BONES

ANCIENT MEN AND MODERN MYTHS

This is a volume in

Studies in Archaeology

A complete list of titles in this series appears at the end of this volume.

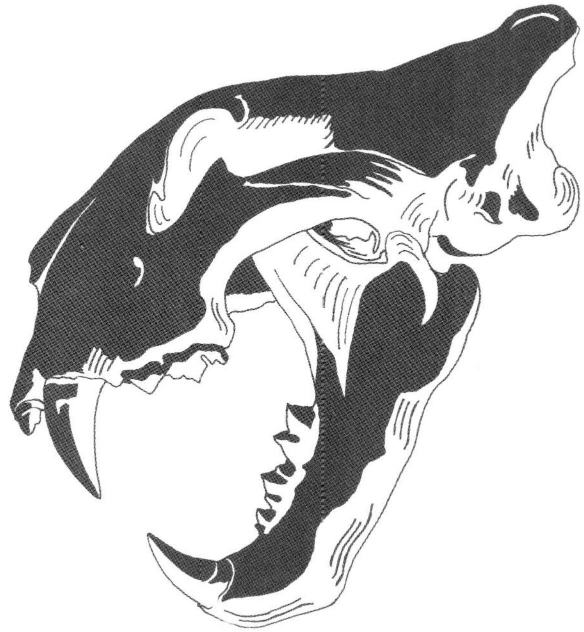

BONES

ANCIENT MEN AND MODERN MYTHS

Lewis R. Binford

Department of Anthropology
University of New Mexico
Albuquerque, New Mexico

With a Foreword by
F. Clark Howell

Academic Press
A Subsidiary of Harcourt Brace Jovanovich, Publishers
New York London Toronto
Sydney San Francisco

Figure 4. 42 is from G. Frison, M. Wilson, and D. J. Wilson, Fossil bison and artifacts from an early Altithermal period arroyo trap in Wyoming, *Antiquity*, 1976, *41*, (1):Fig. 17. Reproduced by permission of the Society for American Archaeology.

Extract on p. 203 is from A. J. Sutcliffe, Spotted hyaena: Crusher, gnawer, digester, and collector of bones, *Nature*, 1970, *227*:1111. Reprinted by permission from *Nature*. Copyright 1970 Macmillan Journals Limited.

ACADEMIC PRESS, INC.
111 Fifth Avenue, New York, New York 10003

United Kingdom Edition published by
ACADEMIC PRESS, INC. (LONDON) LTD.
24/28 Oval Road, London NW1 7DX

Library of Congress Cataloging in Publication Data

Binford, Lewis Roberts, Date.
Bones: ancient men and modern myths.

(Studies in archaeology)
Bibliography: p.
Includes index.
1. Human skeleton. 2. Animal remains (Archaeology)
3. Anthropometry. 4. Forensic osteology. 5. Bones.
I. Title. II. Series.
GN70.B53 930.10285 81-1776
ISBN 0-12-100035-4

PRINTED IN THE UNITED STATES OF AMERICA

81 82 83 84 9 8 7 6 5 4 3 2 1

To the memory of David L. Clarke

It is the duty of theoretical hypothesis to outrun fact so that speculation o'erleaps the present information state and points the way, then careful accumulation of tested data will revise the validity of the theoretical position, which may then leap ahead again.

—CLARKE (1972:239)

Contents

Foreword

Lewis Binford is probably the most influential and stimulative archaeologist in recent years, in respect to his insistence on hypothesis testing and theory building. This book is a continuation of the best of his previous methodological work and will, without doubt, once again redirect archaeological and paleoanthropological researches in important and useful ways.

As always Binford has set a *grand* task for himself, and here he has made a major positive effort to advance method and theory in regard to several aspects of human versus animal subsistence behaviors and their attendant residues. From his observations on Alaskan Eskimo butchery and meat-processing procedures, and his analytical observations on wolf kills and their lairs, he has recognized a series of distinctive ("diagnostic") patterns uniquely common to each set of circumstances. This builds on the earlier work of Binford and Bertram and of Binford (*Nunamiut Ethnoarchaeology*), but extends and amplifies that work very substantially. An increasingly appropriate corpus of comparative data is being afforded by ethnoarchaeological studies in Africa (for example, by J. Yellen, D. Gifford, D. Crader), by various taphonomic studies (for example, by A. K. Behrensmeyer, A. Hill, P. Shipman and associates), and by various studies of carnivore (particularly hyaena den) accumulations (by A. Sutcliffe, R. G. Klein, J. A. Maguire, G. Itani and associates, for instance). A set of analytical procedures has been devised that seems to afford measures of a variety of differences between (nonhuman) mammal osteological assemblages in regard to composition of unitary and fragmented body parts, and their respective condition and preservation.

Ultimately Binford is concerned "with the development of diagnostic procedures for recognizing or identifying the agents which might have been responsible for the faunal facts preserved in an ancient deposit." I believe he has made substantive methodological contributions appropriate toward resolution of this very complex set of problems. These include:

- the recognition and differentiation of animal (carnivore) and human procedures (behaviors) in carcass dismemberment and partitioning
- the elucidation and description of patterns of bone breakage and explication of degrees of destruction ("ravaging") by carnivores (particularly, but not exclusively canids)
- the thorough inventory and systematic description of the disposition and character of traces on bones resulting from human-skinning, butchering, and meat-processing practices
- the development of analytical procedures (measures) for ascertaining the (expected versus actual)

composition of various bone assemblages (drawn from assemblages of known derivation) through the employment of several diagnostic indices of the frequencies of body parts, complete or in various stages of fragmentation/destruction

Binford has asserted, largely justifiably, that archaeologists–paleoanthropologists have been unduly cavalier in their treatment of bone assemblages in general, and conjoint artifact–bone assemblages in particular. He insists that "claims for the use of tools should be supported by the citation of marks produced by tools." Although there are some notable exceptions in archaeological occurrences of late Pleistocene age (for example, Pincevent), and even in Mousterian occupations (for example, La Quina), it is not unreasonable for Binford to take this stance. One can only applaud the view that progress is not "made through discoveries . . . which are treated as self-evident in their meaning," and that "basic research makes possible the reliable assignment of meaning to observations." It is again to the point to insist that the development and testing of "middle range theory" is critical to the realization of a scientific study of the human career. It is, unfortunately, also true that there is substantial "mythology" in paleoanthropology, even "modern" paleoanthropology. Hopefully, Binford's position and efforts will ultimately provoke others to develop and advance methodology appropriate to the resolution of what are clearly real problems in a proper understanding of the past. No one familiar with the state of the art would deny that there is still much to be learned when (and if) there is substantive progress in the development of methodology and analytical procedures.

Binford has sought to analyze Olduvai faunal assemblages, mostly from reputed hominid occupation places, on the basis of published accounts provided by Mary Leakey in *Olduvai Gorge,* Volume 3 (1971). His analyses are bold and imaginative, and the result provocative in the extreme. However, are his sweeping conclusions justifiable and realistic? There are substantive problems concerning this corpus of faunal material; hence there are problems concerning Binford's efforts at analysis, most of which he recognizes. Heretofore these assemblages of animal residues have not been studied and inventoried systematically for breakage patterns, for survivorship of skeletal parts per individual, or even collectively for a particular taxon; and "cut marks" as evidence for butchery–dismemberment have not been recorded. Maps of spatial distributions of such bone–artifact concentrations have been published, but the recording of particular taxa and body parts do not enable a linkage to be effected between these distributions and the faunal inventories.

The difficulty (and hence limitations) of the faunal summary in the *Olduvai Gorge,* Volume 3, is that the overall frequencies of body parts for large mammal species are summarized (p. 276), but there is no full breakdown of body parts, or an indication of their preservation for individual taxa. (However, A. W. and A. Gentry [*Geology, Bulletin of the British Museum of Natural History, 29*(4) and *30*(1), 1978] have provided relatively full, but still incomplete body part data for the Bovidae, and these data often do not match that tabulated in *Olduvai Gorge,* Volume 3.) Moreover the limb side is never stated, and those data are critical to the realistic calculation of MNI values.

Mary Leakey early recognized, and on several occasions has stressed, the several sedimentary environments within which these various occurrences are situated, and has also directed attention to their state of intactness or extent of disturbance. Thus, she distinguished "living floors" (largely intact hominid occupation occurrences), possible (hominid) kill sites, channel fills, and "diffuse" occurrences. Binford's analyses have been directed toward the elucidation of the nature and extent to which a diversity of natural as opposed to (or in conjunction with) hominid factors may have been responsible for these accumulations. He has been concerned with discerning events and formation processes, and thus in the degree of resolution afforded by occurrences relative, in part, to their integrity. Binford concludes that a diversity of agencies were probably responsible for the state of these occurrences. He has discerned various patterns and degrees of bone destruction (and preservation), and employing factor analysis has sought to demonstrate resemblances with modern analogues, and thereby to infer the role of various agencies in producing the various accumulations.

Binford concludes that "the large, highly publicized sites as currently analyzed carry little specific information about hominid behavior." He believes that "arguments about base camps, hominid hunting, sharing of food, etc. are certainly premature and most likely wildly inaccurate." This pretty strong statement is probably arguable. However, he does accept as demonstrated "hominid scavenging [of] the kills and death sites of other predator scavengers for abandoned anatomical parts of low food utility, primarily for purposes of extracting bone marrow." This conclusion, if further substantiated, is itself of major significance for understanding the dietary and behavioral adjustments of Plio-Pleistocene hominids. Recent studies of bone assemblages, with artifact associations, from Koobi Fora (by H. Bunn) and from Olduvai Gorge (by R. Potts and H. Bunn) now promise to document in detail the nature and extent of hominid butchery practices evidenced at these earlier Pleistocene sites. These researches also do confirm the complexity of formation processes for particular hominid "occupation situations," as Binford has concluded.

There has been, and is in progress, a variety of research that should provide substantive data relevant to the resolution of many issues about which Binford has expressed his profound skepticism. Several monographs (under the editorship of J.-K. Woo) on the *Homo erectus* locality of Choukoutien I are in preparation and should afford, at last, relevant detailed data on those bone assemblages, and—hopefully—the extent to which they reflect activities of carnivores, early humans, or other agencies, or a combination of many factors, at that important site. Claims have already been made (by P. Shipman and associates) that a distinctive *Theropithecus owaldi*—rich occurrence at the Olorgesailie locality reflects an unusual form of hominid predation (despite the apparent absence of butchery traces!) rather than some other natural catastrophe. A detailed study of this and other bone assemblages is currently being undertaken by Christopher Koch.

Binford offers some pithy comments relative to our own work at the Acheulian localities of Torralba and Ambrona in central Spain. It is worth mentioning that when those sites were initially excavated (1961–1963), initially under my direction and subsequently jointly with L. G. Freeman, *no one* was concerned with bone assemblages in any of the ways which have become more commonplace 10–15 years hence (in fact the Leakey's became concerned with such matters only in the late 1960s as a consequence of the extensive excavation program at Olduvai Gorge). Unfortunately, and for several complex reasons, those faunal collections from the Spanish sites were never fully analyzed, beyond taxonomic identification, and thus fracture patterns, cut marks, and details of dismemberment and butchery were only superficially recorded in the field. These faunal collections are now relocated and in course of detailed study. Renewed excavation (in 1980) at the Ambrona locality has revealed an abundance of evidence relevant to butchery, dismemberment, and bone breakage, as well as human manipulation and fashioning of bone and ivory. Sedimentary processes at this site are more adequately understood than before, but the situation is an unusual one (for Europe) and there is still much to do.

There is, in fact, a diversity of research completed, and presumably in course of publication, or now in progress, which addresses the deficiencies in paleoanthropological studies with which Binford is so intensely—and rightly so—concerned. These range from the Upper Paleolithic of the U.S.S.R. (and western Europe), the so-called "M.S.A." of eastern and southern Africa, the Mousterian of many parts of Europe and western Asia, the Acheulian (and its facies) of caves in southwestern Europe, and cave and open-air occurrences in Europe which may, it is thought by some, testify to the initial occupation of Europe by hominids. In each and all instances these researches will gain substantially from the analytical procedures developed in Lewis Binford's provocative and stimulating book.

F. CLARK HOWELL
University of California, Berkeley

Preface

This book developed from an article started in January 1979. I had made observations on wolf behavior while conducting ethnographic research among the Nunamiut Eskimo of North Central Alaska. I though that my observations might be of some value to those engaged in taphonomic studies.

As I searched the literature for some comparative material on assemblage composition, I realized that my data on dens, bone breakage, and patterns of attrition were unique, and that there was more to say than originally planned. The paper was then expanded to a short book, in which I described butchering and butchering marks, and compared marrow cracking as performed by man and by animals. I expanded my literature coverage to deal with Dart's arguments, read the fascinating literature treating Choukoutien, and delved into the German literature that clearly foreshadowed the "osteodontokeratic" arguments of both Breuil and Dart. As I treated the problems of pseudotools, butchering patterns as manifest in bone modification, and George Frison's arguments about muscle stripping, I also began to envision some interesting conclusions stemming from my data and research. These conclusions were not related to the composition of wolf kills and dens, which now seemed peripheral to the thrust of the "small technical book" developed thus far. Then Diane Gifford lent me her copy of Andrew Hill's important doctoral thesis, submitted to the University of London (Hill 1975). Hill's thesis contained data on assemblage composition for animal kill and death sites in African settings. Now I had comparative material for viewing my data in a broader context. I was fascinated to discover that the assemblage composition of remains at wolf kill sites in north central Alaska was essentially the same as the composition of assemblages from animals of comparable size in Africa, where the predators were unknown and/or variable and the prey—antelopes and equids—so very different from the caribou of Alaska. I began to work up the assemblage composition comparisons between Hill's data and my own, in the process gaining a new appreciation for many of the facts and their implications.

As I worked on the kill data, I increasingly wished for some comparative den material. It was at this point that Richard Klein arrived in Albuquerque to deliver several lectures. He had previously sent me offprints of many of his articles treating faunal assemblages from South Africa. Naturally when he was here we had some intense conversations about faunal assemblages. Richard responded to my moaning and groaning regarding the lack of data on animal dens. He had dug one! He promised to send me an offprint of his report on his return to Chicago. The day it arrived was full of excitement for me, since I had already formed some ideas as to what a den should look like based on the kill site assemblages and the hints I had from my Alaskan material. On inspecting his data, I realized that the assemblage had suffered heavy attrition, and that what the surviving bones gave me was more a reflection of the relative strength of the anatomical part rather than a clear picture of the parts transported by animals to their dens. Could I reconstruct the original assemblage using the techniques developed earlier for understanding attrition? (See L. R. Binford and J. B. Bertram 1977.) As will be seen, I was successful, and the correspondence between the result and the data from the kills was remarkable. With this last comparative component in place there was a kind of closure and a general methodology seen outlined. Clearly what was needed was an application! I chose to implement my approach on the important faunal materials from Olduvai Gorge. I had played around earlier with some of the models developed in my Nunamiut study (L. R. Binford 1978b) and found that there appeared to be some "fits" between my models and the actual faunal data. My impression at that time, however, was that all the variability in the Olduvai materials was probably referable to nonhominid behavior. That is, the fauna was background natural to geological deposits in Africa, deriving from the continuous action of predators and prey as well as natural deaths in such a fauna-rich environment.

I had completed the book up through the first half of the Olduvai chapter when Jim Hill and Jan Orcutt visited me. They were conducting multivariate analyses of settlement data at the University of New Mexico computer center. Jan had an operational program for carrying out factor analysis. I put together the Olduvai data and Jan made several runs using different assumptions. The fascinating result, which in my opinion did isolate pattened consequences of hominid behavior, was totally unexpected. Consequently the writing of the last half of the Olduvai chapter and the conclusions was sheer excitement, rather than the normal labor of finishing a book whose "outcome" has been known for some time.

The writing of this book has been an intellectual adventure, with several important turning points, as has been indicated. I am convinced that a basic methodology has been outlined and that findings of importance have resulted. The methodology is not "clean," and much more information regarding both kills and animal transport of fauna is needed, as is much better control on the anatomical facts of bone density or hardness so important to reconstructing assemblages from surviving elements. I hope that others will be motivated to join in the research needed to move from this outline to a robust set of methods for giving meaning to the facts of the archaeological record.

Acknowledgments

In many cases acknowledgments are a kind of formal recitation of normal relationships, in which an author thanks the typist, the photographer, acknowledges former teachers for inspiration, and closes by acknowledging a spouse for putting up with the tensions of writing. This format is germane to my situation, but there is in addition a complicated set of intellectual and technical debts that must be acknowledged as extraordinary in the writing of this book.

With regard to my experience with wolves and the data collected, I owe a very special debt to two men who continually prodded me into recognizing that some of my assumptions were perhaps not warranted in working with the tundra fauna of Brooks Range Alaska. One was Simon Paneack, who played skeptic with me as I asked him questions about the faunal materials my crews were regularly recovering during the field seasons between 1969 and 1973. For instance, if I described a collection made or field observations recorded on fauna from a particular place, Simon would smile and say, "How you know it's not Indians or wolves?" He said that to me so many times that I began attempting to find out what wolves "looked like" when viewed from the perspective of fauna. I tried to do the same thing for "Indians" but was not successful. The second man was Johnny Rulland, who in his quiet way pointed out to me my first examples of wolf kills and enumerated all the characteristics that led him to distinguish a wolf kill from one made by his fellow hunters.

While I was working at Anaktuvuk, Bob Stephenson of the Alaskan Department of Fish and Game began his extraordinary study of wolves and the Nunamiut knowledge of wolves. This stimulated considerable interest and discussion of wolves among the Nunamiut hunters with whom I was associated, and in turn I was the beneficiary. I am particularly indebted to David Mekiana, who became interested largely through stimulation from Justice Mekiana, who was working closely with Bob Stephenson. David went out of his way to inform me of evidence of wolf behavior.

Although the collection of wolf data was largely my project done in conjunction with ethnographic work, two members of my crew, Dan and Allison Witter, were untiring in their interest in fauna and did the nasty job of collecting bones from one of the major dog yards at Anaktuvuk. Dan developed the first stages of the classification of destruction and modification presented in this book. He also organized and recorded the marrow-cracking experiment, which serve as important controls on the tool-assisted breakage patterns described here. It is one of my major regrets that Dan could not continue working on the Alaskan materials with me, but academic considerations as well as a job in Australia led him in new directions before this work on fauna could be undertaken from the analytical perspective.

A number of people have been important in conditioning the character of the final form of this manuscript. All of the drafted graphs and charts were drawn by Dana Anderson, who never failed to produce the exact illustration I had in mind. The illustrations of bones used in demonstrating the cut marks (Figures 4.06, 4.11, 4.16, 4.20–4.22, 4.25–4.32, 4.36–4.39, 4.48, and 4.53) were all drawn from actual specimens by Judith Browne of Albuquerque. Judy has made a major contribution to the utility and quality of this book.

Most of the photographs of bone breakage and modification were taken by my daughter, Martha R. Binford, and James Moore. These photographs are crucial to the types of comparative arguments that make up much of the dismemberment and breakage chapters. Martha and Jim spent long hours propping up scales, adjusting lighting angles, building supports for specimens out of all kinds of things so the angle would be right or the alignment between specimen and scale would be on the same plane. For this investment and the fine pictures that resulted (Figures 3.02, 3.10–3.16, 3.18, 3.20–3.30, 3.32, 3.33, 3.40–3.45, 3.51–3.53, 4.50, 4.56, 4.57), I am most grateful. After the initial "spree" of picture taking by Martha and Jim, I kept discovering other subjects that needed illustration and Martha took all the additional photographs unassisted (Figures 3.01, 3.03, 3.05–3.09, 3.17, 3.19, 3.31, 3.34–3.39, 3.46–3.50, 4.14, 4.18, 4.20, 4.24).

A number of people contributed photographs that they had taken and I requested. This assistance is acknowledged in the following list: Robert Stephenson, Figures 5.01, 5.03, 5.04, 5.11; James O'Connell, Figure 4.43; Jean-Philippe Rigaud, Figure 4.07, 4.10, 4.15, 4.16, and 4.41.

Through suggestions and actual research, several people have helped me cover some of the foreign language literature. Michael Jochim responded to one of my cries for help by supplying copies of articles from the German literature that were not available in Albuquerque. Further aid with the German literature was provided by Larry Todd, a student in the anthropology department. In addition, my friend James Finley of the biology department here in Albuquerque also helped out with some translation when my own skills failed. Jean-Philippe Rigaud duplicated two important manuscripts unavailable in Albuquerque, and suggested several other references. Lawrence Straus helped by solving the mystery of several French idioms, and he graciously lent me some interesting and rare articles from the early French literature. F. Clark Howell clarified several points for me regarding Torralba and Ambrona and sent me literature as a further aid. James Judge helped considerably in tracking down some of the literature treating early man in the New World.

Intellectual stimulation and inspiration during the course of writing this book came basically from long telephone conversations with Diane Gifford, intense conversation with Richard Klein, and general discussions of the problem with Bob Vierra, John Pfeiffer, and Stanley South. To those good friends, I am sincerely grateful.

Several typists were involved in the preparation of this manuscript. Lisa Edelhoff again treated my rough draft with gentle and considerate skill. Freddie Height, Marilyn Daily, Ruth Stewart, and Louise McGuff also prepared major sections of the manuscript. Their commitment to a good job is greatly appreciated.

A rough draft of the manuscript was critiqued by six highly qualified readers: George C. Frison of the University of Wyoming, Diane Gifford of the University of California, Santa Cruz, F. Clark Howell of the University of California, Berkeley, Richard E. Morlan of the Archaeological Survey of Canada, Timothy White of the University of California, Berkeley, and a final unidentified reviewer obviously with interests in the New World. These reviewers provided me with a truly impressive body of criticism, that made it possible for me to clear up numerous vague or misleading parts of the manuscript. Their responses forced me to look again at the manner and substance of my treatment of others' data or ideas. In addition, I have added (largely in notes) numerous volunteered pieces of information that have strengthened the presentation considerably. I cannot thank these readers enough. I have never had such a body of thoughtful, constructive, and candid criticism. To the six who took it seriously and did their best to find the weaknesses and points that needed clarification, I am very grateful.

Funding for the fieldwork during the course of which the data on Alaskan wolf behavior was obtained came from the National Science Foundation and the Wenner–Gren Foundation for Anthropological Research. To these agencies, I am most grateful.

List of Figures and Tables

Figures

Tables

Abbreviations

Note: Only the most commonly used abbreviations are listed here.

ANT Antler
SK Skull
MAND Mandible
MAX Maxilla
AT Atlas
AX Axis
CERV Cervical vertebrae
THOR Thoracic vertebrae
LUM Lumbar vertebrae
PELV Pelvis
SAC Sacrum
R Ribs
ST Sternum
SC Scapula
H Humerus
PH Proximal humerus
DH Distal humerus
RC Radio-cubitus
PRC Proximal radio-cubitus
DRC Distal radio-cubitus
CARP Carpals
MC Metacarpal
PMC Proximal metacarpal
DMC Distal metacarpal
F Femur
PF Proximal femur
DF Distal femur
T Tibia
PT Proximal tibia
DT Distal tibia
TAR Tarsals
AST Astragalus
CAL Calcaneus
MT Metatarsal
PMT Proximal metatarsal
DMT Distal metatarsal
PHAL 1 First phalange
PHAL 2 Second phalange
PHAL 3 Third phalange

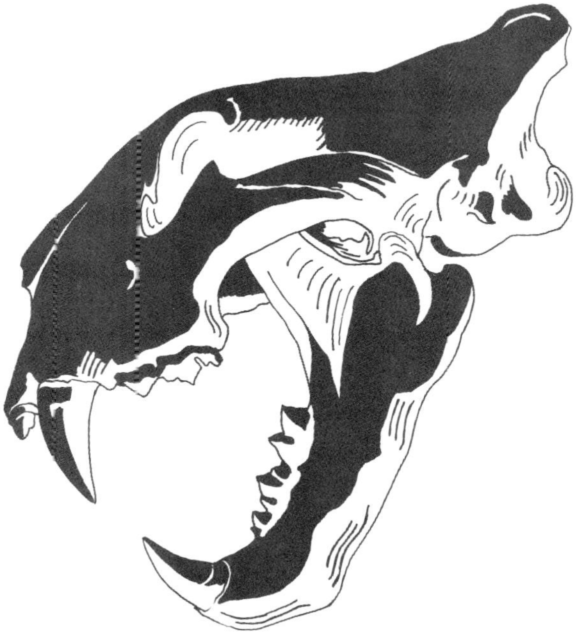

Part I
Documenting a long-term archaeological problem: Concepts and observations

The focus of this book is bones. It may come as a surprise to some that most of the behavioral ideas regarding our ancient past are dependent on the interpretation of faunal remains and depositional context—not, as most textbooks would lead one to believe, stone tools. In fact, in all the arguments I have reviewed regarding the behavior of ancient man, almost all are dependent on inferences drawn from faunal facts or facts of association between fauna and stone tools. Few if any pictures of the past are, strictly speaking, derived from interpretations of stone tools.

A study of bones is, however, not sufficient to clarify our ancient past. We must know how to use information and organize research, and we must increase the reliability with which we construct the past.

I will argue that archaeologists have regularly generated a variety of modern myths by virtue of failures in the inferential process. Many such modern myths have been generated by serious and dedicated archaeologists through the interpretation of faunal materials. I am referring to arguments regarding the character of life at the dim boundary of our humanity as known from the archaeological remains of Lower and Middle Pleistocene hominids. There is a current set of pictures of the past: the mightly elephant hunters of Middle Pleistocene Torralba, the cooperative altruistic sharing behavior of the hominids at the Pleistocene boundary. These ancient men are depicted as living in base camps, hunting, and in many ways resembling a band of modern San speakers of the southern African veldt! Or if you prefer a different view, the myth of the killer ape sees our earliest ancestors as successful predators gaining their livelihood from the flesh of other animals and possessing the "killer instinct." These modern myths about ancient man are based on the interpretation of the bones. Some of these views may be accurate, but in the absence of a reliable methodology for giving meaning to bones they must remain myths based on judgments, opinions, and in some cases clear biases, perhaps motivated by political or humanistic philosophies of the contemporary world.

It will become clear that I do not have new information or controlled data of direct relevance to all the types of interpretative myths archaeologists have generated. For instance, I will only mention in passing the problems involved in interpreting bone destruction of human skeletal material by animals and other natural agencies, yet this is crucial to evaluating the validity of arguments for cannibalism and even to the character of "religious" behavior or the "spiritual life" of ancient man. (See Berounioux [1958] and Marshack [1972] for examples of such interpretive speculation.) Even the simple recognition of intentional burial and the presence of graveside ritual is not an easy inference. For instance, the Neanderthal "burial" of Teshik-Tash (Movius 1953) might just as easily be seen as the consequence of destruction and sorting of skeletal parts by predator–scavengers instead of as ritual behavior by the Neanderthalers. Nevertheless, the arguments to be presented regarding the tenuous contextual assumptions archaeologists frequently make, coupled with the lack of controlled information about bones, render many interpretations strongly suspect. In short, their status as modern myths becomes more apparent.

Chapter 1 focuses on some of the assumptions that have guided many of the reconstructive inferences made from archaeological observations. Chapter 2 focuses on the problems of inference that archaeologists must squarely face if they hope to increase the accuracy with which the archaeological record is used as a means to gaining knowledge of the past. In essence, this part of the book is concerned with two problems: (a) the understanding of the past that we would like to improve through research, and (b) the research strategies that must be employed if we hope to achieve increased understanding of the past.

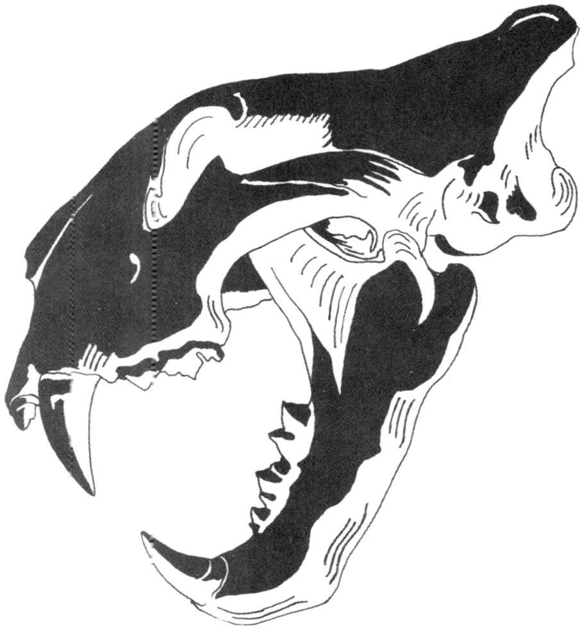

Chapter 1

Relics to artifacts and monuments to assemblages: Changing conceptual frameworks

As is suggested in the title of this book, my focus is the study of fauna and the interpretations archaeologists offer regarding data summarized from observations on fauna. Although the primary interest is bones and the development of a methodology for using fauna to gain knowledge of the past, this work goes beyond a mere technical concern for bones. David Clarke (1972: 238) once characterized much of my earlier work as metaphysics:

> Metaphysics . . . the science of being as such . . . the term science is used in its classic sense of knowledge by causes [Runes 1962: 196].
>
> Archaeological metaphysics is the study and evaluation of the most general categories and concepts with which archaeologists think [Clarke 1973: 12].
>
> The exposure of archaeological metaphysics to critical appraisal allows us the self-conscious capacity to consider the possibilities of altering or rejecting current disciplinary concepts in favour of alternative forms. Thus, at the moment, archaeology is still a discipline in which artifacts, assemblages, sites, and their contexts are identified and related as relics of communities in accordance with rules formulated in terms of artifact taxonomies—the traditional Montelian paradigm [Clarke 1973 in Hammond *et al.* 1979: 94–95].

In the foregoing sense, this book is directly concerned with some aspects of archaeological metaphysics. I will advocate the replacement of some basic concepts, and modifications in the conventional use of others. In addition, I will offer certain methodological suggestions, which are inevitably called for given basic shifts in metaphysical perspective. In fact, most of the book is concerned with the development of methods; however, the need for such new methods is only made clear in the context of changed perspectives regarding the archaeological record and the past. I will be concerned with the concepts *artifact, assemblage, site,* and *habitation*. I will attempt to demonstrate that the assumptions about the archaeological record that stand behind the use of these concepts are frequently misleading; nevertheless, assumptions strongly condition what is considered to be a problem in need of solution as well as what is considered to be a justifiable statement about the past. More specifically, I will be concerned with what the archaeological record is like and how it comes to be so. Given basic conceptualizations about the archaeological record and in turn what the record means for the past, there are linked methods that are generally taken for granted.

The student of a discipline frequently reads and uses the products of his predecessors as discrete contributions from which he picks and chooses ideas and observations to serve his ends. This activity should be carried out with an appreciation of the intellectual history of the field so the writings of earlier workers may be viewed against generalizations about the thought of their time. Some view of history in the latter sense is needed to provide a pause before the user of the works of others begins selecting ideas and observations for personal purposes and/or expounding on the value of past work relative to present points of view. A productive researcher needs a wellfounded set of ideas regarding the state of the art both past and present to understand why past workers chose to invest their time as they did. He or she needs to make judgments as to what is in fact worth seeking through research effort at the present time. If we gain such a perspective on how the state of the art changes, then the past deeds that we criticize need not be seen so much as misdeed relative to our contemporary frame of reference but as important contributions necessary to the genesis of the point of view from which our criticism derives.

There have been several histories (Daniel 1968; Willey and Sabloff 1980) of the archaeological discipline. None have treated either the fundamental or metaphysical propositions about the character of the archaeological record, or the sources or assumed "causes" for the properties of the archaeological record to which meaning was frequently attached by the archaeologists.

I will attempt to characterize our field in fundamental terms, that is, in terms of the basic ideas with which archaeologists have worked. I will attempt to illustrate how these ideas and their use have contributed to the current "state of the art" while also conditioning the character of many problems currently facing archaeologists.

Viewed from a metaphysical perspective, the discipline has had two phases, set apart from one another in terms of the concepts commonly used for treating the phenomena from which archaeologists seek to learn of the past. I term these two periods the *relic and monument phase* and the *artifact and assemblage phase*. If some of the current challenges to the tenets of the artifact and assemblage phase prove productive, a new phase may be in the making.

The Relic and Monument Phase

The relic and monument phase of archaeological investigation came first and in many ways changed our ideas of the past most drastically. It pushed the frontiers of human history back from the 4000-year depth of the biblical perspective to that of man as an ancient resident of the planet. The intellectual procedures used in accomplishing this were based on two propositions: (*a*) the definition of man as a manufacturer of tools and (*b*) the definition of tools manufactured by man as designed or manufactured to a plan—in short, exhibiting patterned redundancy in form.

Given these two propositions and the recognition of the existence of stone tools (as opposed to "thunder stones," etc.), it was possible for a conclusion to be reached regarding whether or not tools had been found, given the evidence presented by such men as Schmerling (1833) and Boucher de Perthes (1849). Demonstrating the antiquity of man rested with the strength of the association between tools and extinct animal remains within undisturbed deposits. The recognition of the "hand of man," as it was frequently referred to in the early literature, required the demonstration that a design or plan had been used in the production of the item (relic) or construction (monument) being investigated. The place where the relic or monument was located was important primarily for demonstrating *associations* with natural objects, such as the bones of animals. The latter provided the basis for chronological assessment as well as some clue to the nature of the environments characteristic of the eras when man had abandoned his items of handiwork. The investigation of relics in these terms remained the primary approach in Paleolithic research for a considerable number of years. The investigation of monuments was largely an activity of classical archaeologists, or at least those interested in Neolithic and more recent remains.

The early procedures were based on assumptions characteristic of the era in which archaeology began to emerge. Central among these ideas was that of the *Creator*. A distinction was made between God's creations (nature) and man's creations (civilizations). The plans of the "joint" creators, God and man, were to be seen in patterning manifest in their separate creations, nature for God, and relics and monuments for man.

However, during these early years of the relic and monument era, the role of God as divine Creator of nature was severely questioned. With the debate over Darwin's unprecedented insight that patterning in nature need not be referable to the prior execution of a divine plan, but instead might result from the normal operation of unplanned processes in nature itself, the equation between God's design or planning and patterning was challenged and largely abandoned. Yet man's role was not questioned. Patterning deriving from the "hand of man" continued to be viewed as a material manifestation of man's plans and creative designs.

It was in this intellectual context as applied to monuments that important advances in archaeological technique were pioneered. Changes in method during this period are generally attributed to two sources: (*a*) *classical archaeology* (Daniel 1968: 65), which emphasized the detailed mapping of items relative to one another, as well as in situ description of what Taylor (1948) would later call "affinities," as sources of information about the past; and (*b*) *monument-oriented research* initiated by Pitt-Rivers at Cranborne Chase (published between 1887 and 1898). The latter served as a model for data-collecting techniques, and formed the basis for a conceptual shift toward the investigation of "places," a shift that would be solidified in the succeeding artifact and assemblage period.

It is not surprising that the development of technique was the product of monument-oriented research. Excavations within monuments were investigations within previously identified human constructions, such a hill fort, a long barrow, or a great palace. Within such complicated constructions the design or plan of the builders could only be uncovered by careful excavation and recording of the many parts making up the overall construction. The techniques were not developed or employed to demonstrate that a monument *existed;* that was assumed. The techniques were viewed as procedures permitting or facilitating a more accurate description of the manifest design or plan in terms of which the monument was constructed.

Researchers of the relic and monument phase worked with several basic ideas and a metaphysical position that viewed the subject of study as things, specifically relics (movable) and monuments (not movable). These things were recognizable as relevant to man's past by virtue of their demonstrable organization of properties betraying a design or plan that must have been in the minds of their makers. For much of Paleolithic archaeology, these views were coupled with the Victorian idea of progress, so that temporal trends in increasing complexity, judgments of refinement, or increased intelligence were taken as directly reflecting the character of the minds or the knowledge available to those minds for designing or creating their products. The early researchers into man's antiquity were concerned with "man" in the generic sense, and tended to have a single view of man as the creator of a progressive, unilinear emergence. This view is well illustrated in the following statement:

> It was about this period—the end of the last century—that a great truth—the inception of which we owe to our

> colleagues of France—began to leaven the researches . . . this truth is simply the recognition that the law of change or progress, which influences all the worldly affairs of man, holds true not only for the present but also for past generations of mankind. Every generation has its own distinctive fashions and ideas; it builds its houses, it tills its fields, it makes its implements, it writes its books, it wears its clothes, and paints its pictures in a manner slightly different from the generation which went before it [Keith 1931: 488–489].

In addition to generation by generation "progress," another idea gradually emerged during the last part of the nineteenth century: the idea of plural creators:

> De Mortillet himself helped to show that one assemblage of type-fossils could be used to define, not only a period of archaeological time in France, but also a nation known to written history. . . . It was however, in Northern Europe and especially in Germany . . . that archaeologists first came to see clearly that assemblages of type fossils might characterize not only distinct periods of time, but also distinct nations or tribes within a single period. And it was German prehistorians who came to term such recurrent assemblages of type fossils "Kulturen"—a word . . . translated into English as "cultures" [Childe 1956: 28].

These "discoveries" set the intellectual stage for the development of the basic archaeological concepts with which we generally continue to labor today. Certainly they provided the basis for a shift from the search for man's place in nature to the search for an appreciation of man's ethnocultural history: "The most profitable inquiry [of archaeology] is the search for the origin of epoch-making ideas in order to comprehend the history of civilization [O. T. Mason 1893: 403]."

The Artifact and Assemblage Phase

A relic was a constructed object, planned, designed, and manufactured by man. An *artifact*, on the other hand, was any object or thing exhibiting properties that appeared to be the results of human action. Relics and monuments were also things that exhibited properties of interest. Classifications and taxonomies had reference to variability in the properties manifest among these individual units or things. With the conceptual shift to artifacts and assemblages a fundamental change occurred in the character of the basic archaeological units. Artifacts were recognizable by virtue of their *association* with other objects, in particular places, as well as by virtue of a distinct design or evidence of planning, as was the case with a relic. Similarly, assemblages were only recognizable by virtue of demonstrable patterning in the *associations* among things and classes of things at *places*.

Although I have not attempted to cover all the literature of the period, it is my impression that the earliest use of the concept of artifact (but not necessarily the word) arose in two slightly different research contexts. The first context is that of Old World archaeologists investigating relatively late remains where items included within human constructions (monuments) could be viewed as being there by virtue of the hand of man, even though they may not have been manufactured to a design or a plan. Under such conditions an item or class of items receives its artifact status not by virtue of formal properties exhibiting "design or planning," but instead by virtue of its association with demonstrable relics or monuments exhibiting unambiguous planning and design within deposits considered to have been constructed or at least formed "by the hand of man." The second context in which the concept of artifact gained some importance was among Paleolithic archaeologists who frequently spoke of recognizing prehistoric "habitations." When such claims were made the investigators tended to treat the contents of "deposits" as if they were monuments constructed by the hand of man. Some of the shift to the metaphysics later associated with the concepts of artifact and assemblage is seen in the controversy between the proponents of the use of fauna as a criteria for defining eras or periods in man's past, and their opponents, who advocated the classification of relics and in turn the recognition of eras by virtue of the constellation of relics systematically found in association at *places*. By the turn of the century this shift to recognizing assemblages was generally adopted. (despite the fact that some particular relic classes were considered more diagnostic than other forms). In short, "habitations" were increasingly treated as monuments, and the systematic investigation of cave deposits pioneered earlier was pursued with great interest.

> It was in the Mousterian age that man first made his home in caves; the period takes its name indeed from the cave of Le Moustier in the valley of Vézère, Dordogne, where its remains were first carefully studied [Sollas 1924: 212].

> Caves have been repeatedly frequented by groups of men; during each occupation broken implements, discarded ornaments, bones left over from repasts, ashes and other refuse will be dropped and trodden into the cave floor, forming a habitation layer or culture-stratum (French foyer). If the cave be then deserted, cave-earth falls from the roof or stalagmite may form a sterile layer. The latter will seal the underlying habitation layer, but may form a floor on which man, eventually returning, will deposit a second occupation layer. . . . Plainly the artifacts and other data in each deposit constitute an assemblage of associated types [Childe 1956: 59].

The shift to the treatment of a habitation as if it were a monument brought an interesting change in the character of the criteria used to identify the habitation. In the earlier period one recognized a relic and/or a monument by virtue of the plan or design manifest in the organization of the thing itself. Habitations were not so easily recognized, particularly in Paleolithic research. Habitations were instead identifiable by a characteristic pattern of association among things at a particular place. That is, recognition of habitations was in terms of a *diagnostic association* among things at particular places. This provided the basis for a shift not only in the units of analysis and synthesis, but also in the fundamental character of the units of observation. What were the criteria for recognizing a habitation?

They were as follows:

1. The aggregated presence of demonstrable relics in the design sense of the term
2. The associated presence of animal bones assumed to be indicative of the meals of ancient men
3. The associated presence of fire
4. The association of the foregoing elements in a "sheltered" setting, as inside a cave or rockshelter

These characteristics provided an operational definition of a habitation that was drawn from our contemporary human experience of living in homes or residences It was considered natural for man to bring food to his home, to increase the comfort factor by warming the place with fire and by either living in sheltered locations or constructing shelter in naturally unsheltered locations. If the relic and monument period worked from a minimal definition of *man the toolmaker,* the artifact and assemblage period worked with an additional definition of *man the homemaker.*

It was in the intellectual context of both of these ideas of man that the problem arose that is addressed by this book. *Namely, how do we recognize the products of human–hominid behavior in faunal remains?* This problem arose early in the history of archaeological research and was addressed both from the perspective of attempting to identify relics and from the standpoint of viewing remains of human activities conducted in man's "home"—that is, from an associational perspective utilizing the concept of habitation.

The two strategies used and argued over have been straightforward. According to the first, the morphological properties of postmortem modifications on bones show evidence of design and planning and the modifications are therefore referable to the hand of man. This strategy is a *pattern-recognition* approach, in which one first attempts to establish the existence of a pattern of modification, and then argues that the form of patterning isolated is referable only to the actions of man. This is the same strategy that has been widely researched and applied to the diagnosis of modifications on stones and in turn the recognition of stone tools.

Early in the relic and monument era, claims were made for bone tools based on the recognition of patterned modifications and the referral of such patterns to human agents. Similarly there had been the recognition that animals modified bones and that many modifications that were uncritically attributed to the actions of man could well be referable to the actions of animals. Both arguments have been present from the very beginnings of modern archaeology. For instance, Boucher de Perthes (1849: 312) reported a phalange with a round hole in the shaft, which was discussed as a "whistle." Dean Buckland, as early as 1823, reported on the similarity between bone modification observed in deposits believed to have been accumulated by hyenas and a bone experimentally given to a spotted hyena at a traveling menagerie. Boyd Dawkins, writing during the formative years of the relic and monument era, commented on Buckland's earlier work as worthy of emulation:

> Dr. Buckland's method of solving the problem of the introduction of remains of so many and different animals into so small a space, is a model of scientific analysis. He argues from the abundance of the remains of the hyaena and from the correspondence of their teeth with the marks on the bones, and from the quantity of their coprolites, that the cave was inhabited by many generations of those animals, and that the gnawed fragments were relics of their prey. . . . He further demonstrated the truth of his conclusions by the crucial experiment of subjecting the leg-bone of an ox to a spotted hyaena from the Cape of Good Hope, in Wombwell's menagerie . . . the exact correspondence of one of the fragments of the tibia of an ox gnawed by the Cape hyaena, with the corresponding bone of the bison from Kirkdale,. . . . The same kind of identity runs through the whole series of bones gnawed by the living and fossil hyaenas.
>
> Dr. Buckland's conclusion that the Kirkdale cave was the den of the spotted hyaena (*H. crouta*) that preyed upon the animals of Yorkshire in ancient times, and that it was undisturbed down to the time of its exploration, cannot be disputed. The tread of the hyaenas in their passage to and fro had polished some of the bones and jaws scattered on the floor [Dawkins 1874:281–284].

The recognition that animals and other nonhuman agents could modify bones to a pattern prompted some early and important research designed to provide recognition criteria for modifications affected by humans as opposed to other agents. This was considered important, since claims for "Tertiary man" had been frequently made during the last half of the nineteenth century, based not infrequently on claims that bones that had been modified to a "plan" had been found in early deposits. In fact, approximately the first 50 pages of De Mortillet's famous work *Le Préhistôrique* (1900) are concerned exclusively with the recognition of bones modified by man as opposed to animals and other agents.

What I find extremely interesting is the *second* research strategy followed in identifying bones modified or introduced to deposits by man as opposed to other agents. The most common strategy was to study the patterning considered manifest on bones recovered from habitations. Since man was responsible for the deposit he must also have been responsible for the patterning manifest in and on its contents!

One of the earliest systematic descriptions of modified bones recovered in an archaeological setting was that by Henri Martin (1907–1910). His work is a mixture of solid observation and description, which are still useful, and a kind of speculation that seems little more than romantic fancy. Part of the "fancy" derives directly from the assumptions and reasoning that underlay the pattern-recognition studies he conducted. Because the material he was analyzing came from the archaeological site of La Quina, which was assumed to have been a habitation, the contents of the site were accepted as being there by the hand of man. Consequently, patterned modifications were thought to be referable to human behavior. In short, one assumption about the character of the archaeological record—that the site was a habitation—was being used as the control condition for referring the patterning observed in the contents of the deposits to human behavior. This approach is well illustrated in one article by Martin (1906) in which he suggested that a reindeer metapodial that bore the unmistakable impressions of animal teeth demonstrated that the occupants of the site of La Quina must have domesticated the dog, for in what other way could the marks made by a beast occur at a human habitation?

Many of the early interpretations of ancient life were the consequences of a similar logic. A good example is provided by the Sirgenstein cave, excavated just after the turn of the century (Schmidt 1912). Of particular interest was Layer C, which yielded a Mousterian assemblage associated with large numbers of cave bear bones. This mixing of tools and bones was cited as evidence that the cave bear bones did not occur in the context of a den deposit but in fact represented bears that had been exploited by man as food (Schmidt 1912: 165). In addition to the large quantities of bear bones, there were at least two "rodent layers" that yielded no cultural materials. On top of the Mousterian materials was a layer yielding Upper Paleolithic tools associated with a fauna dominated numerically by hyena. Despite what appears to me to be provocative evidence of a palimpsest deposit, deriving from a variety of events and actions by both man and animals, the excavators interpreted the entire deposit as the result of hominid behavior. Man selectively hunted bears and later hyenas!

This type of approach was common and was not infrequently associated with the claims for the use of bone as a tool. The important cave of Vogelherd in Würtemberg, Germany, is an example. The site was excavated in 1931 by Gustav Riek, who published a three-volume report (Riek 1934). The cave had not been disturbed by modern excavations prior to investigation by the archaeologists. Numerous layers were recognized, but most of the bone material associated with Mousterian tools was recovered from an area described as oval in shape just to the left and behind the southwest entrance. The largest bones were said to be arranged along the cave walls; the smaller bones, in the center. The Mousterian tools were described as largely restricted to an area 3.5 × 4.0 m, where the artifacts and bones occurred in "close contact." Cave bear, cave hyena, wolf, and cave lion bones were the most common in this "close association," where the approximately 50 stone tools were recovered. In addition to the stone tools, a number of modified bones designated tools were recognized by the excavators. These were listed as seven "bellscrapers," three "hip-joint-socket scrapers,,' two "smoothing" tools, and one awl. Bellscrapers have been described as "the tibias of large animals such as Rhinoceros, horse or cattle. The tibia was broken in half . . . [and] the sponge-like insides of the bones were more or less completely removed so as to produce a hollow space [Zotz 1951:93]." This is exactly what has been described by Kitching as "tankards or scoops" (1963: Plate 25) and flaying tools (1963: Plate 30). Andree (1931:372) reports that these same types of items were found in the Lindentaler hyena den and other German sites.

The "hip-joint-socket scrapers" are described (Zotz 1951:93) as "joint sockets of the cave bear. The apophyses of these sockets were partly removed, and the socket edge shows work traces in the form of smoothness and splintering."

A more expanded list of "bone tools" from the Renis cave was given by Hulle (see Andree 1931:406). Numerous levels were recognized and the lower two "cultural" layers, Renis I and Renis II, yielded Mousterian blattspitzen materials in association with a fauna dominated by cave bear in Renis I and both cave bear and cave hyena in Renis II. Occurring in both levels but reported to be more common in Renis II were a number of "worked bone tools." In addition to the already described bell-scrapers and hip-joint scrapers were so-called "fur-smoothers," and "worked" cave bear mandibles. The identity of these items as tools has been repeatedly questioned (Wetzel 1969; Zotz 1951:136), nevertheless, the description of such "tools" from levels dominated by hyena and cave bear remains is interesting. Some of the more imaginative and fanciful "bone tools" have been described from famous bear cave deposits. Particularly important in this regard is the report on Petershole by K. Hörmann (1923). A total of 20 stone tools are reported from three separate depositional levels in the cave. Large numbers of cave bear bones occurred throughout the deposits in various side chambers These were described as piled in crevices, on stone slabs, and in interesting combinations. In addition to the cave bear there were 55 identified elements of carnivores from Level I. Cervids, bovids, and horses were represented by only 48 identified elements (see Hörmann 1923: 137–138). Stated another way the assemblage was dominated by Carnivora. There were few stone tools, but a relatively large number of "worked bones" were reported. One of the more imaginative forms, "buttons," are illus trated as small sections of long-bone cylinder. In addition, there were a number of "bone knives," described as sections of bone with chipped retouch along ends or lateral margins; "spearheads and arrowheads," identifiable by their suggestive shapes only; and finally a number of "worked bones" exhibiting radial scarring, lateral "retouch," and smoothing of ends—all of which are indistinguishable from fragments of humerus identified by Kitching as "knives, pounders and scoops." (See Plate 17 in Kitching [1963] and compare Plate XLVII in Hörmann [1923].) Many of these same "types" of "worked bone" have been reported from more recently investigated locations such as Repolust-Höhle (Mottl 1951:23, 33–34, Plate XXIII), where the fauna is likewise dominated by Carnivora.

Although the German and other North European literature certainly has many examples of what can only be called anthropocentrism—the assumption that if any evidence exists for the presence of hominids everything present is the product of hominid action—some very explicit examples come from Great Britain. For instance, the famous Kent's Cavern in Devonshire, England, excavated by the Reverend John MacEnery and later by William Pengelly, exhibits all the properties of having been primarily a den of cave bears and then later a hyena lair (see tooth counts in Campbell and Sampson [1971: 15]. Mixed with the faunal remains were Acheulian and Mousterian tools. The spatial juxtaposition was sufficient to lead Campbell and Sampson (1971: vii) to suggest that "specialized butchering of hibernating bears during winter may have been practiced by the so-called 'unspecialized' Acheulian hunters."

The dominant line of reasoning used in defining both patterns "diagnostic" of human behavior and associations of items that served as the basis for behavioral inferences was as follows:

1. The argument was made that an archaeological site was a habitation.
2. A habitation was considered a type of monument.
3. The associated elements within a monument were all there by virtue of the hand of man. Therefore,
   a. Any patterning demonstrable within classes of associated elements was referable to human actions.
   b. Any patterning demonstrable among classes of associated elements was referable to human actions.

This paradigm of reasoning permitted some "reconstruction" of the past based on observations made on the archaeological record, and as a result of its use several ideas about the past were gradually solidified. There was, for example, periodic flirtation with the idea that man may have made and used bone tools before he used stone (see De Mortillet 1900:25–54; Menghin 1931). More important to the future directions of Paleolithic archaeology were three recurrent arguments about man's past:

1. *Bone tools were common and frequent components of Middle and, perhaps, Lower Paleolithic tool assemblages.* They were recognizable by their shapes, which were suggestive of their uses, as well as by superficial modifications and forms of breakage. Advocates of this view were Henri Martin (1907–1910), Schmidt (1912), Riek (1934), Andree (1939), Hörmann (1923), and Bächler (1921, 1940).

2. *There were distinctive bone tools associated with the remains of cave bears as well as evidence for the special treatment of cave bears and their remains.* This view argued for the existence of a bear cult and propositions were advanced regarding the spiritual characteristics of Middle Paleolithic man. These arguments may have developed from views popularized through the work of Schmidt and Wernert (1910). Schmidt's work has been summarized by Sollas (1924), who wrote,

> Of late years Germany has added its contributions, and one of the most precisely investigated caves of modern times is that of Sirgenstein . . . the broken bones of animals were strewn about. The Cave Bear (*Ursus spelaeus*) . . . was by far the commonest; this animal was the universal "caretaker" occupying the cave during the absence of the hunters and receiving short shrift when they returned. It was also a favorite food, as is shown by the great numbers of bears' bones which are massed together near the threshold of the cave as well as plentifully scattered about. No one cave could have supplied so many bears, and the hunters must have ransacked the surrounding district in pursuit of them. Most of the bones belonged to young animals which were, no doubt, an easier prey as well as more delicate eating [pp. 221–223].

In these early works there is the notion of an almost specialized hunting of cave bears. In the mid-1920s this emphasis or suggestion of a special relationship between man and cave bear was elaborated by Bächler (1921, 1940) and Hörmann (1923), and the idea of a bear cult emerged (see Abel 1926). In the words of later writers presenting "general understandings" of the past,

> Why Neanderthal man began hunting the Cave Bear is not certain. It was a formidable animal, standing more than eight feet tall when reared in anger, and must have been a dangerous foe. It also lived in much more inaccessible places than most of the other game. Nevertheless, it was hunted—perhaps to fulfill an early hunting ritual. Discovery of bear skulls stacked in a stone chest in Drachenloch, Switzerland, supports this idea; the heads may have been trophies [Howell 1965:126].

> In high-level caves of Switzerland, Bavaria and Croatia, evidence exists of a cult of bears involving the deposition of bear skulls and bones in certain caves. . . . In a

> chamber of the Drachenloch in Switzerland, a stone cist had been built to house stacked bear-skulls; piles of sorted long bones were laid along the walls of the cave. Another heap of bones contained the skull of a bear through which a leg bone had been forced, the skull resting upon two other long bones, each bone was from a different beast [Coles and Higgs 1969:286–287].

This idea of a special relationship between man and cave bear seems to have struck a responsive note in modern readers, for almost all literature treating the "origins of religion" cites these arguments and places particular emphasis on these "important" developments leading to our "human" status.

3. *Ancient man was a cannibal.* Perhaps one of the more influential early claims for ancient cannibalism was made in the report on excavations at Krapina (see Gorjanovic-Kramberger 1906). Other authors during the same era sounded similar notes (see for instance Rutot 1907). These early suggestions were greatly elaborated both prior to and immediately after World War II. The cannibalism suggestion was gradually linked with the so-called "Cult of Skulls" (Bergounioux 1958; Wernert 1948). For instance, Weidenreich (1928) describes the Ehringsdorf skull in the following terms:

> the frontal area bears undoubted signs of having been hit by stone weapons and it can be assumed that this individual had been murdered, (*b*) the absence of the cranial base, no fragment of which was contained in the block of travertine from which the skull has been extracted, indicates that the skull had been opened to extract the brain [p. 67].

Later Berckhemer (1934) described the Steinheim skull as having been artificially widened at the base around the foramen "for purposes of extracting the brain." This view was further supported by the description of the Monte Circeo skull by Blanc (1939). This important skull was recovered from a small chamber linked to several other cave rooms by narrow passages. In all, the complex can be viewed as seven caves, five of which are reported to have yielded some Mousterian remains. Blanc has reported on this find in numerous places (1950, 1958, 1961); the skull is said to have been lying as an isolated element surrounded by rocks near the center of a small chamber. Bones of other animals are reported concentrated in at least three additional locations within the chamber (see Blanc [1958] for a good description).

The three basic ideas just described largely grew from early twentieth-century excavations of European "cave" and rockshelter deposits referable to men of the Mousterian era and of more recent times. The excitement of a first glimpse of men considerably earlier than the Neanderthalers came with the discovery of the site of Choukoutien in China. This site was remarkable for its great antiquity, and the presence of important fossil hominid remains. It was also extraordinary in its being a cave deposit that yielded tools, evidence of fire, and bones in vast quantities; in short, it was viewed as a habitation site in the sense in which the European caves had been viewed. The methods of interpretation already common in Europe were used to make statements about man's behavior in the remote past as well as to provide the "control" for recognizing patterned products of man's ancient handiwork.

In 1936 in Oslo the Abbé Breuil delivered an influential paper in which he proposed that there had been an extensive bone tool industry at the ancient Chinese site:

> Let's go back in thought, then, to that dark dawn of the industrial phase of man. We find him or his intelligent ancestor surrounded by animals better armed by Nature than himself, lions and bears, with teeth and claws, grass-eaters with horns and antlers. What more natural than to rob them of these weapons to use against them? Ever a hunter, Man had around him the skeletal remains of his victims, quickly unfleshed by himself, by carnivores and by natural agencies.
>
> He also encountered on his wanderings the cast antlers of deer and the carcasses of the carnivores' prey. Some of the completed longer bones made excellent clubs with handles not easily broken; some of the bones could in their natural state be used for pricking–piercing holes; some larger and flat could be used as shovels; others which were longer, as levers; yet others which were short and stout might suggest an anvil or a hammer, while such as were broad, thin and trenchant, when grasped, became scrapers and planes.
>
> When Man, armed with a pebble, broke long bones to extract the marrow, like a hyaena, they broke in different ways. Some had articular ends convenient for holding if man wished to use them as implements, while the other end was pointed. Others, pieces of

> diaphyses, had ends that could be used as points or chisels and cutting edges which could serve as knives or scrapers. *Unlike the hyaena Man often broke the bones lengthwise* [Breuil 1938:57–58; emphasis added].

Here we see the imaginative picture of bone-using hominids that later captured the fancy of many audiences as the "osteodontokeratic culture" of yet earlier man advocated by Raymond Dart during the 1950s. However, the history of ideas dealing with the Choukoutien material is in itself fascinating, because it illustrates opposing interpretive points of view regarding bone tools, views still with us today in the Old World and increasingly becoming an active issue in New World studies (see Friedman 1980:7).

Breuil (1939) studied the animal bones from Choukoutien, and Weidenreich (1941) reported the hominid remains. In the works of these two men, parallel theses were presented: (*a*) that an extensive bone tool industry existed at Choukoutien and (*b*) that "*Sinanthropus*," the "man" responsible for both stone and bone industry, was a cannibal. The argument offered by Breuil was based on the suitability of the shape of a bone for use in a given manner, with the inference that it would, therefore, have been so used, and the modification of bone as evidence that intentional shaping by man was involved. Breuil cited the presence of rounded and "polished" edges, chipped or flaked edges, incisions or abrasions on flat surfaces, and "characteristic" longitudinally broken bone. Weidenreich's argument for the cannibalistic consumption of bone marrow was based on fractures of *Sinanthropus* long bones. "Long splinters cleft off the shaft longitudinally so that the medullary canal lies exposed" were considered characteristic of the actions of *Sinanthropus*: "This manner of splitting is generally believed to be the result of human manipulation carried out with the aid of implements in order to extract the marrow [Weidenreich 1941:75]."

In the work of both Breuil (1939) and Weidenreich (1941) there was the clear belief that some long-bone breakage patterns were distinctive and even diagnostic of human agency. In addition, Breuil interpreted modified bones as resulting from human action.

In this era there was, however, relevant research that provided evidence to the contrary. Weidenreich cites a study by Zapfe (1939:76), who experimentally studied the way in which living carnivores, especially the hyena, chew long bones, and even illustrated how some of the broken *Sinanthropus* long bones corresponded to the properties cited by Zapfe as characteristic of carnivores. However, he concluded that "the possibility exists that the long femur splinters were not the product of *Sinanthropus* himself. This alternative, however, seems to me very unlikely [p. 77]." In other words, despite the evidence, Weidenreich preferred to view *Sinanthropus* as a cannibal!

Nearly the same situation exists with the arguments presented by Breuil. Wen-chung Pei (1938:1) then director of the Cenozoic Research Laboratory in China, published a very important monograph, largely in answer to Breuil, rightfully posing the question, "Comment reconnaitre l'action humaine sur un debris osseus?" Pei proceeded to present the information available to him as to the modifications produced by rodents and predators, as well as the effects of water, soil chemistry, and soil texture or bones. Pei was convinced that many of the modifications cited by Breuil as evidence for human use appeared referable to other agents.

After World War II (1939–1945) many of the prewar arguments were continued (see Battaglia 1953; Ehrenberg 1959, 1962) regarding the existence of the bear cult and associated distinctive "Alpine Mousterian" (Leonardi 1958). The focus of controversy regarding bone tools was, however, taken out of its regional and provincial context with the claims by Raymond Dart (1949) for an osteodontokeratic culture ancestral to the production of stone tools by our earliest hominid ancestors. Dart's arguments have been reviewed by Wolberg (1970). These claims placed the controversy over interpretation of modified bones at the very roots of human history and drew the attention of all regional specialists as well as a widespread lay audience. Dart's claims for bone tools in the australopithecine sites of South Africa provided paleoanthropology with a lively and certainly imaginative controversy, which has continued to the present time.

Dart's argument was very similar to that of Breuil. He cited the potential utility of the shapes of various anatomical parts and then inferred that early man must have taken advantage of this potential. Actual modifications cited as diagnostic of human actions included an allegedly distinctive mode of long-bone breakage, the "crack-and-twist" method resulting in preshaped bones and spirally fractured bone splinters, and (as in the case of Breuil) various abrasions, striations, and incisions, as well as chipped or "worked" edges. In addition, Dart made a slightly different argument, namely, that evidence for man's selective use of anatomical parts rested with the differential frequencies of "useful" parts in the sites of early man. Thus, for Dart two important sets of facts converged: modifications on bones, which were interpreted as produced by man, and the associated differential frequencies of anatomical parts, which were interpreted as resulting from the differential transport of hunted foods to "man's home." For Dart there were *facts of modification* and *facts of assemblage composition,* which he believed could be cited as evidence for hominid behavior. It is not surprising that, given Dart's claims, others would offer as evidence modified and/or broken bones in support of the position that tool-using hominids occupied different places other than Africa prior to the development of stone tools. Such claims have been made for the paleontological remains from the Netherlands (Luttschwager and von Bemmel 1962) as well as Rumania (Nicolaescu-Plopsor and Nicolaescu-Plopsor 1963). Many of these arguments have been seriously challenged (see Feustel 1969, 1970); nevertheless, they persist on the level of "opinion." That is, specific modifications are compared, and then the opinion is offered that clearly only a man could have produced the modifications illustrated!

With Dart's claims we take an important step away from the propositions that permitted the recognition of a habitation during the first half of the twentieth century. Most important was the absence of stone tools as a criterion or concomitant in his sites. He sought to meet the tool criteria by arguing for the presence of bone tools. Dart, obviously knowledgeable about the earlier work that described morphological patterning among bones associated with unambiguous stone tools at habitations, could now appeal to that patterning to support his claims for the modified and broken bones from his sites as tools. This claim lent support to his further claim that the sites were habitations, thereby justifying inferences from included materials about the behavior of *Australopithecus.* Dart's case is of no small significance, since his early claims for the ancestor status of the Taung skull to modern man had largely been rejected on morphological grounds. Dart increasingly argued that the australopithecines were protohuman by virtue of *behavioral* characteristics manifest in the archaeological record. *This was the first important demand for behavioral information from the archaeological record of ancient man.* Prior to Dart's claims there had been some attempts to characterize the behavior of early man, but at no time were the inferred behavioral characteristics considered either crucial to arguments about the sequence of human evolution or central to the important question of the history of "human nature," as was the case for the behavioral claims by Dart 1949, and especially 1957):

> the innominate bone found at Sterkfontein and the occiput from Makapansgat have provided further evidence that these creatures are more closely related to man than they are to the apes. They are proto-human beings and they are the most primitive hominids of whom we have knowledge. *It therefore becomes correspondingly urgent that we should elicit from those deposits in which proto-humans occur all the available information bearing on their manual dexterity and implemental intelligence* [Dart 1949:2; emphasis added].

The questions being put to the archaeological record of early man in Africa were not historical in the sense of those asked by archaeologists working with remains from more recent periods. They were concerned with fundamental questions about the very nature of "protohuman" behavior; that is, how do we get behavioral information from the archaeological record? In the context of this important shift of emphasis and in turn its demands for information about the past, some important changes began to take place in Paleolithic research.

In the era immediately after World War II a set of claims was made by Africanists, which foreshadowed a long and important set of arguments relating to the nature of early man's behavior. These were the claims by Louis and Mary Leakey that they had isolated living floors at the site of Olorgesailie. Both J. Desmond Clark and Merrick Posnansky have expressed the opinion to me that the search for living floors was the contribution of Mary Leakey, who had had experience with "Neolithic" excavations and an interest in the use of archaeological techniques to investigate the "place," rather than just using the place to recover artifacts. Several persons have indicated that the Leakeys' original claims for living floors at Olorgesailie were based on the argument that the sediments were fine and hence the geological processes resulting in burial were "gentle." Under such conditions one could expect items to remain where they were deposited by agents such as the hominids of the early time ranges. This "original" argument regarding living floors is at least partially presented by Isaac:

> Olorgesailie was suitable as a testing ground because artifacts could be recovered from fine-grained sediments that had preserved detailed evidences of archaeological associations. These could be related to stratigraphy and to features of the original environment. This preservation was in marked contrast to the sites that had yielded most of the evidence on which the classic interpretations of Middle Pleistocene culture had previously depended [1977:1].

Despite the fact that Louis Leakey reported the Olorgesailie living floors (L. S. B. Leakey 1946), some believe that he did not "realize their implications" (Vincent 1978:32), as he viewed them as useful for assemblage definition and not so much as sources of behavioral information about the past. The view that the Leakeys treated the surfaces more as "curiosities" is perhaps supported by reference to Isaac's descriptions of their excavations at Olorgesailie (1977:28–79), which cite pitifully few records made by the Leakeys for excavations carried out between 1943 and 1947. More recent work and actual analysis of the facts from Olorgesailie render the identification of living floors less and less likely (see L. R. Binford 1977b; Isaac 1977). Nevertheless, the search for living floors was first prompted by the Leakeys' claims at Olorgesailie. Most clearly it was in the work of J. Desmond Clark at Kalambo Falls where the techniques of recovery that have fortunately become somewhat standard in Paleolithic research were originally pioneered. Clark states,

> The lower horizons containing the artifacts showed little evidence of disturbance by water action before burial and it was apparent that the lithic and other material lying on them represented debris from human occupation sites. We developed, therefore, a technique for excavating these "floors," and the later disturbed surfaces, first cleaning off the overlying sterile sands and clays and exposing the floor, or section of floor, so that the relationship of everything on it could be clearly seen and plotted. In this way we were able to obtain details of the relationships of the tools, factory waste, natural stones, and wood and so of relationships between classes of shaped tools, thus supplementing the information to be obtained from a study of the artifacts themselves [J. D. Clark 1969:11–12].

In Clark's early work geological criteria and facts of site structure such as the association of debitage with cores and anvils appear to have been the criteria for recognizing a living floor. This approach to living-floor recognition is well illustrated by Geoffrey Bond's description of these floors:

> The floors are only one stone thick, and every stone is either an artifact, flake, or anvil. The enclosing sediments are non-pebbly and on the floors themselves there are no pebbles which were transported by natural means; they were all left in their present positions by prehistoric man. They are sharp and unworn, nor were they apparently moved significantly from the original positions by the action of water. They were factory floors in the best sense of the term [Bond 1969:207].

F. Clark Howell searched for living floors at Isimila and later at Torralba, and the Leakeys searched for such floors at Olduvai Gorge. In the latter case a new definition of a living floor was offered: "the occupation debris is found on a palaeosol or old land surface with a vertical distribution of only a few inches [M. D. Leakey 1971: 258]." This definition is very close to that which most archaeologists would offer for an archaeological site—an aggregation of artifacts (relics) on a land surface! The problem, which should be clear, is the degree to which the assumptions governing "interpretation" of a *monument* are justified with respect to such places as living floors or even habitations as defined by Paleolithic archaeologists. It will be recalled that a monument is considered a construction produced by man; hence, all the elements, or components of the construction can be assumed to be there "by the hand of man." The interpretative justification that went with the extension of the monument concept to habitations and later to living floors has in my opinion resulted in numerous myths regarding man's ancient past. "Tools" have been described that were probably produced by hyenas, cave bears, and other predators, and the hunting of such unlikely foods as hyena and cave bear has been considered a regular part of man's behavior.

The most incredible example of this type of myth making must be the analysis of materials from Pin Hole Cave at Craswell Crags, Derbyshire, which was excavated by Armstrong and studied by Kitching (1963). There Kitching worked with 6985 identifiable bone elements, of which 3197 (46%) were from carnivores and 1061 were from the hyena alone. Mixed among this carnivore-dominated faunal assemblage were some telltale stone tools. These were sufficient to lead Kitching to the interpretation that everything found in Pin Hole was in fact the result of hominid activity:

> The transport of entire bodies of both large and small carnivores into the cave or the killing of them while they were marauding the cave accounts not only for the great number of their foetal remains (especially those of the hyaena) but also for the highly eroded condition of a large number of bone flakes and deciduous teeth [Kitching 1963:19].

All of these arguments regarding the subsistence base of the early hominids are based on inferences from associations between bones and stone tools at alleged living sites. To what degree may this assumption be made casually and how likely is such an assumption about the past to represent the past reality? How "realistic" is the assumption that associations between stone tools and bones at a site imply behavioral integration in the past?

Given that predation and death are normal events in any environment, to what extent may the archaeologist expect to find evidence of such events in situations favoring preservation of bones? This question is being investigated by a number of researchers, particularly in Africa. To appreciate the character of this problem, one must become familiar with some basic facts of nature. For instance, George B. Schaller (1972:398) estimates that predators (lion, hyena, leopard, cheetah, and wild dog) in the Serengeti–Ngorongoro study area "may actually account for 150,000 to 200,000 animals [killed] annually." In other places predators remove an estimated 9–10% of the total prey biomass annually (Schaller 1972:397). As a simple exercise, let us see what such a situation would look like if we assumed that predators are equally successful in all habitats. Table 1.01 presents summary data on the amount of secondary biomass that might be expected in different environments, assuming equal success of predators (not very likely—we can expect that as food density goes down predator effectiveness may be reduced somewhat; certainly there would be a dependence on smaller animals). Also the amount of prey (dead animals in kilograms) that might be expected in the different habitats per square kilometer.

What is clear is that we may expect major annual accumulations of bones in tropical savannas, substantial accumulations in temperate grasslands, and minor accumulations in temperate forests, tropical rain forests, boreal forests, and tundra. Just as a demonstration, let us imagine an environment along the grass margins of a boreal forest such as one might find in central Manitoba, Canada, or in numerous mountain settings Let us propose that predator kills would be roughly the mean of expected values for the

TABLE 1.01
Secondary Biomass and Estimated Prey Accumulated for Various Habitats[a]

Habitat	Secondary biomass (kg/km^2)	Amount of prey expected (kg/km^2)
Northern or polar[b]		
Tundra	79.0	7.1
Boreal forest	2.0	.18
Temperate[c]		
Mixed broadleaf forest	450.0	40.5
Prairie grassland	3,450.0	310.5
Tropical[d]		
Acacia savanna	15,760.0	1,418.4
Rain forest	5.0	.45

[a] Data for northern habitats from McCullough 1970; data for temperate and tropical habitats from Bourliere 1963.
[b] Data refer to caribou only.
[c] Data refer to ungulates only.
[d] Data refer to ungulates and primates only.

boreal forest and a temperate grassland (155.34 kg/km^2). If a deposit accumulated over approximately 10,000 years (not an unreasonable estimate for many Pleistocene events), we would expect 155,344 kg of prey during that period. Given an estimated mean weight of 222 lb (100.57 kg) for an adult caribou (*Rangifer tarandus*), we would therefore expect 1544 individual caribou to have been deposited over a single square kilometer during that period of time, if caribou were the only prey animals in that habitat. Obviously, that is not realistic, since many different species would be involved. Nevertheless, as an object lesson for the archaeologist, one can see that an impressive amount of bones is deposited annually, particularly in a tropical savanna. Given the types of geological units archaeologists work with in Pleistocene deposits, we can expect as *normal* a very large number of included bones, since most geological deposits represent appreciable periods of time.

Certainly it must be realized that all places where deaths occur are not equally propitious for the preservation of bone. Recent pioneering work in taphonomy, particularly by Diane Gifford, K. Behrensmeyer, and A. Hill, has begun to illuminate the conditions under which preservation is most apt to occur: a quick burial, and either the perpetuation of moisture on the bone, or complete desiccation. Stated another way, in situations in which it is perfectly wet, or perfectly dry, and the bone is being buried, bones are most apt to be preserved. For all practical purposes this means that areas being deflated (see Gifford 1977:108–257) or eroded are not good candidates, and areas subject to extremes of wet and dry are not good candidates. Lake and stream margins, delta formations, as well as protected areas such as caves and rockshelters, are very likely places. Behrensmeyer (1975b) studied bone distribution and assemblage content in several different geological contexts. She observed that bone densities were greatest in delta environments and least on the floodplain, with channel deposits intermediate. The implications of these observations are that (*a*) death and predation result in considerable numbers of animal carcasses being scattered in the environment annually, and (*b*) these carcasses are apt to be preserved in dry or wet conditions where the bones are apt to be buried shortly after deposition. This means that lake, spring, and stream marsh margin deposits and internal drainage pockets can be expected to yield considerable quantities of dead animal remains as a normal condition. Stated another way, the normal functioning of a biotic community can be expected to yield considerable deposits of animal bones in geomorphological settings conducive to their preservation. This seems like a reasonable proposition; nevertheless, the methodological implications have not been explored by archaeologists. In the New World many death sites of Pleistocene elephants have been recorded and only a few have yielded any evidence of human involvement at the time of death. The vast majority of these elephants have been recorded in the contexts of lake, stream, and spring margin deposits. In fact, this is such a regular pattern that it is quite likely that elephants of all types seek bodies of water when their temperatures go up; hence the most common place of death is in such a marsh–bog setting. This is a *normal* condition.

The conditions favoring preservation must be also present if bones deposited as a result of hominid actions are to be preserved. If there is evidence of hominid behavior in a deposit containing preserved bones, the conditions under which the deposit was formed are also likely to be conditions conducive to the preservation of bones derived from the normal pattern of predator–prey interaction as well as the normal mortality characteristic of the setting.

Perhaps this point is best made with an example: The existence of (*a*) an active biological community with normal death and predation rates and (*b*) a lake margin bog deposit is well documented at the famous site of Torralba in Spain (see Howell 1966). Within this site were approximately 785 "artifacts," of which approximately 174 are lithic debris. These provide exciting evidence for the participation of hominids in the Middle Pleistocene biotic community, a glimpse of which is preserved at the site. Tools at Torralba are restricted to the important deposit designated IIc of the "lower grey colluvium." I can find no estimates by the researcher as to the length of time during which this deposit accumulated. I asked Howell, and he said that he had no good idea but that it could not be more than "several tens of thousands of years." During the unknown span of accumulation, at least seven "levels" were generated with several additional lenses. That is, land surfaces were differentially stabilized, permitting the excavators to recognize sequential depositional situations. No estimates as to the length of time involved in each "stabilization" have been offered. For the entire sequence representing perhaps "several tens of thousands of years" an estimated minimum of 115 (See Howell 1966:121–122) individual animals were recovered from the Torralba levels. Assuming that only 10,000 years is involved, there would be an average of only one dead animal every 87 years. That does not appear excessive as a normal death rate for a normal glacial environment! The fauna is interesting in its composition, in that it is remarkably stable from level to level:

> The greatest estimated minimum number of animals for any level is only 15 (Occupation I). Variations in species composition do occur from level to level, but are no greater than can be expected from sampling error alone. . . . The most abundant remains are those of elephants (31 percent of individuals), although they are always outnumbered by the combined total of individuals of other species. . . . Horses are next most abundant (21 percent of individuals) followed by cervids (18 percent excluding Dama), bovids (13 percent), Dama (6 percent), rhino (4 percent), and birds (3 percent), with about 2 percent each of Felis and other carnivores [Freeman 1975:668].

What role did man (represented by 611 tools deposited during "several tens of thousands of years") play in this Pleistocene biotic community? The researchers' answer is that he controlled it completely, dominating the entire situation so that every animal found in the Torralba deposit (with the possible exception of several birds) is assumed to have gotten there by the hand of man! (See Freeman 1975:667–682.) Man killed the animals while executing game drives—possibly aided by fire—butchered them, and carried the meat away—truly extraordinary!

> Even though much meat might have been wasted, an impressive quantity must have been transported from the site, to judge by the missing skeletal parts. In Occupation 7 alone: animals whose carcasses would have yielded at least 30,000 pounds of usable flesh were recovered. Assuming a daily consumption of 5 pounds per man per day, this represents about 6000 man-days of meat rations. Just bearing away so much weight would have presented a small group with severe logistic problems. If each person carried away 100 pounds at a load, 300 loads would have been required to move all the meat; 50 able-bodied men would have to make 6 trips each to carry the lot to the living site. The amount of meat actually utilized might have been considerably smaller than the total available, but the figures still suggest a relatively high group size. Based on such considerations, I am inclined to think that the total size of the cooperating social groups which provided the personnel responsible for the Torralba occupation residues was very large, perhaps on the order of a hundred individuals or more. Such large population aggregates might have been feasible only periodically or seasonally, but it is quite possible, given the undoubted natural wealth of the region in early mid-Pleistocene times, that large human groups were a constant feature of the landscape [Freeman 1975:682].

This is a truly remarkable set of conclusions to draw from the Torralba data. We are told that the region is one of "undoubted natural wealth" (presumably wealth in animals), yet we must assume that none ever died from predation or natural causes around the Torralba lake-bog! We are told that nature operated only when man entered the picture. When man arrived to hunt animals cooperatively in large groups, nature also cooperated by preserving a record of these heroic exploits. Yet, when man left, nature stopped and no further preservation, or natural deaths of animals, occurred! How extraordinary! We are led to conclude that, because there were so many animals represented in "levels" with tools, human groups were practicing cooperative hunting drives in which all species over certain sizes were taken time and again in roughly equal proportions, butchered, and the meat transported to consumers living in large groups that were "a constant feature of the landscape." Amazing!

The researchers at Torralba have operated with a set of *assumptions* about the nature of the deposits, as well as about the processes in the past that resulted in deposit formation. They have been misguided by the idea of a living floor as a spatial aggregation of preserved matter, all of which results from human actions. Given an aggregation of stone tools—evidence of hominid behavior—it is assumed that all other remains associated with the stone tools are also a by-product of human behavior. The researchers of Torralba have certainly made this assumption.

Pleistocene archaeologists need to abandon such an approach. They must adopt an approach that view deposits yielding stone tools as natural deposits, recognizing the potential accumulation of derivatives from a cross section of biotic activity in the environment of a location favoring preservation. Man is only *one* potential contributor.

My discussion up to this point has focused on Old World materials and on Paleolithic research. I do not wish to imply that my criticism is restricted to such studies. In the New World, arguments over the significance of bone modifications are relatively recent.[1] It is true that Webb and DeJarnette (1942:763) suggested, in their "generalized profile" of the Flint Creek shell mound along the Tennessee River, that there was an early period characterized by worked bone, followed by a lithic period, and finally by a ceramic period. However, they did not imply that bone was the exclusive raw material used in tool manufacture. The latter position, linked with sometimes remarkable claims for the antiquity of man in North America, is a position that has developed relatively recently stemming from work at Old Crow Flats (see Harington *et al.* 1975) and excavations by Dennis Stanford (1979b) at the Selby and Dutton sites in Colorado. (See Canby [1979] for a popular discussion of this material.) In addition to the "osteodontokeratic" claims for man in North America perhaps as early as 60,000 years ago (see Canby 1979: 348; Friedman 1980:7), there has been a stepped-up set of claims regarding the use of bones as expedient tools for butchering. This argument was most forcefully set forth by George Frison (1970) with regard to materials recovered from the Glenrock Buffalo Jump in Wyoming. Since that time this argument has certainly captured the imagination of a number of workers, and excavators are now increasingly "seeing" bone tools in their faunal assemblages (see, for example, Johnson 1978).

Stone tools are exciting traces of hominid participation in an environment, but they in no way may be treated as prima facie evidence that human behavior is responsible for the deposit in which they occur. This is not to say that habitations or living floors do not exist. They certainly do, and as we know from the archaeological record of relatively recent times human behavior may well be, on occasion, the agent responsible for the location and content of materials remaining from the past. However, this is something that requires demonstration, not something to be assumed.

We need to abandon the logic of treating alleged habitations, living floors, and living sites as if they were monuments. The logic is one of identifying an envelope or container, and then uncritically accepting everything within it as referable to the behavior of ancient man. A complete reevaluation of basic archaeological concepts and the logic associated with their use is in order.

Minimally, let us begin by asserting that an *archaeological fact* is an accurate observation that carries unambiguous information relative to human–hominid behavior. The degree to which an association between two or more things can be taken as an unambiguous archaeological fact is at least partially conditioned by the character of the deposit; that is, by the matrix or "container" that bounds the association. Most would agree that archaeological remains are items and arrangements of matter that are derived from the operation of a human–hominid system of adaptation. An association between elements included in a deposit may be uncritically accepted as archaeological fact only when the deposit or matrix containing the elements in association is itself an archaeological deposit, and man can be viewed as the agent responsible for the presence of both the associations and elements within the deposit.

Archaeological deposits are matrices together with included remains that are derived from the operation of human–hominid systems of adaptation. That is, the matrix itself is an archaeological phenomenon. From sites of relatively complex cultural systems, archaeological deposits are common; midden accumulations, mounds composed of collapsed building material, and mounds specifically constructed for the burial of the dead are a few examples. Archaeological deposits may be expected to be rare and infrequent derivatives of hunter–gatherer adaptations, since high mobility and relatively low investment in facilities, housing, buildings, and so forth are characteristic of such systems. It is much more likely that archaeological remains will be found within *geological deposits*. Geological deposits are buried matrices, the formation of which de-

[1]An exception is Merriam, 1906 and Putnam, 1906.

rives from the operation of geological processes. Materials within a geological deposit may derive from a wide variety of circumstances and events that occurred during the deposit's formation, none of which is necessarily representative of a behaviorally interrelated set of conditions. That is, an *association* between two or more elements within a geological deposit in no way implies a systematic, behavioral, or dynamic relationship between the components. All that is implied is a roughly coincidental contextual relationship for the events of which the elements were derivatives. For instance, the finding of a whale bone and a tuna vertebra in a geological deposit in no way implies that whales ate tuna or for that matter that tuna ate whales! Nevertheless, the depositional context of an aquatic environment may be indicated. Geological deposits may be thought of as a kind of grab bag, made up of samples of by-products of events and entities that existed or occurred with a given "catchment" and during a given period of time when the deposit was accumulating. A good example is the preserved land surface at Laetoli where dampened volcanic ash hardened to preserve hominid footprints, depicting a stroll by two "early men" across a land surface some 3.5 million years ago (see M. Leakey 1979). Also preserved in the "ashen ledger" were tracks of elephants, guinea fowl, giraffes, hares, ostriches, and other animals and insects, as well as plant remains. Are we to assume that the association of hominid footprints on the same land surface with those of other animals indicates that the action of the hominids conditioned both the presence and character of the patterning observable in the other footprints? Was early "man" herding elephants, ostriches, or guinea fowl? This is clearly an absurd question. The footprints and other traces of past events and actions must be viewed as roughly contemporary but unrelated in any integrated sense. That is, as far as mutual interactions and "causal" linkages among the events recorded are concerned, none seems indicated. Clearly more justification for assuming a behavioral integration among materials associated in a geological deposit is needed than the simple association itself!

As a way of placing emphasis on this fact, it is perhaps useful to think about aspects of past processes or conditions that might result in different kinds of geological deposits. For instance, it is perhaps profitable to ask ourselves about the historical integrity and resolution of aggregates of materials stabilized in depositional association through the operation of geological processes.

Historical integrity refers to the degree to which inclusions within the deposit derived from the same or different dynamic conditions in the past. For instance, if all the materials in a deposit derived from the action of hominids, we could argue that the deposit had considerable integrity. Similarly, if all the included material was referable to the actions of hyenas at a den, predators in killing prey, or the natural deaths of animals, each would command a judgment that the integrity of the deposit was great. On the other hand, if all the mentioned agents were represented in a deposit, we would say the deposit had little historical integrity.

The relative *resolution* of an accumulation of materials is to the degree to which items and classes of materials may be referable to a specifiable and limited, hence unambiguous, set of events or actions in the past. Assemblages with high resolution are assemblages in which all parts are referable to the set of events or conditions in the past. Resolution of assemblages may vary independently of the degree of integrity. For instance, we might have a deposit with high integrity in that all the included materials are referable to a single agent—hominids. Yet the events and activities that the hominids participated in might span a considerable period of time and represent a wide range of different behaviors, which have in common only the fact that they occurred in the same place. I think it should be clear that *integrity refers to the relative homogeneity of the agents responsible* for materials in a deposit, whereas *resolution refers to the relative homogeneity of the events or situational conditions* whose by-products are preserved in the deposit.

The basic idea as to the relative resolution of a population of depositionally associated items has been conceptualized previously. Archaeologists have recognized that deposits yielding archaeological remains might be multicomponent in character, in that the deposits derive from the independent occupation of the location by populations representing different cultures time periods, or historical eras. Thus, archaeologists could speak of a site having several occupational *components*.

In treating sites that were occupationally differentiated within a system, I adopted the term *grain* (L. R. Binford 1978b:482–483) to refer to properties of assemblages. *Fine-grained assemblages* are those in which all the included items, features, and land surfaces relate to a very few events; that is, all associated archaeological characteristics of the deposit are the consequences of basically the same events. *Coarse-grained assemblages* accumulate over a considerable period of time and/or during periods of rapid "turnover" of events, resulting in the *association* of items, debris, features, land surfaces, and the like that were differential participants in different events during the course of the occupation. Thus, the term *grain* refers to the relative contextual complexity of an assemblage from the perspective of events occurring during the course of a continuous occupation and derivative production of an archaeological assemblage. Recognizing that deposits may be variable in integrity, and in turn in resolution even given high levels of integrity, we must *minimally* have some reliable means of referring observed patterns of association to potentially different formation contexts. We must further recognize that the resolution of an assemblage must be assessed at several organizational levels before meaningful comparisons can be carried out. We may find assemblages to have low resolution because of independent occupational episodes, or because of the relative redundancy in the events occurring within a given occupational episode. These and other differences surely contribute to differences in content and hence meaning.

In later chapters of this book it will be emphasized that carnivores are apt to generate distinctive faunal assemblages in basically two contexts: at kills, and within and around lairs. These two types of assemblage, as well as contributions from natural deaths, can be considered as very likely contributors to the "background" faunal materials occurring in most any geological deposit where bones have been preserved. As suggested earlier, preservation is most likely in caves and rockshelters and in open deposits accumulated where burial was relatively quick and where moisture was either very high or very low.

I have suggested that most if not all, locations yielding evidence of our Pleistocene ancestors are most likely geological deposits, not archaeological deposits as is commonly assumed.

This assessment of the character of our Lower and Middle Paleolithic data demands that we develop means for recognizing the derivatives of different agents and different events as contributors to the geologically associated palimpsests within which may occur some traces of hominid behavior. We can no longer be content with such tautological conditions as identifying a habitation or living site by virtue of the presence of relics and other associated remains—generally bones—and then using the data from such sites to prove the nature of man's home life or other characteristic behaviors!

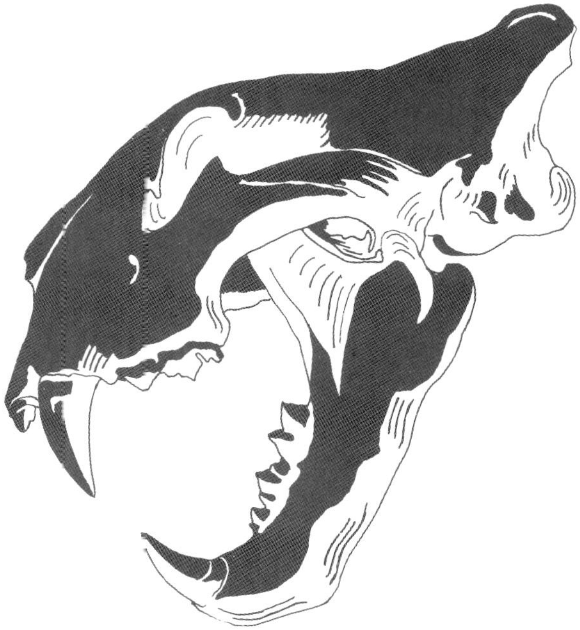

Chapter 2

Middle-range research and the role of actualistic studies

In the previous chapter I showed that the assumptions made regarding the conditions under which the archaeological record was formed directly condition the character of inferences about the contents of the archaeological record. I showed that we may be frequently incorrect or at least highly uncertain about our reconstruction of the past. In this chapter I will explore a somewhat more complicated issue—how we might proceed so as to minimize the likelihood of constructing false pictures of the past. I will be directly concerned with research tactics and how we might use secure knowledge to aid in the development of new knowledge or understanding regarding the past. *How do we carve out knowledge from ignorance?*

The challenge to archaeologists is simply this: How do we proceed? How do we unify the world of archaeological things with our ideas as to the character of the past? How may we use the empirical world of archaeological phenomena to stimulate ideas about the past and at the same time use these empirical experiences to evaluate the resulting ideas? How can we proceed so as to develop confidence that our ideas of the past are informative about the actual past? We face the challenge of science itself—how to keep our feet on the "empirical" ground and our heads in the "theoretical" sky. Basic to the development of a science is a recognition of the domain to which scientific procedures might be profitably addressed—empirical with respect to what? Theoretical with respect to what?

Many archaeologists accept the argument that the discipline of archaeology needs to adopt a scientific approach, yet they are not necessarily in agreement as to the domain of experience to which such an approach is to be most profitably addressed:

> there is in a sense an "archaeological theory" although it might be better characterized as evolutionary anthropology . . . human and cultural evolution is of such scientific and intrinsic interest that there is certainly an essential nomothetic role to be played by archaeologists [Watson *et al.* 1971:164].

In the foregoing view, archaeological theory addresses a domain of past events and conditions. It is concerned with explaining why certain events and systems came into being in the past. It addresses the domain that most traditional archaeologists considered to be their target for seeking understanding: such interesting problems as the origins of the state, the shift to agricultural production, or perhaps the origins of culture itself. Under this view of "doing" archaeology, the act of investigating the archaeological record is viewed as the experimental phase or perhaps the archival phase of investigating the past. I have referred to such interests as *general research or general theory building*. By these phrases I refer to the actions of investigators seeking to explain characteristics of cultural systems past and present. The domain of interest is cultural systems, how they vary, and how they may be modified from one form to another. The domain is interactive, generative, and dynamic.

Important, however, is the fact that all knowledge of the dynamics of the past must be inferred:

> To say that historians construct the past so as not to falsify certain theoretical presuppositions is not to point to a defect in historians or in their method. It is to focus upon the means historians use to find out what happened. It is to say, in a somewhat different way, what has been said before: *that we infer the events of the past from the events of the present by linking them in terms of some general principles* [Kitts 1977:67–68].

If we recognize that science is concerned with developing means for increasing our understanding or observations of nature, this implies that science is simultaneously attempting to generate understanding and to sharpen or increase the informational potential of our observations. The archaeologist investigates phenomena that he has reason to believe remain from the past. These investigations are conducted in the present, resulting in all the observational statements generated by archaeologists being contemporary facts. How does the archaeologist convert these contemporary observational statements or facts into meaningful statements about the past? The first thing that must be realized is that this can only be accomplished intellectually or with reason. Thus there is no way of converting observational statements about the present into meaningful statements about the past in the absence of a reasoning process.

Insofar as archaeology remains a discipline that searches for an understanding of the past through the use of objects and other organizations of matter believed to have been parts of past situations, archaeologists must operate as historians attempting to give meaning to observations on the particular archaeological record being investigated. The accuracy with which we may accomplish the conversion of contemporary observational statements into meaningful statements about the past is a direct function of the character of the reasoning processes employed and the methodology developed for evaluating the products of this process.

Our problem is then twofold: (*a*) We must know the past by virtue of inferences drawn from knowledge of how the contemporary world works, the principles mentioned by Kitts, and (*b*) we must be able to justify the assumption that these principles are relevant—that at least in terms of the properties of the principles, the past was like the present; we must make a uniformitarian assumption.

The point that we must use general principles in giving "historical" meaning to our observations no longer seems at issue:

> History differs from the generalizing social sciences only in that its primary aim is to explain individual situations in all their complexity rather than to formulate general laws for indefinitely repeatable events and processes. That is what is meant by saying that history is idiographic, the social science nomothetic (Nagel 1961:547; Elton 1969:22–24, 41). This does not mean that historians deny the existence of general rules;

> rather they seek to employ them to gain an understanding of individual (i.e., unique) and non-recurrent situations [Trigger 1978:26–27].

Where do such general principles come from, and how can we be assured of their accuracy and relevance to our activities as archaeologists seeking to explicate the past?

Those who claim that archaeologists should be historians and not scientists most commonly advise that we should "borrow" our general principles from other nomothetic sciences. The trouble with this suggestion is that I know of no nomothetic science attempting to understand the archaeological record! Many other "sciences" may be concerned with various aspects of human behavior, history, and sociocultural change in which the phenomena studied are events, behavior, or patterning in communicated thought. However, the basic phenomena with which we work are (*a*) static, (*b*) material, and (*c*) untranslated into symbols or clues to human "thoughts." No other "science" addresses such phenomena. It was the recognition of this fact that prompted the following statement written in 1966 and published two years later:

> Accepting Spaulding's minimal definition of what archaeology is, we can go a step further and specify its aim as the explanation . . . of the order we observe in the archaeological record. Archaeological theory consists of propositions and assumptions regarding the archaeological record itself—its origins, its sources of variability, the determinants of differences and similarities in the formal, spatial, and temporal characteristics of artifacts and features and their interrelationships [S. R. Binford and L. R. Binford 1968:2].

Directing attention to the archaeological record rather than continuing the self-deceit that we were studying the past seemed central to progress. My view was that we could not reconstruct history until we first addressed the problem of how we give meaning to the archaeological record (see L. R. Binford 1968d). Meanings are carried by concepts and arguments and the archaeological record contains only arrangements of matter. If archaeologists are to know anything of the past, they must develop a science. The domain of this science must be the archaeological record per se.

In seeking to develop a science of the archaeological record, are there not some fundamental characteristics of both science and the archaeological record that we must consider to guide the growth of this science? The answer must of course be yes. One characteristic particularly important to the arguments advocating a science of the archaeological record is that science attempts to evaluate the role and utility of ideas for enhancing understanding. Ideas are of course cultural forms:

> if we view culture as at least referring to the particularly human ability to give meaning expediently to experience, to symbol, and in turn, view experience through this conceptual idiom, science is then concerned with evaluating the utility of the cultural tools produced [L. R. Binford 1977a:3].

The reference to the "cultural tools" produced is of course to the concepts and ideas in terms of which we conceive the world of experience. If we gain a "knowledge" of the world through the use of cognitive devices, words, concepts, and ideas, and the world is described in these terms, we must face the problem of the accuracy, utility, and "reality" of such cognitive devices themselves. This is one fundamental problem the scientist must face. The second problem relates to the degree to which we seek knowledge and understanding beyond simple description. We frequently attempt to understand why the world is the way it appears to be, given the description generated. Scientists carry out their work with essentially two sets of intellectual tools: a conceptual frame of reference or paradigm (Kuhn 1962), and various theories that seek to explain the world as "known" through the use of the paradigm.

The Paradigm—One's Guide to Describing the World

The cognitive frame of reference or paradigm consists of the ideas and concepts with which we approach experience. These condition what one considers relevant to describe or chooses to discuss as of interest. One's cognitive frame of reference may be thought of as the culture of a science. It consists of the concepts in terms of which experience is intellectually assimilated. Despite all the definitional controversy (see Masterman 1970), I follow Kuhn (1977) in viewing paradigm as the intellectual terms upon which one meets experience. The character of one's frame of reference conditions what is considered relevant to describe, what is interesting to discuss, and even how we view the world in terms of problems to be solved. In short, it is what we expect the world to be like. Things become complicated when we recognize that we cannot gain a direct knowledge of the essential properties of the world. Our cognition is neither direct nor objective, but may be indirect and subjective relative to our beliefs about the world (i.e., our paradigm).

We generally defend our claims about what the world is like with inferential arguments. I prefer to call these *warranting arguments;* they are arguments advanced that tend to *warrant to others* the beliefs one has about the world. If done in a robust manner, they make one's claims appear plausible, and acceptable to others. Rarely are such arguments formalized in that the premises are rarely explicitly stated, so conclusions are warranted by appeal to a "common body of knowledge or belief." The more comprehensive the alleged knowledge, or widespread the belief serving as the intellectual context for a warranting argument, the more plausible it appears and therefore the greater likelihood it has of being accepted.

Working within a frame of reference is similar to participation in any other culture; we accommodate experience through our shared cognitive devices. The fact that they facilitate this accommodation appears to us as proof that the world is in fact the way we expect it to be. We may be astonished that others do not see the world the way we do. Anthropologists should be familiar with cultural differences and should be fairly comfortable with the idea that the nature of experience does not necessarily determine the nature of culture. Many persons share identical experiences yet ascribe to them very different meanings; this is essentially the message of anthropology. Cultural man has for all time believed that his beliefs were given by "reality" and were therefore "right," whereas those of other cultures were clearly misguided or "stupid" for not having seen the "truth" inherent in given experience.

Archaeology is perhaps in a fortunate position. Although there is much contemporary "culture" or paradigmatic bias regarding the nature of man and the causes of history, there is very little folk knowledge regarding the formation of the archaeological record. This means that there is little explicit prior development of cognitive devices and frames of reference for accommodating archaeological phenomena in the literal, static sense of the word. For the further development of archaeology, the growth of a paradigm, developing cognitive means for identifying properties of the past or diagnosing the archaeological record and thereby giving meaning to the archaeological record, is crucial.

Much of the time use of a paradigm is viewed as an act of identification. Can we identify a habitation, a hide scraper, a matrilineage, a base camp, agriculture? Or can we diagnose the functions of a site, tool, or element of debris? In most cases we are seeking an unambiguous definition, and realistic concepts with which to partition or diagnose the archaeological record and thereby generate meaningful statements about the past. All such interpretations are dependent on a general, accurate, and unambiguous knowledge of the relationship between statics and dynamics, the formal consequences for organized matter that derive from the operation of a dynamic system. In developed sciences, what is being sought here at the conscious level through "middle-range" research may be taken for granted as paradigmatic:

> The distinction between "empirical" and "theoretical" . . . may be only a relative one. It is relative historically. . . . A scientist who undertakes the study of a particular problem, for example of a biological one, and who uses various scientific instruments constructed on the grounds of different physical theories, is quite aware of the fact that together with the equipment he uses he accepts also these theories. In spite of this fact, however, he will treat the statements he will formulate by means of these instruments as observational. The observational language is, for him, something already present and historically given by the development of science and common knowledge [Amsterdamski 1975:86].

An observational language is essentially nonexistent in archaeology. The concepts and hence paradigmatic characteristics of traditional archaeology are believed to be essentially useless for modern archaeology. Today the archaeological record is not being viewed (by most) as a material manifestation of mental phenomena; it is not being viewed as a preserved past; it is not being viewed as uniquely determined by history; its variability is not being viewed exclusively as a manifestation of past ethnic variability, and so on. As is suggested by Amsterdamski, the instruments that permit and facilitate unambiguous meaningful observations must be developed, demonstrated, and tested, using scientific means. Later, as the science of archaeology becomes more mature, these "instruments for measurement" may be taken for granted and results of their use treated as direct observations on the past.

We are a long way from this level of maturity today. We need to recognize very explicitly the current state of the art and address the growth of a new paradigm as basic and fundamental. Recognizing that this is a historical phase in the growth of the "new archaeology," I began using a special term for this endeavor: *middle-range research* or *middle-range theory building.*[1]

What we are seeking through middle-range research are accurate means of identification, and good instruments for measuring specified properties of past cultural systems. We are seeking reliable cognitive devices; we are looking for "Rosetta stones" that permit the accurate conversion from observation on statics to statement about dynamics. We are seeking to build a paradigmatic frame of reference for giving meaning to selected characteristics of the archaeological record through a theoretically grounded body of research, rather than accepting folk knowledge—let alone implicit folk knowledge—as the basis for describing the past.

Theory—One's Guide to Explaining the World

> Theories are the key to the scientific understanding of empirical phenomena, and *they are normally developed only when previous research has yielded a body of information, including empirical generalizations about the phenomena in question.* A theory is then intended to provide deeper understanding by presenting those phenomena as manifestations of certain underlying processes [Hempel 1977:244; emphasis mine].

Given that we have made observations on the archaeological record, offered some generalizations about its properties, and gained considerable experience with the record, I must now ask the crucial question: Why is the archaeological record the way it appears to be? When we seek to reason about the "causes" of the world as known, we are attempting to build theories about the world. "Where it is some event or system of events that is to be explained, explanation has to do with cause [Quine and Ullian 1978:111]."

We are concerned with organizational properties of the world. We seek to understand how the properties of entities and/or events were produced in characteristic ways:

> One very central use of "theory" involves an epistemic device which is used to characterize the state-change behavior of isolated systems within a general class of phenomena . . . one can discover that they [theories] invariably postulate a class of states of systems' change over time . . . and are used to characterize how natural classes of phenomena would behave if isolated [Suppe 1977:658].

Quite literally, theories are the answers to the "why" questions of dynamics. They are concerned with understanding variability and how systems proceed from one state to another.

If we are going to build a theoretically informed paradigm for referring observations on the archaeological record to dynamic conditions in the past, where do we begin? It seems to me we must begin with certain fundamental statements of "being as such." *The archaeological record is a static contemporary phenomenon.* It is structured matter motionless and noninteractive in terms of the properties of historical interest to the archaeologist.

> Only a universe of energy could have no past. If there is matter, structures grow and differentiate and a past can be recognized and partially reconstructed. It is the problem of durationless non-matter versus enduring mat-

[1]This is essentially identical to what David Clarke called *interpretive theory* (Clarke 1973:8) and appears to be what Schiffer (1976) means by *behavioral archaeology.* (See also Sullivan 1978.)

ter. . . . At one end of the spectrum is biblical chaos, a past without a past, because no matter exists to convey information. At the other end there is only information and no decisions—static information forever [Margalef 1968:97].

The archaeologist is of course working with static information preserved in structured arrangements of matter. Since there is no energy remaining, there are no culturally relevant interactive relationships to be observed in the archaeological record. Such relationships existed in the past but ceased when system-serving energy was no longer powering the rearrangement and modification of matter—in short, once a static condition was achieved. In a very essential way the contents of the archaeological record must be viewed as products of a complex *mechanical* system of causation. It was mechanical in that the fundamental genesis of the archaeological structure is a situation of forces acting to modify matter in both its organizational and distributional properties. The archaeological record is a structure of relationships between the distribution and form of matter as caused by energy sources acting on matter in the past. In one very important sense, all properties of matter, whether they be chips removed from a flake of flint, mixing of soil betraying the former location of a pit, piles of debris from meals, or the remnants of a construction such as a mud brick wall, are the mechanical consequences of the actions of forces on matter.

I used the term *causal* in the literal sense, that is, to express the idea of a category of generic connections; it refers to the way of producing things. Or, "something, E, is brought forth by something else, C, in a necessary (constant and unique) manner [Bunge 1979:49]."

Clearly if we can isolate causal relationships between things, and if we can understand such relationships in terms of more general principles of necessity, such as the theories of mechanics or some other basic science, then we have a strong warrant for the inference of the cause from the observed effects. We would be building a strong theoretically informed bridge between properties of the contemporary archaeological record and characteristics of the dynamic past.

Insofar as our inferences regarding the past refer to the causal relationships that obtained between dynamics and its static derivatives, then any attempts to discover the character of such causal relationships must reasonably be conducted through the study of living systems where both dynamics and static derivatives may be potentially observed. Taking as an example the problem outlined in Chapter 1, the identification of the agency (energy source) responsible for generating certain patterns remaining in the archaeological record, we might reason as follows:

First, we must attempt to isolate the different agents or forces that might be expected to contribute to or "cause" a given pattern. Second, we would have to conduct studies of these agents or processes in the contemporary world so as to develop criteria of recognition. In short, we need to specify criteria for recognizing traces, "signature patterns" apt to be preserved in the archaeological record, of the agents likely to have contributed to deposits in which hominid remains might also occur. The procedure is similar to that painstakingly worked out over the years for recognizing lithic materials modified by man as opposed to stones modified through other natural processes. The problem is one of pattern recognition linked with the demonstration that the pattern is redundant and unambiguous, a diagnostic signature that discriminates one agent or set of agents from another.

Such a demonstration must be developed by studying phenomena actively generated in a contemporary setting, since there must be little problem of *inference* regarding the identify of the agent producing the patterning or traces that one is demonstrating as a signature pattern sufficient for the unambiguous identification of the agent. The problem is one similar to the development of a key identification of animals through the study of their footprints. The persons who develop the knowledge that permits the recognition of the track, and hence the identification of the animal responsible, *must* study the footprints of identified animals so that the relationship between animal and track is a controlled or known relationship. Given such a control in the contemporary world, and given that one is successful in recognizing and describing diagnostic criteria (constant and unique) between cause and effect, animal and footprint, then when one encounters the diagnostic footprint in the future the inference of the prior presence of the indicated animal may be considered an inference of high probability.

For an inference about the past to be of high probability, an additional proposition must be met—that the same relationships obtained in the past as obtained in the present between bears and their footprints! Here we introduce the interesting and important, perhaps crucial, problem archaeologists must solve—how do we justify a uniformitarian assumption? This issue is perhaps well illustrated through a discussion of the treatment given the problem by the pioneers of historical geology:

> Lyell's concept of uniformity has four major, and very different, components:
>
> (1) Natural laws are constant (uniform) in space and time. *As John Stuart Mill showed, this is not a statement about the world; it is an a priori claim of method that scientists must make in order to proceed with any analysis of the past.* If the past is capricious, if God violates natural law at will, then science cannot unravel history.
> (2) Processes now operating to mould the earth's surface should be invoked to explain the events of the past (uniformity of process through time). *Only present processes can be directly observed. Therefore, we are better off if we can explain past events as a result of processes still acting. This again is not an argument about the world, it is a statement about scientific procedure* [S. J. Gould 1977:150; emphasis mine].

As was pointed out by Gould, the remaining two senses in which Lyell used the concept of uniformity were in fact assumptions about the world, existential in character. One has been largely sustained by research (i.e., geologic change was largely uniform in rate, slow, gradual and steady, not cataclysmic). The other claim was also existential, namely that the earth has a uniform configuration, or it has been fundamentally the same since its formation. Most would agree that this has been demonstrated to be quite false as a general descriptive statement.

What is indicated here is that we must make uniformitarian assumptions if we are to gain any understanding of the past. On the other hand, when we do so we are making empirical claims about the past and these must be warranted; they must be subjected to evaluation. The degree to which such uniformitarian assumptions are warranted is a measure of the degree to which our inferences drawn from knowledge of the contemporary world and/or our understanding of its processes in the form of theories and laws are relevant to the past.

Insofar as our inferences regarding the past refer to the dynamics of the past, these inferences must be accomplished by appeals to principles or knowledge about dynamics and how static properties preserved in the archaeological record may be derived from dynamics. Since the only access a researcher has to dynamics is through contemporary experience, all research directed toward the development of principles that serve to make possible inferences about the past must be conducted with documented dynamic situations generally in the present. Such knowledge of "connections" between statics and dynamics must derive from experimental research conducted with documented living systems.

Since knowledge of dynamics derives from experience with living systems, observations of linkage between statics and dynamics must be made on living systems. In order to use these principles of linkage for making statements about the past, we must make a uniformitarian assumption with respect to the properties used in inference. In short, we must assume that knowledge gained from actualistic studies is relevant and applicable to the living systems of the past. This basic proposition must be true if inferences employing principles gained through the study of contemporary dynamics are to be used in inferring the past from patterned statics. This means that the assumption is always conditional and may be false; that is, we could be wrong in our judgments regarding the condition shared by systems or entities of the past and the present.

For instance, any number of "correlates" between statics and dynamics might be observed in the modern world. However, the first question we must ask is whether we are observing an incidence of cause and effect, or whether there is simply correlation or coincidence. The second, and equally important, question to be considered is whether the proposed causation was also characteristic of the past. Both questions must be answered affirmatively before an actualistic observation may realistically serve as a premise for inferences regarding the past.

Although I basically agree with much of Schiffer's (1972, 1976, 1978) general discussion of archaeology and the need for understanding of formation processes, I generally disagree with almost all of his suggestions as to how to solve archaeological problems. He fails to make the critical distinction between description and explanation. This is clear as he cites Nagel for "experimental laws" (i.e., empirical generalizations) and Hempel for "covering laws" (i.e., theoretical laws), as if these were the same thing (see Schiffer 1976:4). Schiffer also argues that "the subject matter of archaeology is the relationship between human behavior and material culture in all times and places [p. 4]." I might agree that this is one way of viewing the concerns of middle-range research, but find it hard to accept as the central focus for archaeology since the archaeological record contains no direct information on this subject whatsoever!

How do we know what experiences with living systems are relevant to the past? This question is particularly germane with regard to central issues such as identification. Identification, as mentioned earlier, is a key issue in archaeology, since it is this "act" that establishes the language for discussing the past, and in turn the language carries meanings and provides the units for logical analysis. Identifying things (see Whitehead 1967:144) becomes the act of translating from the domain of matter into the domain of ideas. It is the identities that bridge the gap between the past and the present, that provide, as Whitehead (1967:159) would say, the "eternal objects," the "durables," which serve as the basis for recognizing events, the basis for analyzing events and recognizing transitions from one event to another:

> Whatever passes is an event. But we find entities in nature which do not pass. . . . Factors in nature which are without passage will be called objects . . . recognition is reflected in the intellect as comparison . . . but it is not the events which are compared. For each event is essentially unique and incomparable. What are compared are the objects and relations of objects situated in events [Whitehead 1957:124–125].

It seems to me that uniformitarian assumptions function much like intellectual anchors, for they provide the "points of knowledge" from which we may judge the extent of our ignorance regarding properties of the archaeological record.

What are the durable unchanging characteristics that the events of the present share with the past? As I indicated elsewhere [L. R. Binford 1977a:8],

> We may reasonably ask . . . whether or not there are classes of data remaining from the past which might better support uniformitarian assumptions. In short, are there not classes of phenomena available to us for which a more reliable set of conditions might be projected into the past than for human behavior per se?[2]

I answered the rhetorical question by suggesting that the study of the spatial structure or the arrangement of "objects," in the Whitehead (1957:124) sense of the world, would be a useful area for development. I continue to be of this opinion. On the other hand, I had suggested that ecological and anatomical characteristics of the species still extant with which ancient man interacted were enduring objects for which uniformitarian assumptions might be securely warranted. It is hoped that others will elaborate this list of domains and pursue middle-range research along as many diverse lines as we may be able to justify uniformitarian assumptions.

I began the discussion in Chapter 1 with a demonstration that the "interpretation" of certain archaeological observations was dependent on a basic premise, an assumption about the conditions in the past surrounding the formation of the deposit within which archaeological remains were recovered. I showed that the assumption was generally made that man was the agent responsible for the disposition of all materials found in association with demonstrable artifacts. All the "interpretations"—the postulation of bear cults, cannibalism among early hominid populations, mass killing of elephants at Torralba, systematic hunting of hyena by Neanderthalers at Pin Hole shelter—were inferential arguments consis-

[2]Richard Gould has argued that, since some characteristics appear unlikely to bridge the present and the past, we should avoid uniformitarian assumptions. "The less the archaeologist must depend upon uniformitarian assumptions to infer past human behavior, the more valid his explanations will be [R. A. Gould 1978:255]." This is nonsense, in my view, since any inference, even a simple identification, to the past *must* make a uniformitarian assumption.

tent with the initial assumptions, the premises upon which the inferences rested. There is an important characteristic of all inferential arguments, simply that *we can never reason in a valid manner from premises to a conclusion that contradicts the premises with which we start.* This fact has important implications for archaeologists:

1. All our statements about the past are inferences relative to observations made on the contemporary archaeological record.
2. The accuracy of our inferential constructions of the past is directly dependent on the accuracy of the assumptions or premises serving as the basis of our inferential arguments.

The conclusion we must draw is that we cannot use either the archaeological record or the inferred past to test our premises or assumptions. Quite literally, all our reasoning is "locked in" by our original premises and observational language. Unless we can take our premises to experience and permit experience to pass judgment on their accuracy, we can never gain a critical perspective with regard to our beliefs about the past. "Can we present historical events as instances or confirmation for a law? We cannot if the very law we wish to test has been presupposed in inferring the event [Kitts 1977:79]."

Put another way, since we construct the past inferentially we cannot use our constructions to test the accuracy of the premises that provided the basis for the characteristics constructed.

Since we cannot use the inferred characteristics of the past to test the basis for our inferential procedures, how do we develop reliable means for knowing the past? The answer, as I have intimated, is that we must engage in middle-range research, which consists of actualistic studies designed to control for the relationship between dynamic properties of the past about which one seeks knowledge and the static material properties common to the past and the present, Whitehead's "eternal objects"—in short, the characteristics about which uniformitarian assumptions may be made, those things which the present shares with the past. These common things provide the basis for a comparison of the events of the present with the events of the past or events from different times in the past.

The reason that middle-range research must be basically actualistic is that only in the present can we observe the bear and the footprint together, the coincidence of the dynamic and the static derivatives. In more mature disciplines, where a relatively sound methodology and a sophisticated observational language exist, it may be possible to use inferred conditions about the past as premises for further inferences if the initial premises serving as the basis of the original inference are securely documented and "verified" at the middle-range level of research. As illustrated in Chapter 1, this is probably a very risky strategy, given the lack of sophistication in contemporary archaeology.

The dependence of our knowledge of the past on inference rather than direct observation renders the relationship between paradigm (the conceptual tool of description) and theory (the conceptual tool of explanation) vague, it also renders the "independence" of observations from explanations frequently suspect and commonly standing in a built in relationship, thereby committing the fallacy of "confirming the consequent."

It is this condition that renders it imperative that our methods for constructing the past be intellectually *independent* of our theories for explaining the past. That is, the theories explaining the archaeological record, the work that provides our observational language and conveys meaning to archaeological phenomena, must be intellectually independent of our a priori ideas of the past, or our theories regarding the processes responsible for past events, patterns of change, or stability. *Our middle-range theory must be intellectually independent of our general theory.* Middle-range theory must be tested primarily with documented living systems. Middle-range theory treats the relationship between statics and dynamics, between behavior and material derivatives. General theory may be tested using archaeological phenomena meaningfully operationalized through middle-range research. Stated another way, general theory must be evaluated using instruments for measuring the variables specified in the theory. These instruments must have been developed independently through middle-range research. In the absence of methods for reliably monitoring the variables said to be determinantly operative, no archaeological test of general theory is possible.

The conclusion should be clear: Middle-range research, with particular emphasis of theory building, is crucial to the further development of archaeology. We cannot "know" the past without it, and we cannot evaluate our ideas about the past and why it was the way it appears to have been without means of monitoring the conditions or variables believed to be important. Both of these tasks are dependent upon the development of middle-range research.

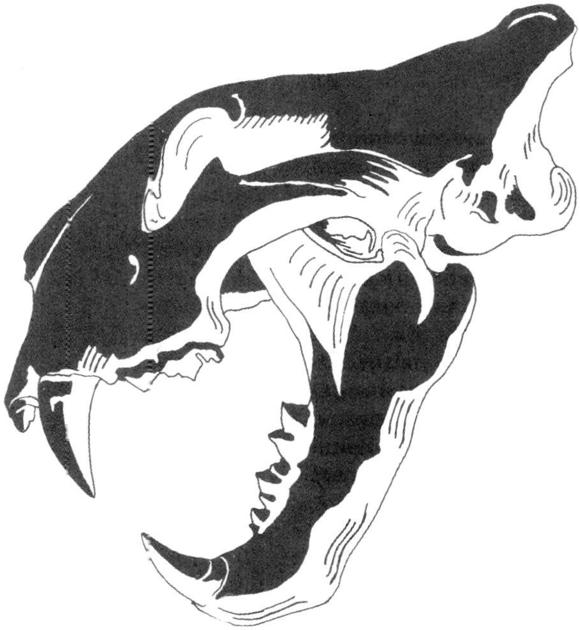

Part II
Middle-range research—In search of methodology

In Chapter 1 I argued that a particular set of assumptions about the formation of deposits that contained archaeological remains served as the warrant for further inferences regarding materials found in association with the archaeological remains.

I attempted to show that Paleolithic archaeology developed in a situation where a specific concern for methodological research was not seen as separate from research conducted for purposes of learning about the past. Interpretations were largely developed post hoc or after discoveries had been made in the archaeological record. These procedures largely consisted of using inferences based on assumptions regarding the formation processes operative in the past or the conditions responsible for morphological properties or patterns of association observed in the archaeological record. Once such post hoc interpretations were offered and judged "probable" or plausible they frequently became conventions whereby additional observations at new sites were interpreted. Gradually a myth was built up about the past. The myth consisted of inferences drawn from unevaluated premises and its very scale of acceptance gradually became further justification for belief in the myth. Unfortunately this remains one of our dominant "methodologies."

In Chapter 2 the point was made that we need to develop methodologies and observational language through middle-range research. This research was envisioned as largely actualistic, where controlled information about causes and effects could be evaluated experientally rather than inferentially, as has been the common practice. Methodological research requires that both the bear and the footprint be observable. Methodological research in service of archaeology must normally be conducted with living systems (ethnoarchaeology) or ones in which the relevant dynamics have been recorded (historical archaeology) or where the relevant dynamics may be replicated (experimental archaeology). The point of view adopted here is that actualistic studies or middle-range research is crucial to archaeology and should be conducted from the perspective of the archaeological record. We are not studying material "residues" (R. A. Gould 1980:250–251) or "behavioral correlates" from the perspective of the ongoing living system. This view contrasts markedly with that of Schiffer (1976), Rathje (1979), and Stiles (1977), who seek a science of material culture, or relationships between behavior and material culture. I am not suggesting that studying contemporary systems from a materialist's perspective should not be done, only that doing so is not archaeology; it is ethnography and as such faces very different methodological and intellectual problems. Stated another way, we are not attempting to specify the relationships between "behavior" in any exhaustive sense and material remains. Instead, we are attempting to understand the determinants of patterning and various structural properties of the archaeological record in order to learn about their past.

It was suggested in Chapter 1 that we need methods for assessing the integrity (the number of identity of the agents that acted in the past to produce the deposit being investigated) and the resolution of a deposit (the number of and redundancy in the events represented by the deposits). The bulk of this book is concerned with the development of methods for identifying the agent and the behavioral context of nonhominid predator–scavenger contributions to paleolithic and paleontological deposits. This is basically a problem in identification.

Following the suggestions developed in Chapter 2 I will describe acquired properties of bones and patterns of assemblage variability that are referable to known agents. I will attempt to justify that some properties are uniquely referable to specific agents, and will suggest some of the behavioral characteristics of the agents that ensure the uniqueness of the properties described.

The majority of materials described in the following three chapters were collected during a 4-year study of the use of animals and animal products by the Nunamiut Eskimo of north central Alaska. (See L. R. Binford 1976, 1978a, 1978b, 1979, 1980; L. R. Binford and J. B. Bertram 1977; and L. R. Binford and N. J. Chasko 1976 for reports on this research.) The focus of the Nunamiut study was the relationship between their hunting and consumer strategies as executed in functionally and seasonally differing sites (L. R. Binford 1978b) and the faunal remains at those sites. The study was middle range in character, aimed at developing methods for recognizing sites of varying function from faunal remains.

Chapter 3 summarizes a considerable body of observations on bones modified by animals. As indicated in Chapter 1, many controversial claims regarding the nature of the past have been inferences from observations on broken and modified bone. I will review some of the "interpretative literature" where faunal remains have been central to arguments about the past. This material is presented because the contents of Chapter 3 are germane to these arguments, but also because my knowledge of these arguments has guided much of my observation on bone. As Darwin is credited (Cohen and Nagel, 1934: 197) with having said; "How odd it is that anyone should not see that all observation must be for or against some view, if it is to be of any service." Chapter 3 treats a number of the properties of bone modification about which a vast number of inferences regarding the past have been prompted. Since there have been so many claims, much of this chapter will appear negative. I will systematically cite properties of bone breakage coupled with surficial modifications produced by either dogs or wolves and compare them to specimens or conditions that others have cited as evidence for human–hominid modification or manufacture. Although it is quite true that much of this chapter appears to be concerned with discrediting the inferences made by others, I am simultaneously presenting descriptive material in sufficient detail to provide a basis for the identification of the agents responsible for modifications that may be of interest in future research.

In a very real sense I have already written a book on the topic covered in Chapter 4—what men do with bones. As you might imagine, I will draw heavily on this previously published material, but some new facts will be presented. My orientation is one of comparing human behavior with animal behavior on the one hand, and evaluating claims other archaeologists have made regarding the meaning to be attached to certain formal patterning in assemblages of modified bone on the other.

Chapter 5 is concerned with the central problem of the book, the diagnostic differences between bones modified by man and those modified by animals. I will present some original observations on wolf behavior and *integrate these observations with those of others regarding the behavior of predator–scavengers. These discussions will then provide the background for a consideration of assemblage variability, that is, the determinants of patterns of association among different anatomical parts in terms of animal behavior contrasted with that of man. From this I will draw some intellectual expectations about the behavior of predator–scavengers, "anchors" that we can project into the past and use to identify the conditions that generated certain configurations of faunal material remaining for us to observe today.*

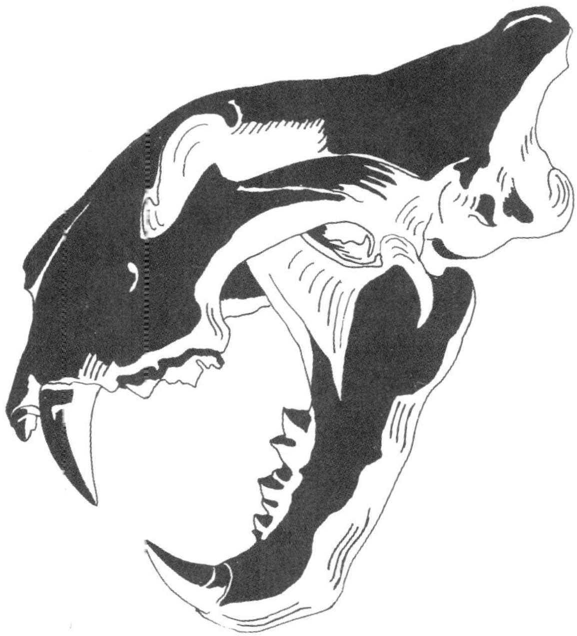

Chapter 3
Patterns of bone modifications produced by nonhuman agents

The observations summarized in this chapter are largely concerned with the modifications that animals, considered strictly as mechanical agents, make on bones. That is, I am concerned with tooth marks and patterns of differential destruction that gnawing animals are capable of producing and in fact regularly produce. In all my descriptions I am concerned with recognizing modifications of bone that are distinctive or otherwise diagnostic of action by nonhuman agents. I cannot emphasize too much that with animals most bone modification is effected by teeth. Teeth are the implements and the jaw is a mechanical device for exerting force; the two constitute a distinctive mechanical system. This is considered to be a very nonhuman mechanical situation, in that man commonly employs a variety of implements, such as knives and hammers, and the mechanical situation is one of directed cutting or impacted hammering. For animals, the teeth are organized in what may be viewed mechanically as a vise. These are real differences in the mechanical basis for human versus nonhuman predator–scavenger modification on bones. Most of my descriptions focus on recurrent properties believed to reflect these mechanical differences.

Obviously animals differ in size, feeding habits, the mechanical strength of their jaws, and the design of their teeth. Nevertheless, they all may be considered animated vises. It is expected that the basic patterns of modification that an animal is mechanically capable of producing are not going to vary a great deal from one species to another. It is more likely that the magnitude of the capability, and the use of it, will vary among species. A very strong animal may be able to accomplish the destruction of large bones but the tactic of destruction is expected to be essentially like that of a smaller animal addressing a smaller bone. This is an operating assumption guiding the presentation of much of the following material. I will remain sensitive to the possibility that the assumption is overly simplistic.

The demand that the bear and its footprint be observed together is only partially met for the materials described. The majority of the bones used in the descriptions and illustrations of modifications produced by animals were recovered from nine dog yards within the village of Anaktuvuk, Alaska. A total of 12,716 bones were observed for patterns of surficial scarring, bone breakage, and destruction. An additional 416 recovered from two small wolf dens served primarily to permit the assessment of the impression that the patterns produced by the domestic dogs were not unique to the dogs.

The samples from the Anaktuvuk dog yards were not collected with this particular research aim in mind. The original motivation was to study in detail the selection of parts by the Eskimo for use as dog food and to monitor any seasonal changes in the feeding strategies (see L. R. Binford and J. B. Bertram 1977: 78–90). These dog yard samples were supplemental to actual records of dog feeding and follow-up studies of the bones remaining at feeding spots after the dogs had been moved away (see L. R. Binford 1978b:262–265). In addition, I had conducted extensive interviews with the families then keeping dogs regarding their dog-feeding tactics.

Thus the sample has a weakness. Man actually introduced the bones to the dogs after he had killed the animal and transported the parts to Anaktuvuk. That is, all the characteristics of the bones cannot be unequivocally referred exclusively to the behavior of the dogs. Nevertheless, the characteristics described can be ascribed to the dogs with the highest levels of confidence. For instance, my information on patterns of dog feeding by the Eskimo permits me to say with confidence that no bones previously broken for marrow within the village were given to the dogs. The Eskimo generally do not feed their dogs "garbage" or table scraps, but when this general pattern is violated it is never with respect to previously broken long bones. When bones that have been previously "consumed" by humans are given to the dogs, they are almost always "women's bones" (see L. R. Binford 1978b:149) or those that have been boiled. These include the neck, thoracic vertebrae, ribs, lumbar vertebrae, pelvis, and scapula. The only modifications that might occur to these bones prior to modification by gnawing dogs would be cut marks from acts of dismemberment, chop marks for sizing the bones to fit in a stew pot, and sometimes "mashed edges" resulting from human chewing at the margins of the bones during consumption. Feeding dogs women's bones is much more common on winter sites where the warm bones are viewed as a treat for the dogs. I never observed this table-scrap feeding in my spring or summer dog-feeding records. Generally dogs are fed complete butchering units, that is, a complete rib slab, a complete scapula with attached meat, a complete pelvis, and so on. In butchering the Nunamiut dismember the body of a prey animal by cutting between the articulator surfaces of the bones. Sometimes multiple bone units will be returned from the field to the village. These may be further butchered at the time of dog feeding, but except in the winter, when the meat is frozen, secondary butchering is the same as primary: dismemberment by cutting between the bones at the joints. This means that single or multiple bone units are what the dogs receive as food. They are almost never presented with previously broken bones. The exception to this generalization is feeding during winter when meat stored outside is always frozen. Secondary butchering under such conditions is accomplished with an ax and it is common to break through the shafts of long bones as a regular dismemberment strategy. For these reasons, three of the dog yard samples (see L. R. Binford and J. B. Bertram 1977:80) listed (samples II, III, and IV) in an earlier report were eliminated from this analysis and only consulted concerning patterns of destruction of articulator ends.

It should be pointed out that I have analyzed over 122,000 bones known to be the result of Nunamiut behavior. I have not reported any type of modification as referable to the dogs that regularly occurred in the known "human sample" (see L. R. Binford 1978b).

The sample of dog-modified bones is considered unambiguously informative about dog behavior with respect to the following properties: (*a*) surficial modifications and (*b*) patterns of destruction or breakage of single bone units. Properties not considered referable to dog behavior are (*a*) the relative frequencies of the bones present and, (*b*) differential destruction due to the accessibility of the different bones—that is, disarticulation had already been accomplished by men before the dogs gained access to the anatomical parts, so patterns of destruction were not generally conditioned by the patterns of anatomical arrangement characteristic of a prey species. Thus the population properties of animal-obtained assemblages cannot be understood through the study of this material. On the other hand, variability in modification and destruction of given anatomical parts is documented in a reliable manner.

Again, I am confident that the properties described are referable to the action of dogs. The bear and the footprint are together with regard to modifications of single bone units. However, I cannot use these data for studying how dogs would modify larger units of anatomical organization, such as a leg or a whole animal.

All the generalizations offered are with regard to caribou bone. I will illustrate sheep bone occasionally to indicate that the generalizations about caribou also apply to sheep. Stated another way, I am talking about the modification of caribou bone by Eskimo dogs—largely malemiuts—weighing between 55 and 70 lb. I will frequently cite literature in which other researchers have described similar modifications of either bones from other species (horses, bison, etc.) or bones modified by other agents (large cats, hyenas, coyotes, etc.). These references serve to warrant the opinion that the modifications produced by dogs may be generalized to other species of predator–scavengers and to the bones of other prey that might be attacked by gnawing animals.

Previous Approaches to Understanding the Significance of Broken and Modified Bone

As was shown in Chapter 1, I seriously question many of the claims Old World archaeologists presented regarding the past. However, I concentrate on New World archaeology in this survey because it provides an interesting case of research moving toward the development of a methodology that, as I will attempt to demonstrate, is largely poorly founded and misleading. It has the added interest that many of the properties cited by New World archaeologists have been cited by Africanists and other Old World archaeologists as diagnostic of human behavior. The very fact that many of the same formal properties have been observed in widely separated places and with reference to differing time periods adds to their middle-range research potential. If these properties can be understood accurately, then they are good candidates for strong methodological tools, since making secure uniformitarian assumptions about their meaning is encouraged by their broad spatial and temporal distributions.

Most claims for man's use and modification of bones as tools have been based on cited evidence of either altered shape or particular modifications such as "flaking," "retouch," "polish," and "abrasion," which were believed to be products of human action. In addition there were claims made that man broke bones differently from animals, resulting in distinctive and diagnostic patterns observable among the fragments of long bone. As was pointed out earlier, Breuil (1938:58) thought that man broke bones "longitudinally" whereas animals broke bones transversely. This view was accepted by Weidenreich (1941: 74–77) and cited as evidence that *Sinanthropus* was a cannibal. Raymond Dart (1960) accepted the suggestion of longitudinally fractured bone splinters, and proposed that this type of breakage was accomplished by wedging bone tools into exposed bone cylinders, thereby splitting the shafts of long bones longitudinally! (for criticism see Feustel 1969, 1970.)

> I have pictured other examples of bone interpenetrations in the memoir on the Osteodontokeratic Culture such as an antelope calcaneus wedged between the Zygomatic arch and the lateral aspect of a split hyaena skull, and flakes of bone or whole bones wedged within other bones. These examples are here shown to be typical and are now sufficiently numerous to constitute irrefutable evidence of a systematic australopithecine technique not merely of smashing up skulls but also of splitting by means of driving other flakes of bone or whole bones forcibly down the shafts of other bones [Dart 1960:136].

The famous crack-and-twist technique of bone fracture was proposed by Dart not as a method of fracturing bones for marrow extraction, but instead as a means of tool production, a position later adopted by Sadek-Kooros (1972).

> we were both ignorant of the means whereby primitive sapient man at Kalbank 15,000 years ago and the still more primitive protoman *Australopithecus prometheus* about 1,000,000 years ago at Makapansgat obtained these spiral blades until one Sunday we preserved the femur from the family's roast leg of mutton.
> The following day he [Kitching] brought it to the laboratory, struck the shaft with the point of a stone implement and holding each end of the bone in a hand, twisted them apart in opposite directions. The beautiful resultant spiral blades and the point of impact of the stone implement are clearly seen. . . . So the next time this Sunday joint provided James Kitching with a further experimental opportunity he simply struck the shaft against the wooden edge of the table and twisted the bone ends contrary wise again and the spiral and pointed results are seen [Dart 1959b:91].

Several points should be cleared up here. Experiments have repeatedly demonstrated (L. R. Binford 1978b:152–157; Bonnichsen 1973:14, 1979:37–53) that spiral fracture results from breaking relatively green bone. It is not conditioned by twisting. Twisting may condition two types of situation; most likely it simply facilitates the separation of fractured bone from the periosteum and in no way conditions the pattern of actual breakage. In situations such as are described by Sedek-Kooros (1972) where a line of minor impacts are made longitudinally along a bone, the twisting may actually ensure a longitudinal crack connecting all the small points of impact. This is difficult to accomplish and most often twisting simply removes or separates fractures that are otherwise held together by the periosteum. (See, for instance, such fractures indicated in experiments 19 and 22 [Figure 5] of Sadek-Kooros [1972:374].) Despite Dart's designation of the crack-and-twist method as one used in tool production, it has been most often cited as a method appropriate to the breaking of bone for marrow extraction. For instance, as a caption to a very imaginative reconstruction of the lifeways of *Australopithicus africanus* in the popular Life Nature Library book, *Early Man,* by F. Clark Howell (1965), we are told,

> Cracking and twisting antelope limb bones, the Australopithecines extract the edible insides. Later the bones, frequently found with prehuman fossils, may have served a different purpose; their sharp edges made useful tools and weapons [Howell 1965:69].

Recent New World literature, although generally acknowledging that twisting is not necessary to the production of spiral fractures, has tended to suggest that spiral fractures themselves are diagnostic of human behavior. This view seems to stem from seminal work by Bonnichsen (1973), which builds on two pioneer research experiences with behaviorally modified bones. Bonnichsen participated in a study with Zierhut (1967) aimed at the ethnoarchaeological study of bone breakage among the Calling Lake Cree. This experience was coupled with the analysis of bones recovered from animal cages at the Alberta Game Farm in an important attempt to develop criteria for distinguishing between bone broken or modified by man and that broken or modified by animals. Research with the Cree involved repeated demonstrations by the local Indians as to how bones were broken for marrow (see Bonnichsen 1973:Plate 1). A standard method was observed among the Cree and Bonnichsen obviously still accepts this (see Bonnichsen 1979:Plate IV–1) as *the* method whereby man break bones for marrow.

Bonnichsen's second actualistic experience was in terms of observations made on bones recovered from the cages of a caracal cat, a North China leopard, a Siberian lynx, and a Siberian tiger (Bonnichsen 1973:14). This list is interesting in that it includes no canid or hyena. Most observers of cats agree that they are fastidious eaters and modify bones primarily in the course of removing meat from the skeleton rather than during the gnawing sessions more characteristic of canids and hyenas. Large cats have, nevertheless, tremendous power in their jaws and are capable of shearing bones during the course of meat consumption. (See the impressive photograph in Beard [1979:96].) All of the bones illustrated by Bonnichsen (1973) and attributable to animals at the game farm were modified by a Siberian tiger, certainly a powerful animal. All of the modifications illustrated are puncture marks on bones of a small animal (sheep), which when broken were simply sheared by the tiger (Bonnichsen 1973: Plate 6b, c, d, and f). In one sheep tibia illustrated, however, the proximal end was sheared off and a spiral fracture originating at the end of the bone was produced.

The last "control" population of bones available to Bonnichsen was a collection recovered from the site inhabited by the Calling Lake Cree, who had served as informants for the marrow-cracking demonstrations. These bones represent garbage, from a local house, that was subsequently scavenged by domestic dogs. Thus, it was not clear which agent, man or dog, produced the modifications within this population. Since man obviously cracked bones for marrow, the spiral fracturing within this assemblage was attributed to man. Only modifications previously described by Sutcliffe (1970) and Miller (1969) as characteristic of animal modification were illustrated as referable to the dogs. Thus the "modern sample" from the Cree was interpreted using previously described criteria for recognizing animal-modified bone, coupled with the ethnographic experience of marrow cracking among the Cree, which was taken as the basis for recognizing bone broken by man. Only a single piece of "ambiguous" evidence existed from Bonnichsen's perspective: the spirally fractured sheep tibia from the Siberian tiger's cage.

> Some bones altered by animals exhibit spiral fracturing such as the sheep tibia from the Siberian tiger cage. . . . As this process has not been witnessed, the following suggestion is conjectural. The bone was probably first perforated by the teeth at the two oppositional points at the epiphyseal end. At the same time the tiger pulled out a segment which terminates in a spiral fracture [Bonnichsen 1973:16].

Bonnichsen then concludes that the long bones are particularly important for recognizing human activity, noting that there is frequently an impact notch and spiral fractures initiated outward from the notched loading point. In contrast, he lists the following characteristics as diagnostic of animal-modified bone: "(1) tooth perforation marks, (2) gnawing and scooping out of cancellous tissue, (3) crunching and splintering, (4) spiral fracturing directed from the epiphyseal ends, (5) partial digestion [Bonnichsen 1973:24]."[1]

[1]How the third characteristic is different from spiral fracturing is not at all clear since this criteria is adopted from Sutcliffe (1970) and he did not utilize the concept of spiral fracturing.

Bonnichsen was led to believe that spiral fractures were characteristic of bones modified by man, whereas tooth marks, scooping out, and partial digestion were characteristic of animal-modified bones. This is important, since Bonnichsen obviously believes that man breaks marrow bones by smashing them in midshaft, the "mid-diaphysis smash technique of spiral fracture [Bonnichsen 1979:36]." Although he admits the possibility that animals might produce spiral fractures originating at the epiphyseal ends, it is quite obvious that Bonnichsen considers spiral fracture to be "artificial." He states,

> The recognition of spirally fractured bones has led to the following hypothesis: spirally fractured bones are indicative of man's presence and reflect patterned human behavior. This hypothesis has been used to guide the analysis of fractured faunal remains from the Old Crow Flats paleontological assemblages. Two additional hypotheses specify how spirally fractured bones can be used for reconstructing prehistoric human behavior patterns. Hypothesis 1 states: The *geographic* distributions of spirally fractured bones may indicate human activity localities; and, Hypothesis 2 states: Spirally fractured mammal bones indicate the general selectivity exploited by man [Bonnichsen 1979:69].

Methodologically speaking, though Bonnichsen refers to his propositions as hypotheses there is no way of testing them using paleontological remains, since they refer to a relationship between a property remaining today (spirally fractured bone) and an agent inferred to have been active in the remote past (man). Since he is attempting to use the spirally fractured bone to prove the ancient presence of man in the New World, only with an accurate and independent means of monitoring the ancient presence of man could he test his hypotheses. In fact his hypotheses are used as conventions for giving meaning to paleontological remains. They serve to justify his belief that man was present in very early times but in no way do they evaluate the belief. That Bonnichsen does not seriously consider predator–scavengers as agents capable of producing spiral fractures is illustrated by his discussion of possible sources of spirally fractured bone:

> The prehistoric occupants of Old Crow Flats apparently processed the limb bones of large mammals for "bone

> grease" and perhaps for tool production material. Five hundred fifty-five spirally fractured limb bones from 50 localities suggest that bone processing was widely practiced throughout "the Flats."
>
> In assessing spiral fractures, the possibility should not be ruled out that occasionally animals broke their limb bones in spiral fractures. . . . Three arguments suggest that the majority of spirally fractured bones from Old Crow Flats are bones broken by man. Of the 1,794 elements collected from locality 89, 237 specimens exhibit spiral fractures. . . . This figure is much higher than would be expected, if only natural breakage were involved. Secondly, bones exhibiting spiral fractures include: mammoth/mastodon, horse, caribou, bison, muskox, elk and moose. Spiral fractures were not observed on the limb elements of bear, symbos, saiga antelope, camel, cat, or ground sloth . . . and the third argument is: the bones exhibiting spiral fractures occur on all the major limb bones. If the bones were broken only by natural causes, a pattern in which the weakest limb bones were broken would be expected [Bonnichsen 1979:77–79].

We see here a classic argument from elimination and a good illustration of its weaknesses. Two causes have been postulated: (*a*) man breaking bones for marrow and possibly for tools and (*b*) animals accidently breaking their legs. The latter is rejected by arguing that accidental breakage of legs should be rare, and it represents at least 13% of the observed bones at one location at Old Crow. I totally fail to understand the logic of Bonnichsen's second point, but the final point argues that all the major limb bones exhibited spiral fracture though accidental breakage would be expected on only the weakest bones. Having rejected animals breaking their legs as a cause, we may accept the remaining "working hypothesis" that man was the causal agent. Such logic only works if one has all the possible causes identified and included in the elimination procedure. As will be shown, Bonnichsen did not have an important cause listed—nonhominid predator–scavengers.

At approximately the same time Bonnichsen's early work was done, George Frison (1970) was studying the faunal material from the Glenrock Buffalo Jump. He excavated in detail the locations yielding stone butchering tools and associated faunal remains, obviously the remains of a mass kill of bison that had been butchered and processed for meat by man. Frison made the assumption, common in many Old World studies, that the characteristics of materials associated with unquestioned tools must also be referable to the action of man. Therefore, all the patterned modifications on the bones from Glenrock were interpreted as the consequences of butchering behavior by man. As I will illustrate throughout my discussions of animal-modified bones, I consider the major modifications illustrated and described by Frison (1970) to have been the products of scavenging animals (wolves and/or coyotes), who presumably feasted at the site after man had taken his share from the kills.

Near the end of Chapter 1, I argued that it seemed extraordinary to me that the interpreters of the site of Torralba made the implicit assumption that nature stopped except when ancient men were on the scene. I pointed out that such an assumption was necessary, since the interpreters referred all the faunal contents of the Torralba deposits to the actions of ancient men! After having reviewed the types of traces carnivores can leave on faunal assemblages, it seems equally amazing to me that of all the mass bison kills and megafaunal remains excavated by archaeologists from the American Plains, not one such location is reported to have yielded evidence of predator–carnivore behavior! All bone modifications and patterns of dispersion have been systematically treated as stemming from human behavior. This situation is truly remarkable, since both wolves and coyotes were common, even plentiful, on the American continent until relatively recent times:

> A wolfer spent each day in the same routine. In the afternoon he would ride out and shoot two or three buffalo and lace the carcasses with strychnine. The next morning he would return to dress out the ten or twenty victims. One wolfer, Robert Peck, left a record of his days in a journal. . . . Peck mustered out of the army in 1861, and he and two friends set up a wolf camp about twenty-five miles north of Fort Larned, Kansas. Over the winter they killed 3,000 animals: 800 wolves and more than 2,000 coyotes, and about 100 fox [Lopez 1978: 178].

If we believe the archaeologists, we must imagine that the North American wolves and coyotes *avoided* the edible remains left by North Americans at such places as animal traps and the bison jumps profusely documented on the Plains. (For some idea of the scale and distribution of such features see Davis and Wilson [1978].) I have searched in vain for discussions of bone modification and dispersion referable to the behavior of nonhuman scavengers at such obvious concentrations of meat.

The majority of the patterns of bone destruction illustrated by Frison (1970) from the Glenrock site may be taken as a clinic in canid-modified bone. Nevertheless, I am getting somewhat ahead of myself. The point is that bones showing patterns of destruction characteristic of canids were mixed with humanly modified bones in the assemblages available to Bonnichsen, and at Glenrock canid-modified bones were misinterpreted by Frison as the product of human actions. Thus, from the perspective of developing methodology, spiral fractures were viewed by Bonnichsen as diagnostic of man and all the modifications produced by canids at Glenrock were advanced by Frison as also diagnostic of man. The latter were interpreted as deriving from a particular method of butchering (muscle stripping) postulated to account for many of the modifications, whereas others were viewed as intentional modification of bones for use as "expediency tools" (Frison 1974).

The work of Bonnichsen and Frison provided the conventions for giving meaning to observed modifications on bones, the principles later summarized as follows:

1. "Spirally fractured bones are indicative of man's presence and reflect patterned human behavior [Bonnichsen 1979:69]." "Bones that exhibit spiral fractures from nonhuman sources are usually those broken after the bone was dried or had been partially decalcified, or borken when the animal fell or twisted its leg [Stanford 1979a:117]."
2. "Bone expediency tools (Johnson 1977 and Frison 1974) are manufactured by producing a spiral fracture and using the sharp edges of the fracture for butchering or hide-working activities. Bones used as tools can be identified by the occurrence of tiny step fractures, polish and striations on the proximal end of the spiral fracture. . . . Expedience tools are usually made from long bones, but ribs, mandibles, scapula and innominate fragments were also used [Stanford 1979a:108]."

Using spiral fractures, and a variety of surficial scarring and bone modifications as diagnostic of faunal assemblages produced by the hand of man, it is not surprising, given the view that most of the "diagnostic criteria" were probably produced by gnawing animals, that a number of modern myths have begun to emerge. Using the Bonnichsen–Frison criteria for recognizing man and tools, many archaeologists have made claims for man's great antiquity in North America (Bonnichsen 1978, 1979; Canby 1979; Friedman 1980; Harington *et al.*, 1975; Irving 1978; Irving and Harington 1973; Morlan 1979; Stanford 1979a).

Similarly, there have been claims for tool assemblages lacking stone or other generally recognizable products of human workmanship such that at least some workers have entertained the idea that early man in the New World manufactured tools primarily of bone:

> Irving further suggested that a sophisticated Upper Paleolithic variety of flint industry is by no means essential to support human life in the far north. In fact, bone forms that could have performed all the necessary tasks of hunting, piercing, butchering, skinning, and perforating have been recovered from the deposits at Old Crow and elsewhere, adding another dimension of possibility to the reconstruction of early lifeways [Humphrey 1979:viii, paraphrasing William Irving].
>
> This study suggests heavy emphasis was placed on bone working in Mid-Wisconsin times. This pattern probably dominated the tool making picture in many areas until Pleistocene extinction began to occur. . . . Man probably responded by developing new technological innovations that emphasized the use of lithic artifacts. In light of these findings, future early man research . . . might well be directed toward reexamining existing Pleistocene paleontological collections for evidence of human modification [Bonnichsen 1979:193].

In the hands of generalizing early man enthusiasts advocating a particular point of view regarding the peopling of the New World, the material from Old Crow Flats is characterized as "a remarkably sophisticated and specialized bone and antler industry [Bryan 1978:323]."

I consider most of the foregoing arguments as having produced a series of modern myths. I will attempt to demonstrate that (*a*) spiral fracture is not unique to man, (*b*) man does not generally break bones by the "mid-diaphysis smash technique," (*c*) spiral fracture when produced by animals is not limited to an origin at the distal ends of the bones, (*d*) the modifications cited by Frison at Glenrock as evidence of muscle stripping are commonly produced by gnawing canids, (*e*) the modifications cited as criteria for recognizing "expediency tools" are commonly produced by nonhominid predator–scavengers, and (*f*) many of the modifications previously cited as evidence for human modification are referable to predator–scavengers.

Skeletal Disarticulation

There have been very few studies of natural skeletal disarticulation. The two that are available are somewhat ambiguous on whether the active agents are decay organisms and chemical processes or whether there has been action by predator–scavengers. In the earlier study available to me (Toots 1965), it would appear that the agents of disarticulation are primarily decay organisms and chemical processes. Toots summarizes his observations as follows:

> From these observations the normal course of disarticulation can be summarized approximately as follows: (1) skull and some limbs become disconnected (possibly also the atlas), (2) ribs become loosened and fall off at least from the upper side, (3) limbs start to disarticulate into progressively smaller segments until only isolated bones are left (sometime during this process the lower jaw becomes disconnected . . .), (4) following this but overlapping its late stages, the vertebral column starts to disarticulate, and (5) after this has continued for some time but before disarticulation of the vertebral column is completed, weathering, splintering and gradual disintegration of the bones set in [Toots 1965:38].

It is interesting that Toots attributes the resistence of the vertebrae to the interlocked type of articulation and the rapid disarticulation of joints such as the atlas–axis to their highly mobile character. He further notes that hide and ligaments retard disarticulation and this is particularly true where desication occurs before decay by virtue of there being very little "meat," as is the case with ungulate lower limbs. In a contrary fashion, where there is a concentration of decomposable tissue, as in the abdominal area, chemical reactions associated with decomposition will hasten disarticulation of such parts as ribs.

The second study of disarticulation is one conducted by Andrew Hill (1975). This investigator actually did a pattern-recognition study based on the comparative degrees of disarticulation manifest at a series of death sites of *Damaliscus korrigum*, or topi. The average adult body weight of topi is 82 kg or about 180 lb. His findings on the sequence of disarticulation are as follows:

> The sequence [given in Table 3.01] is described in terms of the smaller unit, often a single bone, resulting from the disarticulation of a particular joint. Each disarticulation has been given a number, shown on the left, indicating its stage in the sequence. Apparently simultaneous disarticulations . . . are distinguished by supplementary letters. The numbers in the right-hand column indicate the next stage at which, first, the named unit and, second, its complement, are encountered. Thus it makes it possible to trace through this account the course of disarticulation of a particular part of the skeleton, such as the forelimb . . . [Hill 1979:741].

The basic sequence of disarticulation is head and front leg, followed by rear leg, and finally the elements of the axial skeleton. On the basis of these observations, we might expect the following anatomical segments to have slightly independent distributions under natural conditions simply as a function of the sequence of likely disarticulation.

1. Cranium plus atlas vertebra
2. Front leg
3. Rear leg
4. Basic axial skeleton (cervical, thoracic, lumbar vertebrae; sacrum, pelvis, ribs)

For instance, if one can picture competition among carnivores around a dead animal, then it might well be reasonable to anticipate that skull, mandible, and front leg parts would be most likely dragged away from the kill *first* during the early period of competition for parts of the kill among numbers of carnivores.

In addition, we may anticipate that parts will "ride" with others as a function of their probable

TABLE 3.01
Sequence of Natural Dismemberment According to Hill[a]

Stage	Unit (smaller one observed)	Next occurrence named unit	Complementary unit
	Basically appendicular skeleton		
Basically front legs			
1	Forelimb (including scapula)	3	2
2	(Caudal vertebrae)[b]	*	
3	Scapula	*	5
4	(Mandible)	*	6
5	Humerus + radius + cubitus	8A	7A and B
6	(Cranium + atlas)	13	8C
7A	Carpals	*	7B
7B	Metacarpal	*	8B
8A	Humerus	*	10A
8B	Distal phalanges of forelimb	8	9
Basically rear legs			
8C	Hindlimb	10B	15
9	(Proximal phalanges and distal phalanges [front])	*	*
10A	(Radius and ulna) separate	*	*
10B	Femur + tibia	12	11
11A	Tarsals + metatarsals	14	11B
11B	Phalanges (interphalangeal joints as well)	*	*
12	Femur separates from tibia	*	*
	Basically axial skeleton		
13	(Cranium separates from atlas)	*	*
14	(Tarsals) separate from metatarsals	*	*
15	Ribs (average value)	*	16
16A	Thoracic vertebrae (13) + lumbar + innominate + sacrum	17A	16B
16B	Thoracic vertebrae (2) to (12)	*	19B
17A	Innominate	*	17B
17B	Thoracic vertebrae (13)	*	18A
18A	Lumbar vertebrae (7) + sacrum	19A	18B
18B	Lumbar vertebrae (1) to (6)	*	19A
19A	Sacrum separates from lumbar	*	*
19B	Thoracic vertebrae (1)	*	20
20	Axis	*	21
21	Cervical vertebrae (3) to (7)	*	*

[a] Table adopted from Hill 1979:742. (Table originally titled "Sequence of Disarticulation in Damalescus Korrigum").
[b] Exceptions to skeletal groupings are indicated by parentheses.

sequence of disarticulation. For instance, when the forelimb separates, it is most likely that all articulated segments of the front leg will be dragged off, not just some desirable or high-yielded element of the front leg.

I had the opportunity to observe and record the parts remaining on 24 wolf kill sites (reported in Chapter 5). The overall pattern observed was generally consistent with the pattern reported by Hill. In several cases there was clear indication that the wolves had dismembered their prey by (*a*) eating through the proximal humerus, therefore disarticulating the scapula, which was then dragged off (the otherwise unmodified front leg from the middle of the humerus shaft down remained attached to the body by skin in two cases and was scattered in nine cases), and (*b*) cracking the femur in roughly midshaft so that the proximal end of the femur remained articulated to the pelvis (however, the greater trochanter was eaten away); the lower rear leg was missing or widely dispersed around the kill location. Much less common was a break in the lower third of the tibial shaft such that the complete lower leg from the distal third of the tibia down remained articulated, sometimes with part or all of the toes and hooves missing. This means that the proximal femur and the proximal tibia may remain associated with the pelvis, the distal femur and the proximal tibia may remain articulated, and the distal tibia down to the foot represents the final segment of the rear leg typically found articulated. In the case of the front leg, the most common pattern was for the scapula to occur singly, the proximal end of the humerus to be destroyed, and the remaining parts of the front leg to remain articulated. An alternative was for there to be a spiral break through the lower part of the radio-cubitus so that the distal radius was articulated with the carpals, metacarpal, and segments of the foot. Another common segment consisted of the main body of the shaft of the radio-cubitus, and the distal humerus, plus a considerable section of shaft.

I greatly regret that I did not collect all the bones from the wolf kills for return to Albuquerque. At the time I was interested in anatomical part survival and did not make systematic notes on either units found in articulation or the distribution of tooth marks, fracture patterns, or other evidence of carnivore activity. I saved only bones exhibiting extraordinary evidence of carnivore destruction. Gary Haynes (1977, 1978a, 1980b) has informed me that he has currently underway an extensive study of wolf kills of American bison. He expressed the opinion that the pattern of dismemberment was so regular that we might eventually be able to tell how hungry the wolves were and the size of the group making the kill simply from the patterns of disarticulation and bone dispersion. Thus far, the consistency between my observations and those made on very different animals in a very different setting would tend to support the idea of extreme regularity and, in turn, indicate that we might well place some confidence in making uniformitarian assumptions about patterns of dismemberment when contemporary patterns are documented more fully.

Tooth Marks

Viewing animals as denticulated vises, it is obvious that the implements that ultimately modify bones are the teeth. Ignoring for the moment breakage and actual modifications in bone morphology, I will concentrate on surficial scarring caused by the motion of animal teeth on and over the surface of bones.

I recognize four basic types of tooth marking on bone: (*a*) punctures, (*b*) pits, (*c*) scores, and (*d*) furrows. These may be coupled with types of breakage and when distinctive, this will be mentioned.

Punctures (Figure 3.01) are simply where the bone has collapsed under the tooth, frequently leaving a fairly clear imprint of the tooth, as in the dorsal (right-side) view of the distal end of the metacarpal shown in Figure 3.02. Frequently, when the bone is thin and/or porous, the tooth may penetrate the bone leaving distinctive holes in cancellous bone (Figures 3.01 and 3.36). On very thin bone, such as the blade of the scapula, the tooth may penetrate and remove an area of the edge equal to the surface area of the tooth, producing a *crenulated edge* (well illustrated in Figure 3.40). This effect has been illustrated previously by Bonnichsen (1973:19, Plate 3, item A.) and Shipman and Phillips (1976:171, Figure 1, item F).

Punctures were one of the first modifications made to bones noted by early archaeologists. In fact, Boucher de Perthes, the father of modern prehistory, reported a punctured phalange from the site of Abbeville (1849:312). Punctured bones were noted with interest by Lartet and Cristy and a considerable discussion in the French literature followed the suggestion that these were whistles, the first musical instruments. The literature is well summarized by Henri Martin (1907–1910:162–168), one of the first to point out that perforations were commonly made on the weakest surface of a bone and to give many

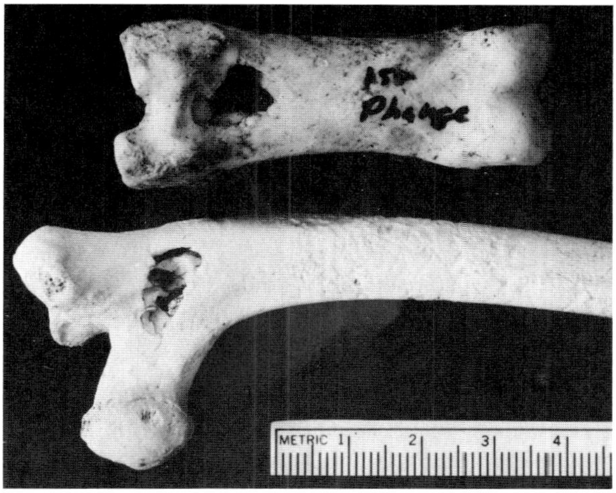

FIGURE 3.01. *Punctures made by animal teeth.*

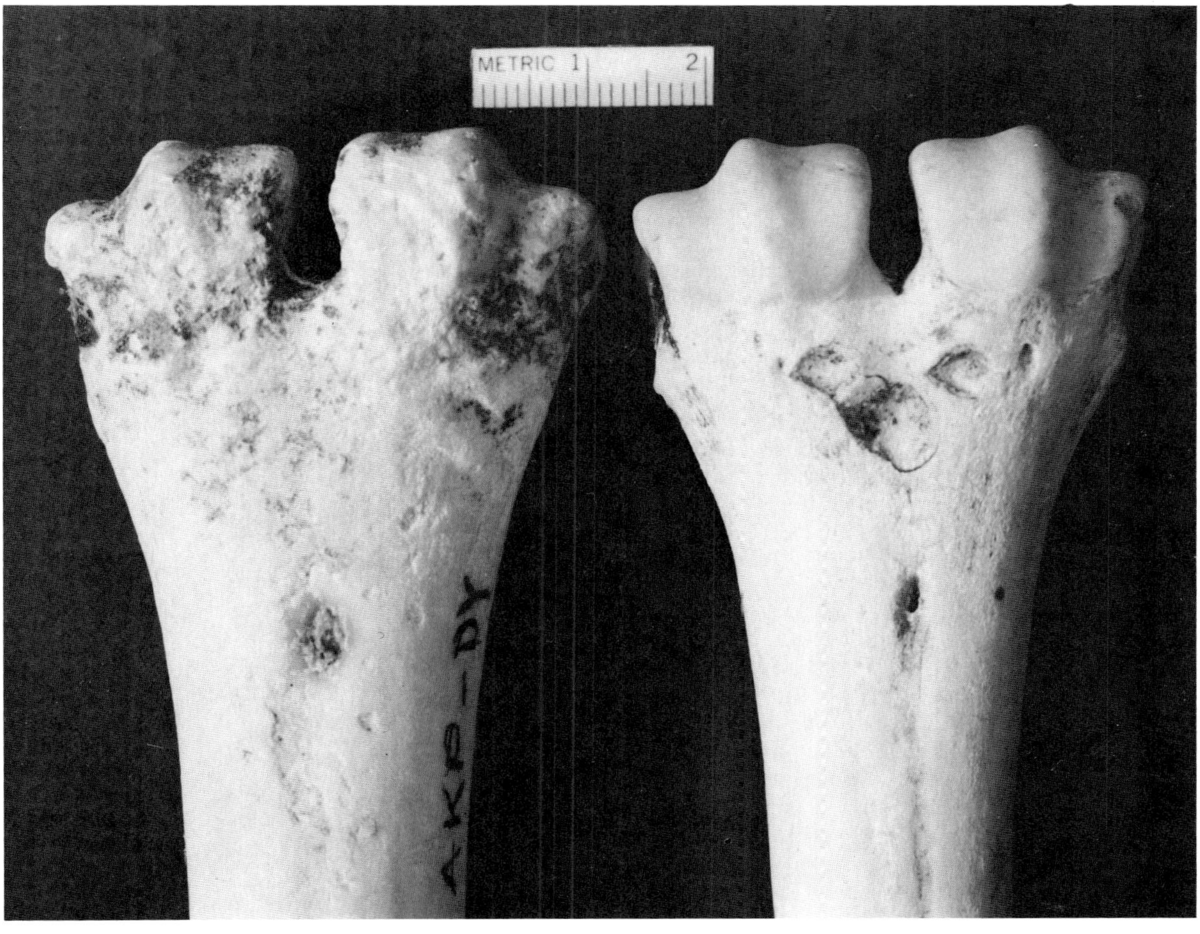

FIGURE 3.02. *Animal-produced pitting and punctures on the distal metapodial.*

examples identical to that illustrated here (Figure 3.01, upper specimen) (see Martin 1907–1910: Plates XXX, XXXI, XXXII).

Since punctures are a function of the strength of the bone relative to the strength of the animal, it is not surprising to find large and deep puncture marks made by lions and other large cats. (See, for instance, Bonnichsen 1973:21, Plate 5, items A and B; Dart 1949:13; Hill 1980: Figure 8.4; Kitching, 1963: Plate 4; Pei 1938: 62, Plate XXI, items 7, 8, 15, and 16.) Puncture marks are really the initial stages of *furrowing*—a term adopted from Haynes (1980). That is, if considerable gnawing occurs, obvious puncture marks are generally obliterated as the cancellous tissue is gnawed away. Gnawing generally proceeds from soft to hard bone; the animal attacks the soft cancellous parts of a bone first. As it gnaws, it encounters progressively harder bone (the bone shafts) and the same actions may result in pitting in that the bone is now strong enough not to collapse under the gnawing action. This *pitting* is illustrated in Figure 3.02; the bone on the left is extensively pitted, whereas the one on the right is punctured. Figure 3.03 illustrates in more detail the effects of extensive pitting. This is a common condition on bones that have been gnawed extensively. I should point out that I have observed this only on bones from wolf dens and dog yards. *No examples* of bones so extensively gnawed were observed on the wolf kills that I will discuss later. Pitting generally results from gnawing bones rather than eating and pulling meat from the skeleton of an animal, as at a kill. Bone gnawing as such is rare at kills but very common at dens and the "sleeping" places of canids that I have observed. It is even less likely at the kill sites of large cats (Diane Gifford, personal communication).

Figures 3.04 and 3.05 illustrate the pitted surface of a compact bone splinter, which if found in the European Middle Paleolithic would be called a tool and designated a "compressor." This category was originally described by Martin (1907–1910), who interpreted it as a compressor or pressure flaker used in stone tool manufacture. The pitting and scoring of the bone was viewed as the result of pushing off flakes of chert during the act of pressure flaking. Bordes (1961:77) has correctly pointed out that the very presence of the technique of pressure flaking has not been established for the Middle Paleolithic, rendering it unlikely that compressors were used in the suggested manner. Nevertheless, Bordes accepts these items as tools with the warning that they were more than likely scarred while being used in percussion flaking. I have examined many of these items (approximately 134 from the site of Combe Grenal) and in most cases there is no tendency for the scarring to be on the end where one would expect it if these items were used as flakers. In fact, I could see no relationship between the placement of pitting (sometimes coupled with scoring as in Figure 3.04) and the morphology of the splinter. It was as if the splinter was produced after the pitting had been done.

These items have been widely illustrated as tools, by Bordes (1961:Figure 108, 1968:100, 111, Figures 34 and 39), Semenov (1964:171–172, 177, Figure 92), Movius (1953:45, Figure 12), and DeLumley (1969:165, Figure 20), among many others.

Scoring is the result of either turning the bone against the teeth or dragging the teeth across relatively compact bone. The result is a scarring of the surface that is commonly linear. Scoring may on occasion look very much like "cut marks" from stone tools in that scoring marks may be parallel and frequently quite close together. They generally differ from cut marks in that cut marks rarely follow the "contour" of the bone, being deeper on the parts of

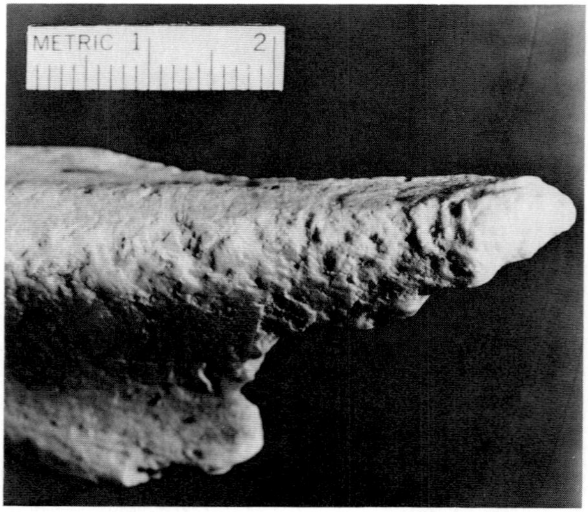

FIGURE 3.03. ***Animal-produced extensively pitted bone.***

FIGURE 3.04. ***Animal-produced pitted and scored compact bone surface.***

short curvature and shallower on surfaces with long curvature or sections of bone tending toward flat surfaces. Another major difference is the placement of cut marks versus scores on bones. Cut marks generally result from three activities: (*a*) skinning, (*b*) disarticulation, and (*c*) filleting. Cut marks from skinning may occur around the shaft of lower legs and phalanges and particularly along the lower margins of the mandible or on the skull. These are the only marks that might on occasion mimic tooth scoring. Cut marks from disarticulation normally occur (*a*) on the edges, or articulator surfaces, of the ends of long bones and (*b*) on the surfaces of vertebrae or pelvic parts. These are all areas of relatively soft cancellous bone and are apt to exhibit puncture marks, pitting, and furrowing. The bone is generally soft enough so that scoring rarely results from normal gnawing, although it may occur on harder adjacent parts of the bone. Filleting generally results in cut marks parallel to the long axis of the bone, whether long bones or flat pieces such as the scapula are involved. Scoring is almost without exception transverse to the long axis of bones, since it results from turning the bone against the teeth rather than "pulling" the teeth along the bone (Figure 3.06). Actually, after some experience with cut marks and scoring, I found few ambiguous situations, except perhaps where pups had been feeding and their fine pinpoint teeth produced a very sharp scoring, sometimes even longitudinal scoring from tugging and pulling in play and competition for bones.

Scoring has been well illustrated in the literature. In Breuil (1939) many scored bones from Choukoutien are illustrated (Plate XII, p. 65, Nos. 1, 2, 3; Plate XXVI, p. 93, No. 5) although these are interpreted by Breuil as evidence of human workmanship. Pei (1938) also illustrates items from Choukoutien but recognizes them as the product of animals, principally the hyena (Pei 1938: Plates X, XIII, XIV, and IV, and XX, No. 2). The foregoing are all good illustrations of tooth scoring by relatively large carnivores. Kitching (1963) illustrates numerous scored

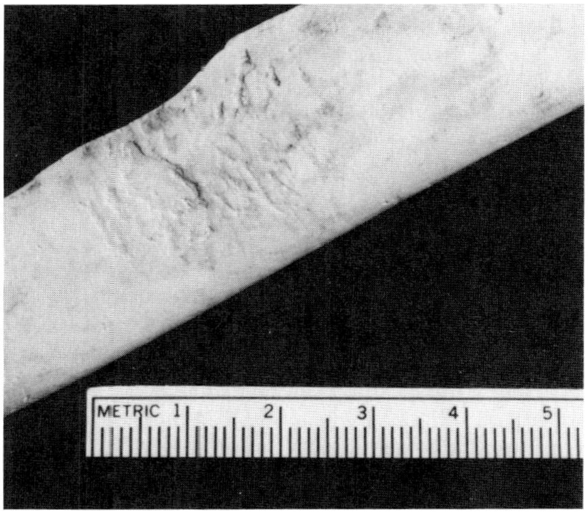

FIGURE 3.05. ***"Compressor" pitted and scored by animal teeth.***

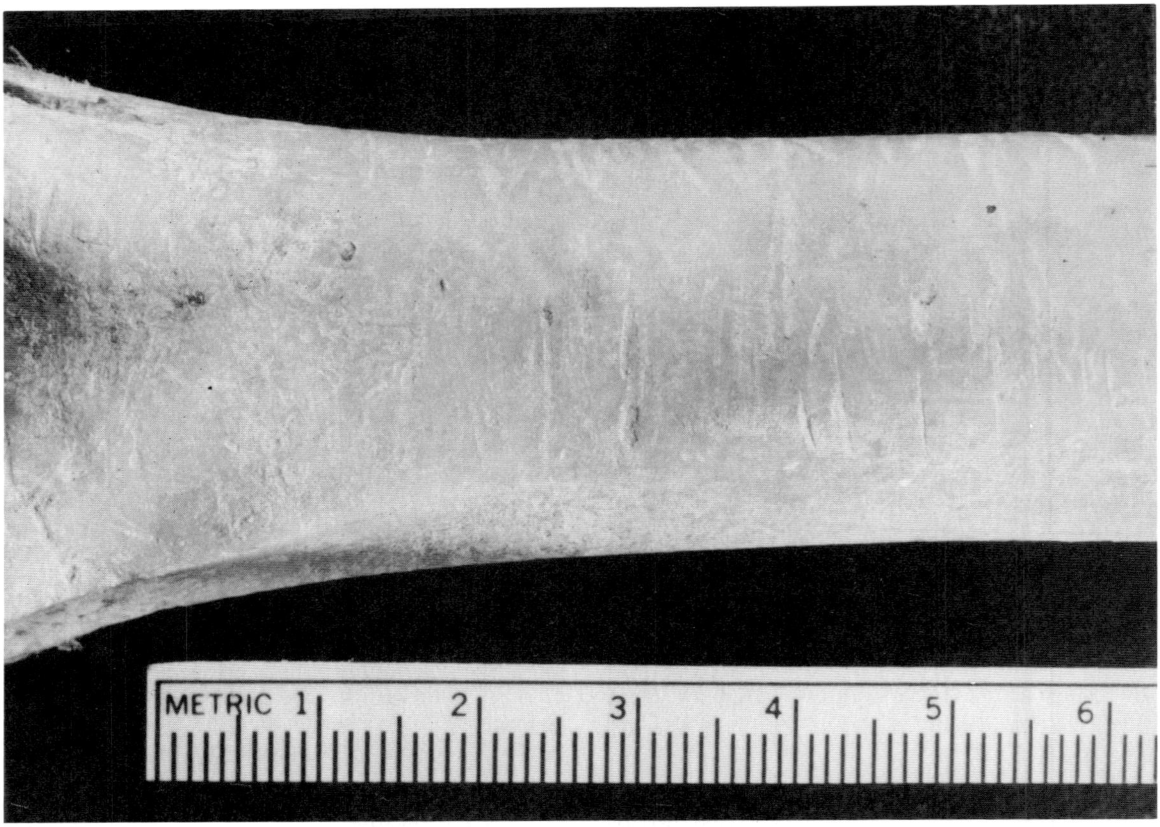

FIGURE 3.06. ***Compact bone scored by animal teeth.***

bones from what was most certainly a major hyena occupation of Pin Hole Cave in England (Plates 5, 6, 7, 15). Miller (1969: Figure 3, bottom, and Figure 4, top) provides good illustrations of tooth-scored bones, as does Haynes (1980: Figure 1 and 2).

Furrowing is a term I have adopted from Haynes (1980) to refer to the effect that repeated jaw action with either canines or carnassials produces on relatively cancellous bone tissue. The result in extreme cases is what Bonnichsen (1973:16), following Sutcliffe (1970:112), has called "scooping out." In furrowing soft tissue is removed, leaving a "hole" which is graduated up against harder, more compact bone. The action of the teeth will sometimes produce undulations on the remaining surface of the cancellous tissue, and there are sometimes steplike irregularities where the teeth have vised down in different spots as the bone is mouthed by the animal. This is by far the most common type of damage produced by animals and is the type that is present even if animals have given only minor attention to the remains under investigation. Examples of furrowing are common in the literature (Bonnichsen, 1973: Plates 6A, 6B, 6D; Haynes 1980: Figures 3, 4; ;Miller 1969: Figure 5; Pei 1938: Plates 12, 20; Sutcliffe 1970;). In the older literature, a number of authors have interpreted furrowing as the consequence of human modification (Breuil 1939: Plate XV, No. 2, and Plate XVI, Nos. 5 and 6; Weidenreich 1939: Plate VIII). In the more recent literature furrowed pieces have been interpreted as the consequences of butchering with choppers and "muscle stripping" (illustrated in Frison 1970: Plates 9, 12; Johnson 1977: Figure 11.2, to mention only two).

Figures 3.42 and 3.43 nicely illustrate furrowing of cancellous tissue by dogs and by wolves.

I made several observations that may aid in distinguishing the behavioral context of different types of scarring by animals. Wolf kills predominantly yielded some furrowing, relatively common puncture marks, and some crenuelated edges; pitting and scoring were much less common. On the other hand, the Eskimo dog yard assemblages and the wolf den assemblages yielded extensive pitting, scoring, and more extreme furrowing. This seems to result from the "boredom" chewing of bones in the dens and the dog yards; meat consumption was primarily at the kills. At kills, furrowed destruction of soft cancellous tissue while meat is still present on the shafts of bones prevents the harder bone adjacent to the cancellous tissue from becoming heavily scored or pitted, as is illustrated in Figure 3.43. Stated another way, when meat is present in considerable quantities, long bones may be furrowed and even polished along the graduated edges *without* correlated puncture marks, pits, or heavy scoring on adjacent denser bone. This may be particularly true when animals are attempting to dismember a carcass, as at a kill.

Other Sources of Surficial Modifications on Bone

In my discussions thus far I have only addressed the issue of Carnivora versus man as an agent of bone modification and surficial scarring. As might be imagined, there are other agents that may systematically scar bones and other natural contexts in which bones may become modified.

Perhaps the most remarkable of the agents whose modifications of bones may be mistaken for human workmanship is the porcupine. Pei (1938) illustrates numerous bones modified by the porcupine and other rodents. Bone "chisels" reported from Hopefield (Singer 1956), and modified bones from ancient deposits in China (Hooijer 1951), although presented as evidence of human modification, appear to have been modified by rodents. The Hopefield bones are particularly interesting, since Singer reports an assemblage of "control" specimens recovered from a rock crevice considered too small and low to have been used by man. Given such conditions, it was reasoned that the bones within could only have been introduced by animals. The control assemblage was then compared to the so-called chisels from Hopefield, and the conclusion was made that animals were indeed responsible for the modifications on the Hopefield specimens.

Less satisfactory claims accompany the studies conducted at Kalkbank (R. J. Mason *et al.* 1958), as well as the Israeli Mousterian cave of Geula (Dart 1967). In both cases rodents had modified the bones considerably; nevertheless claims that humans had modified them were made. Even more remarkable is the fact that the Kalkbank material is considered by Dart and others as a control assemblage providing information on the diagnostic properties of bone assemblages modified by man. No less bizarre claims are made by Dart (1967) for the Geula materials.

Very characteristic patterns of gnawing involving parallel tooth marks, production of "windows" in the shafts of bones, and extensive modifications in localized areas, particularly along breaks, are all diagnostic properties. The relevant illustrations presented in Pei (1938), Singer (1956), Dart (1967), and Bonnichsen (1979) should be consulted for further details.

As in the case of the rodent modifications, Pei (1938) was one of the first to recognize that considerable alterations on bone surfaces could result from the solution of bone surfaces in some situations where dense roots covered the bone. Such etching of bone surfaces in conjunction with the presence of plant roots is common in some environments (see Figures 3.07 and 3.08). Human agency for modifications most likely referable to the latter conditions has been claimed by Bordes (1969) for an "engraved" bone recovered from the Acheulian levels of the site of Pech de L'Aze, and by Freeman (1971) for a piece of "Arte Mobiliar" from Mousterian Level 17 at the Spanish cave of Morin. Both bones are described as having a decorative "macaroni" type motif (Marshack 1976). Bordes' find has been claimed to be the earliest evidence of man's symbolizing and artistic capacities: "If my an-

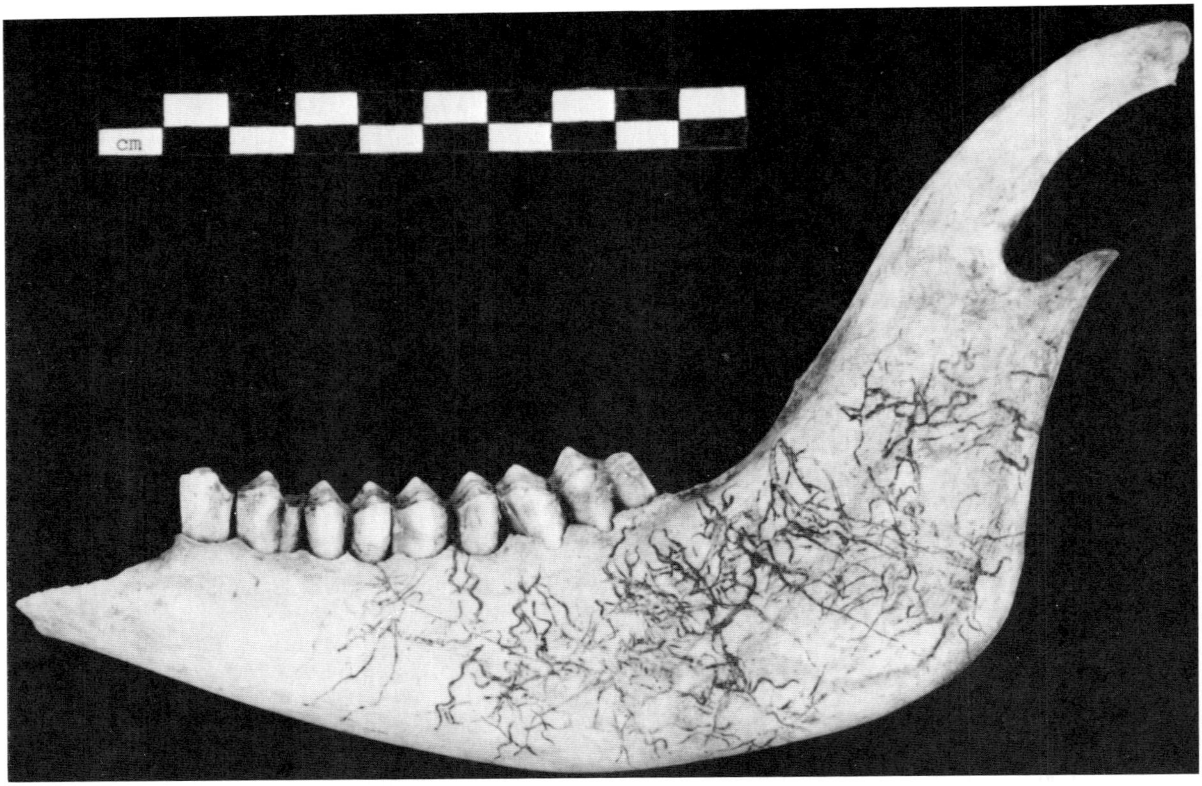

FIGURE 3.07. *"Arte Mobiliar" produced by acids in compact root masses on a sheep jaw from the Bear site, Anaktuvuk Pass.*

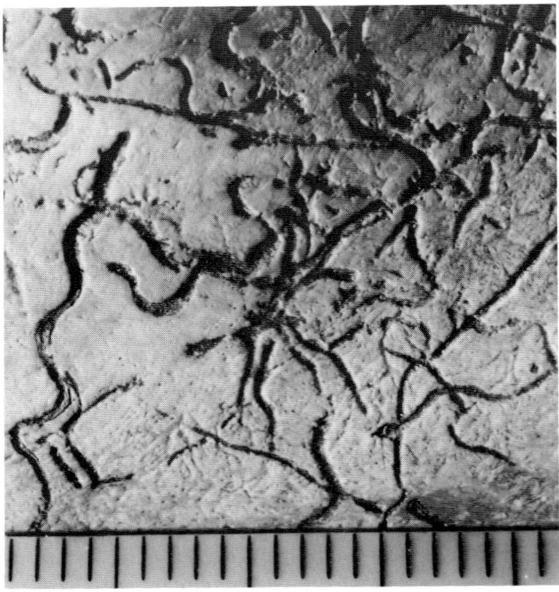

FIGURE 3.08. *Detail of bone in Figure 3.07.*

alytic results [on Bordes' specimen] are correct, they may change our concept not only of the origins of art and symbol making, but also of the intellectual and cultural evolution of *Homo sapiens* [Marshack 1975: 85]." Claims such as these are certainly the basis for modern myths about ancient man.

Bone Breakage and Destruction by Animals

Many hours of observation of both wolves and dogs revealed a very convincing and redundant bone consumption strategy, which leaves a regular pattern in the faunal by-products of animal consumption. Gnawing animals attack the ends of long bones first. They are most successful in chewing away the softest articulator end. The softer the articulator end the faster the animals chew them away and expose the cross section of the medullary cavity. At this point an animal may lick at the end and "scoop out" marrow by turning the canines around in the cavity and simultaneously licking into the cavity, but this is very unsatisfactory. More commonly the animals attempt to collapse the bone shaft by use of the strong cheek teeth.

Given a very resilient bone, the animal most commonly turns it parallel to the orientation of the carnassial teeth and attempts to puncture the crown of the bone cylinder, which results in what I call *channeled bones* (see Figure 3.09). This effect is produced by puncturing the bone back from the transverse edge, leaving a channel running parallel to the longitudinal axis of the bone. The animal may proceed in this manner for considerable distance "down" a bone. In Figure 3.09 the upper bone is the shaft of a long bone, which is clearly scored and chipped along its edges; the channel is evident. In the lower example, a dorsal spine of a thoracic vertebra, the puncture marks are clear. Channeling is unique to animal gnawing and in my experience there are no human processing techniques that result in analogous forms (see Buckland 1823:276; Sutcliffe 1973:Figures 2, 3, 5; Zapfe 1939:151).

In addition to the "direct" bite-down approach, which is most common when a resistant bone is encountered, the animal may distort its face, moving its jaw laterally away from the side where the bone is inserted in the mouth, and turn its head into the bone removing a splinter along the shaft. Success in this strategy results in the removal of a substantial flake parallel to the longitudinal axis of the bone, exposing more of the marrow cavity. Such a flake is generally pointed and scarred; the opposite end commonly shows "step fractures" (Figure 3.10). This pattern is known from many locations of archaeological interest and is commonly interpreted as bone modification by man (Bonnichsen 1978:112, Figures 6a, 6b; Frison 1974:42–43; M. D. Leakey 1971:Figure 36; Stanford 1979a:108, Figures 6, 7). Such flake removal originating at the broken end of a bone may be coupled with polish as in Figure 3.10, or radial scarring as in Figure 3.11, and also has been taken as evidence of tool use. (See Kitching [1963: Figures 5, 6 7b] and also Klima *et al.* [1962:150, 154, 155, 157] for examples of radial scarring from a hyena den.)

Another characteristic form of behavior is what I call "chipping back" an edge. This is done by using the strong carnassial teeth. A bone is placed in the mouth parallel to the orientation of the row of teeth and a ridge or protrusion is placed between the carnassials and "mashed off." When the bone is dense compact bone, the result of this viselike pressure is an effect very much akin to pressure flaking (Figure 3.12). Small chips are pushed off, resulting in an edge along a break that may be continuously chipped, showing all the features of a "microdenticulate" in lithic terminology (see Figures 3.13 and 3.14). Frequently associated with this characteristically microchipped edge is tooth scoring of the external surface of the bone below the chipped edge. As the animal chews on the exposed edge, chips are driven off. When the edge gives way, the large teeth slide along the surface of the bone, resulting in typically parallel and slightly oblique scars below the chipped edge. This is well illustrated in Figures 3.15 and 3.16. Channeling is often combined with chipping back to produce very convincing pseudotools (Figure 3.17).

Chipped edges as illustrated in Figure 3.12 are almost inevitably interpreted by archaeologists as retouch. Dart (1960:5) illustrated eight chipped-back flakes from gray breccia at Makapansgat, about which he comments: "He also prepared feeding

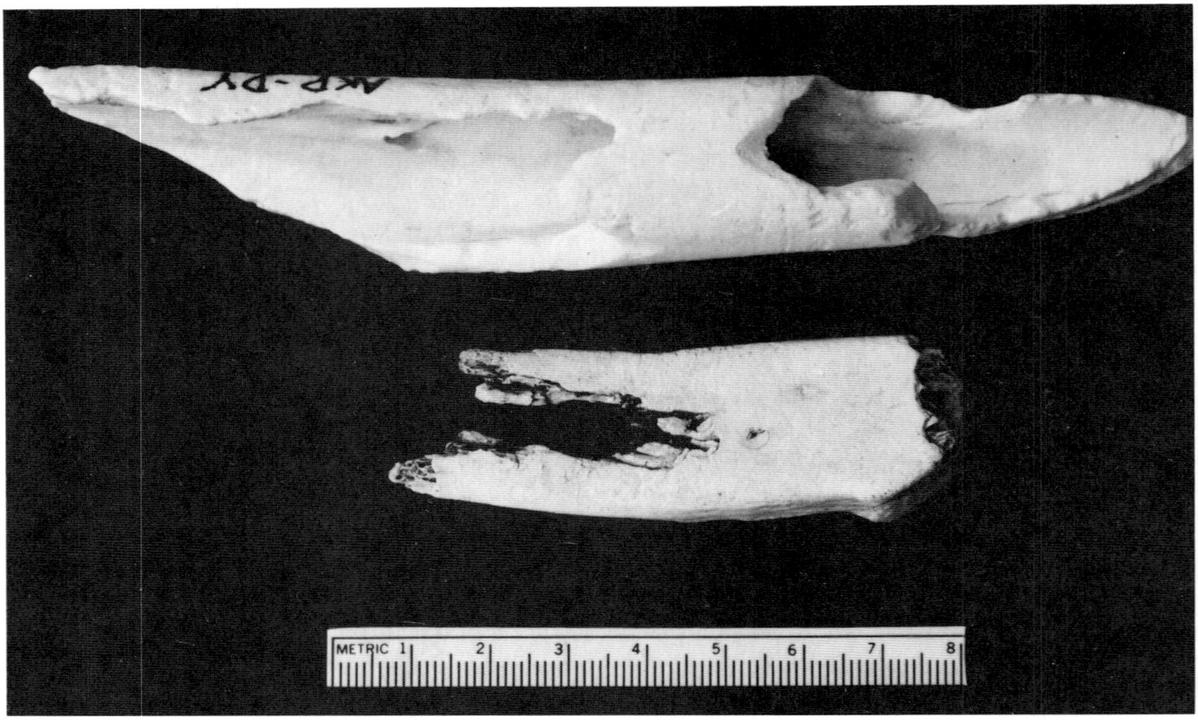

FIGURE 3.09. *"Channeled" breakage of bones by animals.*

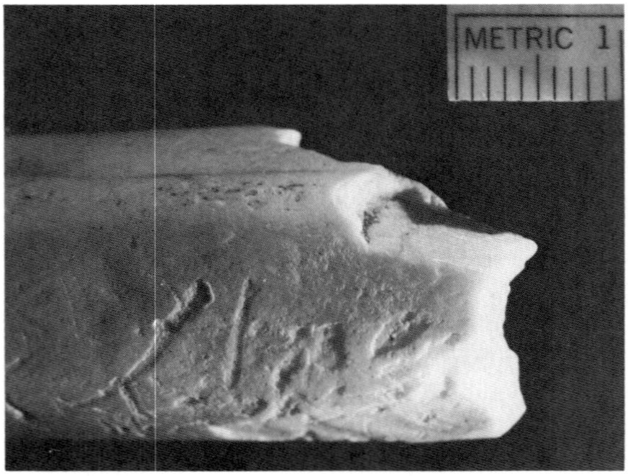

FIGURE 3.10. *Highly polished end of a gnawed bone.*

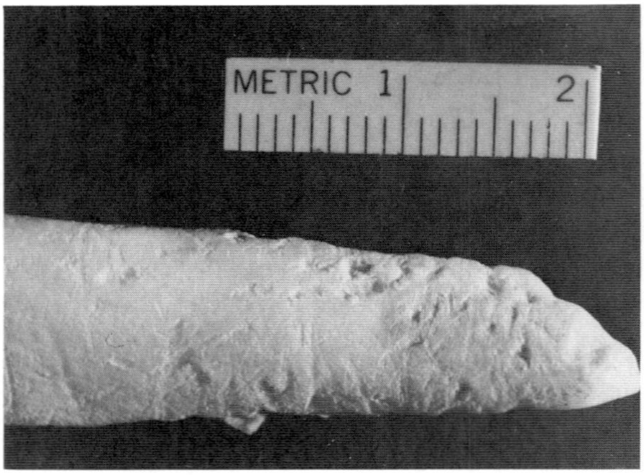

FIGURE 3.11. *Pitted and radially scarred bone, the result of extensive gnawing.*

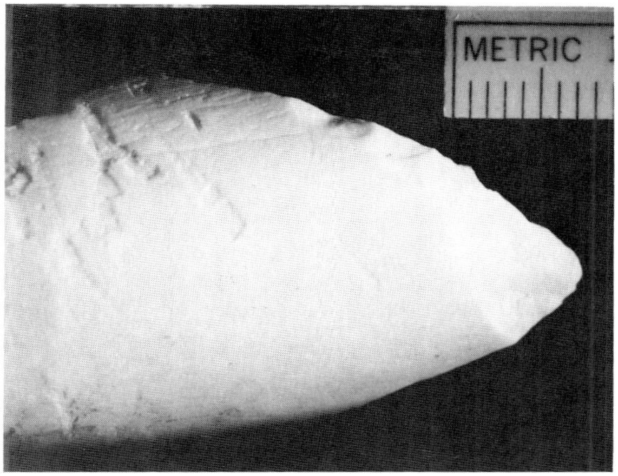

FIGURE 3.12. *"Pressure flaking" on the end of a gnawed bone resulting from chipping back the edge.*

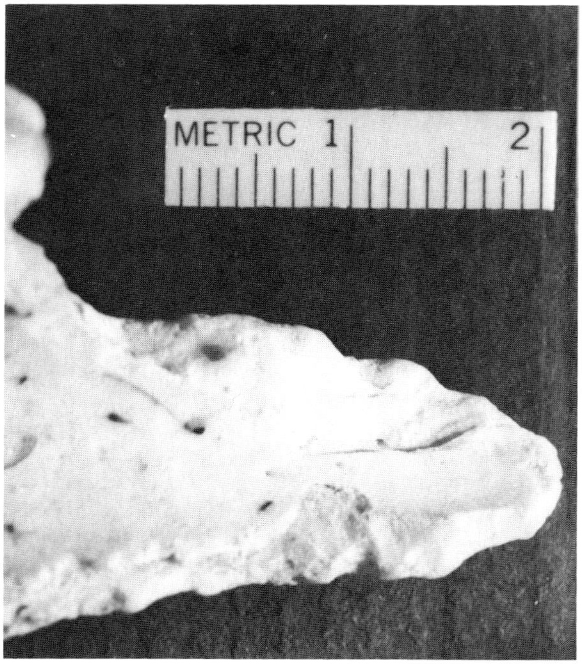

FIGURE 3.13. *Extensively chipped back edges of gnawed bone.*

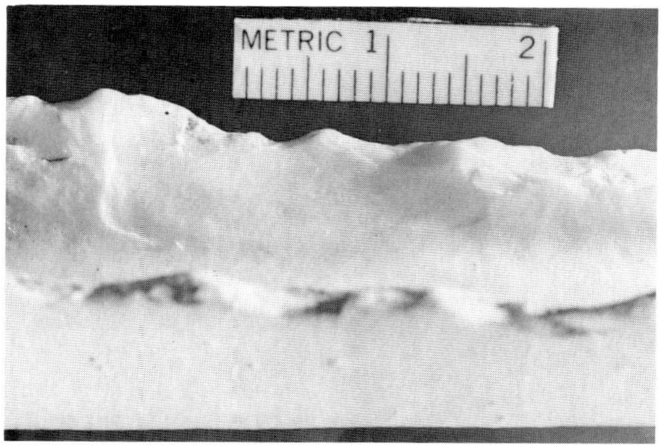

FIGURE 3.14. *Chipped-back edges showing microdenticulated effect.*

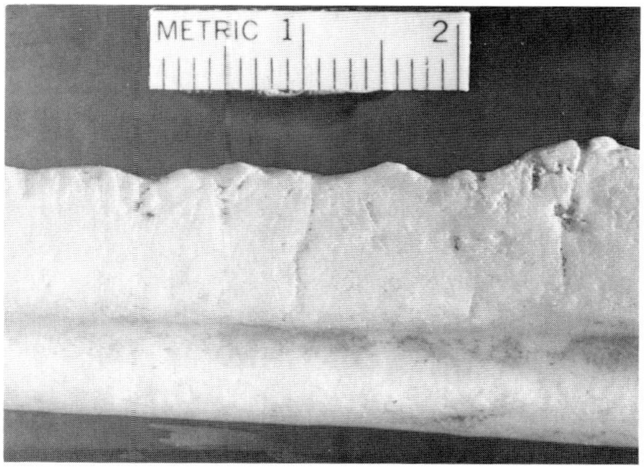

FIGURE 3.15. *External face of chipped-back bone showing oblique tooth scarring.*

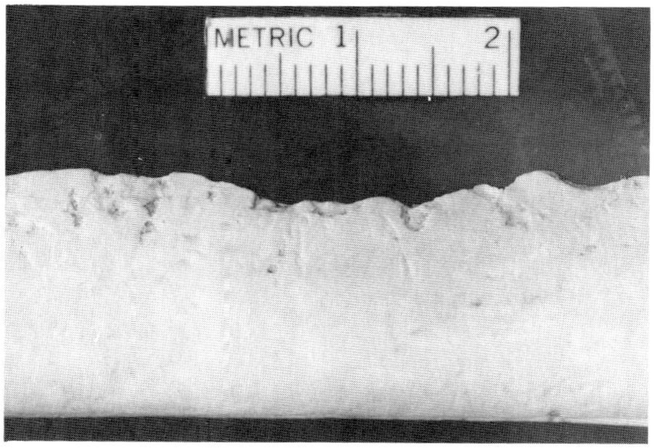

FIGURE 3.16. *External face of chipped-back bone showing tooth scarring from "slipping down" the bone with the teeth.*

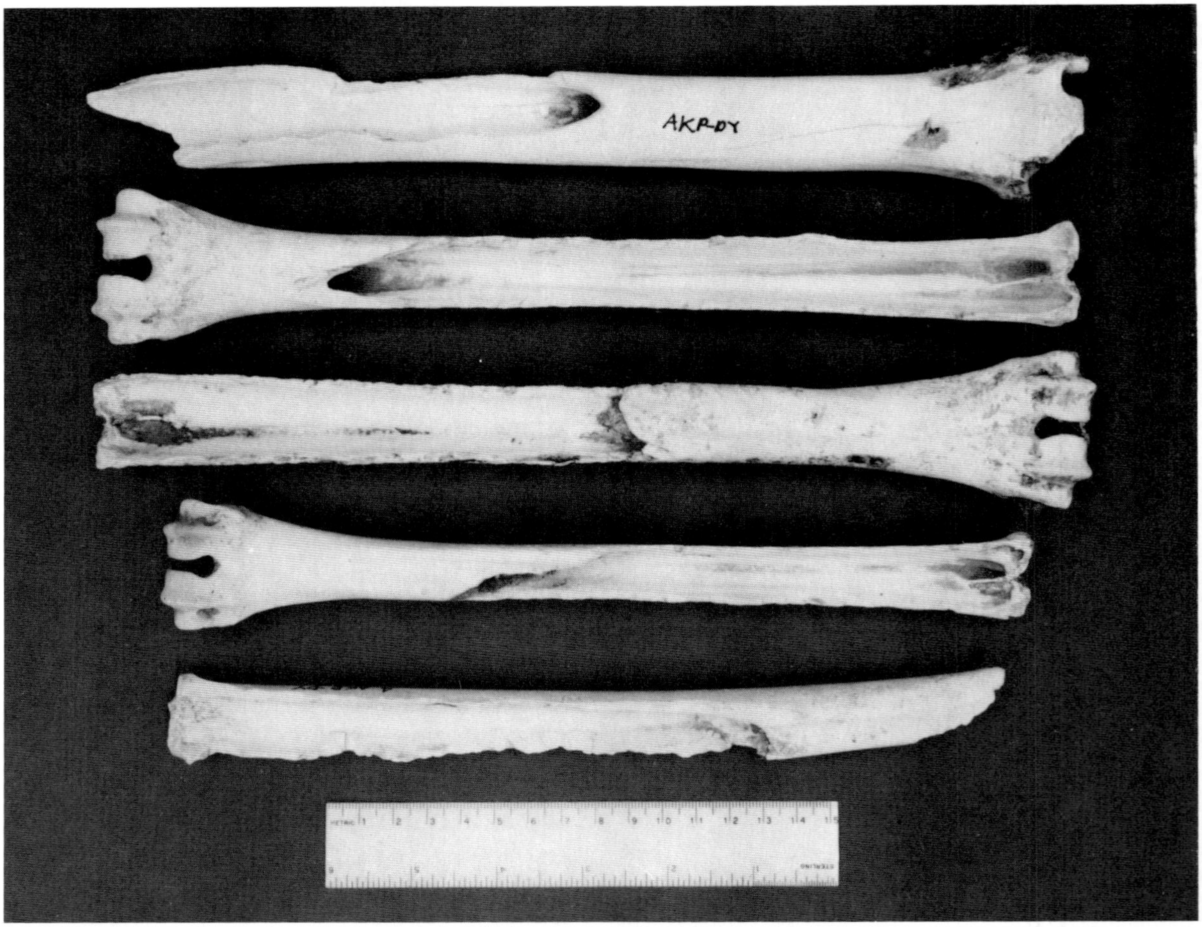

FIGURE 3.17. ***Examples of channeled and chipped-back bones commonly confused with human workmanship.***

utensils and he split bones longitudinally by percussion. He sharpened and pointed by trimming either the broken shaft itself or the flakes derived from splitting the bones. He devised and prepared his tools [Dart 1960:137]." Needless to say, the "he" is *Australopithecus*. For someone accustomed to work in lithics, an encounter with a chipped-back bone produced by animals will almost always convince the observer that he is seeing the work of man. For instance, very solid observers of artifacts such as Veyrier and Combier (1952: 383–385) announced that they had *confirmed* the manufacture of bone tools during the Lower Paleolithic, illustrating what are almost certainly bone flakes chipped back by animals. Earlier researchers had been equally impressed. For instance, Breuil (1939) presents numerous examples of chipped-back bone flakes as evidence for the bone tool industry allegedly produced by *Sinanthropus* at the site of Choukoutien (Breuil 1939:Plates XVIII, No. 3, Plate XXII, Nos. 5, 6, 16, 19, Plate XXV, Nos. 9, 11, 13). Breuil states,

> Long bones of all kinds, chiefly humerus and cannon-bones, were trimmed at the break as chisels or points, the convex distal articulation forming a handle; from Choukoutien to historic times. . . . Flakes may have been

used just as they were, or trimmed at the ends (chisels, piercers, or points) or trimmed along the sides (scrapers). Flakes of very big bones were used after being trimmed like flint flakes of the same size and shape (Choukoutien, lower levels of Castillo including the Mousterian) [Breuil 1939:6–7].

Recently DeLumley (1969) reported retouched bones representing 13% of the total tool industry from the Upper Acheulian site of Lazaret. Chipped-back specimens are well illustrated in Figure 17 (1969:12), where items 3, 4, 5, 6, and 7 are extremely good examples. Figure 18 (1969:164), item 1, is a fine example that also exhibits the frequently correlated presence of both pitting and scoring. It should be pointed out that wolf remains were found in the cave, as were coprolites of predators. I am confident that there was at least a minor occupation of Lazaret by wolves. Characteristically, their remains have been interpreted as the result of human behavior.

In New World studies two chipped-back fragments, both designated artifacts, are illustrated by Bonnichsen (1979) from the Old Crow Flats collections. One is called a "bone core" and the second is designated a "miscellaneous polished and ground bone artifact" (Bonnichsen 1979: Plate VIII–8, and plate VIII–24).

The most enthusiastic recent advocate of "retouched" bone flake tools in the Paleolithic is Leslie Freeman, who comments on his analysis of such "tools" from Level 17 at the cave of Morin: "If the bone pieces were omitted from the study of recovered artifacts, an impressive richness of detail concerning the technology of the Mousterian cave occupants would obviously be lost [Freeman 1978:49]."

Without taking them up piece by piece, I will assert that the chipped pieces illustrated by Freeman (1971, 1978) are classic examples of chipped-back flakes produced by gnawing animals, probably canids. All combinations of telltale properties are represented in the Morin Level 17 assemblage: (*a*) chipping back with diagonal scoring on the external face of the flake, (*b*) chipping back with pitting and/or scoring on the same piece, and (*c*) typically denticulated edges. Such combinations are found with pieces that are described as "eroded," as if by the action of acid (Freeman 1978:46); these are most certainly bone fragments passed in scat. It is probably not without significance that both wolf and hyena were identified (Altuna 1971:380–381). Finally, the "unique" character of Level 17 is commented on by Altuna as follows:

> se nota una clara diferencia entre el nivel 17 (Musteriense) y los nivels del paleolitico Superior, no solo el la compisicion faunis faunistica sian tambien en la composicion de las partes del esqueleto que aparancen. Como hemos indicado un el comienzo de la parte general, en el nivel 17 el munero de esquirlas de diafisis excede con cucho al de epifisis que seria de esperar si no hubiera habido una seleccion intencionada [1971:385].

As will be shown in the following section, the number of bone splinters is inversely correlated with the number of articulator ends in animal-gnawed assemblages. Just the reverse is the case in well-preserved assemblages of humanly modified bone.

Mashed edges are another type of characteristic breakage that animals are capable of producing. On moderately hard bone the animal may vise down with its teeth, at the same time tilting its head while the opposite end of the bone rests on the ground surface. This produces a double set of stresses and may result in the breaking off of an edge of the bone. There may be no apparent tooth impressions in the break side if the leverage is somewhat back of the point where the teeth were impacted. The edge produced in this manner has some of the properties resulting from mashing off the edge of a Styrofoam cup, in that there is a granular character to the surface of the break.

In the event that a mashed edge is then licked repeatedly, it may obtain a rounded polished appearance and may appear as "use wear" as illustrated in Figures 3.10 and 3.18. Such rounded and polished edges are regular elements in both dog yard and wolf den assemblages. However, they are not common—7 in a dog yard population of 886 bone splinters and 267 articulator ends!

When a strong animal is chewing on a long bone, there are always bipolar characteristics in that the chewing occurs between the two sets of teeth. This frequently results in what has been previously described as pitting. For instance, with the metatarsal illustrated in Figure 3.17, when the dorsal edge of the bone has been chipped back, there are accompanying pits on the opposite side of the bone representing the impact of the teeth serving as the anvil for the action of the opposite teeth. Such pitting is

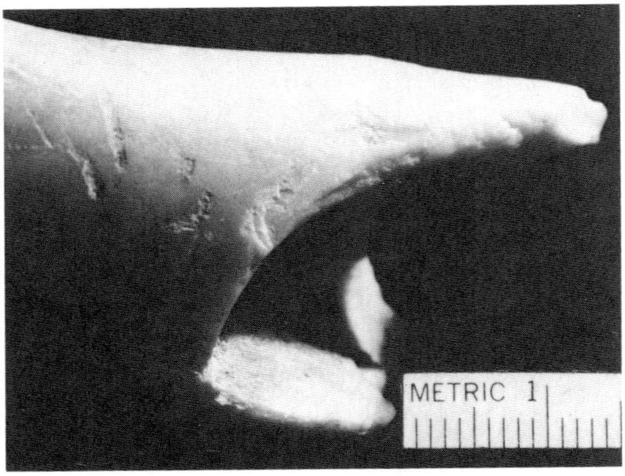

FIGURE 3.18. ***Rounded edge on gnawed bone.***

striking on the more cylindrical long bones of the femur and the tibia.

Pitting occurs typically when an animal is attempting to collapse a cylinder; that is, it has chewed off the articulator end and is working on the cylinder and the exposed marrow cavity. At that juncture it is not uncommon for the animal repeatedly to place the cylinder at right angles to the carnassial teeth and to vise down hard in an attempt to collapse it. If the bone is dense, there is a puncturing or splitting on the short arc side of the bone. Once this occurs the animal begins chipping back along the puncture or split, and the underside of the bone opposite this split becomes increasingly pitted and heavily scarred. Finally, the animal may be successful in collapsing the cylinder and one is left with the impacted scarring on a bone that has been split away from the upper chipped surface.

Under the latter conditions one gets very long, longitudinally split splinters with *chipped* and denticulated edges along one or both ends. This condition is illustrated in Figures 3.19 and 3.20, where "worked" points and edges on exceptionally long splinters are shown. Ironically, what was claimed as distinctive of human behavior by Breuil (1938:38) turns out to be diagnostic of animal gnawing when we have control material.

Mention of bone splinters necessarily introduces the problem of spiral fractures and the degree to which they are uniquely characteristic of human bone breakage, as has been suggested by Bonnichsen (1979).

I have watched dogs and wolves break caribou bones in their mouths. The resulting splinters have all the formal properties of the much discussed spiral fracture. The bone splinters and the broken ends of long bones recovered from the dog yard samples exhibit all the formal characteristics of spirally fractured bone. As was pointed out in my earlier discussion of these samples, the long bones were broken by dogs—Eskimos do not feed their dogs marrow-cracked bone splinters. In the case of my controlled dog-feeding record (see L. R. Binford 1978b: 262–265), there is no question as to the state of the bone at the time of its being fed to the dogs. Long bones were unbroken. On recovery, 44% of the long bones were broken with a spiral fracture; this was done by the dogs. Broken bones and bone splinters recovered from two wolf dens (described in Chapter 5) charac-

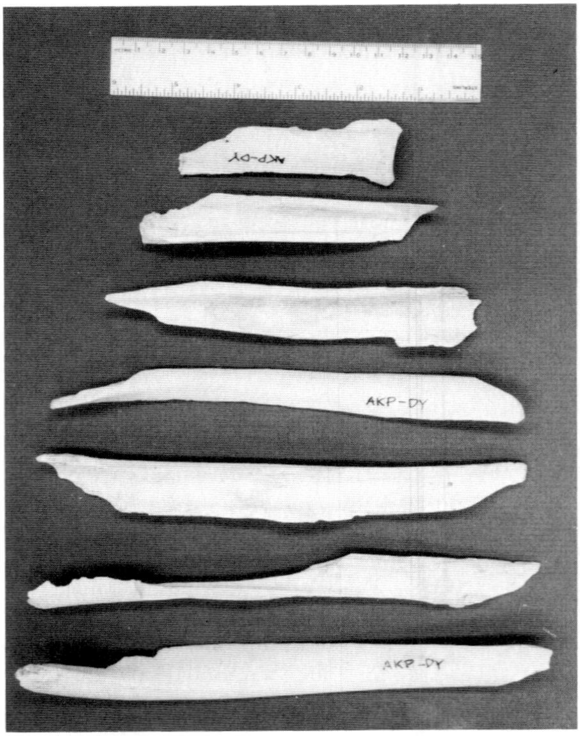

FIGURE 3.19. ***Collapsed cylinders with distinctive chipped-back ends.***

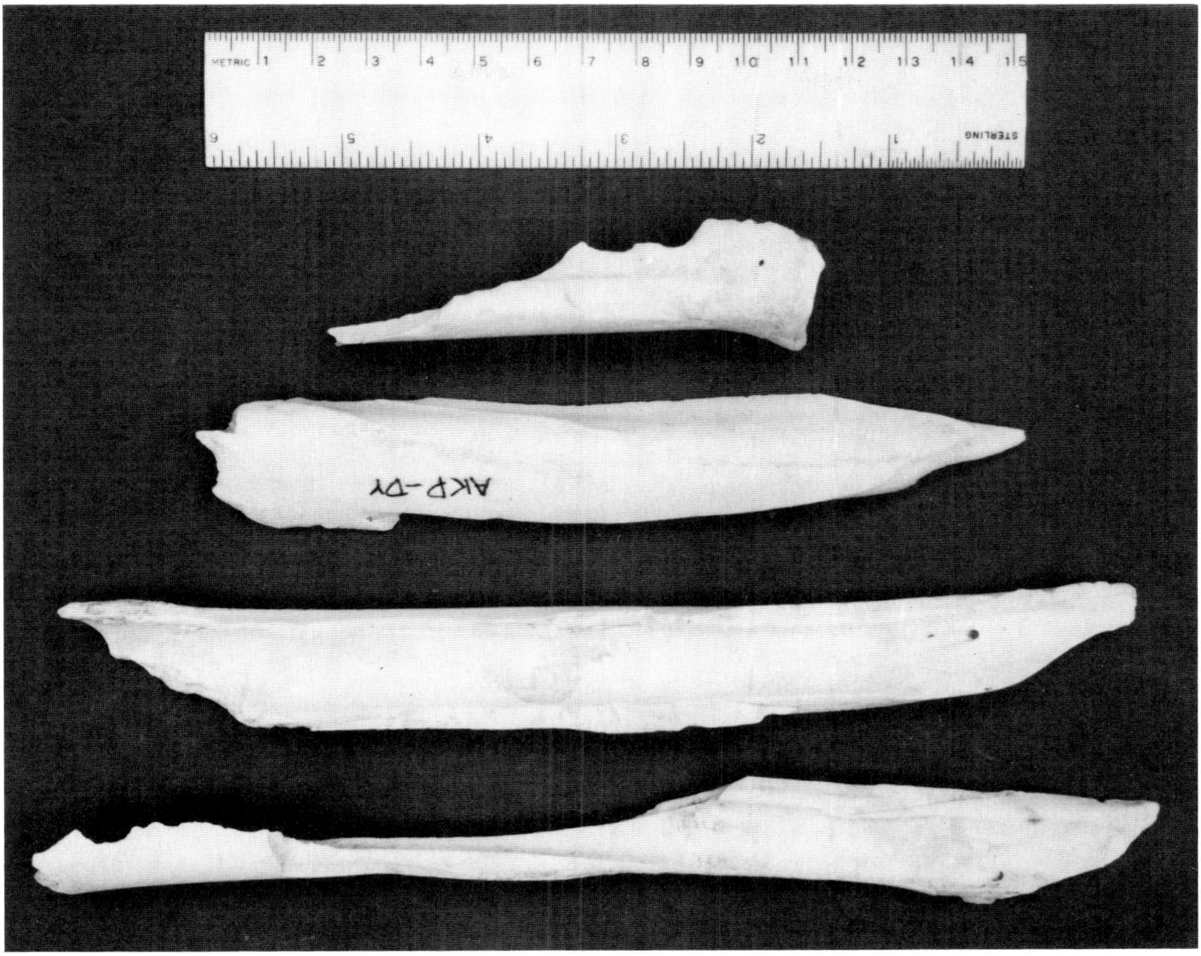

FIGURE 3.20. ***Long "longitudinally" split fragments with denticulated ends.***

teristically exhibited spiral fractures. Broken bones and bone splinters recovered from the 24 separate kills of caribou by wolves (to be reported in Chapter 5) *characteristically exhibited spiral fractures.*

My observations are not unique. Hill has repeatedly reported that spiral fractures were consistently noted in the faunal collections made by him in Uganda and Kenya:

> Spiral fractures are found to occur as a result of natural weathering and carnivore activity that in life are subjected to torsional stress, such as humeri and tibiae. It appears that the internal structure of the bone controls the character of the break [Hill 1976:335; see also Hill 1980:141].

Hill's collections are areal samples in that they represent all the bones remaining on the surface in a given area. Most, if not all, the bone modification is considered to be the result of natural death and subsequent patterns of attrition by scavengers and weathering agents. In most cases the predator is believed to have been a large cat, lion in particular. Since Hill does not know with absolute certainty the agents responsible for the bones, those who wish to believe in the equation of man and spiral fractures tend to dismiss Hill's data as "just a modern sample" where the agent must be inferred as in any archaeological situation. This same criticism may be leveled at spiral fractures recorded by Shipman and Phillips (1976, 1977), who collected areal faunal samples from the Awash area of Ethiopia. In both these cases the bear and the footprint are not, strictly speaking, together, but it is the opinion of the researchers that the bear is not far away. They know the areas in which they worked and find little reason to suspect that man was involved in any significant manner as a causal contributor to the bone assemblages collected.

Zoo studies, where the animals are contained and the parts fed to the animals are known, are excellent experimental conditions. There is no chance that any agent other than the animal fed in the given cage could modify the bone after it was introduced. This strategy was used by Dean Buckland (1823), Pei (1938), Zapfe (1939), and Bonnichsen (1973), to mention only a few. A current study of this type has been under way for several years by Gary Haynes, a graduate student at Catholic University of America (Haynes 1978a, 1978b, 1980).

Paleontological studies—that is, the study of fracture patterns from assemblages where man could not have been a contributing agent—are of utmost importance. One such study (Myers *et al.* 1980) is most provocative.

In this study five New World assemblages of fauna were studied. Four are dated to more than 500,000 years ago; the oldest, to 17 million years. Man could not have played a role in the formation or modification of these assemblages. Spiral fractures were common in these paleontological faunas. The researchers further noted that evidence of carnivore activity was not well documented in the form of tooth marks and other telltale modifications and concluded that still another activity of animals, trampling, is apt to have been the cause. I will return to this point later. Of importance at this point is only that substantial spiral fracturing is illustrated from these localities and no human could have possibly been responsible.

The evidence appears conclusive that animals, both through using their jaws and perhaps by trampling the skeletal remains of other animals, regularly contribute spirally fractured bone to the paleontological record.

The condition of splinters remaining at a wolf feeding area is of some interest. First, the small splinters of bone that are pressured off a bone by the gnawing animal are most commonly swallowed, resulting in few chips and small bone fragments remaining in a feeding area. The latter, of course, appear in the scat, which occurs concentrated in areas of repeated defecation around dens and rendezvous locations, around marking spots, scattered peripherally to kills, and scattered along trails. Defecation is a common marking technique of wolves and is frequently done on prominent rocks around an animal's territory. I have observed considerable bone concentrations around marking spots that result from repeated long-term marking in the same place. Table 3.02 presents summary data on fragment sizes recovered from the defecation and feeding areas around a wolf den. An attempt was made to recover every tiny fragment of bone in both instances.

TABLE 3.02
Size Distribution of Fragments Recovered from Itikmalaiyak Wolf Den [a]

	Defecation area (46 scats)		Feeding area			
			Not gnawed		Gnawed	
Size (cm)	No.	%	No.	%	No.	%
0.0–0.8	37	.29	10	.22	0	
1.0–1.9	29	.23	9	.20	0	
2.0–2.9	23	.18	8	.18	0	
3.0–3.9	17	.13	8	.18	1	.02
4.0–4.9	14	.11	9	.20	6	.14
5.0–5.9	3	.02	1	.02	9	.21
6.0–6.9	3	.02	0		13	.31
7.0–7.9	1	.01	0		6	.14
8.0–8.9	0		0		1	.02
9.0–9.9	0		0		2	.05
10.0–10.9	0		0		1	.02
11.0–11.9	0		0		0	
12.0–12.9	0		0		2	.05
13.0–13.9	0		0		1	.02
14.0–14.9	0		0		0	
15.0–15.9	0		0		0	
	127		45		42	

[a] See Chapter 5 for a description of this site.

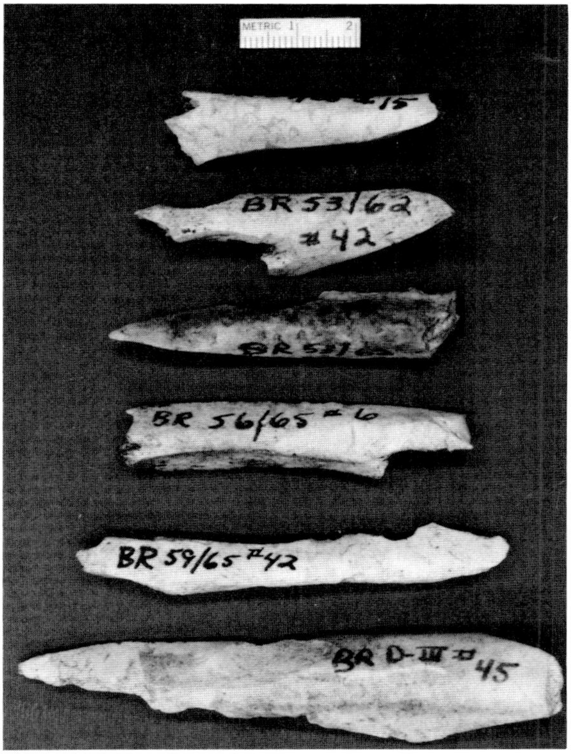

FIGURE 3.21. *Group of "ridge-crest removed" flakes typical of animal gnawing.*

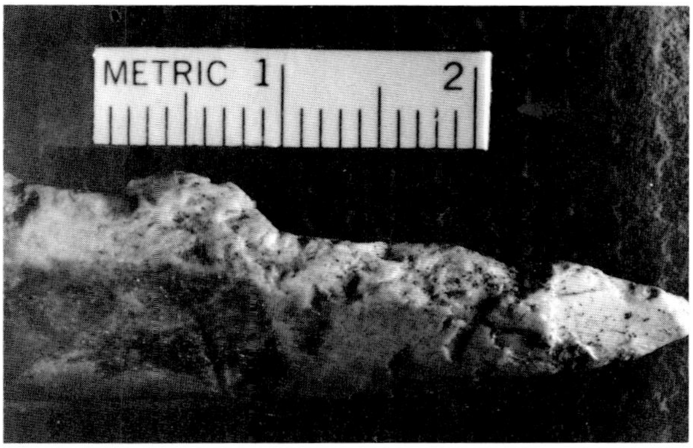

FIGURE 3.23. *Detail of point on a pseudotool shown in Figure 3.22.*

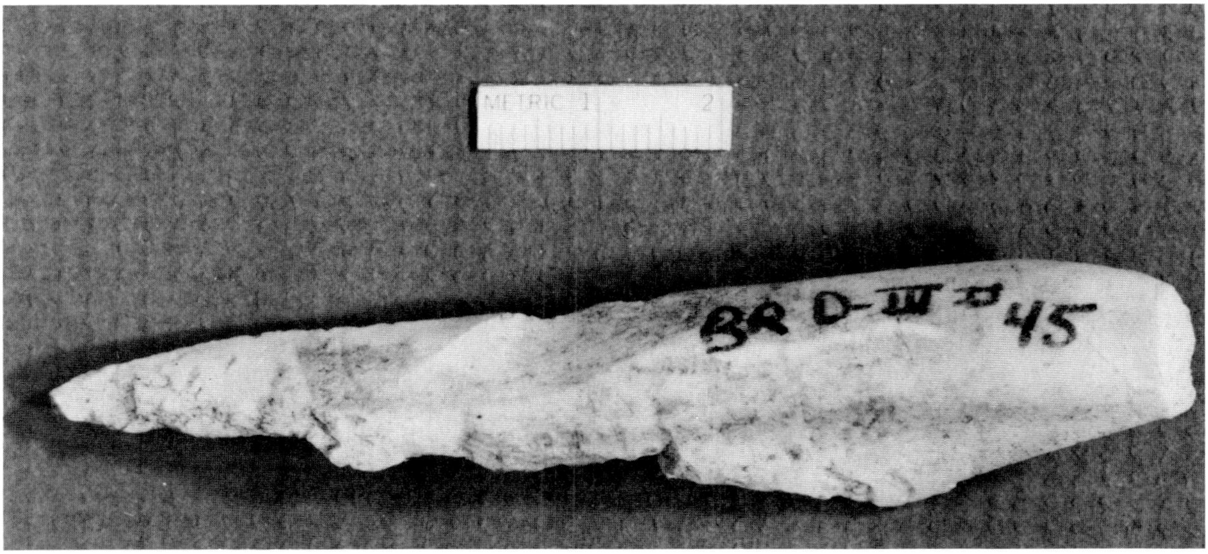

FIGURE 3.22. *Pseudotools produced by animals chewing at ridges.*

Several patterns of interest appear. First, the vast majority of splinters and chips observed in scats were under 4.9 cm in length. Basically the same size range was observed in ungnawed chips and splinters. On the other hand, fragments showing evidence of gnawing are certainly *all* on the large size of the distribution. This makes considerable sense, since the small flakes and chips are pressured off by the gnawing animal during its reduction of the bone, and are generally swallowed. Elements of the bone that show considerable evidence of gnawing, when they are broken, preserve areas of contiguous chip removal and scarring. These are large splinters and therefore will not be swallowed. Thus, when one is dealing with an animal-gnawed assemblage, evidence of gnawing, pressure-flaked edges, incised scarring on the outside of the flake, pitting and abrasion from repeated viselike mashing of a bone surface, and so on will occur on large splinters (greater than 4.0 cm in length); small splinters will exhibit no such modification. When small chips and splinters are present in large numbers, they may show signs of having been eroded by stomach acids (and thus deposited in feces). (See Kitching [1963:19–22] for a good description of the effect of stomach acids on bone splinters.)

One additional characteristic, which appears to be diagnostic of animal-gnawed assemblages, was noted. Animals chew at protrusions of "thin" ridges; a bone is placed in the mouth and irregularities are vised between the carnassials. This action results in pinched-off tuberosities and, more important, in long splinters that are basically triangular in cross section, often with pointed ends. These splinters derive from the biting off of ridges such as the pectoral ridge, the supercondyloid ridge on the humerus, the tibial crest, the linea aspea on the femur, the ventral ridges on the metapodials, and the ventral crest of the cubitus (ulna). When such ridges or crests break, they are generally too long to be swallowed and remain as very common pointed "pseudotools." (See Figures 3.21, 3.22, and 3.23.) Forty-three percent of the gnawed splinters recovered from the den area (Table 3.02) were of this type. I have never observed such splinters in percussion-fractured assemblages; percussion fractures generally cross such ridges rather than run parallel to them. Fine examples of very convincing "tools" are illustrated in Figures 3.21, 3.22, and 3.23.

Modifications by Anatomical Part

As pointed out earlier, one of the more common arguments offered regarding the possible use of bones as tools has been based on the "use potential" of the shapes of various bones, or their shapes as modified through breakage, and/or citations of worked edges or surfaces in conjunction with either normal or breakage shapes. I have described in some detail the types of edge and surface modifications animals are capable of producing during normal bone gnawing, as well as their distinctive breakage patterning.

I will now describe the characteristic patterns of destruction and modification I have observed on caribou and sheep bones from the Eskimo dog yards, as well as from the wolf dens. In each case where false or suspect tool identifications have been made of similar parts, I will attempt to identify the relevant literature and cite the problems I see with the claims for human modification.

I have already demonstrated that gnawing results in *distinctive* patterns of scarring and chipping. These conditions are associated with the pattern of breakage, which has also been shown to be distinctive and diagnostic of animal-modified assemblages. Given these general conditions, it is not surprising that there are distinctive patterns of destruction associated with the different bones as a function of their differences in strength and overall form. The following are observations of patterning in the morphology resulting from animal feeding and gnawing.

Skull

Examples ranged from a complete and unmodified skull to a skull of which only two parts remained—the base around the antlers (Figure 3.24) and the palate with two tooth arcs (Figure 3.25). Animals appear to begin gnawing from the nose inward, removing the face and finally collapsing the cranium, leaving the palate and occiput in two parts. Rarely do the latter two parts exhibit further gnawing. When the skull is eaten in the manner described, few skull fragments remain at the location of consumption; they appear to have been ingested.

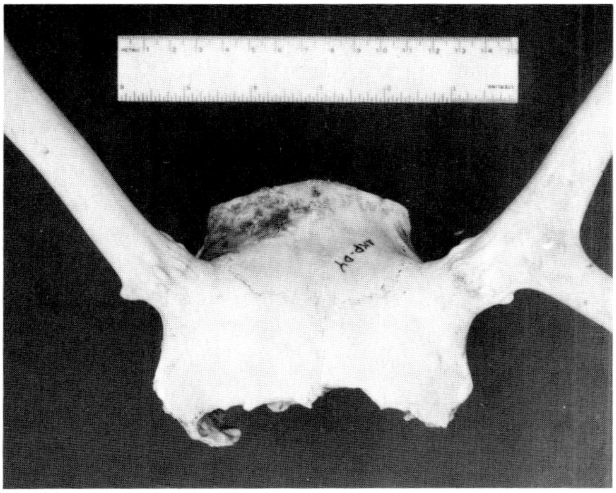

FIGURE 3.24. *Animal-gnawed cranium with attached antlers (caribou).*

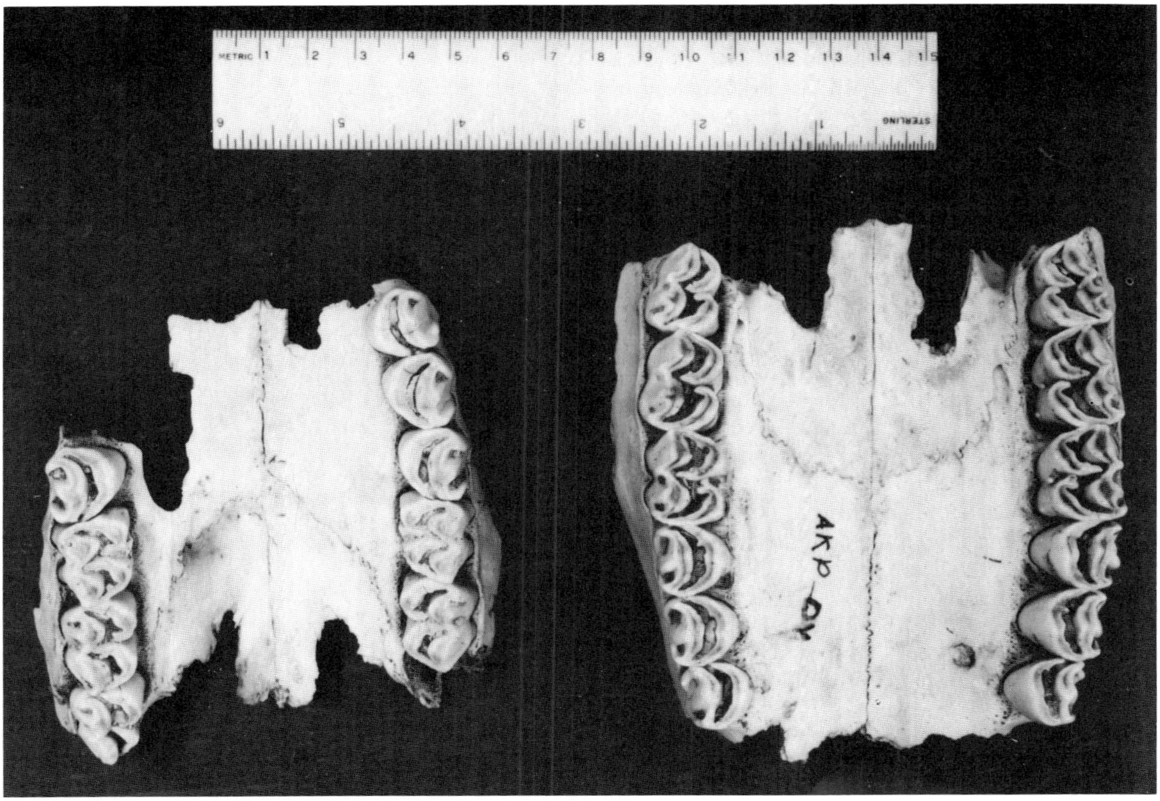

FIGURE 3.25. *Animal-gnawed palates (caribou) showing distinctive channeling and crenulation of thin bone.*

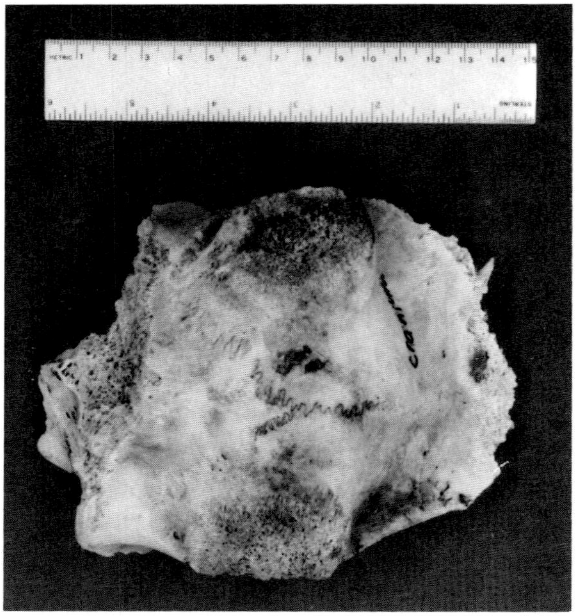

FIGURE 3.26. ***Cranial disc as commonly produced by gnawing canids.***

An alternative pattern, for young animals and those with little to moderate antler development (including large animals in velvet), begins with the chewing back of the face area toward the antler base, but there is also similar breakage and chewing back from the occiput, leaving a "skull disc" as shown in Figure 3.26.

Several points need be made about the observed patterns of skull destruction produced by canids. The skull disc illustrated in Figure 3.26 is a common element on many ancient sites. It is frequently interpreted as a product of human workmanship. For instance, Breuil writes, "The frontal cavity between the severed horns is too often carefully trimmed and worn at the edges not to have been utilized. This is seen in the skull caps and human skulls of la Placard and Laugerie Basse, and perhaps even those at Chou-kou-tien [1939:5]." Kitching (1963:30), writing about the fauna from Pin Hole Cave, where both Mousterian and Upper Paleolithic remains are found, comments, "Amongst the ten skull fragments there are three frontoparietal regions of reindeer that have been neatly removed from the tops of the skulls to form almost perfectly rounded, shallow, saucer like utensils." Confident of the hominid workmanship on such items, Raymond Dart (1962:287) states;

> sufficient of the damaged or discarded specimens are available to display . . . the technique by which skulls were split transversely, i.e., by one or more blows delivered probably with a scapula or lower jaw blade on their bases, into two halves. . . . Each skull half thus formed automatically a cup or bowl.
>
> In creatures sufficiently large such as the buffalo these bowls, especially the anterior one . . . with its horns attached spontaneously afforded a tripod-like basin in which to mash or triturate any tough type of food.

In all these cases, there is no reason to see the "skull bowls" as anything other than the product of normal attrition by animals. It might be interesting in this regard to reexamine the ibex skull–horn units allegedly intentionally placed around the Neanderthal child at Teshik-Tash (Movius 1953:25, 26).

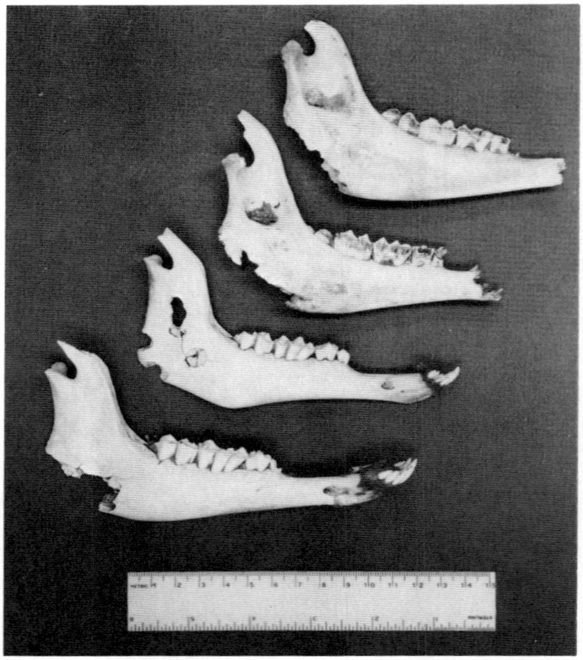

FIGURE 3.27. ***Initial destruction of sheep mandibles by both dogs (lower two) and wolves (upper two).***

Another commonly cited characteristic of skulls is the so-called depressed fracture, which Dart popularized as evidence for the hominid use of tools and the "bludgeon" technique of hunting. I have not personally observed such fractures on the skulls examined from the wolf kills, but others report them to be a regular characteristic of kill and natural death assemblages (Hill 1975; Shipman and Phillips 1976).

Mandible

Animals may make considerable progress in the destruction of the mandible. Exactly where the mandible is attacked by animals is partly a function of differences in mandibular morphology among different species. For instance, Figure 3.27 illustrates patterns of initial or minor destruction of sheep mandibles by wolves (in the upper two examples) and domestic dogs (in the lower two examples). Puncture marks of both canine and carnassial teeth are obvious along the angle of the mandible. In all cases, carnivores were going after the masseter muscle. In Figure 3.28, where caribou mandibles are shown, another condition is indicated. The two lower examples were gnawed by dogs, but the mandibles had first been stripped of meat before the dogs received them. The puncture marks of their teeth ap-

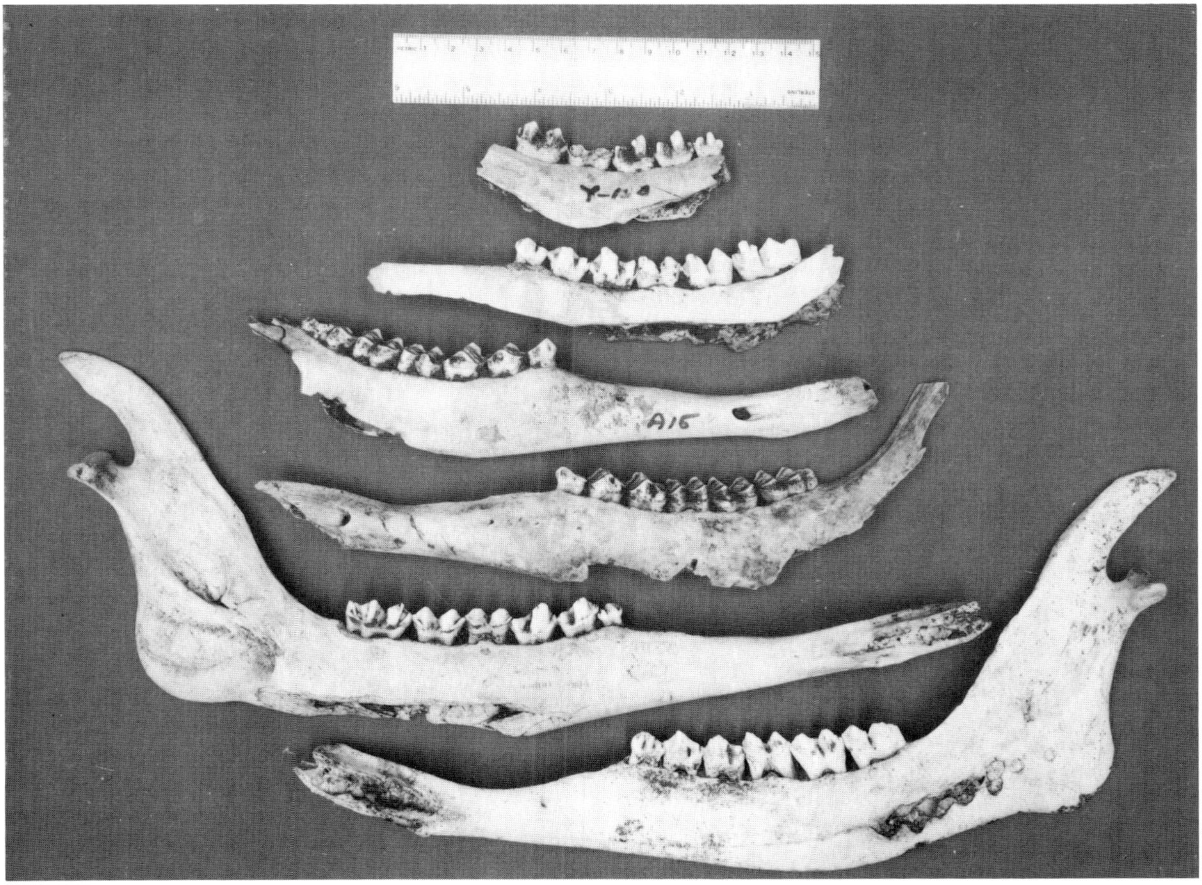

FIGURE 3.28. ***Sequence of caribou mandibular destruction by both dogs (lower two) and wolves (upper four).***

pear directed at the pulp cavity below the molars. The upper four mandibles were reduced by wolves and the pattern seems to be to remove the angle of the mandible and then move into the pulp cavity. This condition has been observed by others:

> Broken lower borders of bovid mandibles need not indicate that anything has been "removed." In the natural "non-hominid" conditions, the lower borders of bovid mandibles are eventually almost always lost. This can arise from the attempts of carnivores to extract the tooth pulp, but more commonly it is the result of natural weathering. At least one deep crack appeared quite early during the weathering process parallel to the horizontal axis of the ramus and the development of this results ultimately in a sharp break. The crack occurs lateral to the mandibular canal, which is at the junction of the basal and alveolar portions of the mandible is also frequently lost. Breaks . . . across the diastemic portion are almost universal in the carnivore-treated remains that I have examined [Hill 1976:335–336].

The foregoing observation, which is consistent with my field observations, should be contrasted with the following statement:

> It would be absurd to imagine that the mandibles received the treatment that they were given, in order to extract the marrow, because there is not enough of this material, if any, in any ungulate mandible to justify the process. The mandibles served in their fresh state as choppers, their anterior or diastemal portions being used as handles and their angles as the choppers. After their usefulness in this capacity had ceased, their lower border still served as pounders, their teeth rows . . . were graters or scrapers [Kitching 1963: 30; see also Breuil 1939:19–31, 62, Plate XI].

Frison has illustrated a mandible "chopper," which was argued to have been used in butchering bison. Its description would be basically identical with that given by Kitching. (See Frison 1970:30, Figure 19; also Frison *et al.* 1978:8, Figures 10 and 12.)

As in the case of "skull bowls," the presence of "worn" edges and polish in the absence of tooth punctures is not conclusive evidence that gnawing animals did *not* produce the modifications, which are frequently cited as evidence for "use" on mandibles.

The mandibles of Carnivora have attracted the attention of many investigators and have prompted some of the more imaginative interpretations of bone tool use appearing in the literature:

> From the similarity between the treatment of these lower jaws of the bear and lion and those of the hyaena and wolf, it is obvious that they were all prepared to serve a common ripping purpose. Their use as tools is confirmed by the all around smoothness of the bone surfaces and their employment as bone tool fabricators by the damage to some of the canine teeth [Kitching 1963:42].

Illustrations in Breuil (1939:60–61, Plates IX and X) and Kitching (1963: Plates 34 and 35) should be compared with Figure 3.29 where a wolf (lower specimen) and an Eskimo dog mandible are illustrated. Both were collected from wolf dens: the lower specimen came from the Itikmalaiyak den reported here and the upper specimen came from a den visited by Eskimo informants who recovered the jaw for me. The interpretation of puncture marks obviously made by Carnivor canines as having been made by man using Carnivora jaws as tools seems to be the ultimate anthropocentrism. Such interpretations have been made on several occasions.

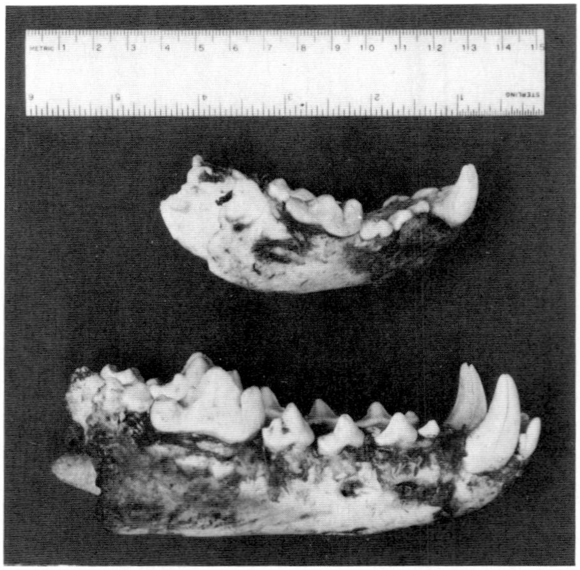

FIGURE 3.29. ***Wolf mandible (lower specimen) and dog mandible (upper specimen) destroyed by wolves. Both specimens were recovered from wolf dens.***

Vertebrae

Vertebrae having irregular shapes are most commonly gnawed on the spines and processes. On the atlas, for instance, the wings and the margins of the atlantal foramen are commonly chewed back to their juncture with the dorsal arch (Figure 3.30). Not infrequently the atlas is collapsed into several fragments. Puncture marks from teeth seem most common on vertebrae, next most common on pelvic elements, and moderately present on skull and toes. Figure 3.31 illustrates nicely the destruction vertebrae can suffer (the upper right specimen is complete). Cervical vertebrae may have transverse, spinous, and articulator processes gnawed away. Thoracic vertebrae have been gnawed somewhat less than cervical vertebrae from the same sites. Animal marks on thoracic vertebrae take the form of broken and gnawed tips of dorsal spines (see upper specimen, Figure 3.32) and, occasionally, punctures on the body of the vertebrae. Lumbar vertebrae are not uncommonly altered by having the transverse processes gnawed away, the dorsal spine attached (see Figure 3.31) and the articulator process gnawed or punctured by canines.

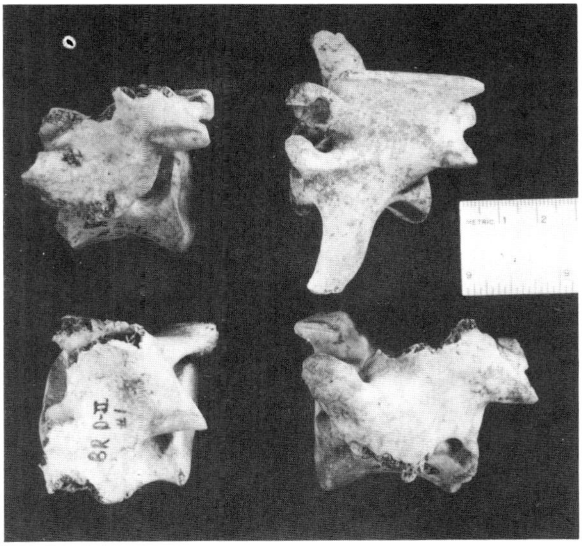

FIGURE 3.31 ***Lumbar vertebrae showing typical effects of gnawing. (Upper right hand specimen is complete.)***

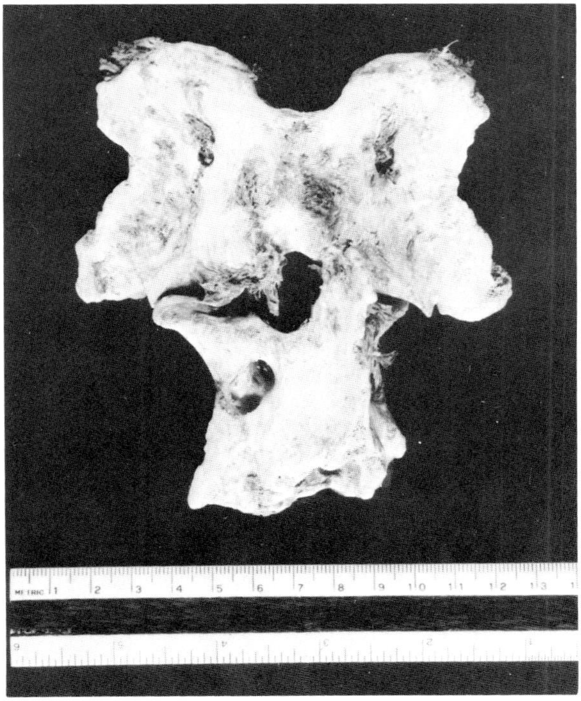

FIGURE 3.30. ***Atlas and axis vertebrae heavily gnawed by dogs.***

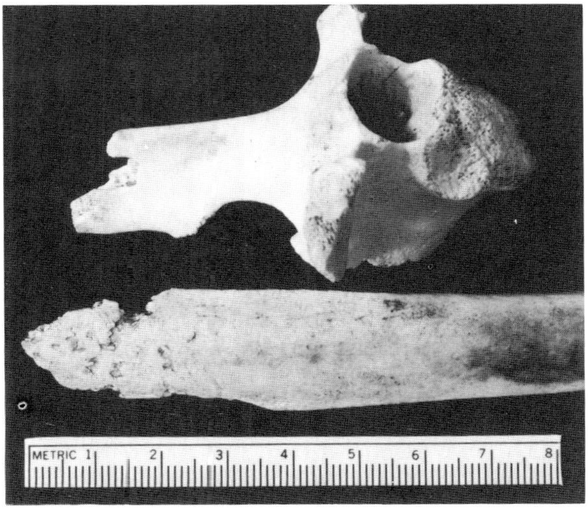

FIGURE 3.32. ***Gnawing on extremities, dorsal spine and distal end of rib.***

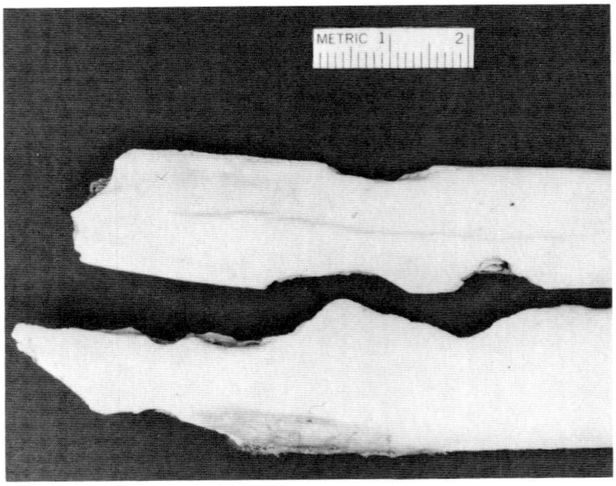

FIGURE 3.33. ***Tooth notching of rib margins by wolves.***

Ribs

Ribs are most commonly gnawed at the distal ends (Figure 3.32), which may appear to have been broken rather than gnawed, although "mashed back" edges are frequent. When carnivores disarticulate ribs, they commonly insert their canines between the ribs and pull. This leaves notches in the ribs and may result in breakage generally an inch or two out from the proximal end of the rib; that is, at the point of maximal curvature as the rib leaves the vertebrae. Figure 3.33 illustrates ribs recovered from a single wolf kill site; gnawing has denticulated the end of the lower example and canine notching is well represented on the upper rib.

Pelvis

The pelvis is one area where carnivores seemingly preferentially destroy much bone. First, it is an "early target" for many carnivores, being one of the first areas attacked during the feeding on a carcass. Most prey animals either fall on their sides or drop from a standing position so that their legs are folded under them. Feeding is therefore frequently initiated at the "ham" from the dorsal surface or, not uncommonly, directly into the pelvis at the anal–genital area.

Look at Figure 5.04, where a recently killed caribou is shown. It is clear that the feeding began in the anal–genital area, resulting in the partial destruction of the pelvis and the subsequent disarticulation of the rear legs, which have been pulled back into the picture by the photographer. Several places on the pelvis are prime targets for chewing—the dorsal spines of the sacrum, the iliac wings or crest, the dorsal margins of the pubic tubercle, and the ramus and tuberosity of the ischium.

Figures 3.34 and 3.35 show the pelvis of a sheep upon which dogs have fed. The entire carcass had been turned over to the dogs by the Navajo owners. This pelvis illustrates the *characteristic* destruction by a canid. Comparison should be made between this pelvis and those illustrated from the Glenrock Buffalo Jump (Frison 1970: Figures 11i and 12). I think

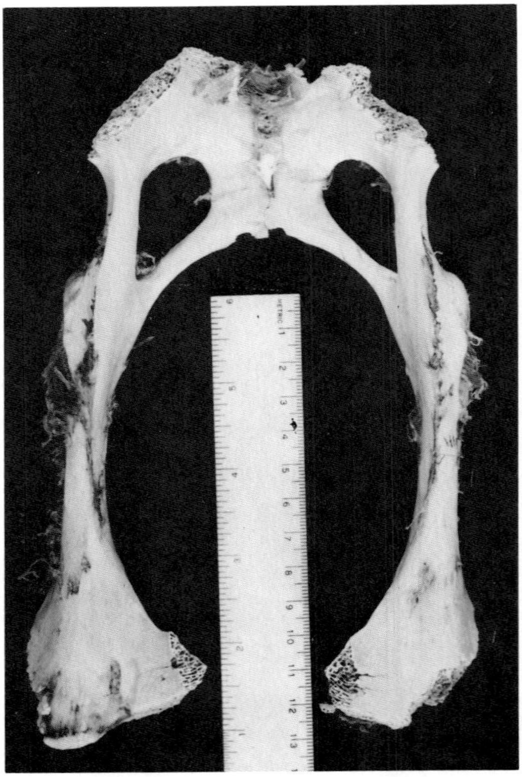

FIGURE 3.34. ***Dorsal view of sheep pelvis chewed by dogs.***

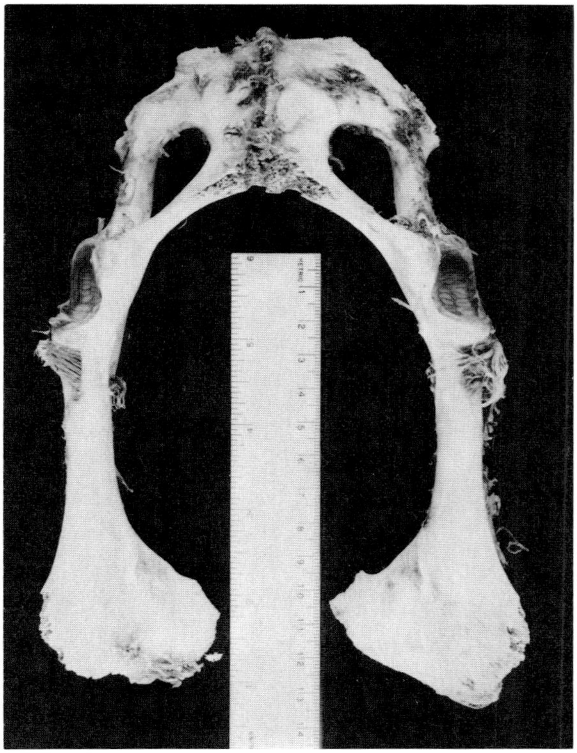

FIGURE 3.35. ***Ventral view of sheep pelvis chewed by dogs.***

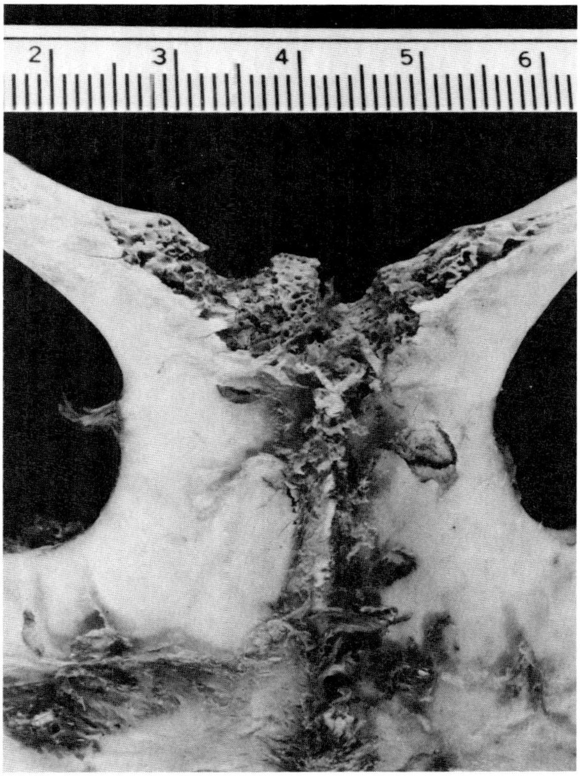

FIGURE 3.36. ***Ventral view of pelvis showing tooth punctures on the tuber coxae.***

very little convincing is needed to conclude that *the pattern of destruction is identical.* The puncture marks so clearly visible on the sheep pelvis (Figures 3.36 and 3.37) are also present on the bison pelves but they are interpreted by Frison (1970:19) as tool marks made by pointed choppers. Most commonly the pelves at Glenrock were broken into halves, and consisted of the acetabulum with both ischium and pubic branches remaining as stubs, the ilium was variously destroyed. Figures 3.38 and 3.39 illustrate parallel destruction of caribou pelves by wolves and by dogs. The irregular denticulated edge where gnawing has occurred is also scarred with puncture marks from teeth (Figure 3.38). Analogous scars were interpreted by Frison as "tool marks from sharp-pointed choppers" (Frison 1970:19).

It is not uncommon for pelvis elements to be broken or cut in places analogous to those illustrated in

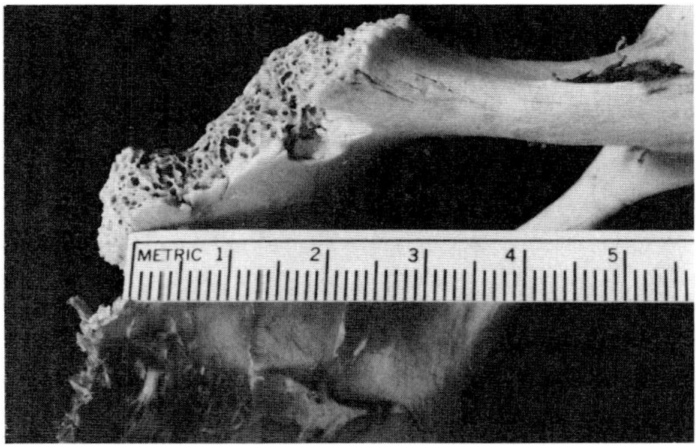

FIGURE 3.37. ***Dorsal view of pelvis showing punctures and mashed-back edges of the tuber ischia.***

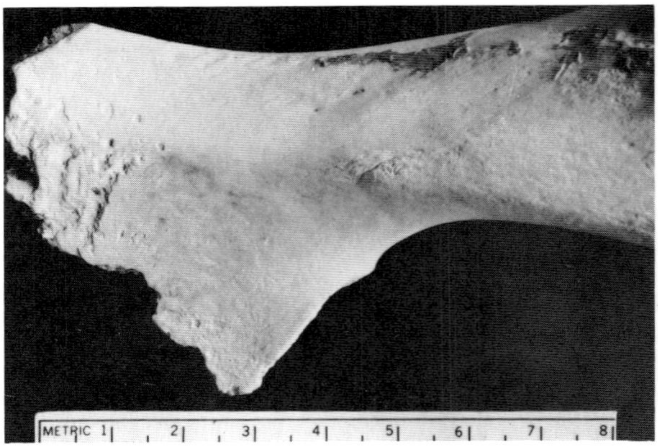

FIGURE 3.38. ***Scoring, pitting, and crenulated edges of ilium produced by dog gnawing.***

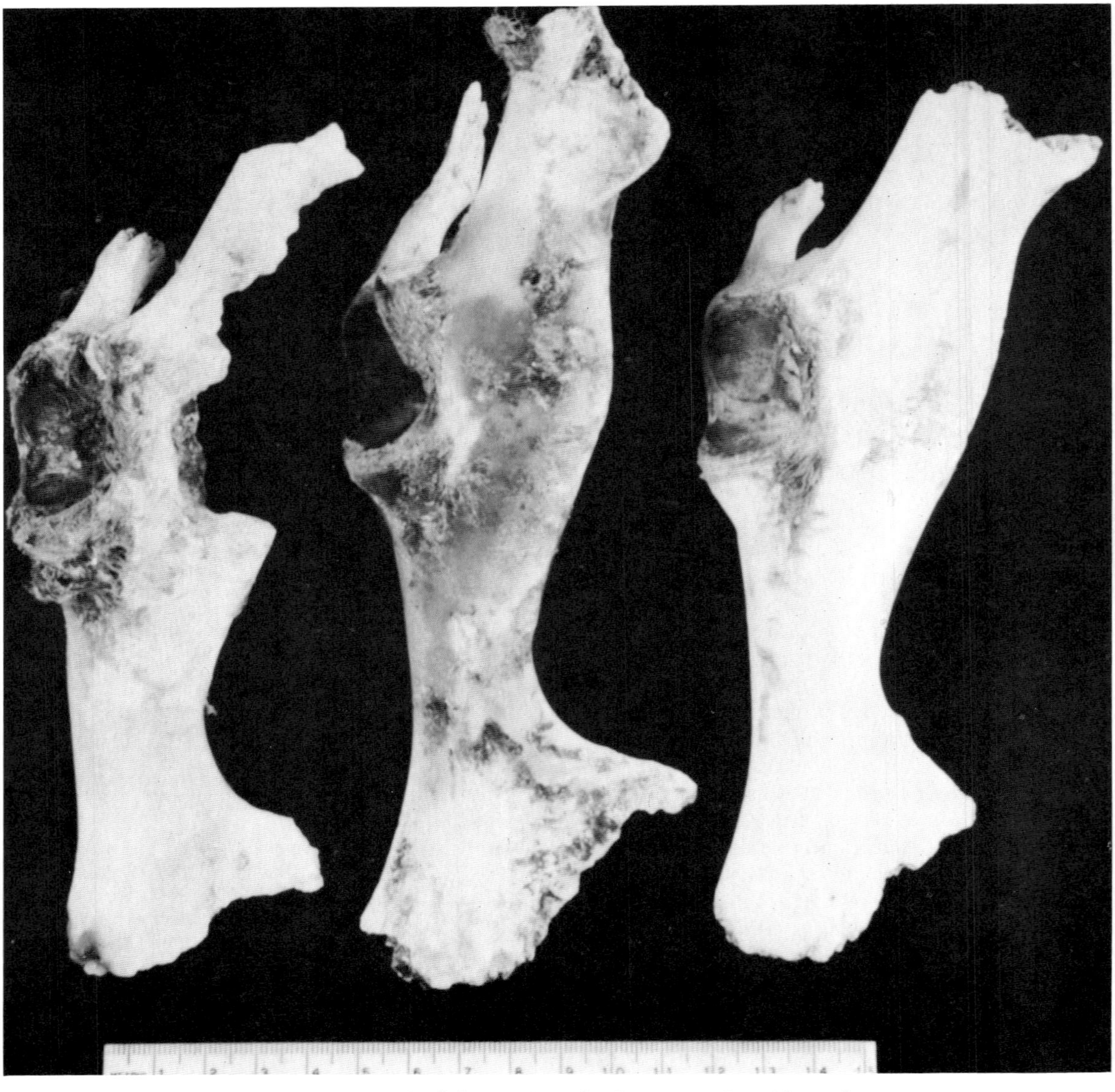

FIGURE 3.39. ***Typical destruction of pelvis as produced by wolves.***

Figure 3.39. Caribou pelvic halves processed by man are frequently broken across the neck of the ilium as well as across the branches of the pubis and ischium. The latter breaks are clean, and although fractures and curves may occur, the fractured bones rarely show the denticulated character seen on those gnawed by animals (Figure 3.39). I think that the pelves from the Glenrock Buffalo Jump, interpreted as evidence of use of choppers and muscle stripping, were simply reduced by gnawing animals.

Pelvic parts have been interpreted as choppers (Frison 1978:307), scrapers (Frison 1970:19), and mortars (Kitching 1963). I have reproduced the statement from Kitching, which becomes informative when it is kept in mind that he is almost certainly describing alterations to bones effected by the Pleistocene hyena of Europe:

> The abraded and grooved condition of the acetabular cavities of the 18 rhinoceros innominate bones shows that some, if not all of them, served as mortars in which food was pounded and crushed. . . . All the innominate bones have had their ischial and pubic portions removed, and the iliac portion remaining forms a handle, each of which show a high degree of polish [Kitching 1963:36].

Scapula

Scapula gnawed by animals have three basic characteristics: (*a*) the acromion (the proximal tip of the spine) is chewed away, (*b*) the coracoid process and the supraglenoid tubercle are generally chewed off, and (*c*) the edges of the vertebral border will be chewed to varying degrees such that on occasion the blade of the scapula will be all but removed. These three features are all well illustrated in Figure 3.40. It should be noted that the typical denticulated edge characteristic of gnawed margins is well evidenced in these specimens. Puncture marks from animal teeth are more common on the margins of the glenoid fossa than any other surface. Such marks produce the crenulated edge of the scapula blade, but the bone is so thin that actual puncture marks are not generally preserved. Instead, "circular" segments of the edge are removed.

Humerus

Fragments of the humerus have received more attention by those attempting to make a case for bone tools and distinctive forms of "human" bone breakage than practically any other bone in the anatomy. Breuil illustrated distal ends of cervid and equid humeri as having been modified by the occupants of Choukoutien (Breuil 1939: Plate XIII). Later Dart (1959b) was to make a grander claim for "daggers" fashioned from the humeri of bovids:

> From the facts presented here it is patent that the antelope humerus owed its prevalence at Makapansgat and its persistence at Kalkbank, as well as its dispersion from South Africa to Peking, in the Far East and to Portugal in the Northwest, to the fact that it was the most generally serviceable domestic appliance discovered by Australopithecus [Dart 1959:92].

In addition to its shape and distinctive pattern of breakage, the fact that the proximal articulator end of this bone generally has the least survival potential and is therefore commonly missing from bone assemblages had prompted many interpretations. For instance, in one of the earliest attempts to recover behavioral information from faunal remains, Theodore E. White states,

> The humerus shows the greatest discrepancy in the number of the two ends of any of the elements, with three specimens of the proximal end and thirty from the distal end. Anyone who has tried to separate the scapula and humerus with a knife knows that it is not easy, even in these days of crucible steel, and the job can be easily and effectively accomplished with a cleaver. *It is clear that a Stone Age cleaver* would demolish the head of the humerus beyond recognition [1952:338; emphasis mine].

Many archaeologists have accepted White's argument and interpreted the low frequency of the proximal ends as evidence of destruction of the joint during butchering. "Replicative" experiments have even been performed, showing that when one destroys the proximal humerus, it is destroyed! Others have butchered animals, showing that if one works at it hard enough, one can even dismember an animal by pounding on articulations such as the proximal humerus. But to be sure, these exercises have

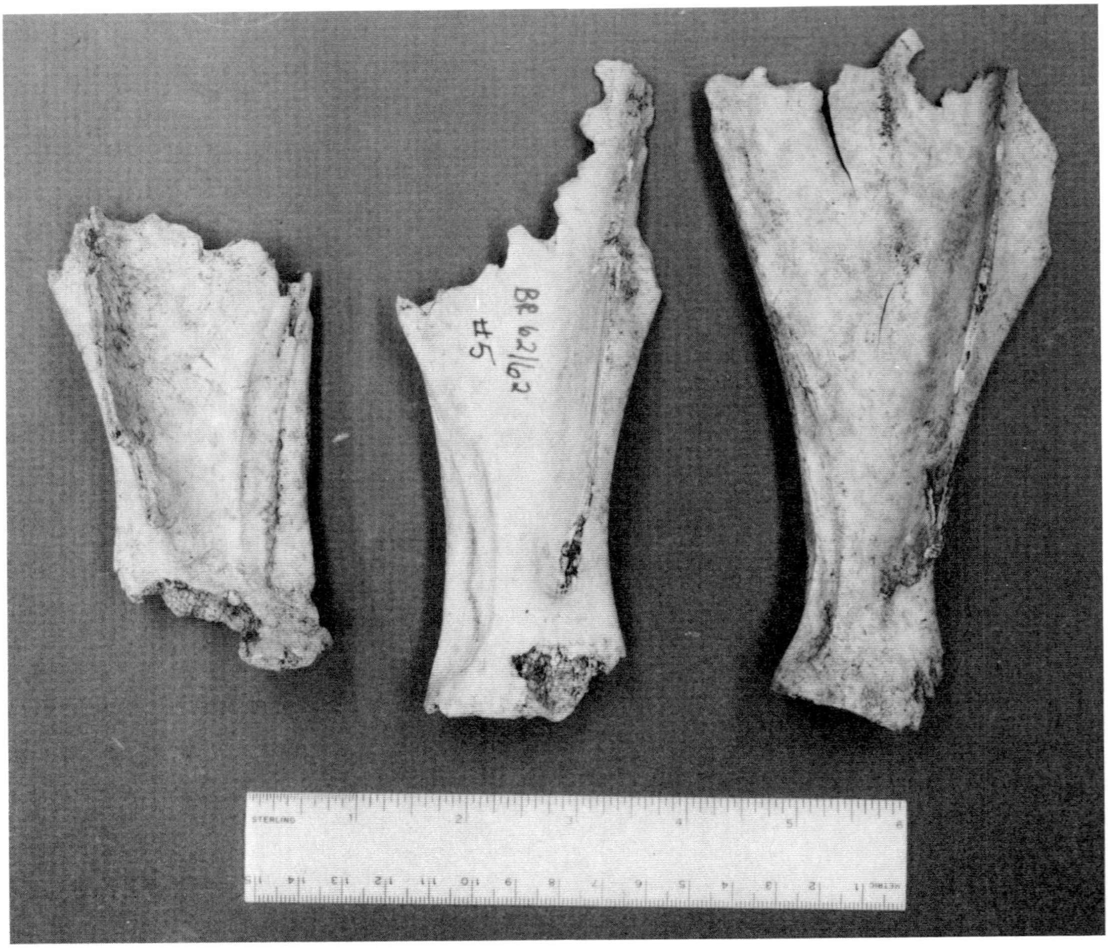

FIGURE 3.40. ***Scapula gnawed by dogs (center) and by wolves—note crenulated edges.***

not shown that men of the past ever did such things. Another argument offered to account for the common absence of the proximal humerus from archaeological assemblages has been that it was intentionally destroyed during the production of a tool manufactured from this bone. George Frison (1970:27) has argued that the humerus "flesher" was a consistently produced and used tool on the North American Plains:

> another use of smaller humeri was to break off the proximal articulator end in such a way that the edges have a scalloped appearance. This end was used as a scraping tool with wear striations and polish parallel to the longitudinal axis of the bone [Frison 1970:27].

In later publications this "tool" is given greater prominence and has been largely accepted by many early man archaeologists as a diagnostic item indicative of human involvement in the formation of deposits and events, even though stone tools and other evidence may be lacking. In the Casper site report (1974) Frison states,

> The articular surfaces were removed, the cancellous bone gouged out and the circumference of the bone

FIGURE 3.41. ***Caribou leg removed from a meat cache and eaten by wolves.***

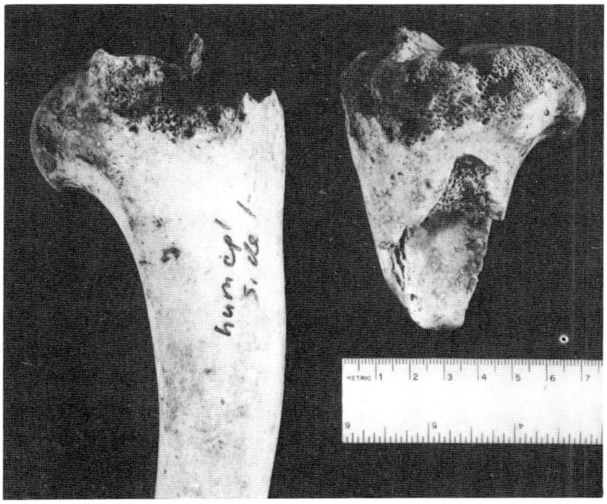

FIGURE 3.42. ***Typical forms of minor destruction (furrowing) caused by dogs on already disarticulated proximal humeri.***

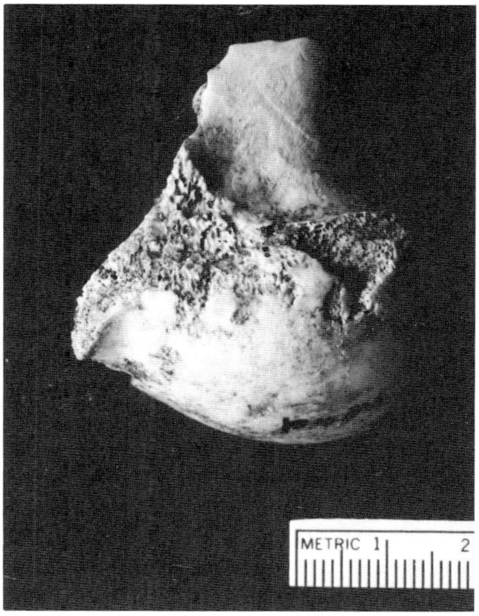

FIGURE 3.43. ***Furrowed and punctured humeral head removed by a wolf while disarticulating the junction with the scapula.***

> around the break given a scalloped effect (Figure 1.14b). They were presumably used in both skinning and fleshing and where bone preservation allowed, wear in the form of a polish appears on the sharp points [Frison 1974:56].

Figure 3.41 illustrates just how rapid the destruction of the proximal end of the humerus by animals may be. The right front leg shown had been dragged from an Eskimo meat cache and "eaten" by wolves in a single afternoon. The proximal end of the humerus is completely gone, as is the proximal end of the cubitus, but the foot and the lower part of the leg are unmodified.

Figures 3.42 and 3.43 illustrate characteristic stages of destruction of the proximal humerus effected by dogs and wolves. Deep furrowing and actual destruction is generally initiated on the lateral tuberosity and the ball head of the humerus (Figure 3.42, left) and often results in the breaking off of the ball articulator surface, as shown in Figure 3.43. In both illustrations, pitting and dragged puncturing marks are visible along the margins of the furrowed areas. As gnawing continues, all the soft tissue on

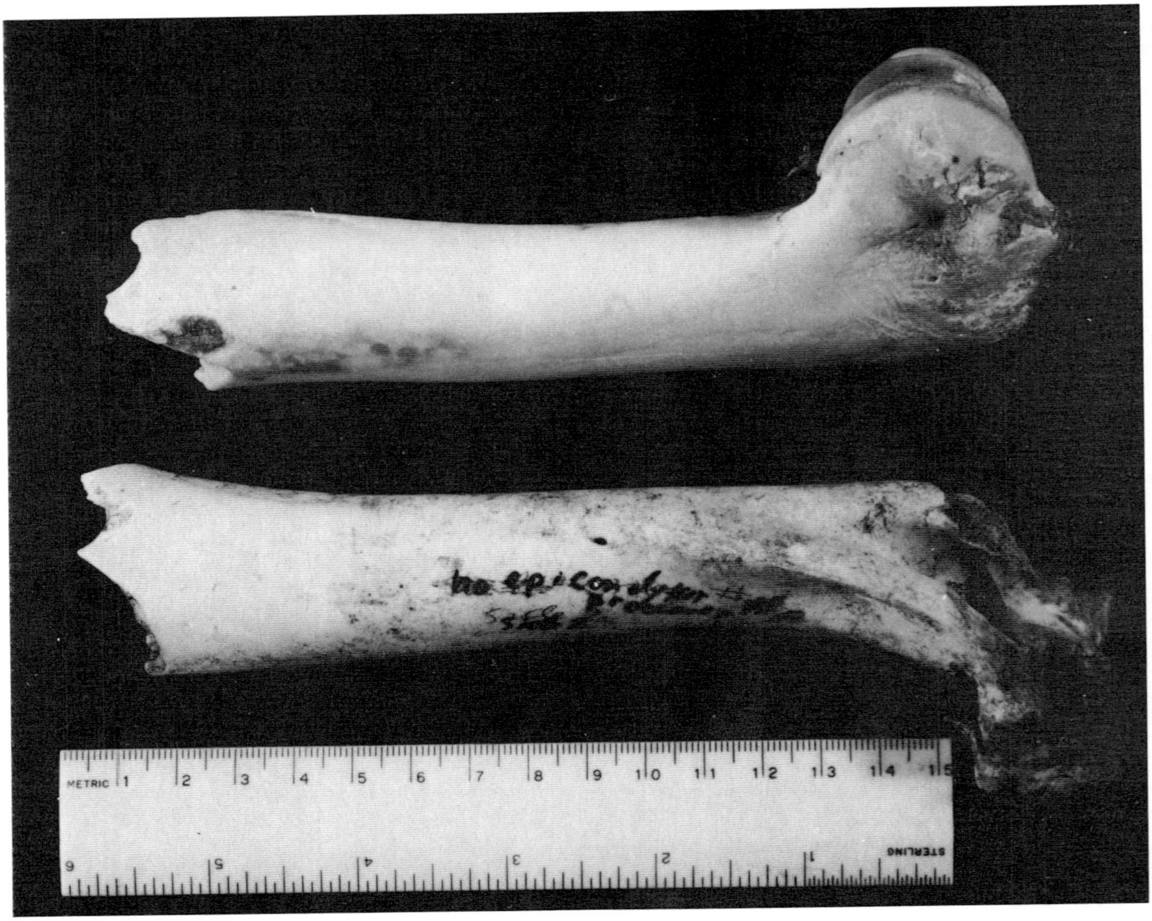

FIGURE 3.44. ***Humerus "fleshers" produced by wolves (upper) and dogs (lower).***

the proximal end may be removed until the animal begins to encounter the harder, denser bone of the diaphysis or shaft, as was the case illustrated in Figure 3.42.

I have discussed the problem of butchering the scapula–humerus joint previously, but it should be emphasized that *all* the so-called diagnostic forms of breakage on the proximal end of the humerus that have been interpreted as evidence for use as choppers and a "muscle stripping" technique of butchering (compare Figure 3.42 with Figure 1.12 in Frison [1974]) are in my opinion normally produced by animals gnawing through this joint. As we will see in chapter 5, animals customarily gnaw through this joint on moderate to large prey, so that the scapula is normally associated with the axial skeleton. In contrast, human butchering commonly results in the removal of the entire front quarter, and the disarticulation of the scapula–humerus joint is a secondary dismembering operation. Even in large animal butchering, as I describe for moose in chapter 4, the disarticulation of the scapula–humerus joint is done first, because the scapula may be used for leverage while it is still attached to the carcass. The scapula, however, is the target element and is then removed from the carcass; it rarely remains with the axial skeleton.

Figure 3.44 illustrates characteristic "humerus fleshers" produced by wolves and dogs. Similar forms have been identified from paleontological assemblages. Good examples of humeri that have been modified by nonhuman agents are given by Miller (1969: Figure 5) and Myers *et al.* (1980: Figure 2).

Radio-cubitus

Figure 3.45 illustrates a fairly typical series of modifications produced by gnawing animals. It will be recalled that the articulation between the distal humerus and the proximal radio-cubitus is a rather durable one, in that it comes apart relatively late in the sequence of normal disarticulation. This is reflected in the somewhat abnormal survival of the proximal end of the radio-cubitus and the typical breakage pattern illustrated (Figure 3.45). The olecranon is frequently chewed off and the proximal radius may be punctured and pitted; the distal end is commonly chewed off. As can be seen in Figure 3.45, the distal end frequently shears off, leaving an oblique and pointed section of shaft. These may be gnawed and chipped back and even polished, resulting in pseudo tools of a rather convincing form. Most often these are interpreted as daggers or picks. Fri-

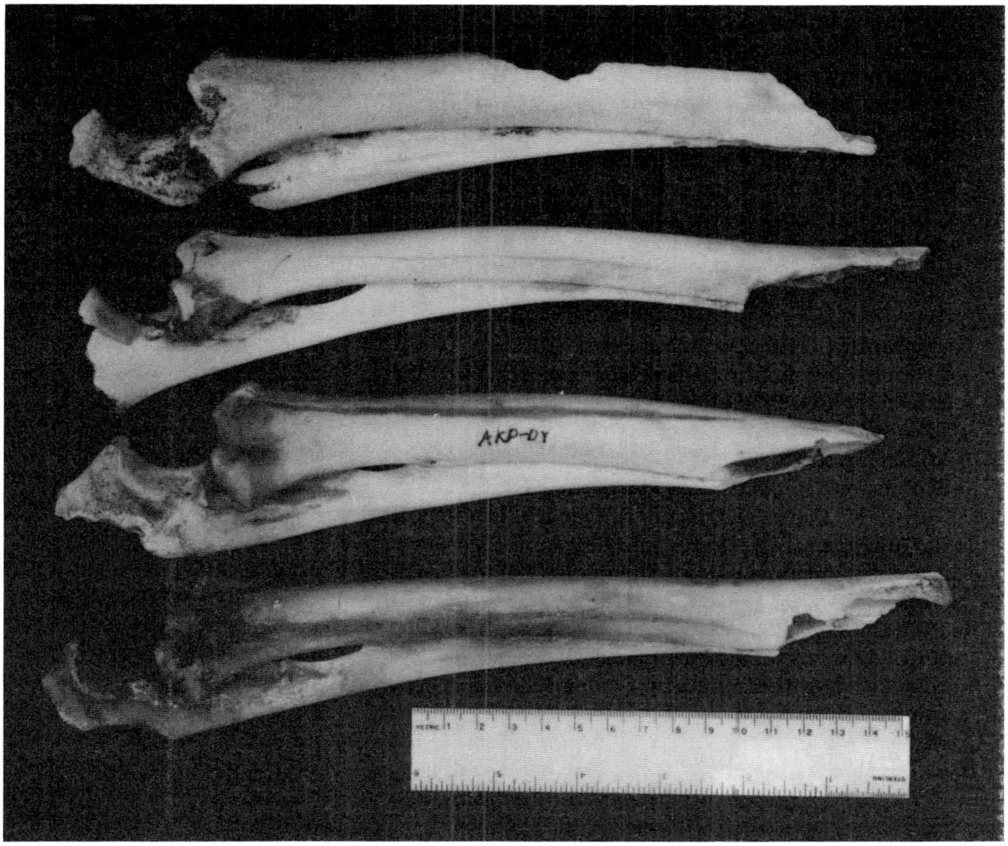

FIGURE 3.45. *Typical animal destruction of the radio-cubitus.*

son (1978:104) notes an analogous item from the Colby site—one of the few convincing mammoth-butchering sites in North America.

Carpals

In animal-gnawed assemblages, carpals will be either present in largely complete form or absent, having been completely consumed. The pattern of gnawing is to address the articulations where the joint bends. If this is done, it results in the ingestion of the carpals, destruction of the distal radio-cubitus, and marginal tooth scoring of the proximal end of the metacarpal.

Metacarpal

Metacarpals are generally chewed in two ways: (*a*) They may be approached as described for the carpals. When this is the case, the metacarpal is commonly abandoned since the proximal epiphysis is dense and contains little grease, whereas the exposed distal end of the radius contains relatively high-quality marrow. (*b*) Gnawing may also begin from the distal end; this is normally initiated by first attacking the foot with attendant puncturing (Figure 3.02) and consumption of phalanges. The articulation between the phalanges and the distal metacarpal is short in most ungulates, and therefore a bent joint is not the approach of most gnawing animals. They typically begin by chewing off the skin and unqual phalanges and then work their way up to the toes, gradually approaching the distal end of the metacarpal. Once the cartilaginous area between the phalanges has been exhausted, the chewing is generally stopped in favor of another bone. The metacarpal is thus commonly found as a complete bone with only minor damage from gnawing. This is particularly true at kills; however, in dens, lairs, and the dog yards of Anaktuvuk, less discriminating chewing was obvious. Such den-related destruction in the caribou, where the metacarpal is not a particularly sturdy bone, is through the shaft, resulting in approximately equal numbers of proximal and distal ends.

Fragments of ungulate metapodials have stimulated the imagination of many investigators of the Paleolithic. Such fragments have been cited as probably used as daggers and scoops (Dart 1959a, 1959b, 1961), and they have been illustrated from Olduvai Gorge FLK North Level 1 (M. D. Leakey 1971: Figure 35) with abraded fractures. Use for digging or perhaps in chopping has been suggested for analogous bones from the American Plains (see Frison 1970). As in the majority of other tool claims, the evidence consists of suggestive breakage and, in the case of the Olduvai material, abrasion and rounding of the edges, which is interpreted as use. As we have seen repeatedly, these are properties that may be developed in a faunal assemblage *without* the participation of man.

Femur

Figures 3.46 and 3.47 illustrate several femoral fragments from wolf kills and from Eskimo dog yards; Figures 3.48 and 3.49 illustrate sheep femora gnawed by Navajo dogs after they were fed an entire carcass during the winter of 1971. Several characteristics that are very regular are illustrated: almost without exception the greater trochanter is chewed off the proximal end by the gnawing animals (see

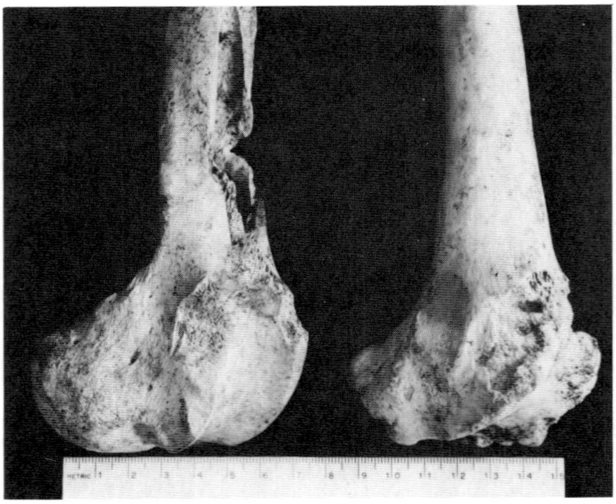

FIGURE 3.46. ***Distal femora from wolf kills showing typical puncturing and furrowing of the trochlea.***

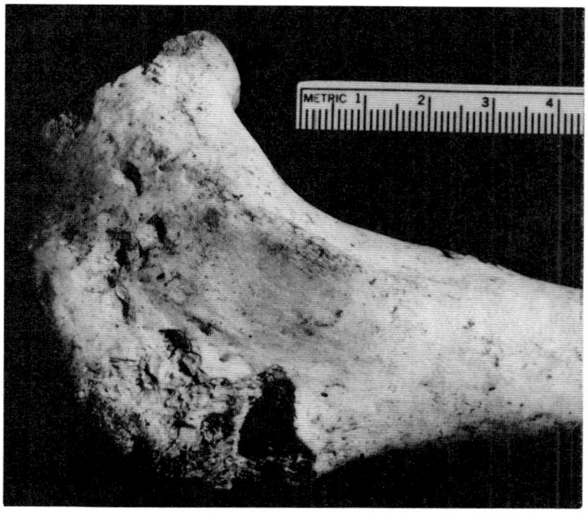

FIGURE 3.47. ***Distal femur showing tooth punctures along medial surface of the distal condyle.***

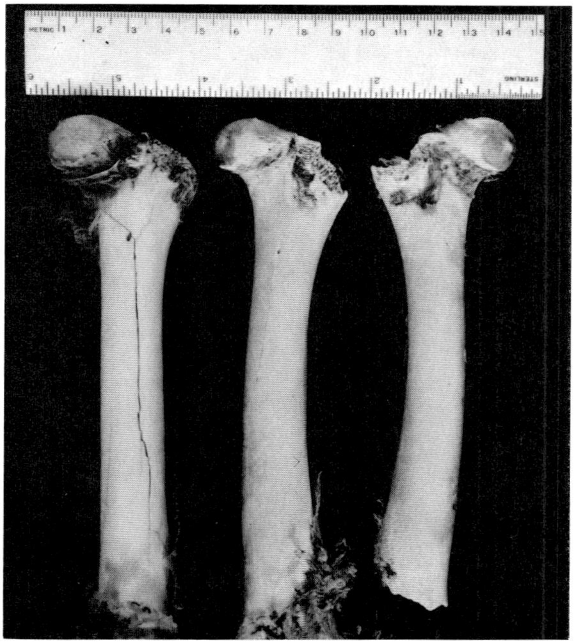

FIGURE 3.48. ***Proximal sheep femora showing consistent removal of the greater trochanter by dogs.***

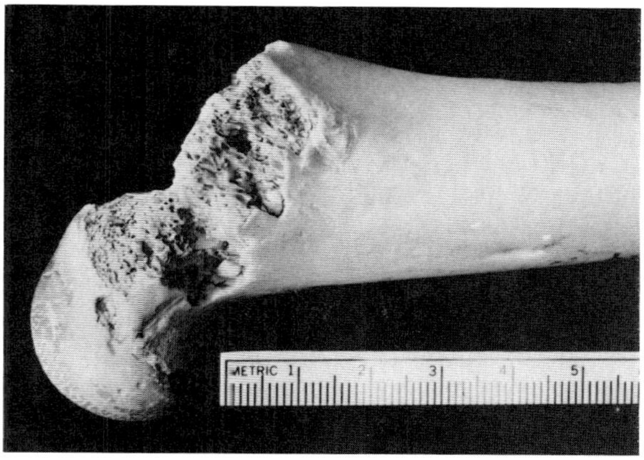

FIGURE 3.49. ***Distal view of proximal sheep femur showing tooth punctures and furrowing.***

Figure 3.49). Animals may then proceed to chew through the neck of the femoral head and finally furrow out the entire proximal articulator end. Figure 3.46 illustrates the distal ends of femora removed from wolf kills and these without exception (that is, of damaged bones) exhibit the furrowed removal of the medial ridge of the trochlea. (Compare this picture with Figure 8.6 in Frison [1978:311].) Both characteristics have been cited as evidence for muscle stripping as a human butchering procedure. For instance, the following "reconstruction" of a butchering procedure, largely based on the ideas of George Frison, is given by Johnson:

> Frison postulated that processing of the hind leg began with loosening of the patella and removal of the tuber calcis of the calcaneum and the trochanter major of the femur. This may also have been the pattern on the Llano Estacado as this type of damage was common, with the exception that the tuber calcis was generaly not removed. Damage to the calcaneum (distal projection cracked or broken off) appeared to be from attempts to sever the lateral and middle annular ligaments. After this was accomplished, the calcaneum would then be free to serve as a handle with which to strip various muscles [Johnson 1978:103].

It should be noted that all the "diagnostic" patterns of destruction cited as evidence for muscle stripping are common in my canid-gnawed assemblages.

Tibia

By this time the reader can probably anticipate the types of destruction seen for the various bones, since the pattern is redundant: biased destruction of the soft bone parts, particularly the articulator ends of long bones. The tibia is no exception and this is well illustrated in Figures 3.50 and 3.51. The tibial crest is generally the first projection to exhibit gnawing, and since it is such a prominent projection it is frequently removed by aligning the bone parallel to the tooth row and vising down with the carnassial teeth. This produces a semicrenulation when viewed from the side, as is the case for both specimens on the right side of Figure 3.50. The left specimen shown in Figure 3.50 illustrates a relatively advanced state of destruction—the entire proximal end is removed. It should be noted that the proximal ends of the tibial shaft are chipped back along the channeled edges, producing a "denticulated" and "retouched" series of edges (Figure 3.51). Typically this level of destruction precedes any major alterations on the distal end, which is commonly left articulated with the tarsals. If destruction continues, the shaft may be channeled back and collapsed so as to leave the distal end with a minimal segment of shaft attached (note such patterning in Table 4.07). When such an advanced stage of destruction is reached, there is an almost invariable pattern of chipping back coupled with polishing, as shown in Figures 3.10 and 3.12, occurring on segments of the broken ends of the attached shaft. Very similar forms are described by Frison as tibia choppers (see Figures 1.17 through 1.19 in Frison [1974]). Given my knowledge of animal patterns of bone destruction, I believe that it would be unusual for an animal to attack the shaft of the tibia before effecting some modification on the proximal end. This observation renders it more likely that Frison's tibia choppers manufactured from the proximal tibia are not animal produced. Nevertheless, the tibia choppers illustrated in Figures 1.18 and 1.19 (Frison 1974) would certainly be suspect in an assemblage where there was other evidence of animal gnawing. Examples of "tibia choppers" given by Stanford (1979a, Figures 6 and 7) are textbook examples of animal-modified bones.

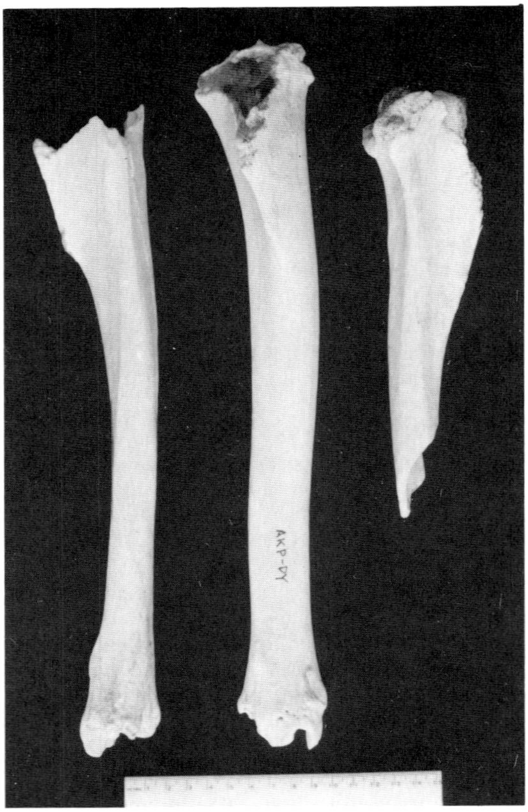

FIGURE 3.50. ***Different views of animal destruction on the proximal tibia.***

Tarsals

The most common form of destruction seen on tarsals is the gnawing away of the distal end of the calcaneus (tuber calcis). Puncture marks are occasionally seen on other tarsals but by far the most common traces of animal gnawing are on the calcaneus.

Metatarsal

Patterns of metatarsal destruction are very similar to those noted for the metacarpal. In the dog yard sample from Anaktuvuk, there were numbers of

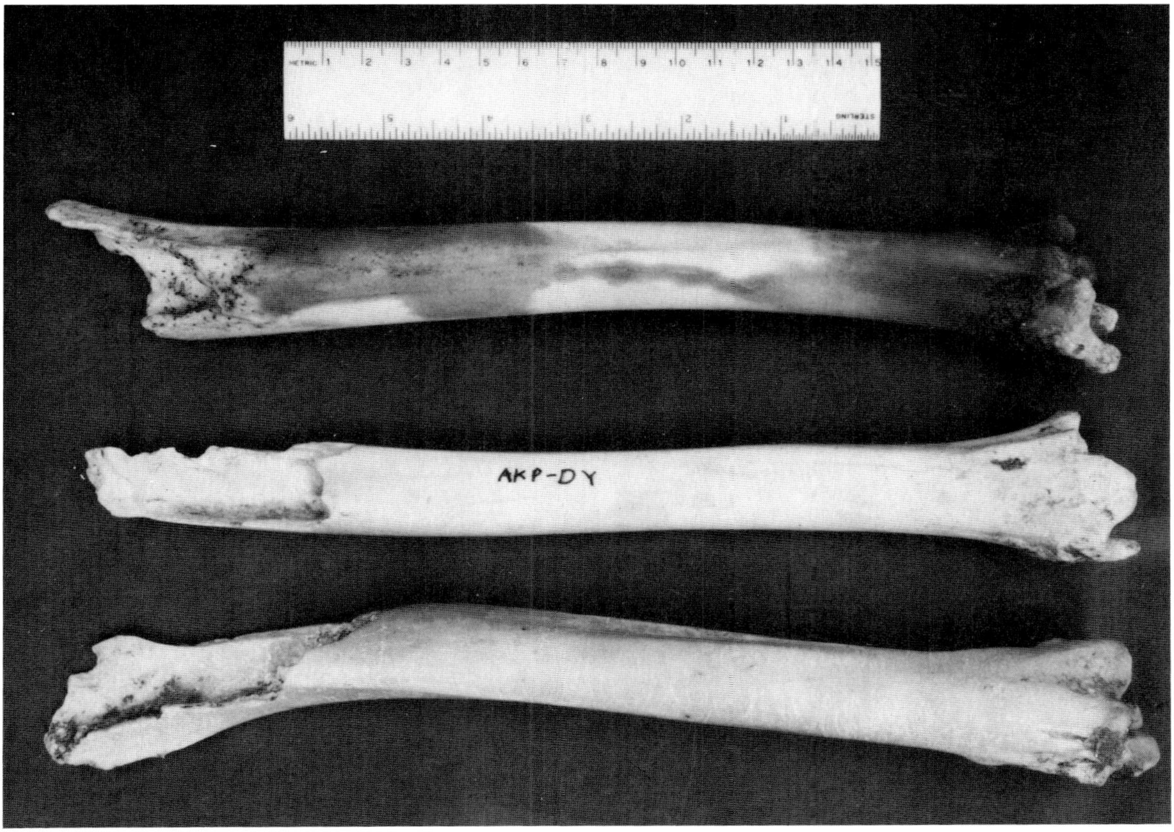

FIGURE 3.51. ***Typical advanced destruction of the proximal tibia showing channeling and chipping back.***

metatarsals that were channeled almost the complete length of the bone, with the channeling having started at the proximal end (the lower four specimens in Figure 3.52). I have not seen such systematically channeled specimens from either the wolf dens or the kills (see Figure 3.17). I suspect that the dogs gnawed the bones in this fashion largely in response to having been fed already disarticulated lower legs, because the only area not covered by skin is the proximal end, and I have observed the dogs preferentially start eating at the exposed end. The pattern of consumption described for the metacarpal is believed to be more common among "wild" animals consuming their own prey. Nevertheless, Dart (1967:120) illustrates some nearly identical specimens from a Mousterian assemblage recovered in Israel. Figure 3.53 shows the distal ends of metatarsals recovered from wolf dens and the channeling is obvious. The specimen at the lower right is channeled and there are puncture marks and pitting on the opposite side. As pointed out for the metacarpals, these are classic examples of "scoops" as illustrated by Dart (1959a, 1959b, 1961).

Breakage Unrelated to Consumption: Trampling and Bone Manipulation

Thus far I have discussed bone breakage as if it occurred primarily during consumption and carcass

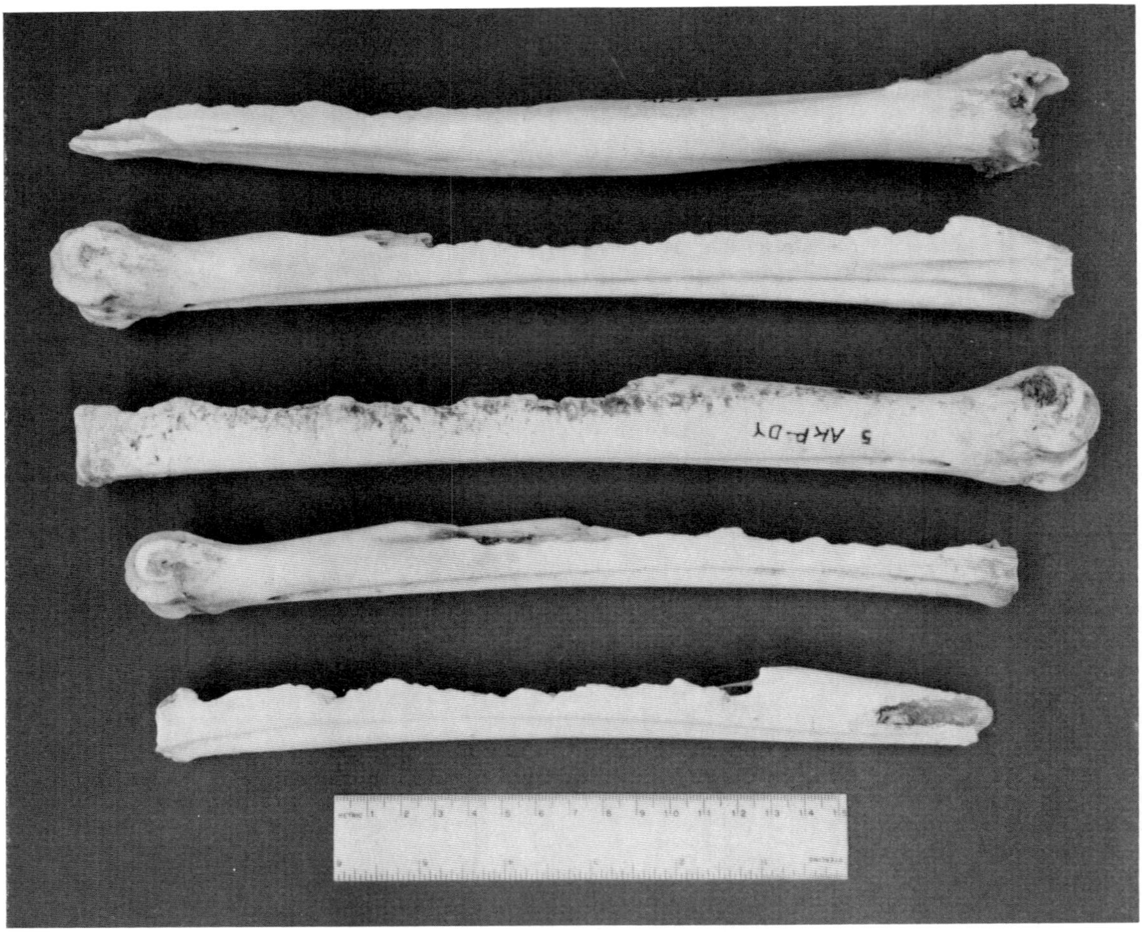

FIGURE 3.52. ***Extreme channeling of metatarsals by tethered dogs.***

dismemberment by animals. Of course this is not the case, and the archaeologist must be continuously aware of the possibility that other agents and contexts of breakage may condition what he or she observed. C. K. Brain (1967b) was one of the first to point to the important role that trampling might play in both breakage and the production of pseudotools. Diane Gifford (1977) has made important observations on the role of trampling by large animals, particularly for bones deposited along the margins of bodies of water. This is of considerable importance when behavioral claims such as those made by L. S. B. Leakey (1968) are considered. In describing his materials from Fort Ternan, Leakey states,

> there are in the same deposit, and at the same level, small areas of fossils where the bones have been broken up, and where the damage includes excellent examples of depressed fractures of the types usually associated with "a blunt instrument" [1968:528].

Leakey then goes on to say,

> The available evidence presented there, therefore, strongly suggests that the Upper Miocene hominid

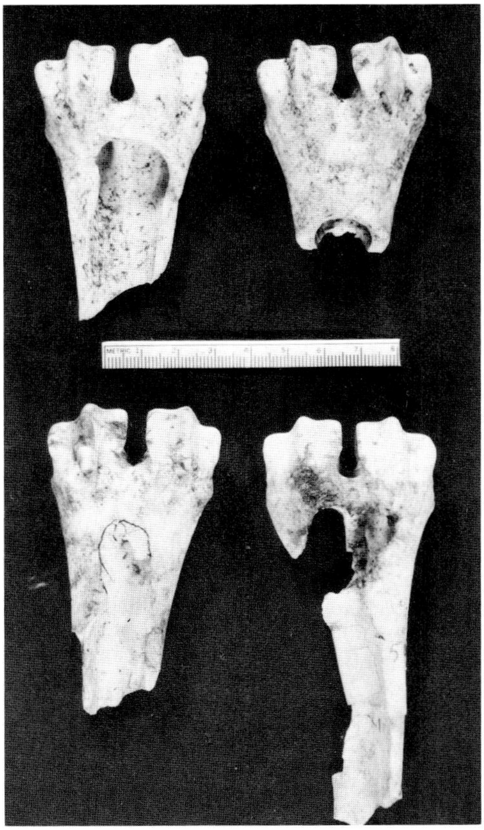

FIGURE 3.53. *Wolf destruction on distal metatarsals, commonly cited pseudotools.*

> *Kenyapithecus wickeri* was already making use of stones to break open animal skulls in order to get at the brain and bones to get at the marrow [1968:520].

All the breakage described by Leakey is essentially of the same type observed by Diane Gifford as derived from the trampling of exposed bone by living animals. The interested researcher should compare Leakey's photographs, particularly Figure 3, with those presented by Gifford (1977: Figures 68–71).

From both the Old World (Biberson and Aguirre 1965) and the New World (Bonnichsen 1978; Irving and Harington 1973), there have been claims that "worked" or "modified" elephant bone represents tools manufactured or used by man. A recent "experimental" butchering of an elephant was conducted by several persons interested in bone tools (see Park 1978) as a demonstration that tools manufactured from elephant bone could be used in butchering an elephant This team of experimenters, led by Dennis Stanford of the Smithsonian staff, concluded that an elephant could indeed be butchered with tools manufactured from its own bones (Park 1978:94).

Such claims must be weighed against research findings where actual controlled conditions have been sought; that is, where the conditions of formation or production of traces or remains are known and one seeks to recognize *diagnostic* characteristics of those conditions. Such controlled research has been provided by two sources (Coe 1978, 1980; Douglas-Hamilton and Douglas-Hamilton 1975) regarding elephants and the possible conditions of breakage and modification directly referable to the taphonomic conditions of elephant death, disarticulation, and burial. Of particular importance is the work of Malcolm Coe (1978), who actually took the pains to observe what happens to an elephant carcass after death. He observed in detail the progress of decay and the scattering of the remains of elephants that died natural deaths. The death sites of elephants are under certain conditions sites of violent activity. Coe observed that "the large patch of disturbed ground around most corpses observed (diameter) indicated that death was usually protracted after collapse [1978:77]." Falling and protracted thrashing (Coe 1978:77) by an elephant is certainly a condition that could result in green-bone fractures.

The death sites of elephants are apparently frequently visited by living elephants, who trample exposed bone and disperse such bones by knocking them about with their tusks and even kicking them. Coe (1978:76) observed crushing of rib bones from trampling, presumably by elephants, approximately a year after the control animal's death. Another interesting comment by Coe is as follows:

> The skeletal remains still persist on all sites studied five years after the animal's death. Long bones and skulls have been scattered up to 100 m. from the death site *by other elephants* while smaller bones have been chewed into small fragments by hyaenas. . . . Skeletons lying in

> the open and subjected to great diurnal temperature extremes are severely fissured and flaked [Coe 1978:79; emphasis mine].

In order to gain some appreciation for the magnitude of dismemberment and scattering that can occur at an elephant death site, the interested reader is referred to the remarkable collection of photographs of elephants in various states of decay and dispersion published by Beard (1979).

Of particular importance to those interpreting breakage of elephant bone is the mysterious excitement elephants exhibit over the bones of dead elephants. The following description illustrates the behavior of elephants toward the bones of their dead:

> They all began their detailed olfactory examination. Some pieces were rocked gently to and fro with the forefeet. Others were knocked together with a wooden clonk. The tusks excited immediate interest; they were picked up, mouthed, and passed from elephant to elephant. One immature male lifted the heavy pelvis in his trunk and carried it for fifty yards before dropping it. Another stuffed two ribs into his mouth and revolved them slowly as if he were tasting the surface with his tongue. The skull was rolled over by one elephant after another. To begin with only the largest individuals could get near the skeleton, such was the crush. Boadecea [a named elephant] arriving late pushed to the center, picked up one of the tusks, twiddled it for a minute or so, then carried it away, with the blunt end in her mouth. The rest of the group now followed, many of them carrying pieces of the skeleton which were all dropped within about a hundred yards [Douglas-Hamilton and Douglas-Hamilton 1975:239].

This behavior, coupled with direct trampling as noted in the studies by Gifford (1977), could certainly be expected to generate bone breakage. Depending on the condition of the bones being trampled or manipulated, green-bone or spiral fractures might be regularly expected.

Particularly germane to this point is the previously cited work by Myers *et al.* (1980). They confirmed the presence of spiral fractures, as well as polished, chipped, and otherwise modified bone in deposits where man had certainly played *no* role in the accumulation of materials. Of more interest, however, is their observation that evidence for predator–scavenger related modifications was relatively infrequent. They suggest that trampling may well have been a major cause of the relatively large number of spiral fractures and other modifications. They experimented with a weathered bone: "We have found that slightly weathered bones break quite easily but still exhibit the characteristic spiral breakage pattern of fresh "green" bone [Myers *et al.* 1980:487]."

These researchers illustrate convincing spiral fractures produced by trampling the bones of a cow that had been weathered under natural conditions and approached weathering stages designated No. 1 or early No. 2 in Behrensmeyer's (1978) study of bone weathering. This research opens up a major possible source of bone modifications unrelated to human activity that has received very little research attention. This may be particularly important to a number of claims regarding the human involvement in the modification of elephant bone. Biberson and Aguirre (1965) attempted to duplicate patterns of breakage among elephant bones recovered from the important site of Torralba. Bonnichsen's (1978, 1979) analysis of alleged tools from the Old Crow Flats locality places strong emphasis on modified elephant bone. He cites a number of elephant bone flakes that have all the properties of having been removed by percussion techniques analogous to lithic reduction strategies. Similar forms have been reported by Stanford (1979a, Figures 8, 12). Noted as companions to these finds are a number of elephant bone "cores" from which flakes have been removed (Bonnichsen 1979:101–129). Morlan (1981), after restudying the faunal remains from Old Crow Flats, has basically accepted the argument that the modifications of bones (including spiral fractures) from animals *smaller* than elephants may well be referable to the actions of predator–scavengers. He has dismissed the role of trampling and bone manipulation by elephants as not likely to produce "point loading," that is, localized impact scars. He concluded that the carnivores are too small essentially to address effectively elephant bone, and that fractured elephant bones, particularly specimens exhibiting "point loading," are the product of human modification and therefore tools. No criticism of this argument is yet available. Nevertheless, the dismissal of trampling and bone manipulation remains an opinion totally unsupported by empirical material; thus Morlan's argument is not a valid argument from elimination, a point I will take up somewhat later.

Bone Modifications and Methodology

Throughout Chapter 1 I illustrated case after case of imaginative interpretation for observations made on the archaeological record. Most of the time my aim was to demonstrate a hidden assumption—namely, that man was the agent responsible for the particular modifications or associations being "interpreted" in behavioral terms. Both the premise as to the identity of the agent and the interpretations offered as to the conditions in the past responsible for the character of the things observed in the archaeological record were, in Chapter 2, shown to be *inferences*. In Chapter 2 it was also argued that all our observational language and the means available to us for inferring the past from observations on the archaeological record had to be rooted in experience:

> There are two ways of regarding concepts, both of which are necessary to understanding. The first is that of logical analysis. It answers the question, how do concepts and judgments depend on each other? In answering it we are on comparatively safe ground. . . . But this security is purchased at the price of emptiness of content. Concepts can only acquire content when they are connected, however indirectly, with sensible experience . . . no logical investigation can reveal this connection, it can only be experienced [Einstein 1947:471].

It was argued that middle-range research with the aim of developing reliable methods for inferring the past had, therefore, to be rooted in actualistic studies.

This chapter has reviewed certain claims that others have made regarding the meaning to be attached to properties of modified bone. Categorically speaking, the claims centered on the degree to which there are distinctive forms of bone breakage referable exclusively to the actions of man, and on the degree to which patterned modifications observed on various bone could be referred to modifications produced by man in order to use the bones as tools.

The advocates of "osteodontokeratic" (Dart 1957), "protolithic" (Menghin 1931), "pre-paleolithic" (Nicolaescu-Plopsor and Nicolaescu-Plopsor (1963), a distinctive bone-using culture "Alpines Palaolithikum" (Bächler 1940), or more recently an early and continuously important bone technology in the New World (Irving and Harington 1973) generally imagine bone modified by man as having many features in common with lithic technology (Bonnichsen 1978). This analogy has served archaeology poorly for several reasons. It is true that many of the modifications observed in bone are the result of fracture dynamics, or the mechanics of impact loadings or pressure loadings, as are the basic dynamics of lithic fracture. Both materials may be studied in terms of the same general questions regarding the physics of fracture. However, in nature there is a fundamental difference: Lithics are apparently regularly used by only man. It is true that lithics may become fractured in a variety of inanimate, mechanical contexts in nature: pounding of stones on beaches, in friction "cauldrons" in rivers, as a result of sluffing off and falling of rocks from cave ceilings, cliff faces, and the like. In all these contexts the stones being subjected to fracturing forces are being modified as forces act on objects and the orientation of the objects is generally free to vary independently of the directions of impact forces. If not, the coincidence of orienting forces and fracture is relatively random, producing rare and irregular combinations of fracture patterns that statistically form a component of a more comprehensive pattern of random breakage configuration. This is in marked contrast to stones modified by man, who orients the stone and directs the impact so that a regular pattern of orientation for both the piece being acted upon and the direction of action or impact loading is maintained, resulting in distinctive and patterned modifications in the stone being acted upon. Such patterning generally permits the discrimination between tools produced by man and stones modified in nature. *Transferring this approach to bones does not work.* For unlike stones, bones are regularly manipulated mechanically by animals other than man. The result is a pattern of regular, mechanically produced modification in bone that shares the configuration of redundant impacting or loading relative to redundant orientation, resulting in chipped edges, pieces that have multiple flakes originating at a common impact surface, multiple flakes removed in redundant orientation, both with regard to the direction of impact and the orientation of impacting to the prior shape of the piece. In fact, all the patterning configurations of tools are seen. These are true analogues, in that the formal configurations have many things in common with stone tools but the causes may be quite different. Not seriously considering this, many have argued that they are homologues; that is, man was using his documented knowledge of mechanics on both stone and bone, or in some cases only on bone, and producing artifacts. Arguments from analogy *assume* the causes of analogous characteristics are the same. This is not true for fracture patterns on bone versus stone.

Failing to recognize the strong analogical component to research reasoning, some researchers have thought that they could gain a greater understanding of the significance of modified bone by conducting experiments with the mechanics of bone fracture. A number of detailed and "controlled" studies have been aimed at increasing our understanding of the properties of bone that condition its fracture potential when mechanically manipulated (see particularly Bonnichsen 1973, 1979). Although these are interesting and the knowledge gained can be used, they in no way inform us about the *causes*, in the proximal sense, of the loading agents responsible for bone breakage. Nor do they provide us with any criteria for recognizing such agents. *Yet these were actualistic studies.* They were "experimental archaeology" of a relatively elaborate form. How could they not provide us with methods? The answer, of course, is that the overall procedure one uses does not ensure success. One must ask productive questions and one must succeed in gaining adequate controls over experimental observations to be able to relate them accurately to "causes."

Making irrelevant observations, in terms of either the accurate isolation of causes or characteristics not observable in the archaeological record, will not contribute toward the development of usable archaeological methodology. This is a major problem with many so-called ethnoarchaeological studies. They are simply materially oriented ethnographies and do not attempt seriously either to isolate causes or demonstrate the necessary relationships between "bears" of interest and "footprints" actually recoverable from the archaeological record.

My approach has been very consciously aimed at not falling into the trap of believing that if a study is actualistic, it must be useful. I sought to obtain control collections of bones known to have been modified by animals and to describe and identify the effects when the causal agents were known. I have used the specimens whose properties could be unambiguously referred to animal manipulation to demonstrate the ambiguity of many inferences made by others.

I think I have demonstrated that spiral fractures are not unique to man; that spiral fracture when produced by animals is not limited to an origin at the distal ends of the bones; that modifications cited by Frison at Glenrock as evidence of muscle stripping are commonly produced by gnawing canids; and that the characteristic modifications commonly cited for recognizing "expediency tools" are regularly produced by nonhominid predator–scavengers; and that, finally, many of the modifications cited as evidence of human modification and tool use are referable to predator–scavengers.

My interests go beyond the evaluation of specific ideas to archaeological methodology in general. Are there guidelines for research that, if more explicitly understood, could reduce the numbers of myths generated by archaeologists? I think this must be answered affirmatively since archaeologists have, in fact, had very little experience in the research area of methodological development. For the most part, traditional archaeology did not recognize middle-range research as a domain distinct from general research. Methods grew up largely as a series of accepted conventions for giving meaning to the archaeological record. Methods "accumulated" out of the most common research situation faced by the archaeologist, in which he recognizes patterning in the archaeological–paleontological record and must face the intriguing problem of what the pattern means. For only with a solution to this problem can the archaeologist convert contemporary pattern observations into meaningful statements about the past.

Most often this problem is faced by the archaeologist after he or she has described the patterning (frequently in a section labeled "typology" or "statistical analysis"). Once patterns of association, covariation, and so on have been established, the archaeologist may then write an interpretative section. This is where the archaeologist may build a model of the past or offer opinions as to what conditions in the past "caused" the properties summarized. This is what I refer to as a *post hoc accommodative argument*. Summarizing then, this is a situation where one has isolated patterning (the effects) and seeks to use imagination to envision the conditions that, if they had happened, would account for the properties seen (the causes). The way one warrants this construction of the past may be complicated and draw on considerable knowledge, even theory, from other fields. *Nevertheless, the degree of fit between the imagined conditions and the observed properties in the archaeological record is not a test of the accuracy of the argument.* Most persons are knowledgeable enough to invent a situation that, if true, would account for the observed facts.

In order to convince others, a number of warranting arguments are frequently proposed. Some are designed to demonstrate that the suggestion is plausible and not totally unrealistic. This is the role in which ethnographic analogy is cast by Thompson (1956:329): "He must test his conclusion by demonstrating that an artifact–behavior correlation similar to the suggested one is a common occurrence in ethnographic reality." However, *the degree of plausibility is not sufficient to establish probability.* Probability is frequently argued in the form of an argument from elimination.

ARGUMENT FROM ELIMINATION

Arguments from elimination have two basic premises: (*a*) All the potential causes are known and listed and (*b*) all but one of those listed are not the cause of the phenomenon in question. Therefore, the remaining cause is considered to be the correct one. If one could not justify the premise that all the possible causes were listed, then the conclusion that the cause not eliminated through rejective argument is the correct one could not be sustained. For instance, I might argue that the causes of a particularly interesting pile of rocks could be (*a*) that it had been made by spacemen, (*b*) that it had been constructed by God, and (*c*) that it had been produced by geological action. I then might proceed to eliminate alternatives a and b through warranted argument, and therefore conclude that the pile of rocks was produced by geological action. I would be very apt to be wrong, since there is no reason to believe that all of the potential causes had been listed in the first place. Of course, in order to construct such a list we would have to understand causation, and if we had such understanding we could certainly devise a more powerful and informative evaluative procedure.

Argument from elimination was the form implied by Bonnichsen's rejection of the alternative that the spiral fractures observed at Old Crow Flats had been caused by animals breaking their legs. His is a very common way of warranting an opinion regarding a preferred post hoc accommodative argument. Normally one uses imagination and an accumulated knowledge to think of all the possible situations that might have accounted for the observations made. One then offers an opinion as to which is considered most likely by denigrating all but the preferred argument. This seems to be what many researchers mean by the "method of multiple working hypotheses." This is a form of warranting argument for one's opinions, but it is in no way sufficient for providing us with methods of inference.

I must add that for an argument from elimination to be valid, one must have available unambiguous means for monitoring the alleged "causes" and therefore a way of actually determining the degree of participation by a suggested cause in a system of past determinacy. In traditional archaeology the argument from elimination was used as a form of warranting argument for one's beliefs about the past. In the new archaeology, it should function more to warrant one's methodological research: "A valid proposition can only be refused through hypothesis testing. However, when faced with valid alternatives, one can evaluate in probabilistic terms the relative strength of alternatives and make decisions as to how to invest research time [L. R. Binford 1972a:57]."

ARGUMENT FROM WANT OF EVIDENT ALTERNATIVES

A particularly interesting form of argument from elimination is one where only one possibility can be imagined. Most commonly this argument takes the form that "only man could do that." It is generally executed by citing complex or regular patterned modifications in nature, which seem self-evident in their manifestation of design or planning: "I suppose that these bone breaks could come about without human intervention. . . . But the broken mammoth leg bones we've found—nothing could do that except humans hurling rocks [quote from Dennis Stanford published in Canby 1979:354]." Commenting on this form of argument, two distinguished professors of logic state,

> Argument from want of evident alternatives is, however, one of the most abused argument forms. Often when we "jump to conclusions" we are abusing it; we leap at the first explanatory hypothesis that comes to mind without duly surveying the field [Quine and Ullian 1978:120–121].

This form of argument is so easily abused because only one alternative is recognized. The only positive form of argument is offered with regard to alternatives, which are rejected, so there is no direct evidence that can be offered in support of the conclusion. This is the form of argument used by Von Daniken (1969) to claim extraterrestrial involvement in the history of culture. Quine and Ullian (1978:121) comment germanely on this point: "Indeed arguments of this kind (argument from want of evident alternatives) are a favorite device of charlatans, if not absolutely indispensable to them."

I think that it should be clear that, although these forms of argument may appear impressive and even convincing to many, unless they are actually executed with means for measuring the alternative causes or actually evaluating the relative contributions of differing variables said to have been active in a "conditioning" sense, the entire exercise remains a form of polemics and is in no way an epistomological procedure. It stands as a set of ranked opinions, not a statement of probable determinacy.

The techniques frequently employed in demonstrating the plausibility of a post hoc accommodative argument are usually suspect. As mentioned earlier, Thompson suggested the citation of ethnographic precedent. What if there is no ethnographic citation relevant to the patterning you observe? This is particularly important when considering the very early eras of hominid experience. It is also common with respect to the types of patterns archaeologists may observe. Such things rarely prompt comments from ethnographers and historians. We have a fine example of this situation in the work of George Frison at Glenrock (1970). Frison observed a large number of patterned modifications in bone. There were no detailed descriptions of such features in the ethnographic record. He made the assumption that man was the causal agent. He then invented a set of conditions that, if they had happened in the past, were believed to accommodate all the "facts" of bone modification. Using his exceptional knowledge of animal anatomy and manual tool use, Frison imagined a procedure for butchering using "expedient bone tools," largely choppers and fleshers, for muscle stripping the animals. The patterns of bone modification noted at Glenrock could then be understood as resulting from chopping loose muscle attachments on bones, as well as wear patterns and intentional modifications made in shaping and using bones as tools in this operation.

Frison then engaged in actualistic research, *replicative experiments.* That is, Frison (1974) actually butchered a bison using the procedures he had imagined and concluded that his model was well conceived in that the properties expected to result from his imagined operations did in fact result. Others then adopted his model of butchering as a methodological device for giving meaning to specified properties of the archaeological record. Other researchers began finding expedient tools and evidence for muscle stripping. A methodology was being molded by the adoption of a set of conventions for giving meaning to properties of the archaeological record based on the faith placed in an individual's work. The replicative experiment showed that Frison's knowledge and judgment used in developing his model of the past was sound. It showed that his model of the past was plausible. It in no way demonstrated that it was accurate!

As pointed out earlier, Dart conducted replicative experiments together with Kitching, concluding that the crack-and twist method of breaking bone was responsible for spiral fractures, and he was wrong. Sadek-Kooros (1972) conducted replicative experiments and concluded that man intentionally broke bones to a pattern for use as tools, and she was most certainly wrong. George Frison conducted replicative experiments with regard to his ideas about muscle stripping as a mode of butchering and concluded that in fact he could muscle strip a bison and simulate the patterns of bone breakage and destruction noted in prehistoric assemblages. I argue that he is most likely wrong. Recently there have been several "replicative" elephant butcherings (see Huckell 1979; Park 1978). One of these was led by Dennis Stanford, who is very interested in illustrating that an elephant could be butchered with tools manufactured from its own bones (Park 1978:94).

Under the epistomology being criticized, replicative experiments provide an evaluation of the judgment of the person who proposed the original model of the past. If one can produce the effects by following a procedure proposed in explanation of certain observed properties, then the proponent's knowledge and skill in thinking are demonstrated. We could then place some confidence in the argument because of the demonstrated skill of the advocate in anticipating the consequences of certain operations.

The demonstration of plausibility, either through the citation of ethnographic precedent or through replicative experiment coupled with warranting arguments by elimination as a linked battery of tactics, was used by many archaeologists to convince people that their views of the past were correct. These tactics primarily serve to demonstrate *competence:*

> The final judgment of any archaeologist's cultural reconstruction must therefore be based on an appraisal of his professional competence, and particularly the quality of the subjective contribution to that competence. Our present method of assessing the role of this subjective element by an appraisal of the intellectual honesty of the archaeologist who makes the inferences is certainly inadequate. But, there does not seem to be any practical means of greatly improving the situation. . . . We can only hope for improvements in the methods of measuring the amount of faith we place in an individual's work [Thompson 1956:331–332].

I suggested long ago (L. R. Binford 1968b: 16–17) that there must be ways other than those adopted by traditional archaeologists for evaluating archaeological argument. Here I add the suggestion that there must be research endeavors designed specifically for the development of inferential methods, and these cannot be easily evaluated using archaeological data nor can they be simply adopted from "respected" archaeologists. This means that inductive arguments developed post hoc to account for patterning observed in the archaeological record must be treated as *provocative ideas in need of evaluation.* Only after they have been evaluated can they be cautiously adopted as observational language, or interpretative conventions, or methods for giving meaning to the past.

I suggest that middle-range research progresses by virtue of the accumulation of knowledge that in fact permits one to use an argument by elimination properly. *If actualistic research can be cited showing that what is observed in the archaeological record is not referable to a suggested cause, then one basis for myth making will be eliminated.* We must continuously focus our attention on the properties of the archaeological record and, in turn, focus our imaginations on those observed properties. Once we have generated a post hoc accommodative argument, our research task is specified. How can we investigate the idea advanced as to the context of causation for specified properties of the archaeological record? Where in the contemporary world or the domain of recorded dynamics can I seek experience facilitating the modeling of causes for a particular pattern observable in the archaeological record? This is the first question. The next question one must ask is this: Even if one can isolate such a source of enlightenment, how ambiguous are the properties for which one is seeking a causal understanding? That is, are all the possible causes listed? This is a necessary condition for an argument from elimination. Not only must the researcher face the possibility of the ambiguity of the patterns as observed in the archaeological record, but he or she must also face the equally crucial question of the relevance of actualistic understanding to the past. Can we make uniformitarian assumptions from our contemporary understanding regarding the production of properties that appear to be common to the past and the present? These are all complicated problems that must be faced in order to develop a strong methodology and create an unambiguous observational language for the science of archaeology. Failure to address them when making methodological claims will generally contribute to the building of modern myths rather than to the growth of archaeological science.

The task of giving meaning to observations is very difficult. We saw in Chapter 1 how "control" was obtained by inferring the identity of a causal agent from an association with identifiable tools. That assumption provided an intellectual anchor. Patterning observed in associated things was then referred to the behavior of the *inferred* common agent. Bones became food, frequencies of species became statements of dietary preferences or hunting strategies. Patterned modifications on bones became tools.

In this chapter we have seen how "control" was gained by using argument from analogy for forms of patterning observed among stone tools and bones and assuming that analogous patterning was referable to the same causal conditions. That was wrong. We saw how replicative experiments and inadequate attempts to investigate alternative suggestions coupled with eliminations by "opinion" served to justify the use of conventions, such as spiral fracture equals man, or expediency tools equal man, or destroyed trochanters and proximal humeri equal muscle stripping. We must explore ways of increasing our accuracy and reliability in making inferences to the past. Traditional archaeology did not recognize the need for, nor even the possibility of developing, middle-range research. Traditional approaches are demonstrably inadequate.

> It seems to me that the method of most rapid progress . . . is going to be to set down explicitly at each step just what the question is, and what all the alternatives are, and then to set up crucial experiments to try to disprove some. . . . When whole groups of us begin to concentrate like that I believe we may, see . . . increases in the rate of scientific understanding [Platt 1964:352].

Postscript

Since this chapter was written I have learned of an argument starting in Australia regarding the agents responsible for a bone assemblage unaccompanied by stone tools (Archer, Crawford, & Merrilees 1980).

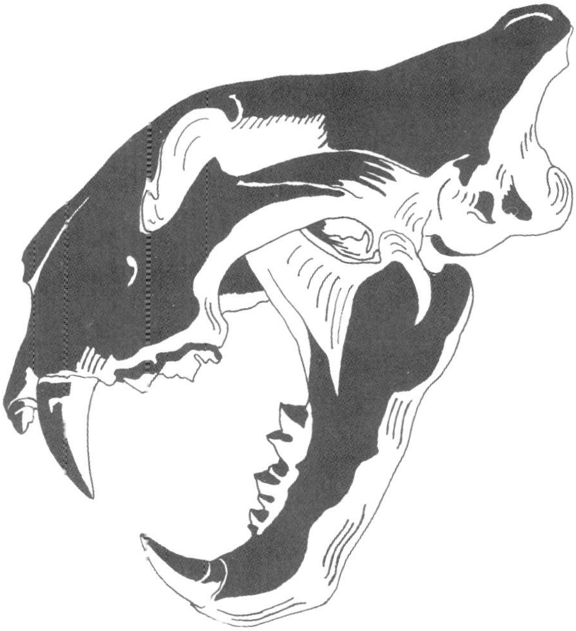

Chapter 4
Human modes of bone modification

Assuming that one can attribute to the hand of man all remains in a deposit associated with accepted artifacts was shown to be a common methodological error in my review of literature presented in Chapter 1. Incorrectly identifying the agent responsible for modifications on bones was a problem emphasized in Chapter 3. In this chapter I will demonstrate that the assumption of monocausality has been a common error associated with the interpretation of bones modified by man. Here the archaeologist observes properties of the archaeological record, for instance modified bone, and assumes that all the modifications arose from a single behavior, such as butchering, marrow cracking, or tool production.

In most cases we may expect that bones and other archaeological remains had a life history during which they were components of a cultural system. Differing actions were taken with respect to the materials and in turn they were altered differentially, depending on the character of their life histories. We must view an item, feature, or site as being the accumulative result of numbers of different actions and events such that the overall morphology of the material remains carries information about many different modifying conditions—not a single causal event but a sequence of causal processes. With respect to bone this view has not generally characterized the analysis of archaeological remains.

Aside from this ontological point regarding archaeological materials, I will continue with my methodological interests. I will explicate criteria for recognizing the agents responsible for materials that might be found either in association with unquestioned tools or remains of hominids themselves, or independently. That is, although my goals are methodological they are also focused on the generic differences between men and other animals in their patterns of modifying bone. For this reason, in the materials to be presented I will make no attempt to discuss the designs or the techniques actually used by man in the production of tools from bones. I am solely concerned with modifications inflicted on bone as a by-product of the exploitation of animal foods by man. In one sense I am only interested in those actions man performs in solving problems that must also be solved by other animals in the utilization and consumption of prey species. I am further limiting myself to a concern with the techniques used in dismemberment and meat removal, and the techniques used in breaking bones for the recovery of bone marrow.

Within this domain I will address the problem of making uniformitarian assumptions about the material derivatives of human behavior that may be observed actualistically. What is the role of ethnoarchaeology and ethnographic analogies in middle-range research?

With respect to the specific subjects covered, the available information is by no means balanced. Following a survey of previous research and a discussion of the data I will be appealing to in my descriptive discussions of human use of bone in a food context, I will begin the descriptive sections with a discussion of dismembering strategies. This is, regrettably, a very short section despite much discussion in the literature of butchering techniques. I will then turn my attention to cut marks on bones, which are more commonly referred to as butchering marks. This is a long, detailed section summarizing the descriptive literature as known to me and presenting new material. It will be painfully clear that little information is available on this subject. I will add to the available material by presenting the grouped data from the Mousterian levels at Combe Grenal excavated by François Bordes (see Bordes 1972), as well as the data from the Kakinya site excavated during my work with the Nunamiut Eskimo (see L. R. Binford 1978b:374). The relatively few cases of described materials are drawn from a remarkably wide geographical and temporal range: two Mousterian cases, a case from eastern North America during the Early Historic period, a Mississippian site from Missouri, several Plains bison kill sites, a Plains antelope kill site, and my materials from the Eskimo. Over all this material I will be able to demonstrate a kind of remarkable redundancy in the placement of cut marks and their particular forms. Thus, I will venture some empirical generalizations about human patterns of butchering and the telltale marks remaining from the use of cutting tools in accomplishing this task. "The plausibility of a hypothesis depends largely on how compatible the hypothesis is with our being observers placed at random in the world [Quine and Ullian 1978:74]." The diverse nature of the samples gives some plausibility to the generalizations offered.

Finally, I will discuss marrow cracking as it has been described by others and as I have witnessed and documented it among the Nunamiut Eskimo. Much of my discussion will center not so much on the production of generalizations but on the degree to which premature generalizations by others have misled analysis in the past. I will make a plea for the collection of more information on this subject and offer some suggestions regarding formal attributes that might well prove useful in future methodological developments for distinguishing the hand of man from the jaws of animals.

Previous Approaches to Understanding Broken and Modified Bone

Reviewing earlier research on human modifications in bone shows that some of the earliest research was on cut marks and surficial modifications resulting from man's use of tools in butchering (Martin 1906:10). This early work was not, to my knowledge, followed up until relatively recently (Guilday n.d.; Guilday *et al.* 1962; Parmalee 1965). Most of this work is sound and provides usable case studies of butchering marks.

Most of the research oriented toward the "reconstruction" of prehistoric butchering practice stems directly from the seminal work of Theodore E. White (1952, 1953a, 1953b, 1954, 1955). White suggested a number of post hoc accommodative models to explain the differential frequencies of anatomical parts revealed through the careful anatomical identification of bones recovered from archaeological sites. These models took two basic forms, postulating (*a*) the differential transport or abandonment of anatomical parts in the context of hunting logistics, and (*b*) the destruction of anatomical parts during the butchering process, which was heavily modeled on the assumption of regular use of large choppers and heavy cleaver-like tools and hafted mauls. Butchering practices were generally considered to vary ethnically (see Wood 1962) and to be a characteristic that could be used for tracing the culture history of different ethnic groups.

Typical of the post hoc modeling of the butchering procedure is the following statement from one of White's early papers on this subject:

> The humerus shows the greatest discrepancy in the number of the two ends of any of the elements . . . Anyone who has tried to separate the scapula and humerus with a knife knows that it is not easy, even in these days of crucible steel, and the job can be easily and effectively accomplished with a cleaver. It is clear that a Stone Age cleaver would demolish the head of the humerus beyond recognition. Further cutting up of the forelimb appears to have been accomplished by smashing the radius and metacarpal near the middle as the ends are usually entire [White 1952:338].

Here we see the post hoc accommodation of a butchering model to the patterns of breakage and bone destruction observed in an archaeological assemblage. It is assumed that the bone destruction resulted from butchering activities. White's post hoc arguments were generally adopted. For a number of years reading an archaeological site report that included a summary of faunal remains was like reading a litany. White's post hoc arguments were invoked to "explain the archaeological facts" and thereby interpret the past It is interesting in this regard that Thomas Kehoe reported that his informants as well as historical accounts indicate: "The limbs were cut up at the joints, not through the bones . . . there was cutting through the scapula–humerus joint [Kehoe 1967:69–71]."

As mentioned earlier, there is running through the Plains literature a very strong "normative" bias. Such discrepancies as the one Kehoe reported, ethnohistorical information indicative of a procedure contradictory to the post hoc model, are frequently treated as evidence for "cultural differences" between the peoples responsible for the two sources of information. There has been little attempt to determine experimentally, ethnoarchaeologically, or from historic records if there are any situational correlates of alternative butchering strategies (see L. R Binford, 1978b:87–90). Little attention has been given to determining if there are diagnostic properties that might permit the archaeologists to discriminate differing strategies, or if the archaeological facts for which post hoc interpretations have been offered did in fact derive from butchering activities!

I will return for a moment to one of my initial points, namely that the bones recovered from a site may have undergone a number of differing event sequences during which modifications could have been made prior to their having entered the archaeological record. In the last chapter I argued that the bison bones excavated by George Frison at the Glenrock Buffalo Jump had been modified by scavenging animals, presumably after the site was abandoned by man. In that situation some of the properties of the bone probably derived from the butchering activities of humans and some from the scavenging activities of the animals. If we recognize that bones may be differentially acted upon by man in various contexts, and that the same acts conducted in different contexts may be carried out in different ways, the "interpretative" picture becomes even more unclear.

Ethnoarchaeological research (L. R. Binford 1978b; Yellen 1977a) that provided new information on butchering practices generally also included information on more than just butchering. Indeed the studies also included descriptions of "animal processing" such that the life histories of bones and their accumulation of properties were empirically illustrated. This knowledge has not yet prompted research leading to refinement of methods for analyzing faunal remains relative to butchering practices.

Before discussing the potential of such research as I see it, it is necessary to review studies aimed at understanding properties of bones derived from marrow cracking and other processing steps such as bone grease manufacture, bone juice production, and the rendering of bone into directly consumable form.

I have already mentioned the replicative experiments by Dart and Kitching (Dart 1959) as they relate to the "proof" of arguments in favor of a distinctive mode of human bone breakage, the crack-and-twist method. It is hard to believe that there appear to have been no ethnographic accounts of bone breakage by man available to researchers at that time. To my knowledge, the first specific study aimed at understanding marrow-bone breakage was the study mentioned earlier by Zierhut (1967) and participated in by Bonnichsen (1973). In terms of published accounts the work of John Yellen (1977a) among the Dobe !Kung provided another account of bone breakage, but no detailed descriptive information regarding the morphology of the derivative fragments. Little analytical attention was given to discriminating fragments produced during different processing phases of !Kung bone use. My own researchers among the Nunamiut (1978b) presented some descriptions of marrow-bone breakage but I did not appreciate at the time the need to obtain controlled samples for purposes of recognizing properties that derived from one phase of bone use and breakage versus another. I was interested in assemblage composition and the differential use and survivorship of bone elements in different contexts where bones were used and manipulated. That is, I viewed my research task as the identification of the function of a site from the overall configuration of an assemblage. It was not until later that I began to worry about morphological properties of broken bone as clues to specific behaviors.

One experimental study (Sadek-Kooros 1972, 1975), one replicative exercise (Bonnichsen 1979), and two pattern-recognition studies of archaeological bone (Lyman 1978; Noe-Nygaard 1977) basically complete the list of research available to the student of patterned bone breakage.

What emerges from this brief survey is that with the exception of research on "butchering practices" there has in fact been very little argument regarding the meaning to be attached to patterns of bone breakage. As will be pointed out in the various parts of this chapter treating different bone-modifying actions, these properties have generally been viewed monocausally. We need methods for discriminating different bone-modifying situations one from another; many such situations may well be reflected morphologically in a single bone assemblage.

The Control Data

The new materials to be presented in this chapter suffer from having not been collected with the specific questions addressed in this chapter in mind. With only minor exceptions the ethnoarchaeological observations were all made between 1969 and 1972 while I was carrying out fieldwork among the Nunamiut Eskimo.

As part of the Nunamiut study much time was invested in observing butchering and the dismemberment of caribou and moose especially but mountain sheep and bear as well. In all, the bones remaining on the sites of over 400 butchering episodes were recorded. Detailed step-by-step observations of butchering procedures were made on 37 separate acts of butchering animals conducted by the Nunamiut Eskimo. I personally field butchered 13 animals under Eskimo supervision using different tools and strategies so I could learn the problems associated with different tactics. Much of this work has been presented (L. R. Binford 1978b) and will only be discussed here in relation to animal behavior or in the context of illustrating variable patterns of bone breakage associated with different butchering tactics.

I will make use of several well-documented Eskimo bone assemblages to illustrate relative frequencies of cut marks, and as comparative data relative to animal-produced assemblages. All these assemblages have been previously described and reported and the reader will be referred to the appropriate pages in my earlier book for the behavioral documentation.

Important to my Eskimo research was a knowledge of the techniques and procedures used in breaking marrow bones and processing bone for other purposes, such as rendering bone grease. At the time the observations were made from which data will be drawn, I was not thinking of the marrow cracking in terms of comparisons with bones broken by animals nor was I thinking in terms of animals other than caribou, which was the target species of my research among the Nunamiut. That is, I was trying to generalize about caribou, not about the variety of animals relevant to interpreting the remains at Old Crow Flats or Olduvai Gorge or other interesting Pleistocene sites. My marrow-cracking investigations were designed to provide a baseline for evaluating whether or not articulator ends from caribou long bones had been either destroyed or removed from sites. I was concerned with establishing an estimate for the expected number of splinters and chips produced by bone-breaking Eskimos per long-bone articulator end. Thus a series of marrow-cracking events were conducted during which an Eskimo woman and a man broke a series of long bones using the repertoire of techniques normally employed by the Nunamiut. The aim was to establish some estimate of the differences between splinter to articulator end ratios for different bones of the caribou anatomy. All splinters, chips, and articulator ends were saved and catalogued from two such controlled situations. In four other available samples all the breaks were produced during marrow processing, but I do not know the exact anatomical composition of the population of bones broken. What is lacking in all these samples is control over the actual breaking tactics and the resulting bone splinters. I was asking population questions regarding the properties of aggregate assemblages in my fieldwork. What we would like to know is the character of any diagnostic morphological properties referable to different methods of bone breakage. Unfortunately, we do not know these things.

Dismembering Strategy

Man using tools is capable of actually taking an animal apart to suit his purposes. That is, he is not dependent on the natural strengths of the articulations and the amounts of connective tissue to separate parts that then may be differentially used or consumed, as in the case with predator–scavengers. This is not to say that the amounts of muscle and the character of connective tissue do not condition man's selection of parts or the tactics he used for dismemberment; they most certainly do. The point is that the particular anatomical sets that man manipulates are conditioned by how he dismembers a carcass, which in turn is conditioned by its intended use.

Several workers have recorded the way different ethnic groups partition the anatomy of animals, particularly ungulates. Diane Gifford (1977) recorded the basic units of the anatomy into which four separate African groups partitioned the anatomy and I have recorded actual butchering data for the Navajo and the Nunamiut Eskimo (L. R. Binford 1978b; L. R. Binford and J. B. Bertram 1977) John Yellen (1977a, 1977b) has observed butchering and hence the character of segments generated by Bushmen. Figure 4.01 compares the ways in which these groups partition the anatomy of animals during butchering.

Several points are of interest:

1. All groups ideally separate the head from the neck between the occipital condyle and the atlas vertebra.
2. All groups except the Navajo separate the neck from the remainder of the vertebrae.
3. All groups separate the front leg from the axial skeleton. Those that further partition the front leg into upper and lower segments generally disarticulate between the carpals and the distal radio-cubitus.
4. All groups separate the rear leg from the vertebrae; however, there is considerable variability in the degree to which half the pelvis is left attached to the rear leg as opposed to the axial skeleton. In all the ethnographic cases recorded where either axes, adzes, or large cleaver-type knives were used in butchering, the pelvis and/or sacrum were butchered off with the rear leg. Where small knives were used, the pelvis and/or sacrum were butchered off with the lumbar vertebrae or as a separate unit distinct from the rear leg.
5. All groups generally treat the spinal column with ribs and brisket distinct from the other

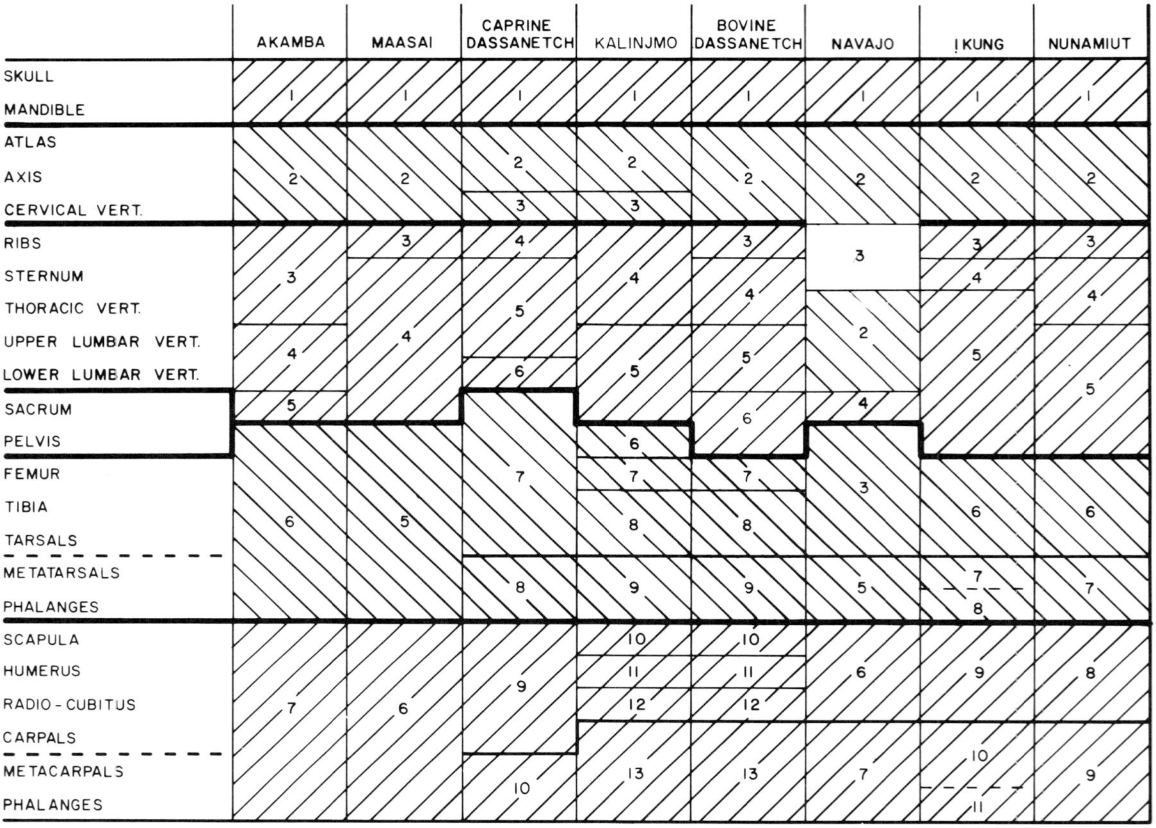

FIGURE 4.01. ***Dismemberment practiced by various ethnic groups during butchering.***

> major portions of the anatomy, but there is considerable variability in the way the thorax and spine section is butchered down into smaller units. Most, but not all, butcher the ribs and sternum off as independent units.

Some of the variability illustrated in Figure 4.01 is likely to be related to degrees of cognitive specificity—that is, the level of discrimination at which informants perceived that the question was directed. In the case of observed butchering, there is a kind of normative "leveling," in that under some conditions the Nunamiut and the Navajo discriminated down to elements even finer than those indicated for the Kalinjmo. Taking this problem into consideration, there is a fair degree of correspondence among the ethnic groups in the elements into which the anatomy is partitioned. Although provocative, these data are either normative or based on informant information and as such are suspect as guides to actual variability that might be generated under specific conditions.

To obtain a clearer picture of the elements of the anatomy that might be expected to be treated as units and as such perhaps to have independent histories under conditions of human utilization, a factor analysis was carried out on 64 unedited assemblages of anatomical parts collected among the Nunamiut Eskimo. These were basically all the assemblages reported in Binford (1978b) minus dog yards. When I say the assemblages are unedited, this means that assemblages were not broken down into subcomponents such as house, meat rack, and dump; only site totals were utilized. Thirty-one variables representing the anatomical parts listed in Table 4.01 were used in the analysis. Values used were percentages of maximum MNI, or the ratio of the variable tabulated in MNIs divided by the maximum MNIs observed for any variable in each assemblage. This procedure ensured that there would be no variance contributed to the matrix by differences in population size among the cases. A BMD factor analytic program

TABLE 4.01
Factor Loadings for Analysis of Nunamiut Faunal Assemblages

	Factors						
Anatomical part	1	2	3	4	5	6	7
1 Antler	−.13	.02	−.11	.09	*−.85*	−.08	.23
2 Skull/maxilla	−.12	−.03	.01	−.02	*−.89*	−.10	.04
3 Mandible	−.39	−.04	−.19	.07	*−.63*	.38	−.11
4 Atlas/axis vertebrae	.08	−.40	.22	−.18	*−.42*	*−.40*	−.27
5 Cervical vertebrae	.26	−.26	.30	.00	*−.51*	−.36	−.15
6 Thoracic vertebrae	*.94*	.01	.07	−.03	.05	.05	.01
7 Lumbar vertebrae	*.91*	−.05	.19	−.03	.09	−.02	.02
8 Sacrum	*.85*	.02	.06	−.13	.02	.02	.23
9 Pelvis	*.81*	−.12	.15	−.02	.05	.30	−.03
10 Sternum	*.64*	.05	−.28	−.12	.21	−.32	.19
11 Ribs	*.72*	.31	−.08	−.22	.10	−.10	.06
12 Scapula	.39	−.21	.41	−.09	.14	.30	−.19
13 Proximal humerus	.35	−.31	*.73*	.12	.03	−.05	.21
14 Distal humerus	.13	.11	*.87*	−.07	−.01	−.01	−.03
15 Proximal radio-cubitus	.03	.08	*.85*	−.10	.07	−.09	−.14
16 Distal radio-cubitus	−.04	.33	*.63*	.08	.06	−.30	.26
17 Carpals	−.16	.08	.28	*.44*	.11	.08	.52
18 Proximal metacarpals	−.18	.14	.41	.18	−.14	*−.66*	.05
19 Distal metacarpals	−.07	.24	.24	.37	−.22	*−.60*	.04
20 Proximal femur	.12	*.86*	.22	.05	.00	−.11	−.12
21 Distal femur	.07	*.83*	.17	.16	.01	.05	−.08
22 Proximal tibia	.06	*.78*	.21	.17	.13	−.06	−.12
23 Distal tibia	−.21	*.69*	−.02	−.10	−.04	−.01	−.36
24 Tarsals	.09	.15	.12	.42	.02	*.53*	*−.41*
25 Astragalus	−.26	.39	−.03	.23	.18	.17	*−.56*
26 Calcaneus	−.20	.29	.10	.25	.21	.14	*−.75*
27 Proximal metatarsal	−.17	*.50*	−.08	.11	−.04	−.14	*−.51*
28 Distal metatarsal	−.15	.36	−.05	.41	.09	−.12	*−.63*
29 First phalange	−.13	.08	−.04	*.89*	−.07	.09	−.06
30 Second phalange	−.12	.05	−.05	*.94*	−.01	−.03	−.09
31 Third phalange	−.08	.08	.01	*.90*	−.01	−.17	−.11
Communality	82.5 Thorax and spine	82.7 Upper rear leg	74.6 Upper front leg	66.7 Feet and podials	65.2 Head and neck	88.9 Metacarpal and upper neck	88.2 Tarsals and metatarsal

NOTE: Factor diagnostics shown in italic.

was employed using orthogonal rotation option. The results as indicated by factor loadings are summarized in Table 4.01.

There is a seven-factor solution with a clear eigenvalue "jump" between the seventh factor (eigenvalue 1.56) and the next factor (eigenvalue 0.97). The cumulative percentage of variance accounted for by the seven factors is 75.2%. The communalities for each factor are given in Table 4.01. This solution appears very clean and is technically a robust pattern-recognition statement.

This solution is provocative and most informative, given what is known about Nunamiut treatment of animals. Of even greater interest are the differences between the anatomical segments that behave independently in the Nunamiut data and the natural disarticulation sequence (see Table 3.01) documented by Hill (1975). Of particular interest is the fact that the thorax and vertebrae group behave together (factor 1) *with the exception of the cervical vertebrae.* The latter are demonstrated to integrate with the skull and mandible (factor 5). This pattern of covariation is most unlikely among assemblages where "natural" processes of disarticulation dominate. Reference to Table 3.01 illustrates that the cranium with atlas vertebra is disarticulated from the remainder of the neck quite early (step 6) in the sequence of decreasing organization, and the neck separates from the thoractic vertebrae at about the same time the ribs begin to separate (step 16). This means that under natural conditions elements that disarticulate at about the same time are most apt to end up associated and such a pattern is in strong contrast with that noted in the Nunamiut data. Other

FIGURE 4.02. ***Johnny Rulland stripping tendon bundle from metapodial.***

FIGURE 4.03. *Johnny Rulland holding stripped unit: tarsals and foot.*

interesting patterns of organized or parallel "determinance" are the grouping of phalanges with the carpals and tarsals (factor 4). This pattern derives from characteristics of Nunamiut processing of bones for marrow (see L. R. Binford 1978b:148). The foot is removed by servering the attachments on the dorsal face of the metapodial between the articulation of its distal end with the phalanges. The foot is then held in one hand and the distal end of the metapodial is grasped in the other. The sheath of tendons running in the ventral channel of the cannon bone is stripped up to the proximal end of the bone. Cuts are then made between the sheath and the bone, resulting in the removal of the carpals or tarsals attached to the tendon sheath, which in turn is attached to the phalanges. The entire unit is then disposed of. (See Figures 4.02 and 4.03.) Finally, the independent distributions of the metatarsals and the metacarpals as well as their independence from the upper leg segments of both the rear and forelimbs clearly reflect the independent processing of marrow as opposed to meat, and the differential quality of front versus rear quarter "marrow" during periods of poor nutrition among the caribou.

I do not offer the Nunamiut data as an example of human patterning only to illustrate that human use of anatomical elements is apt to have certain redundant patterns, to evidence some specialized usage, as in the differences between marrow- and meat-yielding parts, and to be segmentally grouped in terms of use consideration. The contrast must be kept in mind with processes of "natural" dismemberment as documented by Hill (1976:151). In the contrasts rest diagnostic criteria.

Butchering Marks

When man butchers an animal, he leaves traces of his cutting and chopping activities. Few recent studies have been made of these traces, although such marks were a subject of early Old World studies.

In a study seemingly "before its time," Henri Martin (1907–1910) considered in detail traces of skinning, evisceration, disarticulation, and filleting, as well as the fracture and burning of bones. He illustrated nicely the traces made by stone tools on the fauna from the famous Mousterian site of La Quina. These photographs remain some of the best published examples of cut marks made by stone tools. (See, for instance, Martin 1907–1910: Figures XLIV, XLVII, XLVIII, L).

In North America, the first systematic study of butchering marks that I have been able to locate was done by Guilday *et al.* (1962) on faunal remains from an Early Historic site from Pennsylvania (the Eschelman site). This study provides some guidelines for investigating butchering marks: "To qualify as a butchering mark two criteria were applied: (1) repetition in specimen after specimen at precisely the same location on the bone; (2) that there was some anatomically dictated reason why a particular mark should occur at any given spot [Guilday *et al.* 1962:63]."

Several quite useful conventions of presentation and description were introduced in Guilday *et al.* (1962). For instance, the use of skeletal silhouettes as a graphic way of presenting data on butchering marks is notable. These authors were fairly systematic in presenting quantitative data from their observations on the American black bear. Fairly complete quantitative data were given for whitetailed deer and practically none were given for elk. Figures 4.04 and 4.05 are reproduced from this seminal study and illustrate the anatomical "hot spots" where butchering marks were systematically observed.

Considerable attention has been given to traces of butchering by George Frison (see Frison 1970). Most of Frison's work has been with the remains of bison at kill sites and hence should reflect the activities of primary butchering and processing for transport. Frison pioneered the study of bone breakage as a clue to butchering behavior and has argued that

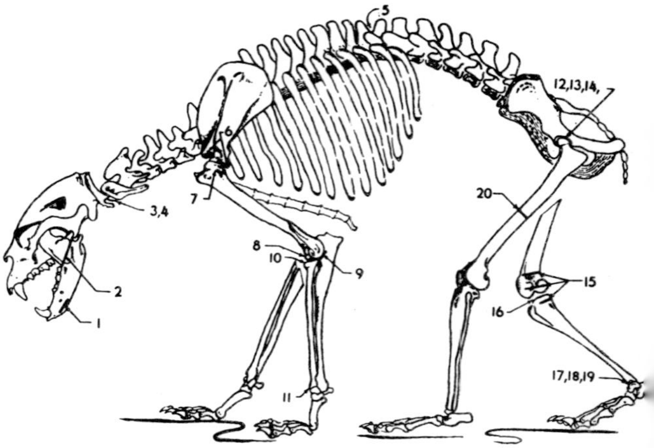

FIGURE 4.04. ***Cut marks observed on black bear at the Eschelman site. (Reproduced with permission from Guilday et al., 1962.)***

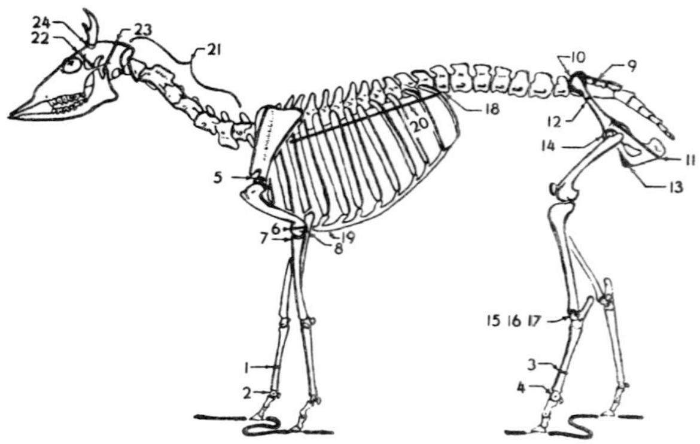

FIGURE 4.05. ***Cut marks observed on whitetail deer at the Eschelman site. (Reproduced with permission from Guilday et al., 1962.)***

muscle stripping, primarily with bone choppers, was a basic dismemberment strategy.

I have made primary observations on two sets of faunal data for the purpose of isolating and tabulating cut marks or butchering traces on bone. The earliest observations were made in 1968 on the fauna from the Mousterian levels at the rockshelter of Combe Grenal excavated by F. Bordes and located in the Dordogne Valley of south central France. The faunal material from this site represents the behavior of Neanderthal man between roughly 90,000 and 40,000 years ago. The second set of observations was made in Alburquerque on five assemblages of bone collected in Alaska from behaviorally documented Nunamiut Eskimo sites. I will describe the latter observations first, since I consider them to be the most behaviorally documented of the descriptive material available.

Cut, Saw, and Chop Marks from Nunamiut Eskimo Sites

I have observed the faunal materials from three residential sites for two species, Dall sheep (*Ovis dalli*) and caribou (*Rangifer tarandus*). These are not

TABLE 4.02
Distribution of Cut Marks on Bones in Five Nunamiut Assemblages[a,b]

	Kakinya caribou			Kakinya sheep			Rulland caribou			Bear site caribou			Bear site sheep		
Anatomical part	MNE (1)	MKD (2)	%[c] (3)	MNE (4)	MKD (5)	%[c] (6)	MNE (7)	MKD (8)	%[c] (9)	MNE (10)	MKD (11)	%[c] (12)	MNE (13)	MKD (14)	%[c] (15)
Antler[d]	1	0	.00	0	0	.00	0	0	.00	3	1	.33	2	0	.00
Skull	4.5	4.5	1.00	3	3	1.00	0	0	.00	6	6	1.00	4.5	4.5	1.00
Mandible	6	2	.33	6	2	.33	2	1	.50	14	5	.36	8	3	.33
Atlas	3	2	.67	1	1	1.00	0	0	.00	7	3	.43	5	4	.80
Axis	3	0	.00	1	0	.00	0	0	.00	3	0	.00	3	0	.00
Cervical vertebrae	14	2	.14	5	1	.20	0	0	.00	6	1	.17	5	0	.00
Thoracic vertebrae	14	9	.64	0	0	.00	2	1	.50	33	24	.73	13	6	.46
Lumbar vertebrae	5	2	.40	0	0	.00	0	0	.00	14	6	.43	0	0	.00
Pelvis	7	3	.43	2	1	.50	1	0	.00	12	5	.42	6	3	.50
Ribs															
Sternum															
Scapula	4	0	.00	0	0	.00	1	0	.00	31	0	.00	3	3	1.00
Proximal humerus	8	0	.00	1	0	.00	3	1	.33	18	4	.22	6	1	.17
Distal humerus	13	6	.46	1	0	.00	4	3	.75	26	11	.42	7	2	.29
Proximal radio-cubitus	13	3	.23	2	1	.50	8	2	.25	22	6	.27	5	1	.20
Distal radio-cubitus	11	2	.18	2	0	.00	5	1	.20	21	2	.10	7	1	.14
Carpals	12	1	.08	10	0	.00	8	0	.00	744	9	.12	28	2	.05
Proximal metacarpal	5	0	.00	1	0	.00	0	0	.00	21	2	.10	10	1	.10
Distal metacarpal	4	1	.25	1	0	.00	1	0	.00	24	3	.13	12	0	.00
Proximal femur	18	1	.06	2	0	.00	13	1	.08	26	3	.12	0	0	.00
Distal femur	20	6	.30	2	0	.00	15	5	.33	22	10	.45	3	1	.33
Proximal tibia	14	7	.50	3	1	.33	7	3	.43	31	9	.29	2	1	.50
Distal tibia	12	1	.08	5	0	.00	7	0	.00	25	1	.04	4	1	.25
Tarsals	21	2	.10	15	1	.07	8	0	.00	43	2	.05	13	0	.00
Astragalus	10	4	.40	6	0	.00	3	0	.00	11	2	.18	7	0	.00
Calcaneus	10	0	.00	4	0	.00	4	0	.00	10	0	.00	7	0	.00
Proximal metatarsal	9	1	.11	4	0	.00	3	0	.00	10	1	.10	8	1	.13
Distal metatarsal	8	1	.13	3	0	.00	2	1	.50	20	5	.25	8	0	.00
First phalange	42	0	.00	8	0	.00	6	0	.00	100	0	.00	36	0	.00
Second phalange	35	0	.00	8	0	.00	6	0	.00	95	0	.00	33	0	.00
Third phalange	31	0	.00	6	0	.00	3	0	.00	77	0	.00	32	0	.00

[a] MNE = minimal number of elements in each anatomical category; MKD = number marked by cut marks.
[b] Values for pelvis, scapula, and vertebrae may well be deflated, since there was much dog gnawing on these parts, which destroyed anatomical properties and could have obscured butchering marks.
[c] Values were italicized if over 20% of the elements examined showed butchering marks.
[d] Elements were only listed when base of antler was still attached to skull.

all the assemblages I have available for study but they are all that have been studied in detail under good lights and laboratory conditions. Table 4.02 summarizes the observations on butchering traces from the five populations observed. This summary treats only the frequency of such marks, regardless of form, on different elements of the anatomy. The information provides a clue to where to look for butchering marks, and also indicates the areas of the anatomy that were most commonly altered or preserve traces of butchering activity. Italicized percentages highlight such anatomical areas (values were italicized if over 20% of the elements examined showed butchering marks).

Two factors are of interest here. First, there are some discrepancies between the butchering mark distribution and the descriptions of butchering that I reported in some detail. (L. R. Binford 1978b:48–61). In none of the field-butchering demonstrations did I observe any cutting or dismemberment of the skull, yet in these assemblages *all* the skulls show some evidence of having been butchered into subunits. Since the assemblages are all from *residential* sites, this further processing of the skull must have taken place during preparation for consumption. This same condition is true for the pelvis. In the field-butchering episodes I observed, the pelvis was *never* reduced to subelements, but in all five assemblages 40–50% of pelves showed evidence of being reduced to subelements. The actual percentage is probably larger, since most pelvic parts ended up in dog yards on these sites and hence suffered considerable gnawing damage, which may well obscure butchering marks.

The second item of interest is the presence of longitudinal cut marks in the super- and infraspinous fossae of all sheep scapulae from the Bear site (Table 4.02 and Figure 4.06). It seemed clear to me that these marks derived from filleting the meat from the scapula, an activity normally associated with drying meat. (See the fillets hanging on the drying rack in L. R. Binford [1978b:101, Figure 3.9].) I then examined some of the scapulae from spring sites, such as Tulugak Late Site 2A (Binford 1978b:205–213), where I knew drying was taking place, and found that 83% of the scapulae remaining with blades exhibited such marks, lending strong support to the relationship suggested between this type of cut mark and the drying of meat.

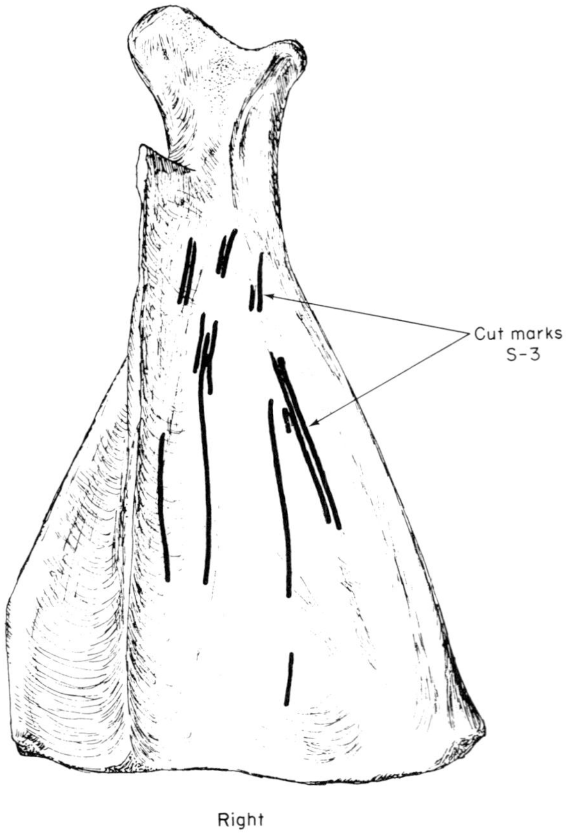

FIGURE 4.06. ***Marks typically produced during the removal of meat from the scapula.***

This observation moves us in the direction of discussing the actual form and placement of cut marks, a subject I wish to delay further discussion of until additional comparative data have been presented.

Marks on Stone Tool Butchered Animals from Combe Grenal

The Nunamiut assemblages represent animals dismembered with metal tools. It is quite reasonable to suspect that butchering strategies would be responsive to the characteristics of the tools available for use. A good Pleistocene comparison is provided by the fauna from Combe Grenal. The MNEs reported should not be taken as an unbiased representation of the site contents, since during 13 years of excavation Bordes followed traditional procedures relative to the collection of faunal materials, instructing his crew to save only teeth, articulator ends, and phalanges in the belief that ribs, long-bone splinters, vertebrae, small fragments of skull, scapula blades, and pelvic parts other than the acetabulum were not suitable for species identification. In some cases, I have listed vertebrae or ribs but the absolute MNE should not be treated seriously only the percentage modifications. Summary data on butchering marks are given in Table 4.03. For additional comparative data see Parmalee (1965:25).

There is clearly a regular *pattern* to the distribution and frequency of cut marks on bones at the site of Combe Grenal. For the three species examined

TABLE 4.03
Distribution of Butchering Marks on Bones from the Mousterian Site of Combe Grenal[a]

	Bos primigenius (aurochs)			*Equus caballus* (horse)			*Rangifer tarandus* (reindeer)		
Anatomical part	(1) MNE	(2) MKD	(3) %[b]	(4) MNE	(5) MKD	(6) %[b]	(7) MNE	(8) MKD	(9) %[b]
Antler[c]									59
Maxilla				2		0	4	—?	?
Mandible	3	3	*100*	7	6	86	27	23	85
Atlas				2			2	+	50
Axis				2			4		
Innominate	2			10			7	3	43
Ribs				4			61	6	10
Scapula	1			6	2	33	44	4	9
Proximal humerus	4			4			1		
Distal humerus	10	1	10	7	1	14	25	3	12
Proximal radio-cubitus	10	2	*20*	11	3	*27*	39	3	8
Distal radio-cubitus	3			2			25	1	4
Carpals	3			2			76	0	0
Proximal metacarpal	1			4			30	0	0
Distal metacarpal	0			6			43	3	7
Proximal femur	2			17	3	18	27	1	6
Distal femur	3			4			14		
Proximal tibia	2			2			9		
Distal tibia	7			9	1	11	59		
Tarsals	7			6			43		
Astragalus	5			3			34	3	9
Calcaneus	2						40		
Proximal metatarsal	2			8	1	13	48	1	2
Distal metatarsal				3			46	2	4
First phalange				12			93		
Second phalange				6			90		
Third phalange	1			4			23		
Total	68	6	9	147	17	12	931	63	7

[a]MNE = minimal number of elements in each anatomical category not converted to MNIs; MKD = number marked by cut marks.
[b]Values were italicized if over 20% of the elements examined showed butchering marks.
[c]All are young "spike" antlers with attached sections of skull. All exhibit cut or "saw" marks around the base of the antler on the skull, as I have observed resulting from the removal of the skin of the face intact generally for use in clothing manufacture.

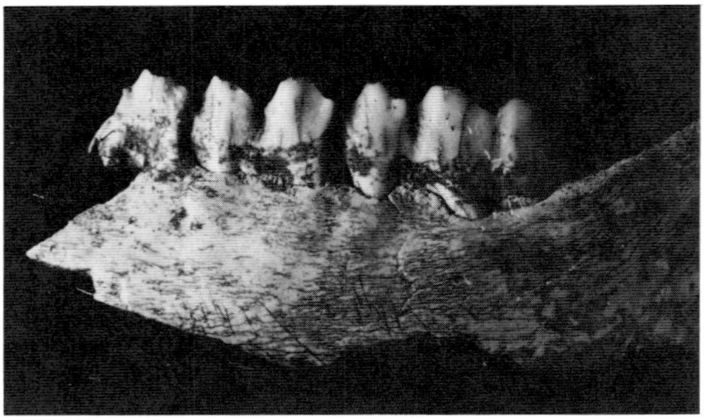

FIGURE 4.07. *Marks by stone tools on the face of reindeer mandible. (From the Mousterian site of Abri Vaufrey, excavated by J. P. Rigaud.)*

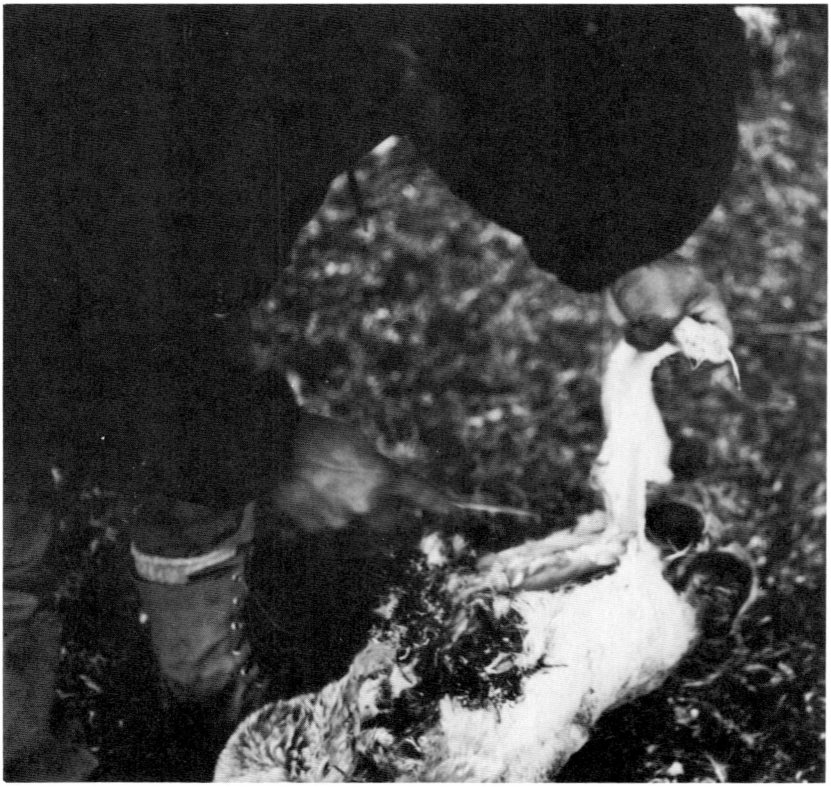

FIGURE 4.08. *Butcher removing the tongue from a caribou.*

(*Bos primigenius*, *Equus caballus*, and *Rangifer tarandus*), the part most commonly yielding cut marks is the mandible. Marks on the mandible tend to be slightly oblique incised marks on the *inside* of the mandible generally opposite the M2 tooth (see Figure 4.07). The marks are believed to originate from the underside of the mandible and to be related to the severing of the mylohyoid muscle during the removal of the tongue. Figure 4.08 illustrates a Nunamiut butcher going in underneath the mandible and removing the tongue of a male caribou prior to the removal of the mandible from the skull. The next most common cut mark is around the lip of the acetabulum (Figure 4.22). Marks encircling the acetabulum are presumably made during the cutting of the iliofemoral and ischiofemoral ligaments, which tend to encase the coxal articulation. A minor type of cut mark, observed is at right angles to the lip of the acetabulum and is believed to have resulted from the insertion of the cutting instrument into the acetabulum to cut the ligament of the fomoral head after the coxal articulation is dislocated. Another type of cut mark was consistently observed on the distal humerus and on the proximal radio-cubitus

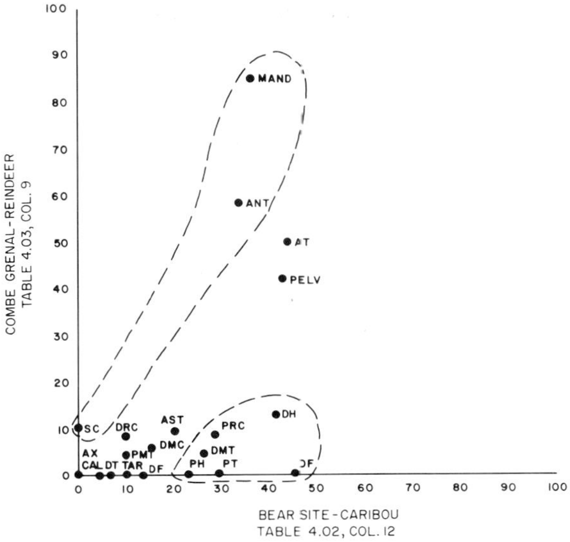

FIGURE 4.09 ***Relationship between frequencies of cut marks on bones from the Mousterian site of Combe Grenal and those from the Bear site of Anaktuvuk Pass, Alaska.***

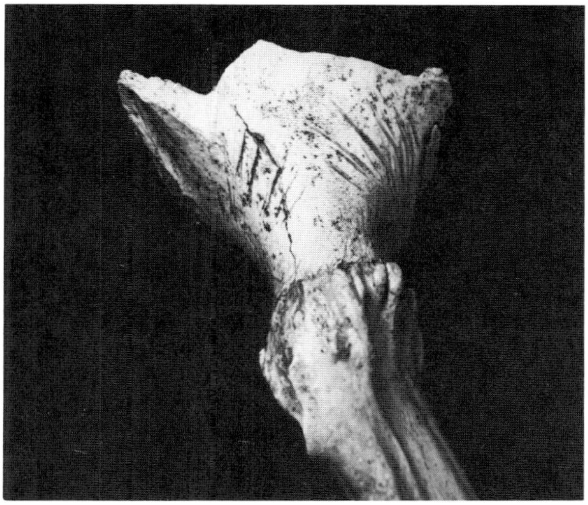

FIGURE 4.10. ***Marks of stone tools on base of reindeer antler. (From the Mousterian site of Combe Grenal, excavated by F. Bordes.)***

(Figure 4.31). This was a transverse cut mark across the lower part of the coronoid fossa and the olecranon fossa on the distal humerus. Cut marks were common on the olecranon, generally running across the process. I will not describe all the cut marks observed at Combe Grenal here, since the point of presenting the material is simply to demonstrate that there is anatomical clustering of marks.

Figure 4.09 compares the reindeer from Combe Grenal with the caribou from the Bear site as to the relations between frequency and placement of butchering marks. Elements from Combe Grenal that exhibit high frequencies of marks relative to the Bear site material are the mandible, antler bases (Figure 4.10), and the scapula. I have already mentioned the Mousterian mandibular cuts. As for the cuts around the base of the antlers, it should be pointed out that all the antlers observed in the Mousterian material were spike antlers of very young individuals. Circular cut marks were seen around the base of the antlers in over half of the cases (See Figure 4.11). I have previously noted that such cuts were most common on Nunamiut sites occupied by special task groups engaged in fall hunting of young animals for use as winter clothing. It should also be pointed out that the Nunamiut skin the head with such care only

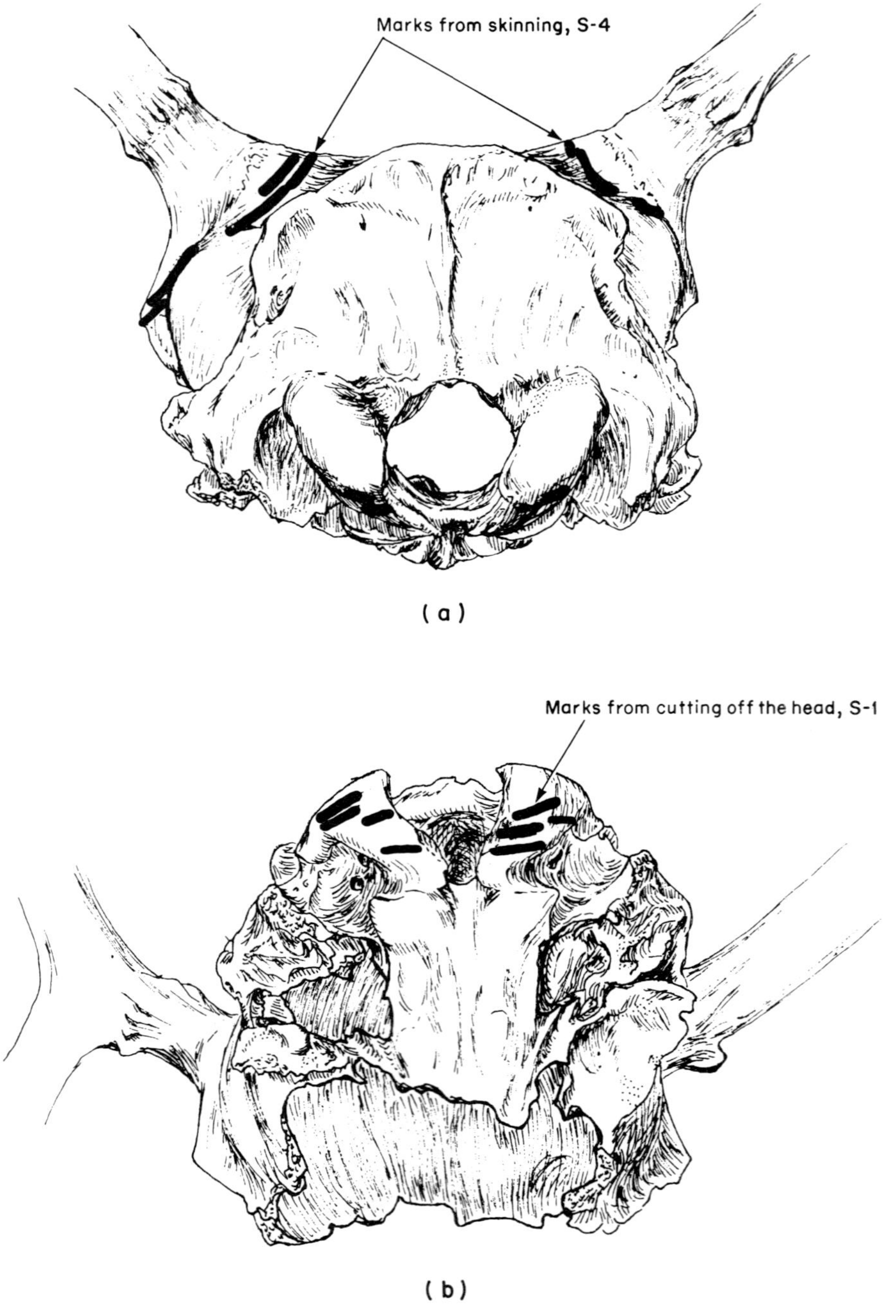

FIGURE 4.11. *Cut marks observed on skulls: (a) posterior, and (b) ventral views.*

when the skin is destined for the manufacture of parkas, where the head of the caribou is used for the hood of the parka.

In a similar vein the very early study of cut marks by Henri Martin (1907–1910:229) reports considerable numbers of marks described as "*incisions circularie des members,*" a cut mark encircling a bone that was interpreted as having arisen in the context of skinning animals. This same type of mark has been consistently recognized by modern workers. The interesting difference is that from the Mousterian materials reported by Martin most cuts occurred on phalanges, whereas recent studies of bison (Frison 1971; Wheat 1979) remains and other forms show this to be common on metapodial bones and relatively rare on phalanges.

In my experience with the Nunamiut, starting to skin an animal by making an encircling cut around the metapodial or sometimes the lower tibia and/or radio-cubitus was almost always associated with butchering activities at a multiple kill, where the primary concern was with meat, not skins. On the other hand, when late summer–early fall hunts were conducted specifically with an eye to obtaining yearling calves or calves of the year for their skins, it was not uncommon for the butcher to take pains in skinning out the foot, initiating the inside leg cut from the hoof. This would have the effect of rendering cut marks on the phalanges rather than the shafts of lower leg bones. The butchers explained that the lower leg skin served as essential components in the manufacture of mukluks (skin boots) and skin socks. The maintenance of the natural conformation of the skin at the animal's foot was essential to providing

FIGURE 4.12. ***Butchering school. Dan Witter being instructed on how to disarticulate the foot from the metapodial.***

FIGURE 4.13. ***Butchering school. Dan Witter being instructed in the use of leverage as a butchering aid.***

the proper shapes for use in forming the instep area of the sock or boot. Perhaps there is one such similar concern standing behind the time-consuming actions of skinning out the foot indicated by the Mousterian remains.

Other marks characteristically noted on the Nunamiut bones related to the removal of the head from the neck. Typically there were marks across the ventral surface of the occipital condyle (see Figure 4.11) as well as related transverse cut marks across the ventral surface of the anterior wings of the atlas vertebra. Both marks derive from severing the head from the neck (see L. R. Binford 1978b:354, Figure 7.11) and are very common on the Nunamiut materials. The frequency among Mousterian materials is unknown since the only parts regularly saved from the skull were the maxillary dental arcs, rendering examination for this cut mark impossible.

The Mousterian material showed high frequencies of cut marks around the head of the scapula. Such marks were rare or absent from the Nunamiut material, but common on bear bones from the Eschelman site reported by Guilday *et al.* (1962: Figure 2). The meaning of this difference is not clear to me. Bones in the Nunamiut material that exhibit higher frequencies relative to Combe Grenal materials are distal femur, proximal tibia, proximal and distal humerus, proximal radio-cubitus, and distal metatarsal (See Figures 4.25, 4.26, and 4.32.) At least some of these differences are in my opinion related to the use of metal tools in butchering. For instance, the procedure for disarticulating the femur from the tibia involves inserting the point of a knife from the posterior face into the joint between the articulator surfaces of the two bones and severing most of the ligaments connecting the two bones (oblique and arcuate popliteal, anterior and posterior cruciate, anterior and posterior meniscofemoral ligaments). This strategy leaves a pair of telltale traces: a set of semicircular cut marks around the lateral intercondylar tubercles of the proximal tibia (cut Tp-1, Figure 4.26) and cut marks on the medial surface of the lateral condyle of the distal femur. The latter marks sometimes extend partway into the intercondylar fossa (cut Fd-3, Figure 4.25). In a similar way, the disarticulation of the foot from the distal metapodial is accomplished by the Nunamiut by inserting the point of a metal knife into the flexed joint, as shown in Figure 4.12. Nicks on the posterior condylar spines result. This technique of inserting a knife point into the joint is probably inappropriate for stone tools since stresses are set up on the flat sides of the blade. Such stresses would render a snap break likely with stone tools but not with metal.

It is my impression that with stone tools the most common strategy couples more superficial cuts with the application of leverage to dislocate a joint, as is illustrated in Figure 4.13. By using a dislocation technique, less scarring of the articulator ends of bones is likely. The ligaments are stretched and exposed, so they may be cut directly. Perhaps this contrast accounts for some of the other inflated cut mark frequencies on articulator surfaces of long bones in the Nunamiut materials.

Thus far, I have been speaking primarily of the distribution of butchering marks on the skeleton; however, the foregoing discussion leads naturally into a discussion of the formal properties of the cut marks and their specific placement on bones. Before taking up this issue, I should emphasize that a population of moderate-sized animals butchered by man using stone tools can be expected to yield evidence of such butchering in the form of cut marks, regardless of the types of cutting tools used.

Cut Marks: Their Form and Placement on Specific Bones

The morphological setting and the form of a cut mark will vary somewhat with the type of work being conducted when the mark was produced and the implement used to accomplish the task. In both the form and placement of marks I hope to document variability that may be reliably referred to known forms of butchering activity as well as to known forms of tools. Of course what is desired is the specification of diagnostic characteristics of cut marks that may serve as the basis for robust inferences about the character of past activities and types of tools.

METAL VERSUS STONE TOOLS

The faunal materials I collected among the Nunamiut Eskimo were from animals butchered with metal tools. The data I collected from the Mousterian site of Combe Grenal are referable to butchering activities conducted with stone tools. Unfortunately, at the time I collected the Mousterian data I was not very knowledgeable about butchering and was not aware of the properties of cut marks made with metal tools. This means that a well-informed and comparably collected body of data that might elucidate differences between stone and metal tools is not available at present.

I have made certain observations that may aid in differentiating the two but certainly more reliable criteria of recognition could be developed if control materials were available (see Walker and Long 1977).

Most of the cut marks made on bones with metal tools are almost hairline in size. They often appear to have been cut into the bone from the side, or obliquely, leaving an overlapping small "shelf" of bone that remains in place. For this reason cut marks produced by metal knives are very hard to see when one looks directly down on a bone (see Figure 4.14). Of course this visibility varies with the pressure applied and is probably an inverse function of how well honed the knife might be. In addition to having this diminutive "sliced" look, the marks are generally long, resulting from cuts running across tissue for considerable distances. Cutting with stone tools requires a much less continuous action, more of a series of short parallel strokes. Also, most stone tools, particularly ones that are retouched, do not have straight or single-plane cutting edges. Marks from stone tools tend to be short, occurring in groups of parallel marks, and to have a more open cross section. They also have a more ragged appearance when viewed from the top. Cut marks undoubtedly

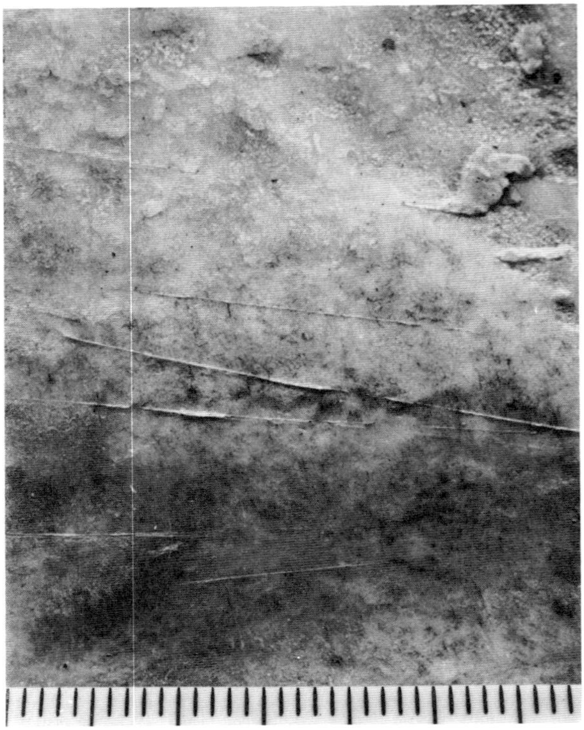

FIGURE 4.14. ***Characteristic marks produced by metal knives used in butchering.***

produced by stone tools are shown in Figure 4.15, where a bone from the Mousterian site of Abri Vaufrey (excavated by J. P. Rigaud) is shown.

Placement of Cut Marks

Many have recognized that cut marks derive from different stages of processing an animal. Since such stages normally result in an accumulation of marks, some traces of earlier and more fundamental stages of processing remain on the bones as they are subjected to more extensive processing during the sequence of actions beginning with the procurement of the animal and terminating with the abandonment of the last elements of the skeleton.

This sequence is almost always (*a*) skinning, (*b*) dismemberment, (*c*) filleting for either consumption or storage, which normally involved still further dismemberment, and (*d*) marrow consumption. It is true that marrow consumption may accompany skinning and initial dismemberment, but this is recognizable as consumption of parts that yield marrow but little else, such as lower limb bones. Marrow removal from meat-yielding bone normally takes place after the removal of meat for consumption or storage. This normal sequence implies that cut marks accumulated on bones will vary with the stage of processing reached prior to abandonment of the bones and will vary in *diagnostic* ways regarding the placement of an assemblage in a logistical–consumption sequence.

Since I am concerned with documenting properties that may be investigated for their diagnostic value in seeking a behavioral knowledge of the past, it is perhaps useful to describe in generic terms the types of cut marks that can be expected from different stages of processing of an animal by man.

Skinning Marks

There are actually very few places on the anatomy where the manipulation of the skin brings the

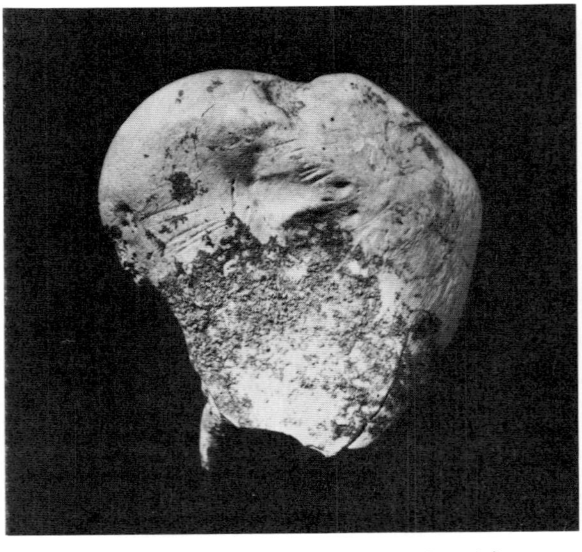

FIGURE 4.15. ***Characteristic marks produced by stone tools used in butchering.***

butcher in direct contact with bone. The two places where this is most likely are the lower legs and the head. I have already mentioned that cuts derived from skinning activities have been noted encircling the shafts of lower limb bones. Such cuts have been observed on the lower tibia, the shaft of the metatarsal (Figure 4.16), and the phalanges. Analogous placement has been noted on the front leg, the distal segment of the shaft of the radio-cubitus, the metacarpal, and the phalanges. This encircling cut is generally one of the first to be made and is basically the "point of entry" for the skinning of the animal. Subsequent cuts normally originate at such an encircling cut and extend down the medial face of the leg to the body of the animal. Cut marks may result from this action.

The second place where the skin is essentially stretched over bone is on the head. There are several places where one might expect cut marks if the head is in fact skinned out: around the base of the antlers (Figure 4.11) or horns, somewhat less commonly around the ears, and around the mouth, particularly in the "chin" area of the mandible.

As noted previously, skinning for skins differs from skinning as a stage of butchering. When the skins are destined for use in clothing manufacture, the animal is skinned out fairly completely, resulting in cuts around the phalanges, around the antlers, and on the chin area of the mandible (see Frison 1970:11). These basic marks are indicated as No. 1 in Figure 4.04 and Nos. 24, 1, and 3 in Figure 4.05. Good illustrations of the cut marks in the phalangeal areas are given by H. Martin (1907–1910): Plates XLV, No. 1, Plate LIX, Nos. 1, 2, 3, 4, and Plate LIX, No. 6 show an encircling cut around a metapodial.

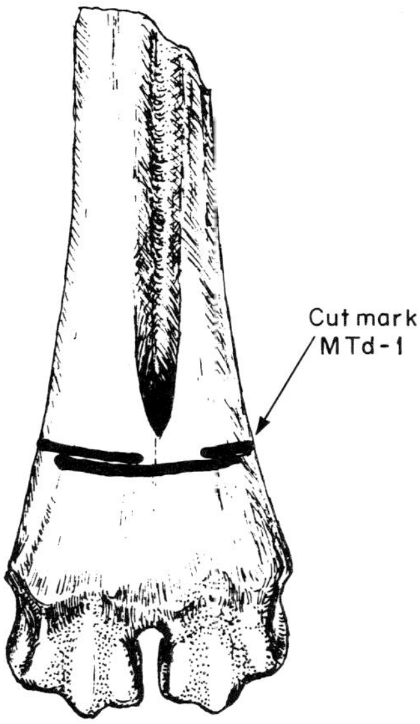

FIGURE 4.16. ***Marks indicative of skinning activities.***

Dismemberment Marks

By far some of the more distinctive cut marks derive from the dismemberment phase of processing an animal by tool-using man. In most cases dismemberment consists of disarticulation; hence, cut marks are associated with points of articulation.

SKULL

I have observed only two types of cut marks that may be exclusively referred to the dismemberment stage of processing. The most common, as well as the most reliably referable to the dismemberment stage, are marks remaining from the removal of the head from the neck and from the removal of the mandible and/or tongue from the skull. Figure 4.11 illustrates the characteristic transverse cut marks on the ventral surface of the occipital condyles (S-1). Corresponding transverse cuts occur on the anterior ventral surface of the atlas vertebra (Figure 4.20).

In some cases the head seems to have been removed by acting on the articulation between the atlas and axis vertebrae. With this strategy marks may occur on the posterior ventral face of the atlas and on the anterior ventral face of the axis. This pattern is well illustrated by Frison for antelope (1971:264 Figure 3aa–dd) and is indicated by Henri Martin (1907–1910: Plate XLVII, No. 6) for the Mousterian site of La Quina. In my experience this method was employed by the Nunamiut on frozen carcasses or on animals that had remained in the field for some time between the time of death and the initial butchering and had therefore become very stiff. Under the latter conditions the most mobile joint is that between the atlas and axis, and "break-

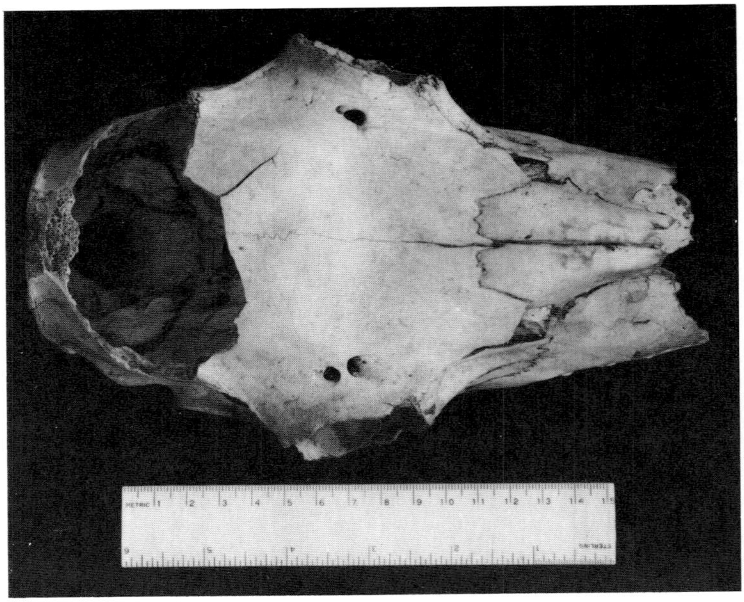

FIGURE 4.17. *Caribou skull showing the most common method of removing the antlers.*

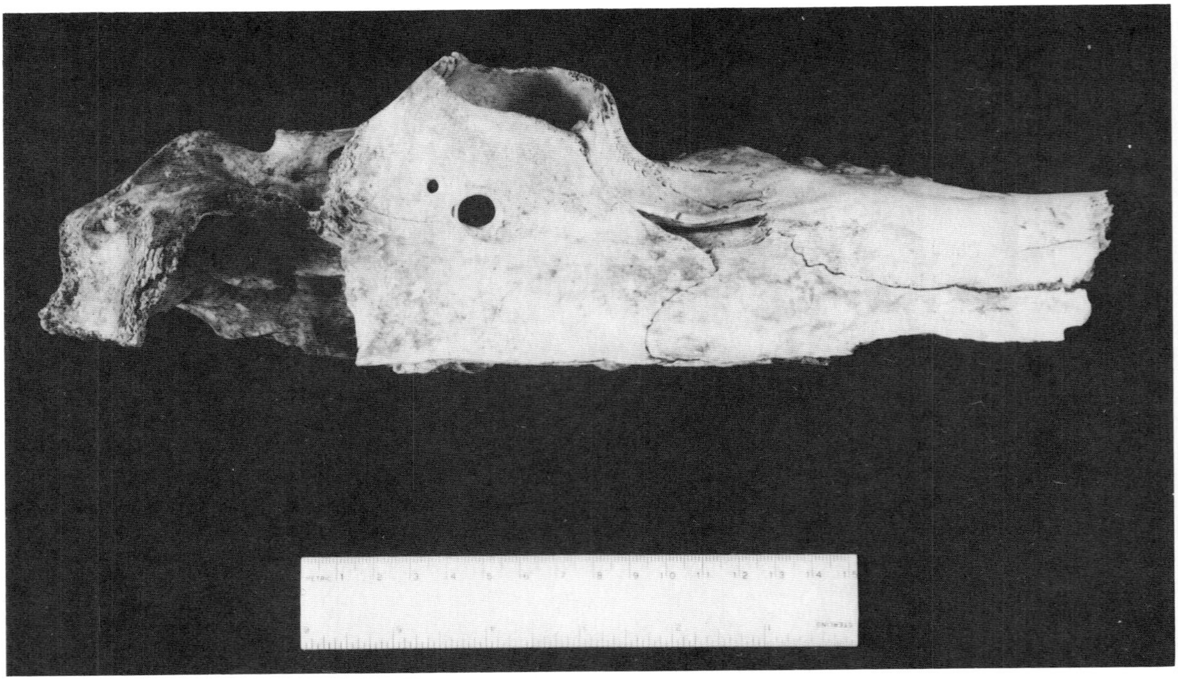

FIGURE 4.18. *Caribou skull cut in half and with the nose removed.*

ing the head loose" by pulling it around using the antlers as a lever is sure to be easier when the axis–atlas joint is penetrated.

On occasion, primary butchering may include the removal of the antlers, particularly if the head is to be returned to camp and limited means of transportation are available. This task can be rather formidable if one does not have axes, saws, or other heavy-duty tools. In the absence of such tools the cranium is usually broken around the base of the antlers and the antlers are removed. Breaking the skull case is much easier than attempting to cut through the antlers. Figure 4.17 illustrates the results of this strategy. *All* the antler bases recovered from the Mousterian site of Combe Grenal exhibited adhering segments of the skull, indicating that they had been removed this way. The Nunamiut frequently place heads in caches and segment them into parts appropriate to cooking on removal from the caches. For fall-killed caribou, the head is normally frozen when it is butchered, so the Eskimo use a saw to split it into two parts. The nose of a large bull might be cut off before the head is split and prepared independently and differently from the split halves of the head. Figure 4.18 shows a bull head sawed in half right down the medial line from front to back.

Since many heads are secondarily butchered when frozen, there is a distinctive mark left on the skull from a particular cut made during the removal of the mandible when the entire unit is frozen. Since the head with attached mandible is completely stiff, a deep and long cut is made from the insert of the masseter muscle along the upper lip area above the upper molars directly back across the ascending ramus of the mandible, severing the masseter muscle completely. Once this is done the mandible may be manipulated slightly and the task of removal is made much easier. Figure 4.19 shows a bull caribou skull with cut-off nose, and the distinctive deep cut above and parallel to the upper molars.

MANDIBLE

There are two sets of cut marks that may occur somewhat independently of one another. Strictly speaking, the disarticulation of the mandible from the skull is the only dismemberment act; however, removal of the tongue is commonly executed during initial butchering rather than during meat distribution or at the stage of preparation of parts for consumption. The larger the animal, or the more rigid the carcass has become, the more difficult it is to remove the mandible. Therefore, more cut marks are generally associated with larger animals or with animals of almost any size that have not been butchered immediately after death.

For instance, the moose jaw shown in Figure 4.14 is from the same animal shown during butchering in Figure 4.44. I observed in great detail the removal of the mandible from this animal as compared to analogous removals from caribou. What distinguished the removal of the moose mandible was the process of *filleting in situ*. The masseter muscle was cleaned off the mandible as well as stripped off its insertion points on the skull. This resulted in cut marks around the edges of the masseteric fossa and longitudinal cut marks completely across the fossa. In conjunction with these cuts was a diagonal cut on the maxilla just behind the third molar (see Figure 4.19 and Frison 1971:264, Figure 3ff). Once the major muscle had been removed, cuts were made against the inferior surface of the mandibular condyle. Then the mandible was mechanically manipulated and essentially pried loose from the skull with small cuts of connective tissue accompanying the pulling away of the mandible. This seems to be the same procedure described by Frison (1971:265, Figure 3, w–z ff) for butchered antelope from the Eden–Farson site, Wyoming: "The extent and depth of the cuts do not appear to have been necessary to cut through the jaw muscles alone [Frison 1971:263]." The seemingly "unnecessarily" deep cuts, the concentration of the cut marks at the axis–atlas articulation of the neck, and the large-animal strategy of removing the mandible lead me to suspect that primary butchering took place after rigor mortis was advanced and/or the animals were partly frozen.

When the tongue is removed while the mandible is still attached to the skull marks are usually inflicted on the medial margins of the mandible just below the third and fourth premolars. These are generally produced in conjunction with severing the mylohyoid muscle, which tends to free the lateral margins of the tongue so it may be cut free at its "root" (see Figure 4.08).

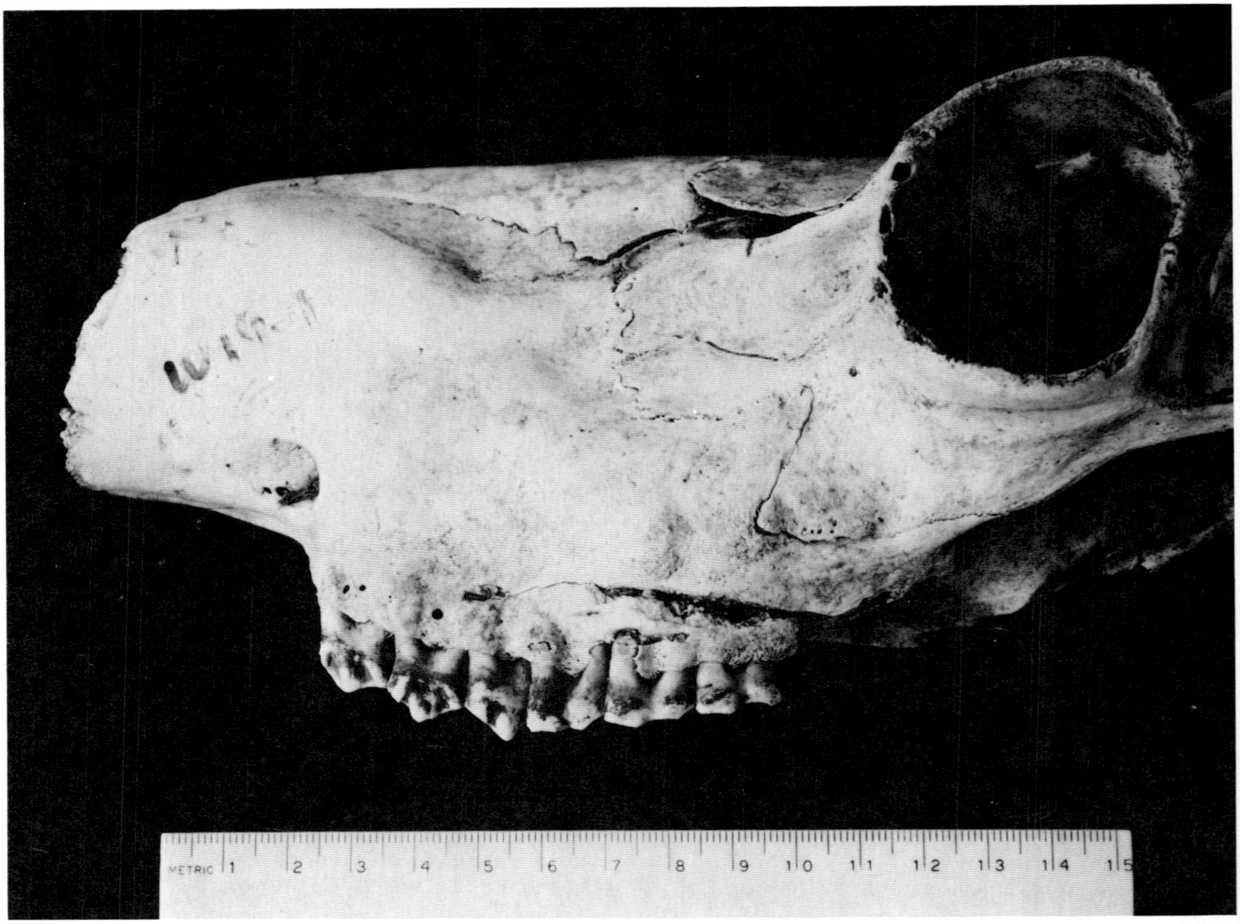

FIGURE 4.19. ***Caribou skull with nose removed and distinctive butchering mark above the molars.***

CERVICAL VERTEBRAE

Aside from the cut marks inflicted during the removal of the head from the neck (Figure 4.20), little dismemberment of the neck proper is conducted during the primary butchering of most animals. Some marks may be inflicted on the sixth cervical vertebra if the neck is butchered off from the spinal column (see Henri Martin 1907–1910: Plate XLVI, Nos. 6 and 7). When frozen carcasses are being butchered, an ax or other massive tool may be used resulting in chop marks sometimes at the axis vertebra and/or the sixth cervical vertebra. Most such marks are inflicted during secondary butchering accompanying meat distribution or during food preparation.

THORACIC VERTEBRAE

There are basically three operations that may result in marks on thoracic vertebrae (Figure 4.21) during primary butchering: segmenting the vertebrae, filleting the tenderloin (TV-2), and removing the rib

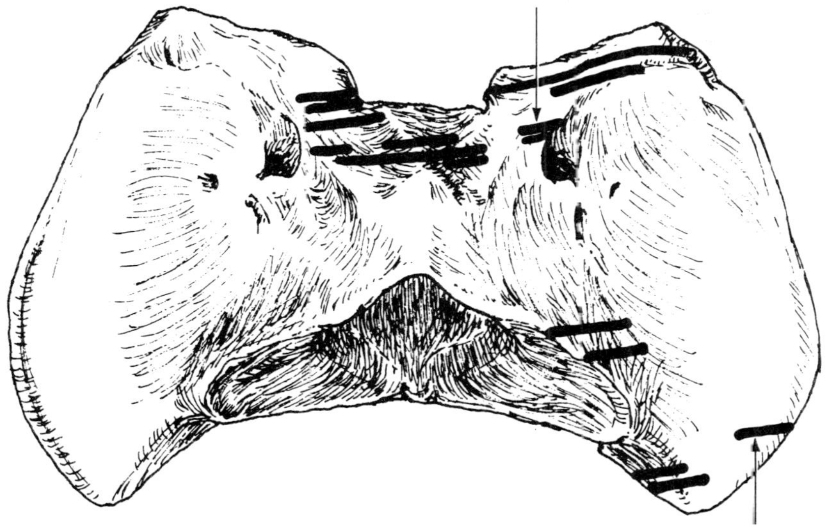

FIGURE 4.20. ***Cut marks on the atlas vertebra produced when severing the head from the neck.***

slabs from the spinal column (TV-5). With moderate-sized animals the segmentation of the vertebrae normally takes place between the second and third thoracic vertebrae and the thirteenth and fourteenth thoracic vertebrae (cut 18, Figure 4.05). In most cases this segmentation takes place after the removal of the tenderloin or eye roast. (See L. R. Binford [1979:49 Figure 2.1] for an illustration of the tenderloin being removed.) The stripping back of the tenderloin is made possible by long cuts made along the body of the muscle, freeing it from the dorsal spines (Figure 4.21, TV-2) of the thoracic and lumbar vertebrae, as well as from the superior surfaces of the ribs (Figure 4.05, cut 20) and their articulations with the vertebrae or the transverse processes of the lumbar vertebrae. Cut marks resulting from the latter operation are commonly oriented transversely or slightly obliquely to the dorsal spines of the thoracic vertebrae. These transverse cut marks are shown in Figure 4.21, as are a minor set of modifications that result from inserting the knife and inflicting marks on the dorsal spines between the vertebrae during the segmentation of the spinal column.

Arthur Spiess (1979:291) has suggested that this method of butchering is made possible by the use of metal knives with long (6-inch) blades. This opinion is based on an incorrect idea of the butchering process, since the roast is not cut loose from the rib attachments but is instead stripped along the sheath of sinew as I have illustrated (L. R. Binford 1978b: Figure 2.1). It is further based on an incorrect idea of the tools used by good field butchers. The Eskimo use the terms *young men's knives* and *old men's knives*. Young men's knives are considered nonfunctional for butchering and rather silly, being selected by young boys returning from boarding school who have been impressed by Euro-American culture in the "lower 48" but know nothing about butchering. These are the long-bladed knives Spiess imagines in use by the Nunamiut!

Given Spiess's opinion it is rather ironic that the cut marks originating during the course of removing the tenderloin are some of the most common marks recorded. I have observed them on thoracic vertebrae from Levels K, L, and 52 at Combe Grenal; Henri Martin (1907–1910) reports at least one specimen from La Quina. They are cited in almost all the reported assemblages where cut marks were noted from prehistoric bison hunters' sites on the American Plains (see Frison 1970:22, Figure 14; Frison *et al.*

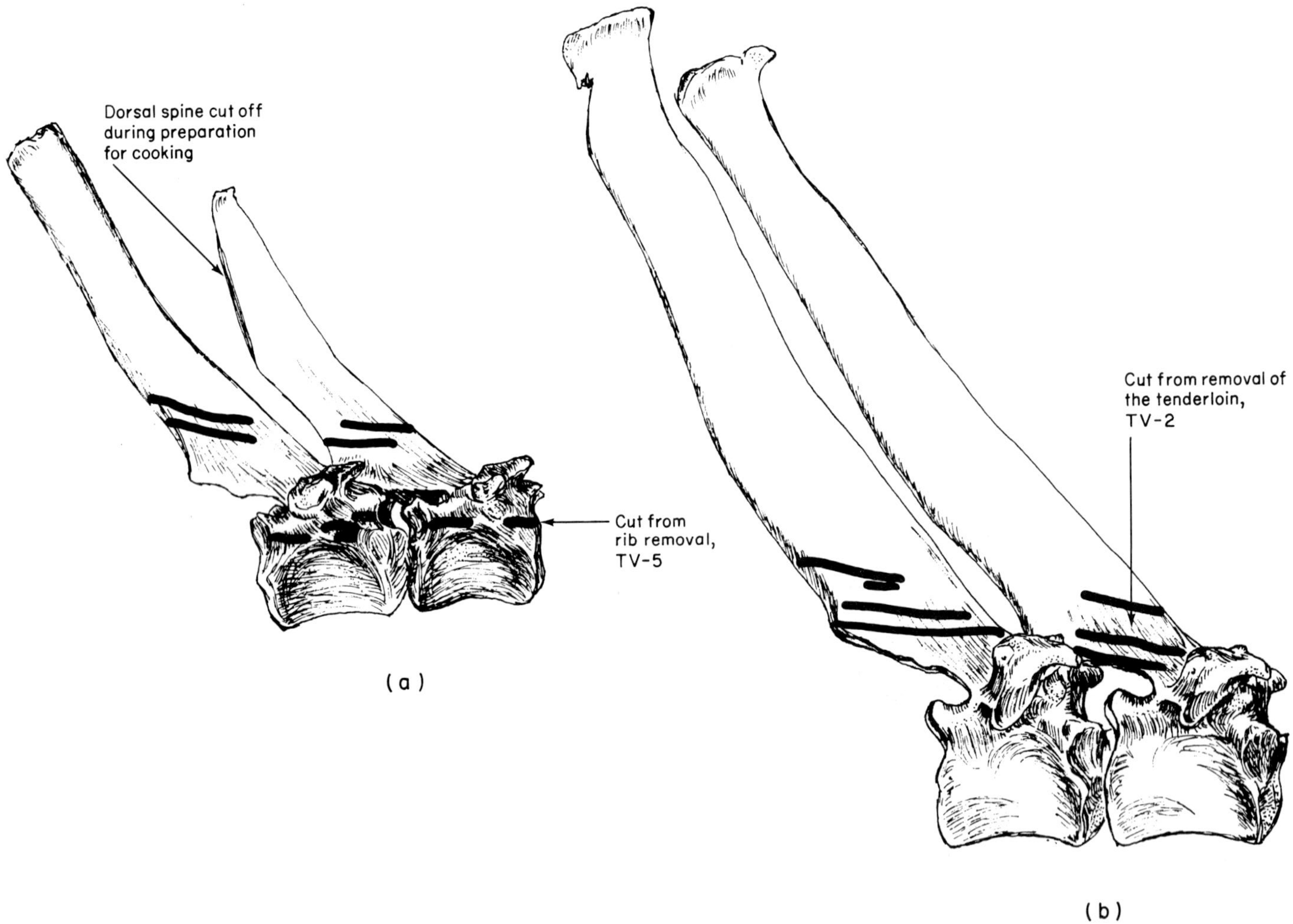

FIGURE 4.21. *Cut marks observed on the thoracic vertebrae: (a) Dall sheep, and (b) caribou.*

1976:53; Gilbert 1969:290; Wheat 1979:67, Figure 33g).

LUMBAR VERTEBRAE

As previously indicated, marks sometimes occur along the transverse processes or at the base of the dorsal spines. These marks derive from the removal of the tenderloin (see Wheat 1979:67, Figure 33h).

RIBS AND STERNUM

There are generally three locations for cut marks inflicted on the ribs and sternum during primary butchering. Transverse marks, derived from the removal of the tenderloin, occur along the dorsal surface of the rib just to the side of the proximal end of the rib (cut 20 in Figure 4.05). The second most common mark results from cutting off the distal end of a rib during the dismemberment of the sternum from the ribs. The last place where cut marks may occur is across the ventral surface of the rib, close to the proximal rib head or sometimes across the articulator head. This cut derives from the removal of a rib slab from the spinal column. Normally the method is to enter the rib cage by severing the distal ends of the ribs from the brisket. Attachments are relatively soft and cartilaginous in character and may be readily cut with even a dull instrument. A cut is then made down between the second and third ribs. For a detailed description of this butchering step see L. R. Binford (1978b:53, 54, 95). Next the slab of attached ribs is pulled up sharply, generally resulting in breaking off the rib heads against the vertebrae. Once the ribs are snapped back, the knife is run along the break severing tissue along the ventral surface. It is this action that results in characteristic transverse or slightly oblique marks on the ventral surface of ribs very close to the proximal articulator end. This is an ancient technique; Henri Martin (1907–1910:233, Plate LVIII, Nos. 10, 11, 12) described in detail this exact cut on many ribs recovered from the Mousterian site of La Quina. I observed this cut on 10% of the rib heads saved at Combe Grenal. It is interesting that this mark is not reported from the North American Plains sites.

PELVIS AND SACRUM

In my experience, most of the marks that regularly occur on the pelvis or the sacrum (Figure 4.22) derive from secondary butchering carried out during

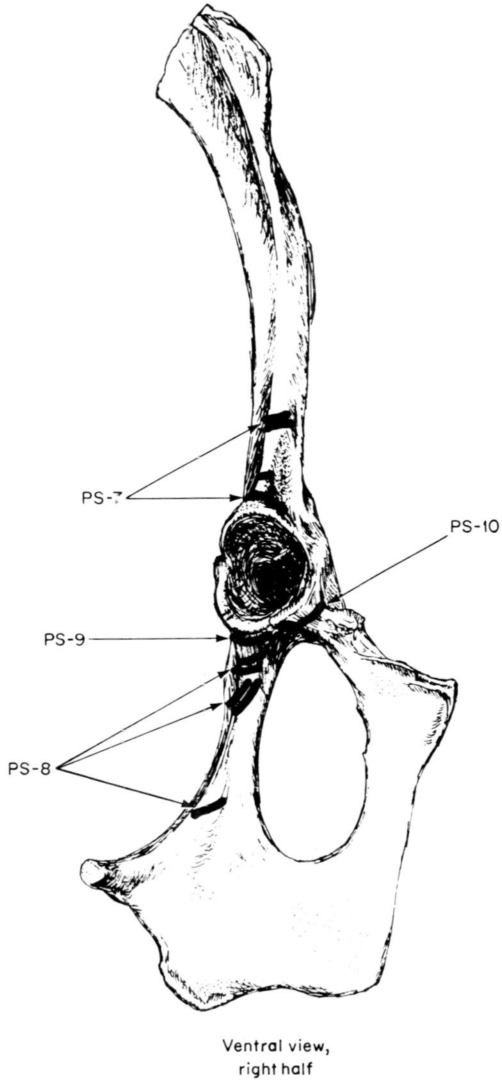

FIGURE 4.22. ***Cut marks characteristically produced on the pelvis during dismemberment.***

meat distribution or during the processing of parts for either storage or consumption. Marks inflicted during primary butchering are generally concentrated just anterior or posterior of the acetabulum socket on the ventral face of the pelvis. These marks are produced during the removal of the rear leg from the axial skeleton. The animal is commonly resting on its back or slightly turned to one side. The foot of the butcher is placed in the crotch of the animal and the leg is then pulled and twisted over against the foot, resulting in the dislocation of the articulation between the head of the femur and the ball socket of the pelvis (acetabulum). Once the leg is dislocated, the knife is used to cut obliquely into the flesh from the obvious fossa between the tissue of the abdominal wall and the muscles of the leg. The knife is run obliquely down and to the rear, frequently contacting bone, while oriented transversely to the ventral surface of the pelvis just anterior to the acetabulum (the move results in cut PS-7, Figure 4.22). An analogous cut commonly begins in the rear along the "point" produced by the extension of the ischial tuberosity and runs obliquely down and anteriorally over the "arm" of the ischium, impacting the bone transversely on the ischial body just below the margin of the acetabulum (cut PS-8) (See Figure 4.23.) These two initial cuts (Figure 4.24) may then be supplemented by tissue-severing cuts around the dislocated femoral–pelvic articulation, resulting in encircling marks around the acetabulum and/or on the head of the femur, or more commonly on the greater trochanter of the proximal femur. Sometimes the cuts may be made inside the acetabulum in order to sever the femoral ligament.

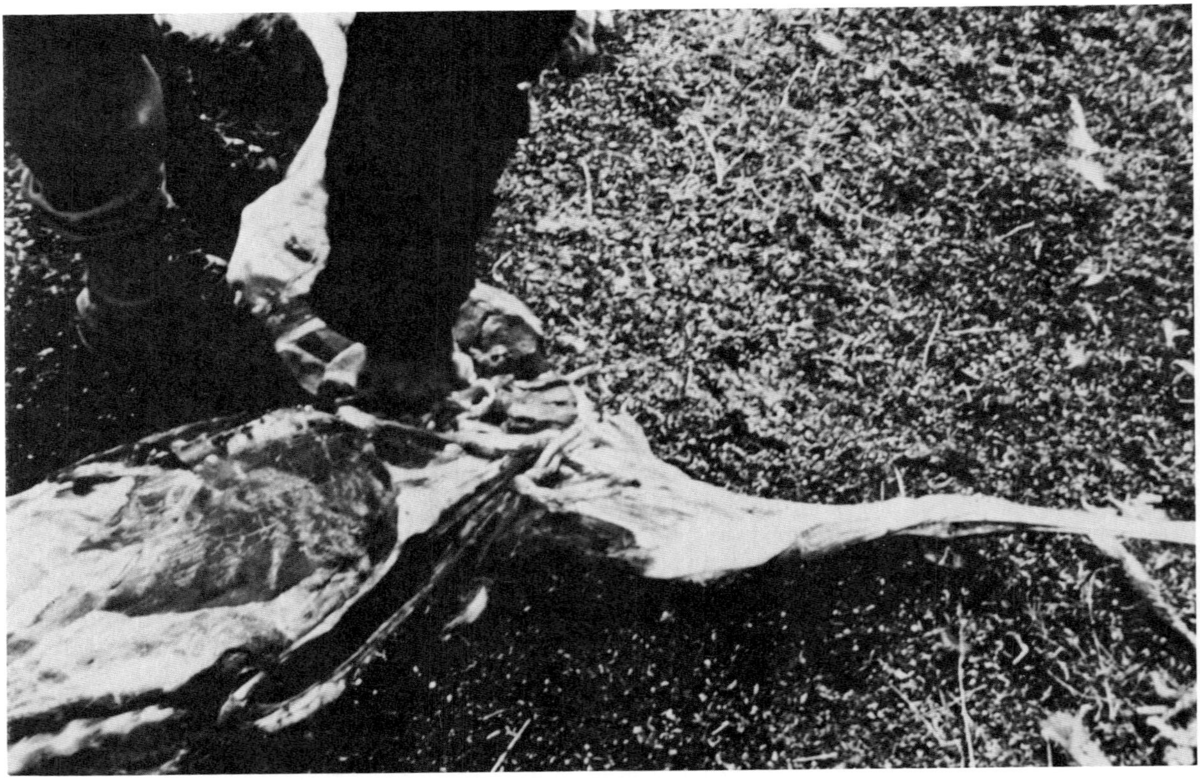

FIGURE 4.23. ***Eskimo butcher removing the rear leg of a caribou calf in the manner that leaves the marks shown in Figure 4.24.***

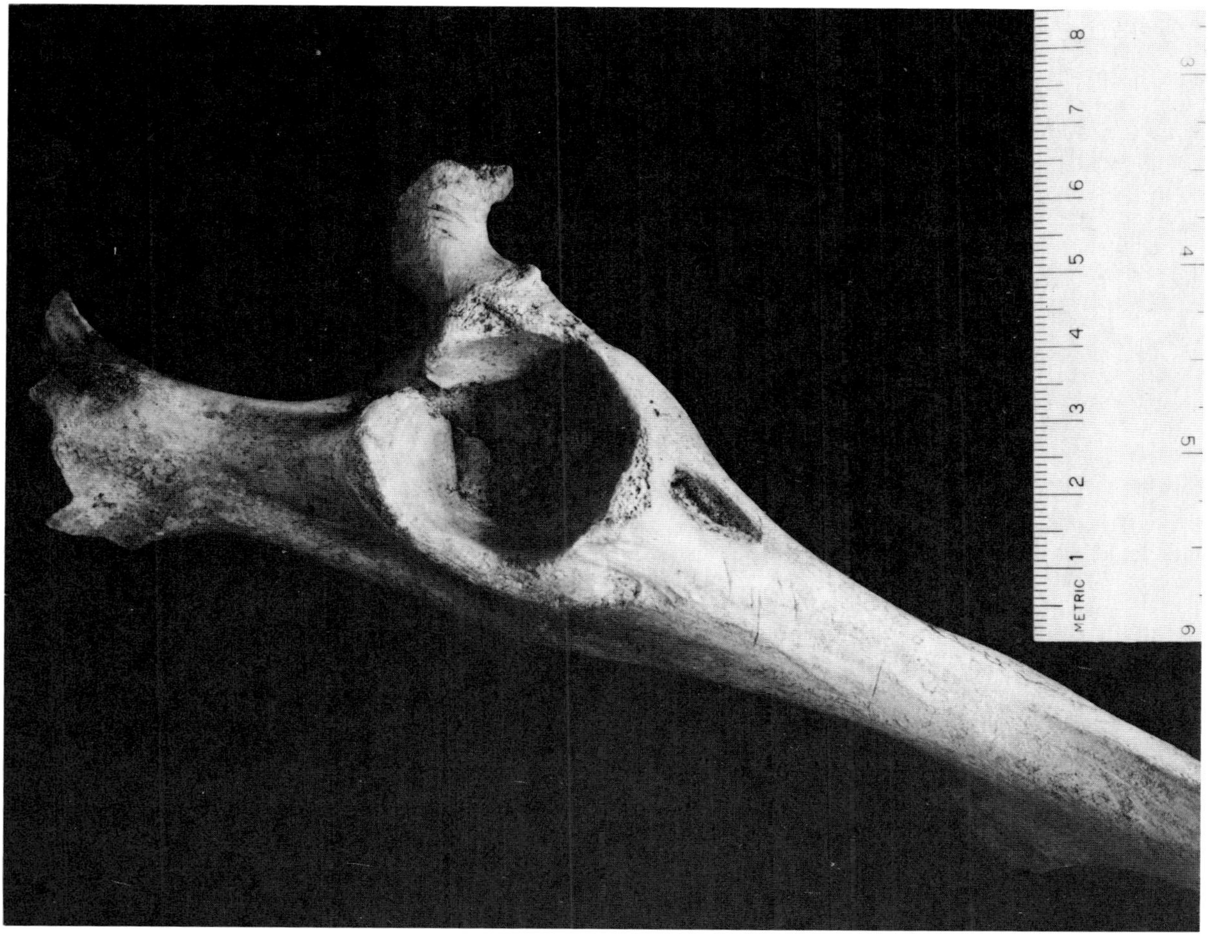

FIGURE 4.24. ***Cut marks on the pelvis produced while dismembering the rear leg from the pelvis.***

When the sacrum is disarticulated from the pelvis during primary butchering (most commonly done with large animals), there may be a trimming of the lateral margins of the sacrum, resulting in cut marks longitudinally down the iliac wings (PS-1 Table 4.04).

There has recently appeared in the Americanist literature considerable discussion of butchering procedures, the most influential of which is certainly the work of George Frison (1970, 1971, 1978). Frison has argued for a muscle-stripping procedure, in which choppers are used to free the insertions of muscles from the bone; the insertion is used as a handle to pull the body of the muscle from the animal. Little cutting with a knife is imagined when this proposed procedure is followed. The argument has been very widely accepted (see Agenbroad 1978; Johnson 1977, 1978; Wheat 1979) as a documented "fact" of North American Plains Indian behavior. The procedure as described by Frison involves the stripping of muscles from the rear leg and the chopping off of insertions on the pelvis prior to the disarticulation of the femur–pelvis articulation. If that was the case, there would be no need to make the cuts that, as I have indicated, would yield marks PS-7 and PS-8 as well as PS-9, since all these marks derive from severing the muscles of the rear leg during the act of disarticulating the head of the femur from the acetabulum. In this regard it is interesting that these marks are documented at a number of the sites where muscle stripping is said to have been the dominant method of butchering (Frison 1970:16, Figure 8a–b; Wheat 1979:66, Figure 32, lateral view). Frison comments on the lack of articulation between the pelvis and the femur at the Casper site as follows:

> With only two exceptions, the head of the femur was removed from the acetabulum. This is not too difficult with the flesh stripped away. The leg need only be given a sharp jerk using the hump formed by the chopped-off trochanter major as a fulcrum to lever the head of the femur from the acetabulum socket [Frison 1974:41].

In all fairness, the marks discussed previously are not identified at the Casper site. Nevertheless, Frison's model for accounting for the disarticulations of the femur is what he had in mind at the sites where the marks are present. Needless to say, I consider these marks conclusive evidence that muscle stripping as described by Frison was not being conducted. I will treat in more detail the evidence for muscle stripping in discussions to follow regarding chopping techniques and breakage patterns.

FEMUR

In many cases the removal of the rear quarter coupled with the removal of the lower limb is the only action taken during primary butchering. In all my experiences with the butchering of animals, even relatively large animals, the rear leg is dislocated from the ball socket at the acetabulum prior to the insertion of the knife and the severing of the connective tissue between the meat of the upper leg and the muscle attachments in the pelvic area. It is during the latter operation that marks may be inflicted on the neck of the femur (Figure 4.25a, b, cut Fp-1), on the ball of the femoral head (Fp-2), encircling the margin of the femoral head (Fp-3), as well as on both trochanters. Marks on the greater trochanter are more common (Fp-5).

Cut marks on the distal end of the femur are most often produced during secondary butchering actions. Generally, the proximal tibia and the distal femur remain articulated after initial field butchering and prior to meat distribution or the preparation of meat for consumption. Under some specific conditions, where there are either large body sizes or where multiple kills increase the bulk of field-butchered meat that must be transported, disarticulation of the femoral–tibial articulation may be accomplished during early butchering stages.

There are basically two general locations where cut marks occur on the distal end of the femur. There may be a mark oriented across the posterior surface just above the lateral and medial condyles (Figure 4.25d). A second and somewhat more common location is across the trochlea or the patellar surface on the anterior face (Figure 4.25e). These marks arise from the insertion of a knife behind the patella with a cut generally directed downward, so that the patella is removed with the proximal tibia instead of the distal femur as one might imagine.

Related to the disarticulation of this joint with metal knives as described earlier, a characteristic mark (Fd-3) is produced on the medial face of the lateral condyle (Figure 4.25f). This is generally produced when the point of a metal knife is inserted into the joint and turned around the intercondylar tubercles of the proximal tibia.

TIBIA

Circular parallel marks around the intercondylar tubercles (Figure 4.26d) of the proximal tibia are produced at the same time as the marks on the internal face of the distal femur (Fd-3). That is, a knife is inserted from the rear of the tibial–femoral joint and rotated around the intercondylar tubercles (Tp-1), simultaneously producing marks around the tubercles and on the internal face of the lateral epicondyle of the femur (Fd-3). This mark (Tp-1) is shown in Figure 4.26, b and d. It is most often produced when butchering is done with metal knives, but at least one example has been reported from a prehistoric North American site, the Jurgens site (Wheat 1979:56). This occurrence can be taken as clear evidence that disarticulation can be accomplished with stone tools by inserting the cutting implement between the articulator surfaces of the tibia and femur.

The only other cut mark regularly noted on the

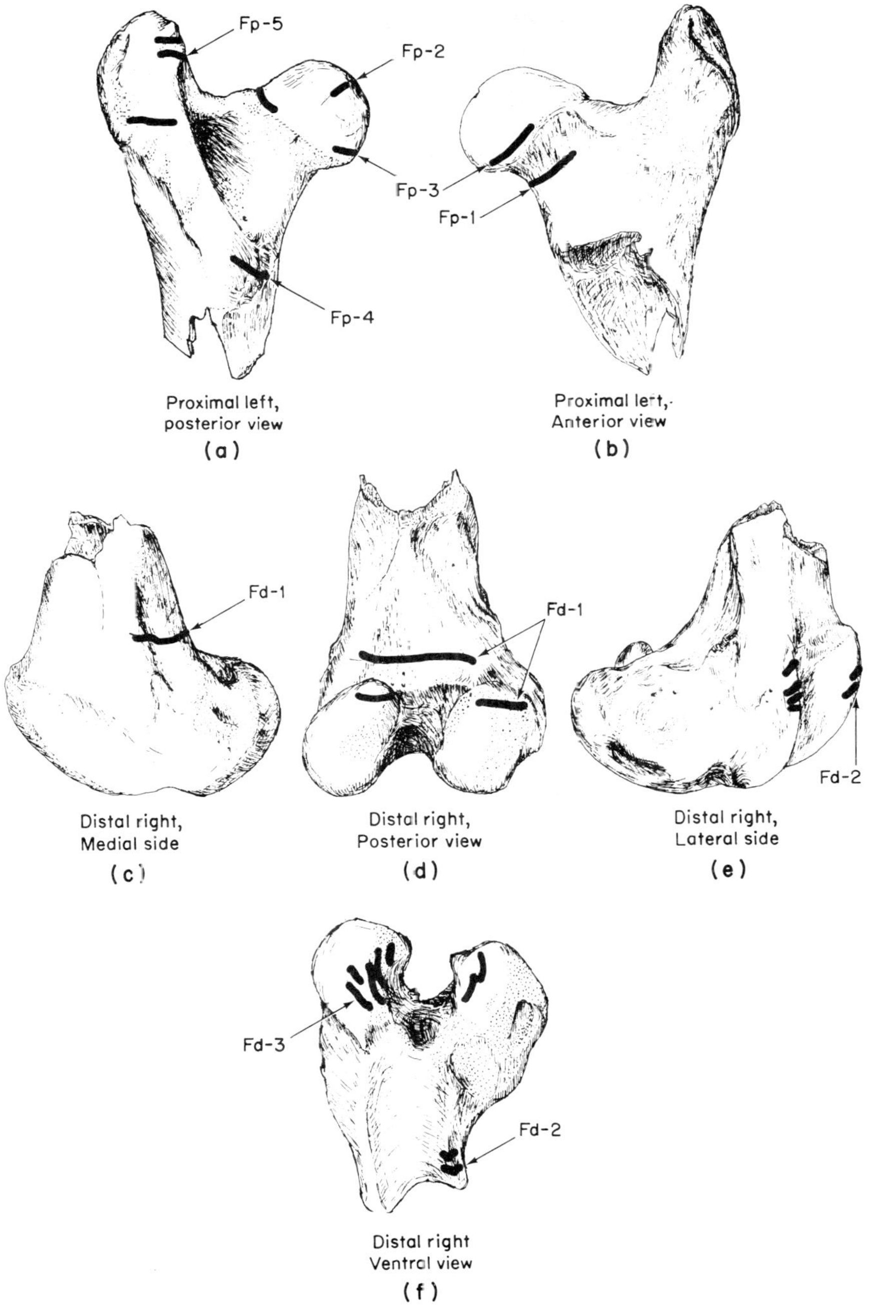

FIGURE 4.25. *Marks produced on the femur during dismemberment.*

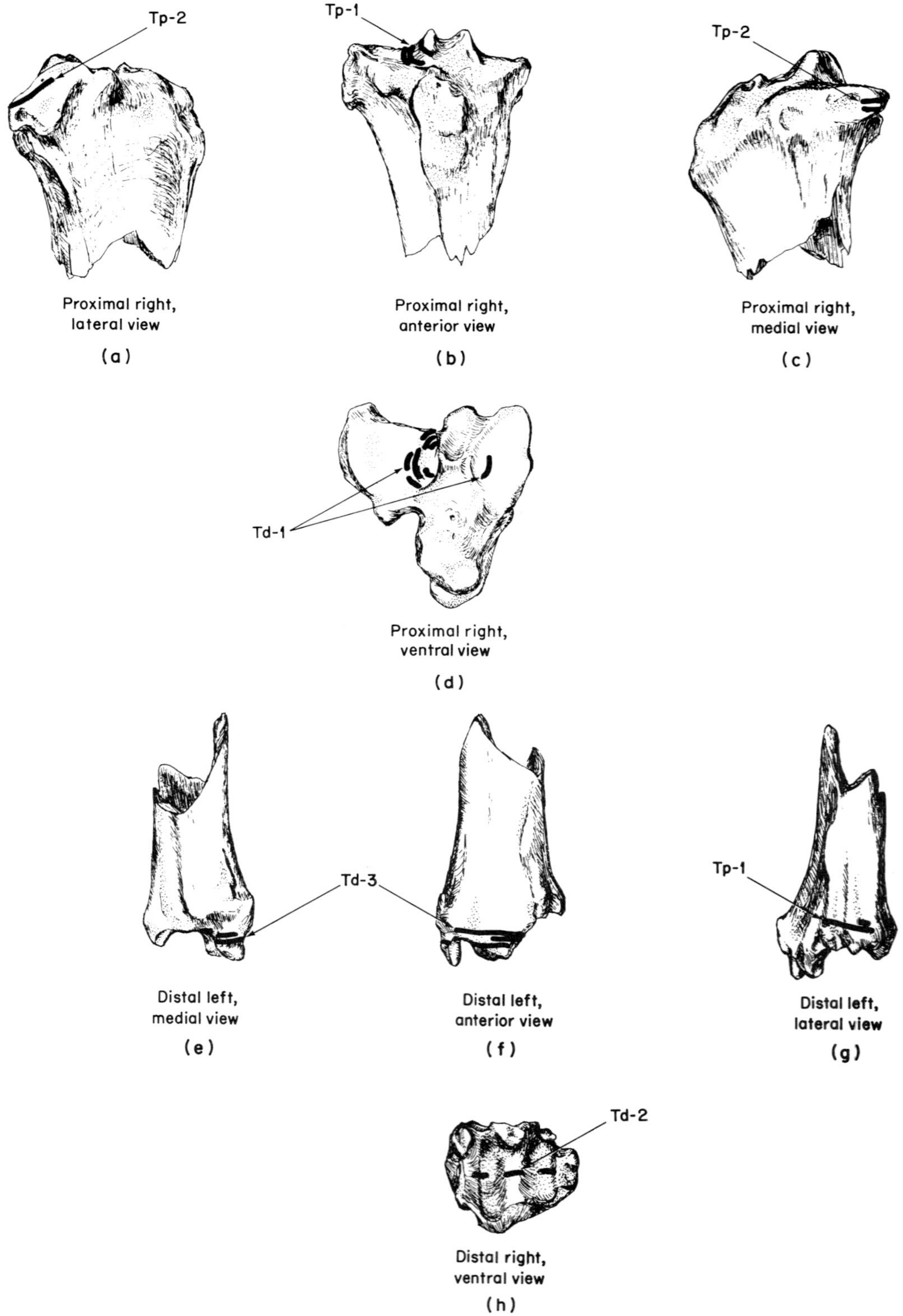

FIGURE 4.26. *Marks inflicted on the tibia during dismemberment.*

proximal tibia runs transversely across the posterior margin of both the lateral and medial condyles (Tp-2). This mark is observed relatively rarely and it may be considered as an alternative to the cut that runs across the condyles or just above them on the posterior face of the distal femur (Fd-1, Figure 4.25d).

The distal tibia appears to yield a very redundant pattern of inflicted marks regardless of time period or geographical location. The marks I will describe have been reported from every faunal assemblage I have consulted, from the Mousterian to the Historic period and from Europe as well as the New World. By far, the most common mark is illustrated in Figure 4.26 (e and f, cut Td-3). This mark is inflicted by cutting across the anterior face of the distal tibia when the leg is outstretched or straight. The result is a mark across the anterior face of the tibia (Td-3) that generally continues across the face of the lateral maleolus, finally intersecting the calcaneus (Figure 4.26, a and b, Tp-1). If this movement is somewhat lower, the marks may intersect the anterior face of the astragalus at essentially midpoint, as shown in Figure 4.27e. In such a situation the knife may "turn the corner" and mark the medial face of the astragalus (Figure 4.27f) at the same time marking the medial side of the distal tuberosity of the tibia (Figure 4.26e). Rarely there may be a prominent mark across the articulator surface of the distal tibia on the smooth prominent ridges that contact the astralagus. This results from a kind of swiping cut as the joint is disarticulated after the cuts just described have been made.

The dismemberment of the tibia–tarsal joint was one of the anatomical areas given special attention by Henri Martin in his seminal study of butchering marks. He supplies detailed descriptions of marks observed in the Mousterian fauna as well as accurate reconstructions of the dismemberment procedure (Henri Martin 1907–1910:251–274).

TARSALS

The foregoing description touches upon all the common marks observed on tarsals (Figure 4.27) except one, cut TC-3, which appears on the dorsal surface of the calcaneus just posterior to the point of articulation with the astragalus. This mark has been frequently interpreted as deriving from the cutting of the tendon attachment of the gastrocnemius muscle at the tuber calcis or the posterior end of the calcaneus. In my experiences with the Navajo, where the carcass is hung (see L. R. Binford and J. B. Bertram 1977:92–93), and with the Nunamiut where the completed upper rear leg is commonly hung on drying racks in cool seasons, another action is believed to produce cut marks on the dorsal ridge of the calcaneus. In both ethnographic cases the tissue between the shaft of the tibia and the attachment of the tendon at the tuber calcis is cut with a knife to facilitate inserting a rope or a gambrel for hanging the rear leg or the carcass for further butchering. This cut "for hanging" was observed to result in marks illustrated here as TC-3 (Figure 4.27, b and c). In light of this observation it is interesting that this mark is regularly reported on bison bones from the Jurgens site (Wheat 1979:66, Figure 32, medial view), and antelope bones from the Eden–Farson site (Frison 1971:284, Figure 3r).

An alternative method of disarticulating the tibia from the metatarsal is to make the cuts at the interface between the tarsals and the proximal metatarsal, which results in marks transversely oriented around the joint, scarring the medial surface across the internal cunieform or on the medial margin of the proximal metatarsal. On the anterior face, marks may occur on the navicular-cuboid (Figure 4.28, TNC-1) as well as the anterior face of the internal cunieform (TE-1) and the margin of the proximal metatarsal (MTp-1). Laterally, marks may occur on the proximal articulator margin of the proximal metatarsal and the lateral face of the navicular-cuboid. These marks may frequently occur *together* with those just described, since the removal of the metatarsal for marrow processing commonly involves the removal of the attached tarsals, as is shown in Figures 4.02 and 4.03. The combination of marks from both points of disarticulation is likely to be a good clue to at least some of the functions of the site, since the processing of marrow bones is normally associated with served meals rather than snacking in hunting camps and the like as is the case among Nunamiut. Similar differences in processing may well characterize other peoples as well.

METATARSALS

As has been indicated, encircling marks along the margins of the proximal end of the metatarsal can be

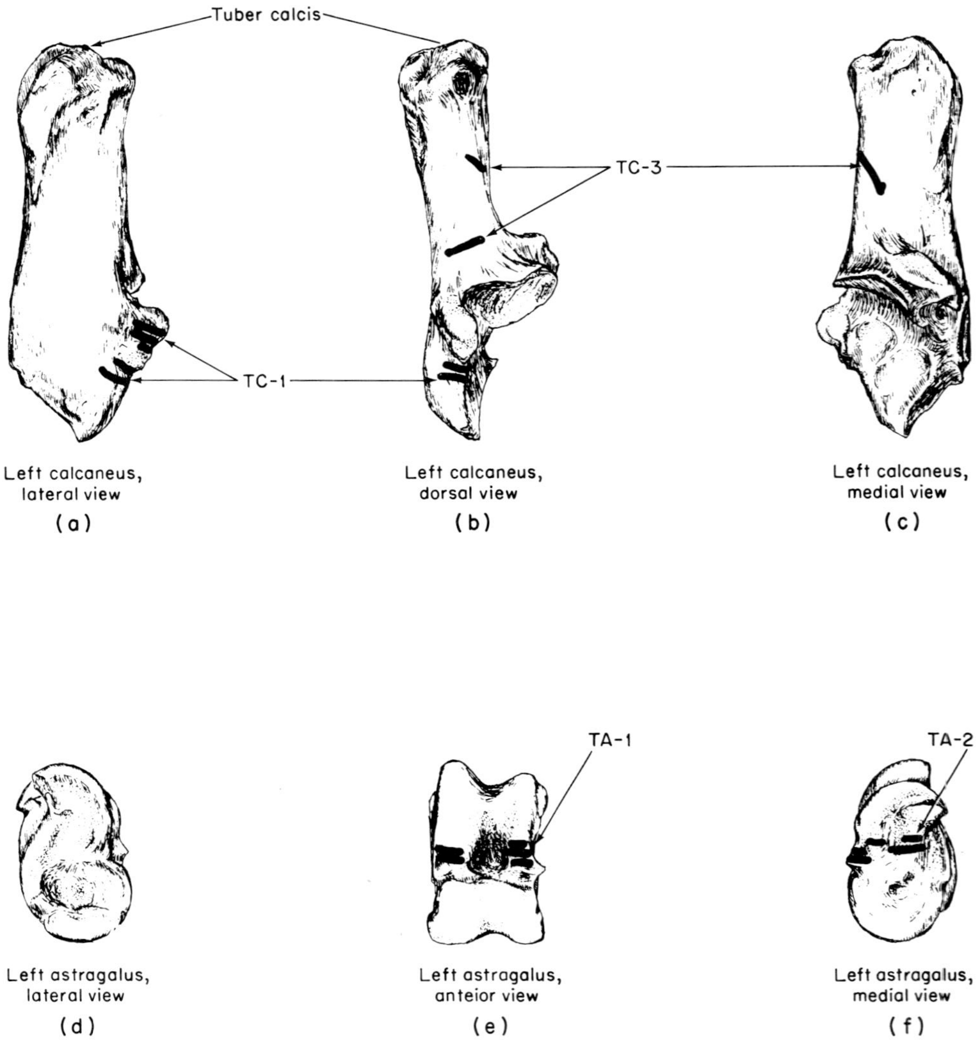

FIGURE 4.27 ***Marks produced on the tarsals and the metatarsal during dismemberment.***

expected on the lateral (MTp-3), anterior (MTp-1), and medial (MTp-2) faces. These may be produced during primary and or secondary butchering.

On the distal end of the metatarsal, marks relating to dismemberment can be expected across the ventral face of the condyle, as is shown in Figure 4.27. Some nicks may occur along the margins of the intercondylar channel (MTd-3). Such marks are produced when a knife is inserted directly into the joint (see Figure 4.12). It is my impression that this method of disarticulation is more common with metal tools and that marks in other locations can be expected if stone tools were being used. Most likely, cuts with stone tools should be concentrated on the anterior as well as the lateral and medial faces just above the epicondyles. In most cases such marks can

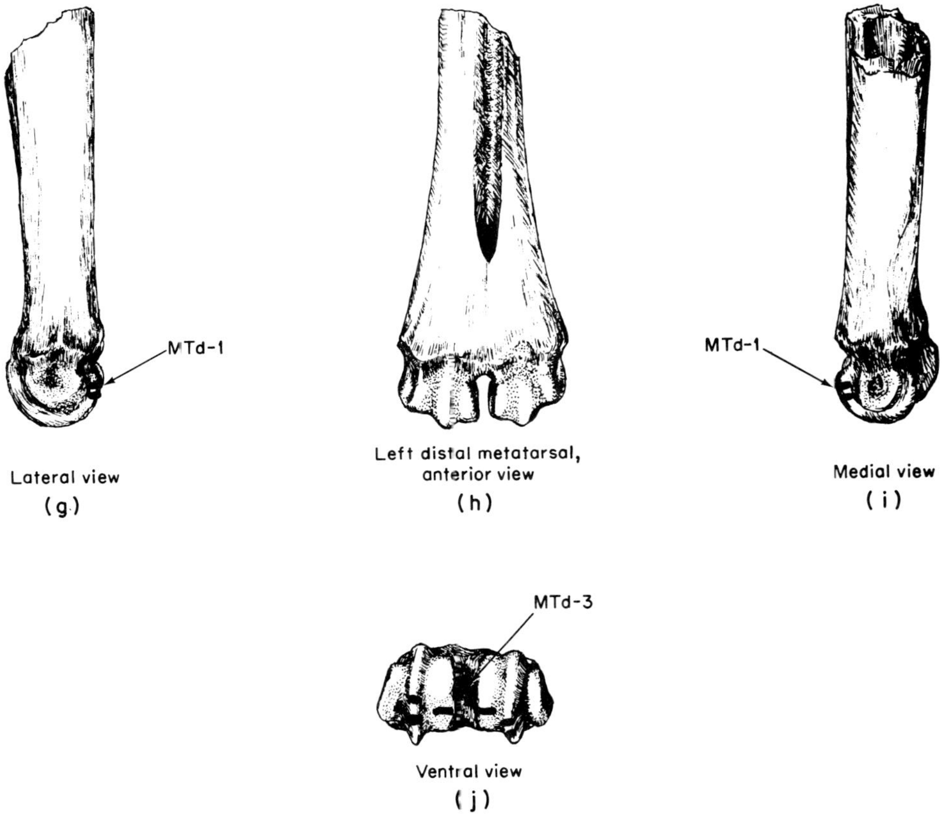

FIGURE 4.27. ***(continued)***

be expected to have been inflicted during secondary butchering and processing for marrow rather than during primary dismemberment.

SCAPULA

Primary butchering traces on the scapula (Figure 4.29) seem to be restricted to marks that encircle the glenoid cavity. There is a tendency for the marks to be on the lateral face and to exhibit some concentration at the origin of the triceps brachia. Analogous cuts may occur up the scapula around the neck. Dismemberment at this joint is most commonly a secondary butchering operation, unless the animal is quite large. This means that considerable numbers of inflicted marks on the scapula in the area around the glenoid cavity are most likely to be seen in locations of consumption, unless there is processing for drying, or in situations where relatively large animals are being butchered and the parts are destined for transport rather than processing.

HUMERUS

As in the case of cut marks on the scapula, few marks on the proximal humerus (Figure 4.30) can be related to initial butchering. When marks do occur they are concentrated in basically two locations and occasionally in a third. Most common are short marks on the lower "lip" of the condyle (Hp-1), as seen in the posterior aspect. The next most frequent is a mark on the apex of the lateral tuberosity com-

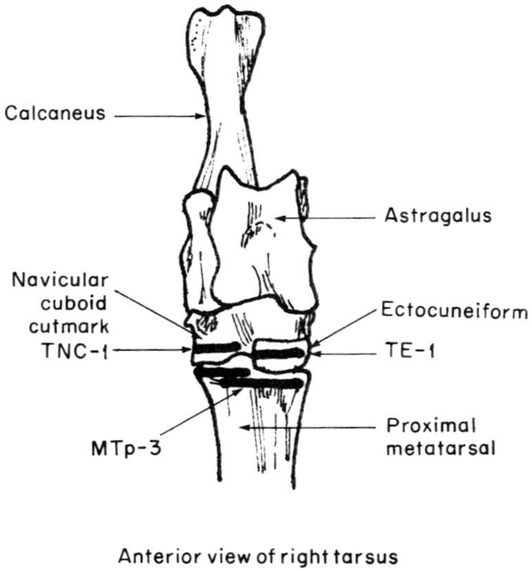

FIGURE 4.28. ***The articulation of the tarsals with the metatarsal.***

monly located on the posterior face. The marks on the lateral tuberosity and the marks on the scapula at the anterior margin of the glenoid cavity are believed to be made at the same time by a simple cut originating on the anterior side of the joint with the joint articulated in a fairly straight fashion. A transverse cut is simply made into the joint from the front, resulting in the knife impacting bone on the lateral tuberosity and along the anterior lip of the margin to the glenoid cavity. After the joint is cut into from the front, an analogous cut is made from the rear, resulting in marks on the lip of the humeral condyle (Hp-1) as well as along the posterior margin of the head of the scapula. These cuts can be made very quickly. The dismemberment of this joint is relatively easy, since the condyle is large and the joint is rather flexible. Leverage is used to great advantage in separating the scapula from the proximal humerus. As will be described later, this joint is commonly separated from the carcass prior to the removal of the scapula when large animals are being butchered. This strategy permits the use of the scapula as an anchor and provides leverage for dislocating the joint. Arguments about the difficulty of disjointing the scapula from the humerus seem odd to me, as do proposals about the use of choppers. Destroying the proximal end of the humerus with choppers seems a strange waste of time and effort when two simple cuts and a little leverage are all that are needed to disjoint the bones in question.

If the proximal humerus exhibits relatively few cut marks, probably because of the ease with which

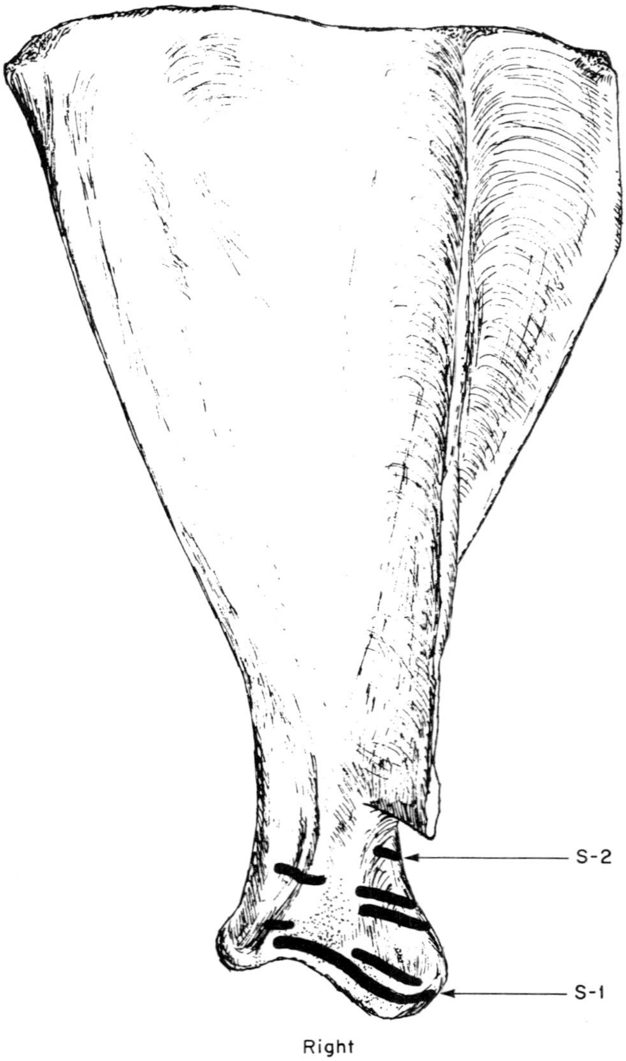

FIGURE 4.29. ***Marks produced on the scapula during dismemberment.***

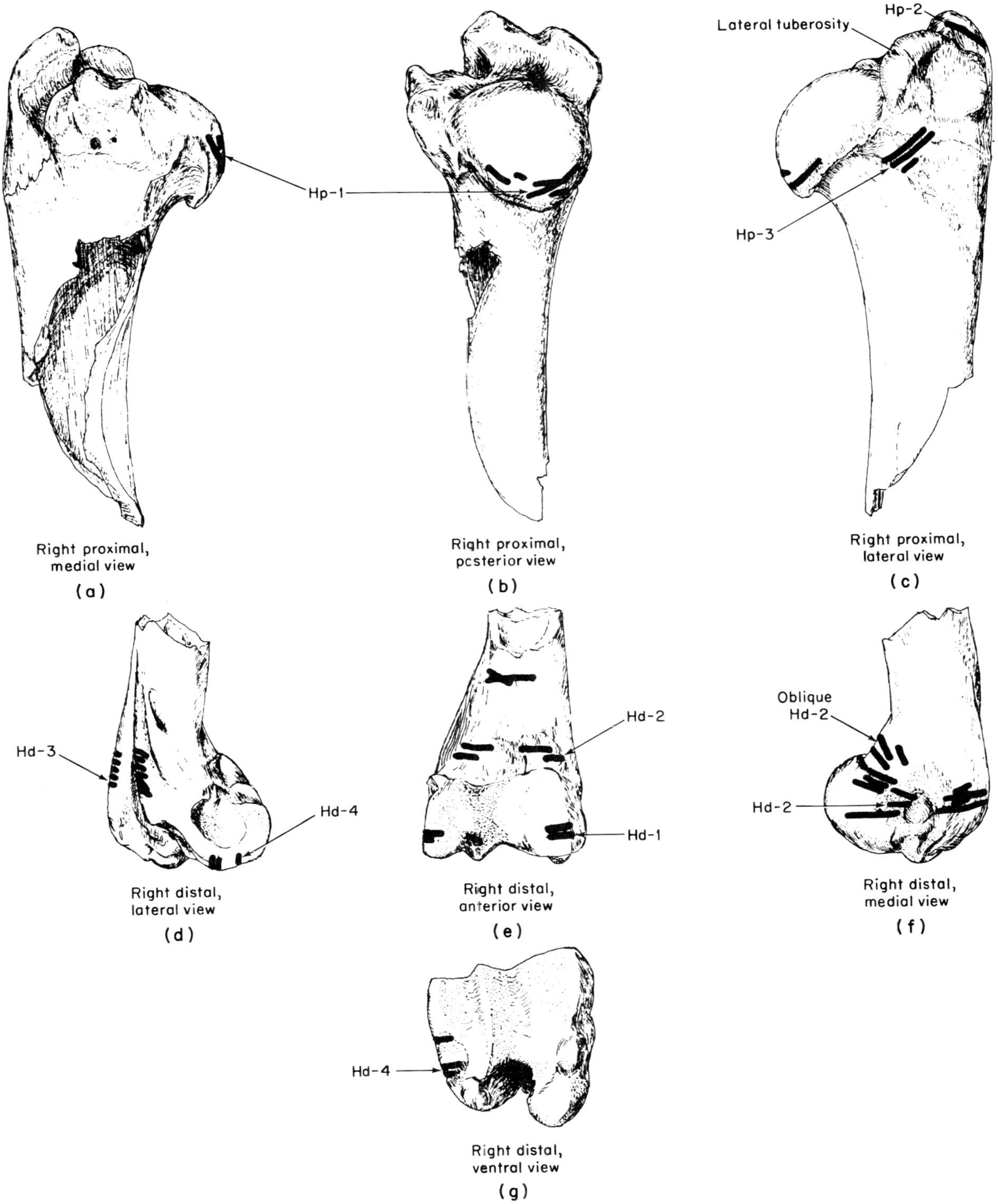

FIGURE 4.30. ***Marks produced on the humerus during dismemberment.***

the joint may be disarticulated, the distal humerus generally sports a consistent and numerous collection of cut marks. The most common mark is on or across the medial face of the distal humerus (Hd-2). The second most common mark is a short mark on the prominent ridges of the anterior face of the articulator surface (Hd-1). These two cut marks are believed to result from the same cutting motion, which also produces the marks along the margin of the radius (Figure 4.32, RCRp-5). The knife is run transversely into the joint and turned inward across the medial face of the distal humerus, sometimes even penetrating to the medial face of the olecranon (RCCp-3). This is generally done with the leg fairly well extended or straight. The second cut commonly employed in dislocating the humerus from the radiocubitus is a diagonal cut made beside the lateral face of the distal humerus, impacting bone on the face of the olecranon just beside the semilunar notch (RCCp-2). This cut makes possible a twisting of the articulation inward toward the body and may also be accompanied by slight twisting of the lower leg or at least the shaft of the radio-cubitus. This has the effect of "rolling" the articulation out of the socket obliquely toward the body. Once this level of dislocation is achieved it is a simple matter to cut the remaining strings of connective tissue and separate the bones. This technique assumes that the body is warm and the joints are flexible. When the body is stiff and/or frozen, a slightly different method may be employed. The first task, making the joint flexible, is accomplished by cutting down from the rear between the olecranon and posterior surface of the distal tibia. This cut is frequently associated with an attempt to flex the joint as much as possible at the same time the cuts are being made, resulting in a series of short cut marks along the dorsal crest of the olecranon and along the margins of the olecranon fossa, which is, of course, on the posterior face of the distal humerus (see Figure 4.31). Once these cuts are made and the joint is flexed, butchering may proceed as before. However, I have observed butchers following this procedure and there is something of a tendency to make the cuts across the anterior face just above the articulator condyles of the distal humerus rather than below them, as is the case with warm animals. When this is done the leg is still flexed, and if the "corner is turned" the knife runs across the medial face of the distal condyle, resulting in obliquely

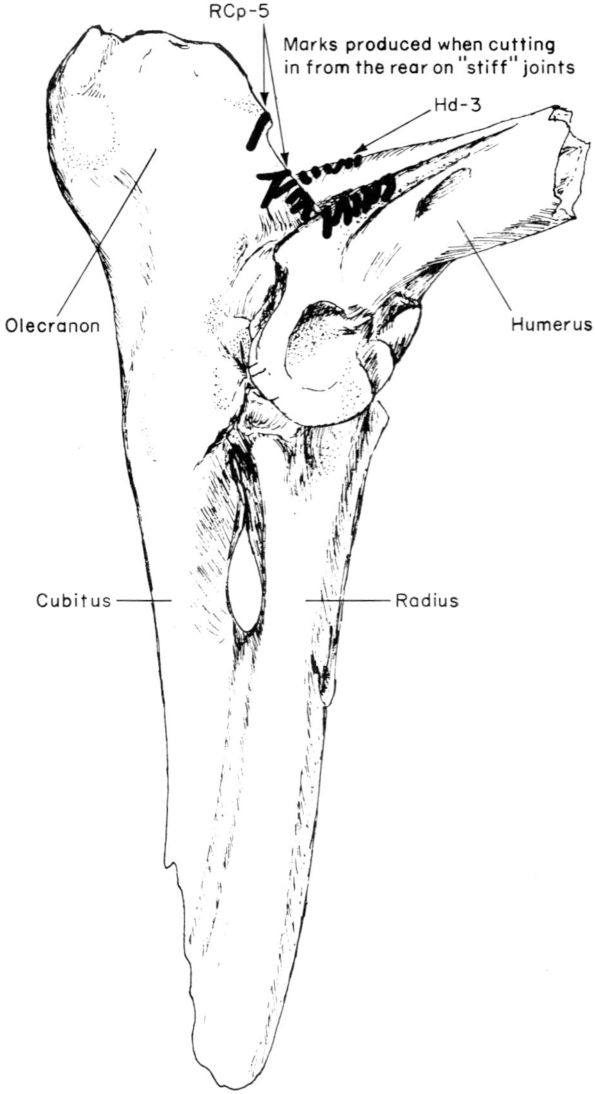

FIGURE 4.31. ***The articulation between the distal humerus and the proximal radio-cubitus, showing marks produced during dismemberment.***

oriented cut marks (see Figure 4.30f). These oblique marks may be somewhat higher up on the condyle, almost at the "neck." Such a placement of marks, coupled with marks on the dorsal ridge of the olecranon and along the margins of the olecranon fossa,

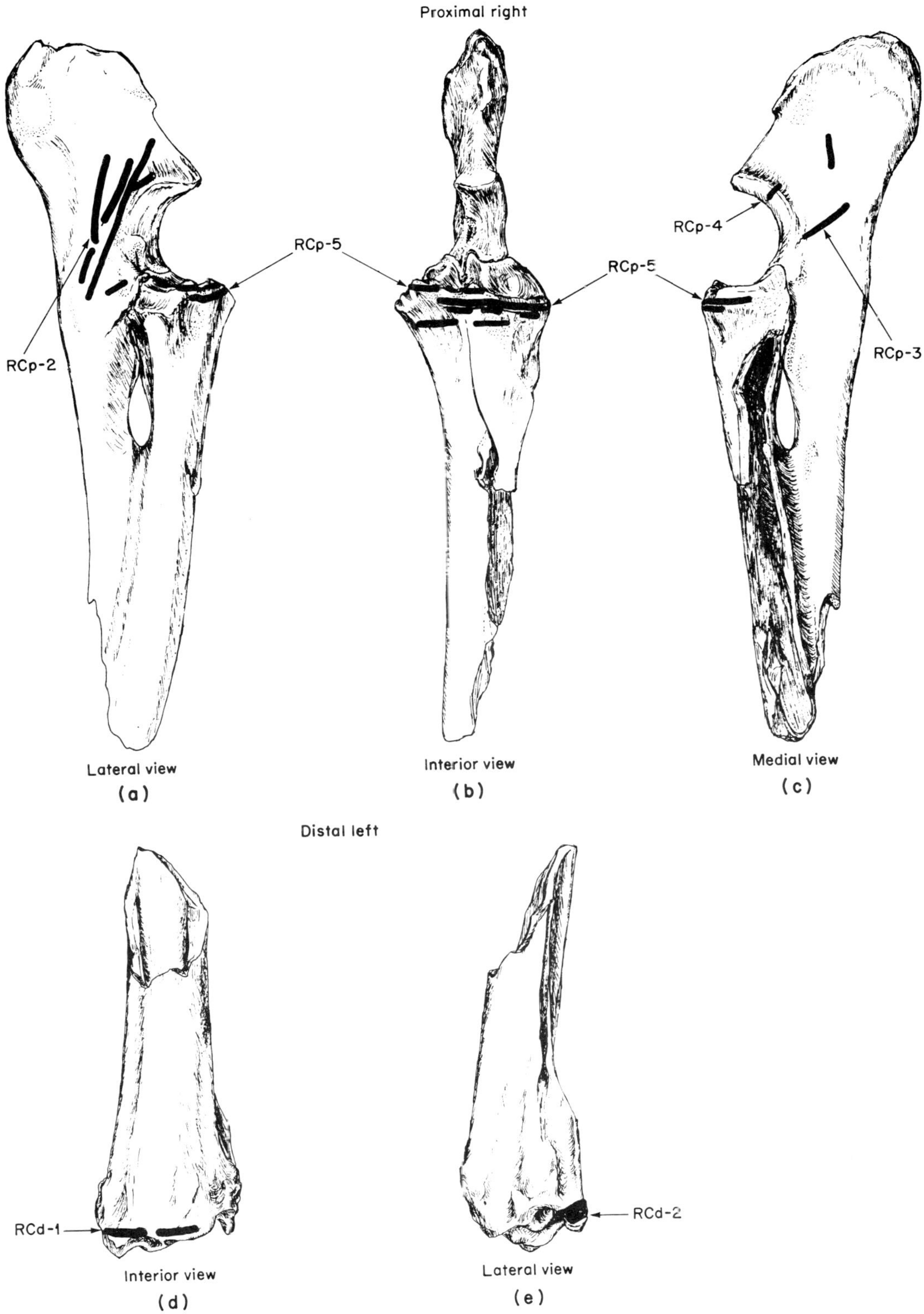

FIGURE 4.32. *Marks produced on the radio-cubitus during dismemberment.*

betrays the butchering of this joint when it is flexed, which is most common when the animal is stiff. Stiffness may be related to the dismemberment being part of secondary butchering activities, or to scheduling problems at mass kills. In this regard it is interesting that the marks illustrated from the Eden–Farson site (Frison 1971:264, Figure 3a–d) all exhibit rather oblique cut marks high up near the neck of the distal humerus. Earlier in this chapter I pointed to other evidence that seems to indicate the butchering of stiff or frozen animals at that site. It is equally interesting that the "stiff joint" pattern of butchery seems indicated at the Jurgens site (Wheat 1979:64, and Figure 31). I would guess that secondary butchering took place, with disjointing occurring after meat was filleted from essentially complete front legs. Disjointing was most likely related to marrow procurement after the joints were both stiff and dry.

RADIO-CUBITUS

Most of the cut marks occurring on the proximal end of the radio-cubitus (Figure 4.32) have been discussed in the course of considering the distal humerus.

The distal radio-cubitus frequently exhibits marks across its anterior face, right along the edge of the articulator surface. Alternatively, marks may occur only on the styloid process, indicating that the cut was made lower down and along the lateral face, impacting the carpals rather than the edge of the distal radio-cubitus. The latter marks are more common. On the inferior face of the articulator surface there may be transverse marks on the prominent ridges. These derive from the insertion of the knife into the joint after it is partially dislocated.

CARPALS

Cut marks may occur on any of the exposed faces of the various carpal bones but are much more common along lateral and medial faces, with far fewer occurring on the anterior and posterior faces. George Frison (1970:12) describes the sources of marks on the bison carpals from the Glenrock site, a very good description of butchering of this joint: "Of the ulnar carpals and radial carpals, 43% of the former and 39% of the latter bear cut marks. This is still a common method of butchering in the field and after cutting in this manner . . . the foot may be snapped off [Frison 1970:12]." What Frison is describing is a simple pair of cuts, one across the articulation on the inside marking the medial of "radial carpals," and another across the lateral face marking the "ulnar carpals" and not infrequently the styloid process of the distal radio-cubitus. Once such cuts are made the joint is "snapped" in half by pushing in or pulling out on the joint, as is illustrated in Figure 4.13.

METACARPAL

Essentially the same marks noted for the metatarsal can be expected on the metacarpal.

PHALANGES

I have rarely observed marks on the phalanges. This experience is held in common with most of the other North American researchers observing the properties of cut marks. On the other hand, Henri Martin (1907–1910) reported numerous cut phalanges from the Mousterian site of La Quina. I have suggested earlier that these cuts derive primarily from skinning when pains are being taken to skin out the foot in great detail. This is done among the modern Eskimo only for obtaining skins from which socks and shoes are manufactured. This is almost exclusively a fall and early winter activity among the Nunamiut Eskimo.

Filleting Marks

Strictly speaking, the marks previously described for the thoracic and lumbar vertebrae associated with the removal of the tenderloin are derived from filleting (Figure 4.21). One might also argue that the removal of the tongue (Figure 4.08) is an act of filleting. Both of these actions are commonly conducted as part of primary butchering activities. Filleting for transport as illustrated by Binford (1978b:229) and described among the Dobe !Kung (Yellen 1977a, 279), and as practiced among groups carrying out mass kills for purposes of generating large quantities of storage meat (as was probably the situation at the

Olsen–Chubbuck site—Wheat 1972) are all instances of primary butchering activities. We might think of primary butchering activities as those that are timed or scheduled with respect to the killing activities, whereas secondary butchering and/or processing activities are timed or scheduled with respect to achieving the goals of placing meat into storage and/or distributing and preparing it for consumption. Secondary butchering normally consists of further segmentation of the parts generated at the time of primary butchering, and/or filleting primarily appendicular parts. I have never heard or seen documented filleting activities conducted with regard to any axial skeletal parts except the lower lumbar vertebrae and pelvis. Thus we can generally expect an overlay of marks deriving from filleting to be restricted to the lumbar vertebrae, the pelvis, and the leg bones. Even in those cases where filleting is conducted at locations of primary butchering, as at a kill site, it is a *second-stage* activity; animals are field butchered into basic anatomical segments, which are then filleted. This means that the character of the segments conditions to some extent the degree to which filleting marks will overlay or, alternatively, be the only marks remaining from butchering on certain parts.

I have observed numerous acts of filleting carried out by the Nunamiut Eskimo and the first impression is that it is done very quickly and with seemingly little effort. For instance, Figure 4.33 shows the bones of the rear leg lying beside the meat removed during filleting. This filleting was done by butcher in the field so rapidly that I did not have time to adjust my camera and take a picture before he

FIGURE 4.33. ***Filleted femur lying beside the meat removed from the rear leg.***

finished making his cuts and removed the fillets. The situation illustrated is rather common—the butcher is aware that filleting will be the next step, and no transport problems exist (e.g., filleting is to be done in the same place). Segmenting of the legs is rarely done during primary butchering; that is, the lower legs are rarely removed separately. Instead the entire leg is removed and it is then filleted and disposed of, a complete or nearly complete leg remaining largely articulated. This situation is well illustrated in Figure 4.34, where articulated front legs minus the scapula (see L. R. Binford [1978b:100] for a discussion of the treatment of the scapula by the Nunamiut) are shown just after having been filleted. The consequences of this behavior at processing locations is frequently a pile of nearly complete articulated legs, as is shown in Figure 4.35. Depicted here are the parts that were filleted for drying from 11 caribou killed earlier (see L. R. Binford [1978b:223–228] for a description of this episode). I am strongly suggesting that there will be diagnostic patterns of articulation, patterns of part association, and spatial features that will covary with distinctive patterns of inflicted marks when filleting is a major or dominant activity at a site. This is an argument to be developed at another time, since my major concern here is to describe the marks that derive from filleting.

The first fact to emphasize is that filleting marks are almost exclusively longitudinally oriented with respect to the bones on which they appear. The very act of filleting dictates this pattern, since the butcher is essentially removing the bone from the mass of meat around it and this is best achieved by cutting

FIGURE 4.34. *Front legs shown shortly after completion of filleting.*

FIGURE 4.35. ***Mass of legs remaining at a processing location after filleting.***

along the bone and rolling back the meat so the bone can be pulled free from the meat. There are generally two types of cuts: (*a*) initial long, longitudinally oriented, bone-exposing cuts, and (*b*) shorter, more oblique cuts made to the underside of the exposed bone to free it from the mass of meat and/or sever muscle insertions. Given such a strategy, it is not surprising to see longitudinally oriented cut marks and shorter, more obliquely oriented marks along the posterior or anterior surfaces of long-bone diaphyses. Good examples of long, longitudinally oriented filleting marks are seen on the scapula in Figure 4.06. Similar marks are shown on the pelvis illustrated in Figure 4.36. The filleting marks indicated in Figure 4.36 were observed on known filleted specimens collected among the Nunamiut.

Filleting marks on bones of the leg are much shorter and are rarely exclusively longitudinal, as is shown in Figures 4.06 and 4.36. Long, longitudinally oriented cuts are made but these tend to be made along the sides of the bone relative to the orientation of the part during filleting operations. Most commonly, the legs rest flat during filleting operations, either on the lateral or medial "sides" relative to the living animal. This means that cuts will appear along the anterior and posterior faces of the bone when viewed in anatomical orientation. Knives are run along the "edge" of the bone and do not, strictly speaking, produce cut marks; irregular longitudinal shallow scratches may be produced, but are relatively rare. Much more common are short cut marks frequently made obliquely to the longitudinal aspect of the bone and concentrated on both the anterior and posterior faces. Such marks will be clus-

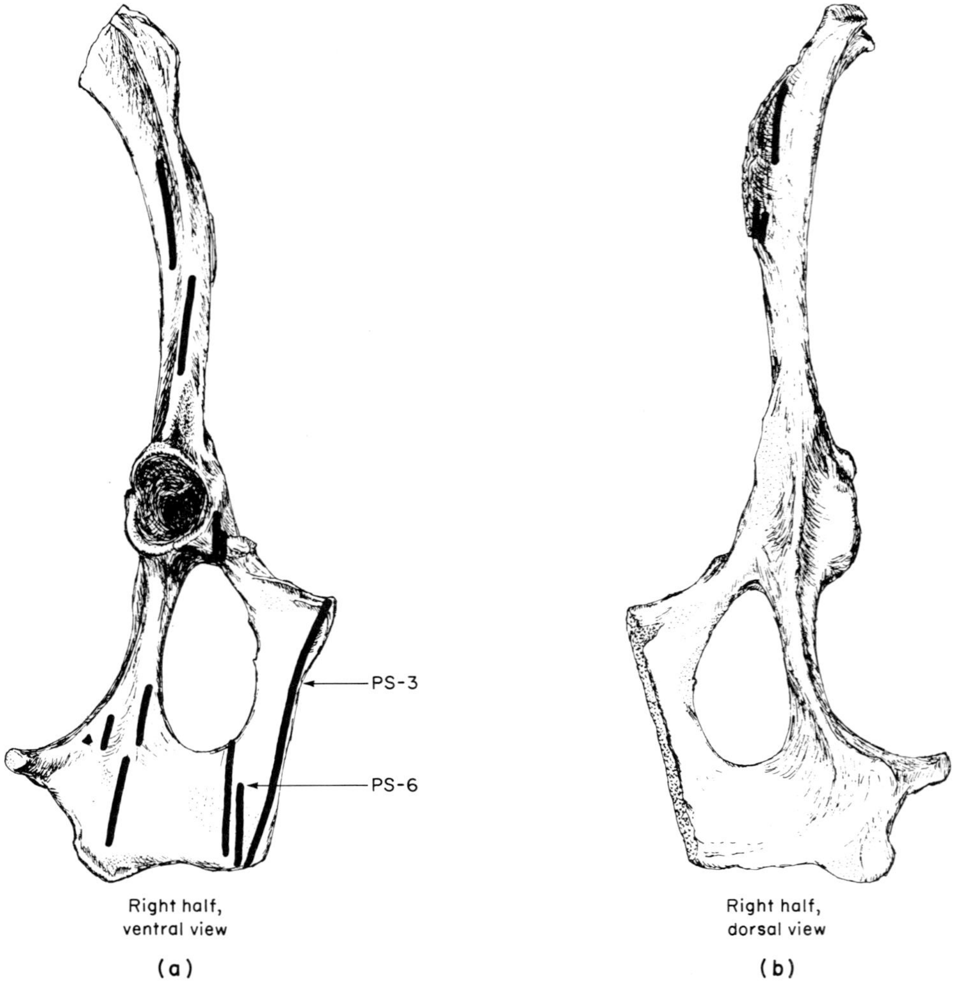

FIGURE 4.36. ***Marks produced on the pelvis during filleting.***

tered where the shape of the bone is irregular and where there are numerous muscle insertions. Irregularity in shape is most common in the immediate area of the articulator ends, hence we can expect a number of short cut marks in the area of the "neck" between the epiphysis proper and the linear shaft of the diaphysis. As anyone who has carved meat knows, it is where bones are irregular in shape that there is the most problem for the carver. This is equally true for filleting, hence we can expect many marks concentrated in fossae or around indentations in the bone such as the area along the lateral side of the tibial crest. Figure 4.37 illustrates marks known to have been produced on the proximal ends of upper rear leg bones during filleting by Nunamiut Eskimo. It should be clear that the marks are (*a*) generally oblique, (*b*) generally located on the "neck" of the bone, and (*c*) commonly in recessed places where one would have to cut the meat out rather than strip it behind a smooth cut.

Figure 4.38 illustrates the distal ends of rear leg bones with marks derived from filleting operations. A very diagnostic pattern, one I call *short chevrons*, is clearly indicated. These are paired clusters of short parallel cuts obliquely oriented and originating from opposite sides of the bone. These are generally more common on the distal ends of bones where there are muscle insertions, as in the case of the distal tibia and particularly the distal radio-cubitus. They are less common on the bones of the upper leg such as the humerus and femur. The bone that tends to exhibit most such cuts is the radio-cubitus, and as is clear this is the bone where the articulation with the humerus is characterized by the most irregular and crevice-like areas, making filleting very difficult. Figure 4.39 shows the location of marks inflicted on front leg bones during known situations of meat filleting; the pattern is generally the same on all the bones. The marks tend to be very superficial, rarely deep; on freshly filleted bones they are only visible as cuts through the periosteum that did not penetrate the bone. Unfortunately, I did not return to

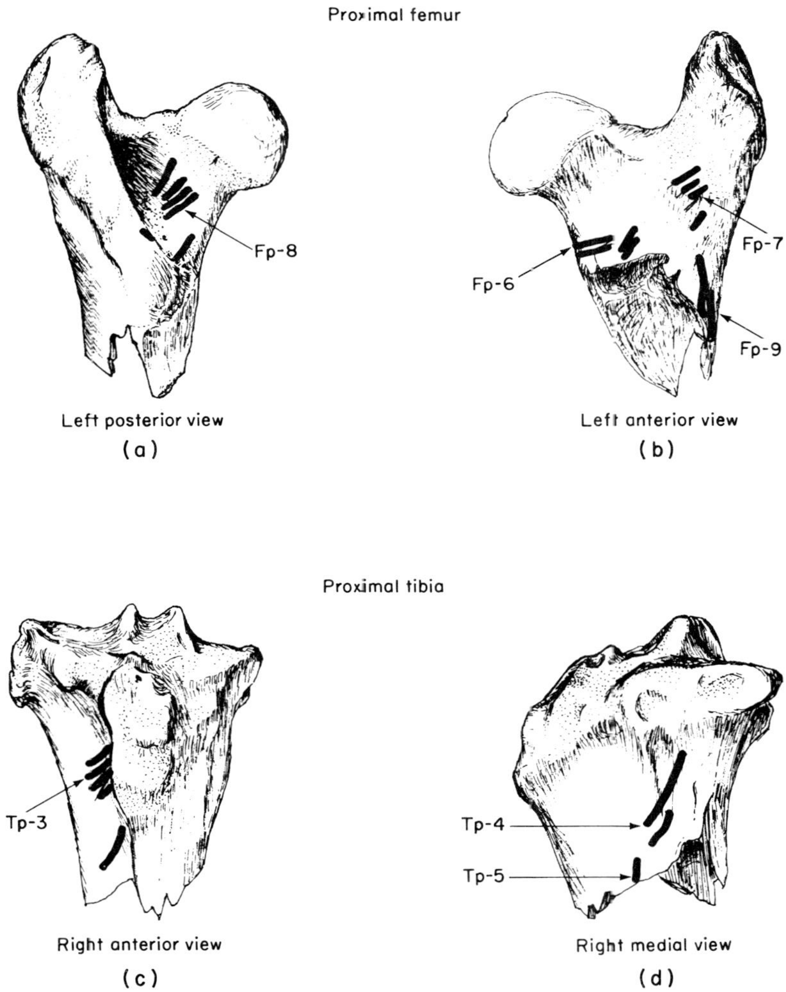

FIGURE 4.37. ***Marks produced on the proximal ends of upper rear leg bones during filleting.***

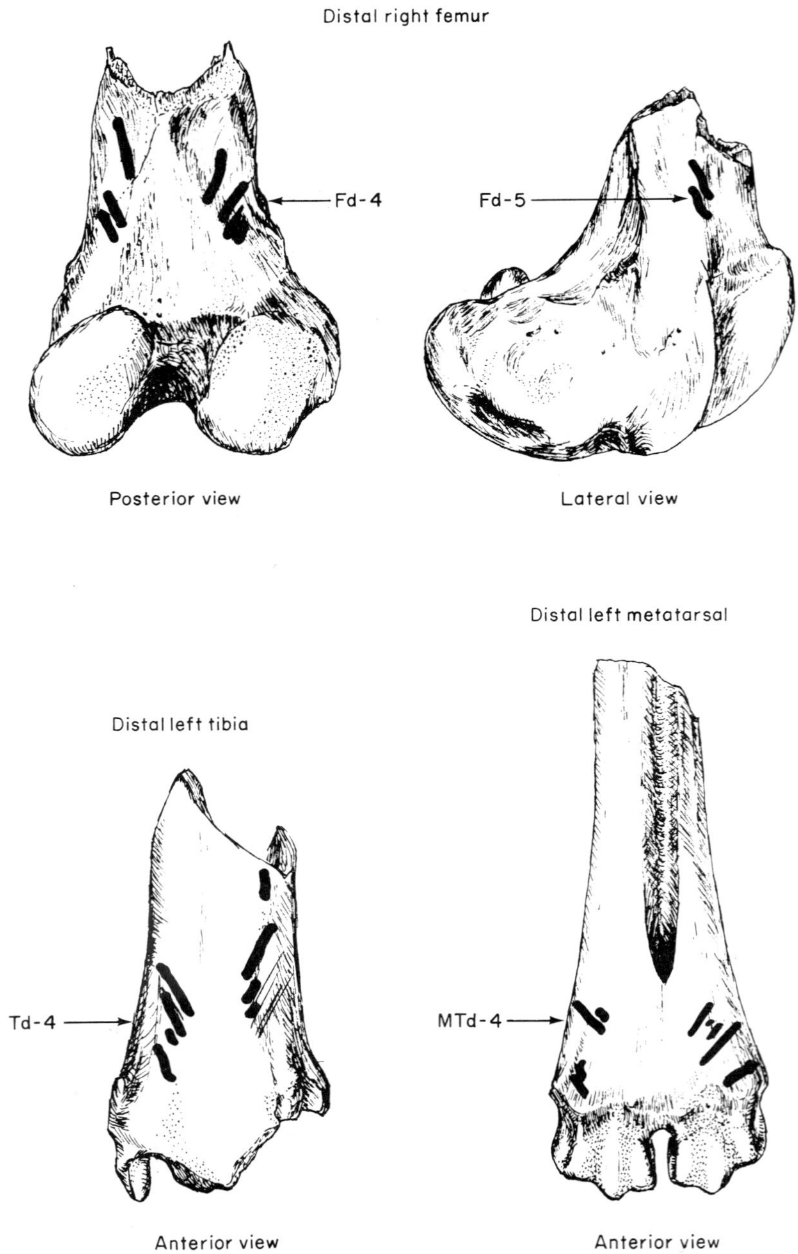

FIGURE 4.38. ***Marks produced on the distal ends of rear leg bones during filleting.***

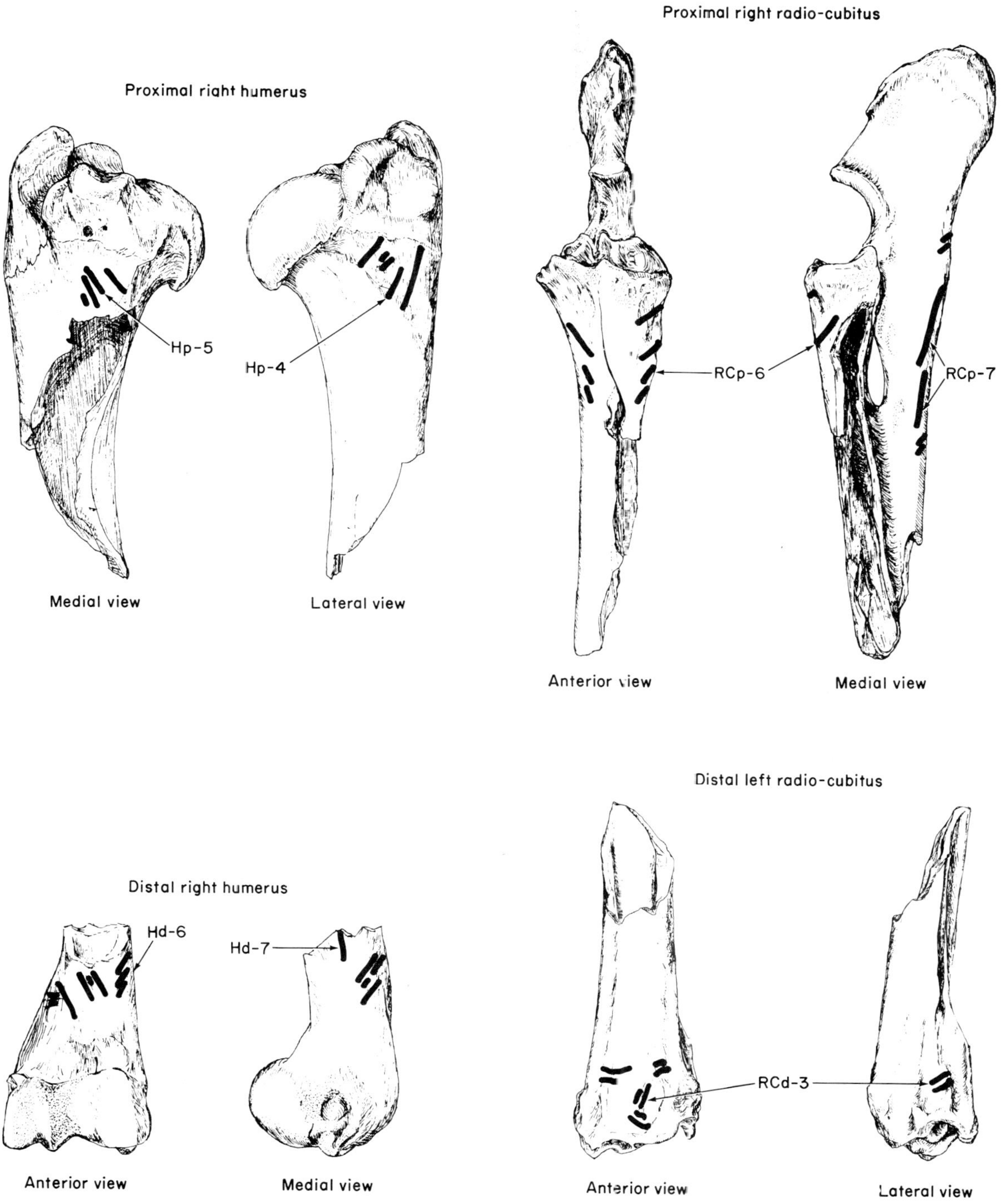

FIGURE 4.39. *Marks produced on front leg bones during filleting.*

Albuquerque a total assemblage of bones from a known filleting operation, so I cannot provide reliable information on the frequency of such marks. However, it was my impression that they were produced on a relatively small number of the bones actually filleted and that the frequency increased as the limbs being filleted were superficially dried, making the "thin" ends such as the distal radio-cubitus and the distal tibia tougher and more difficult to penetrate for starting the filleting operation.

The literature is relatively silent on these types of marks, and when they are described they are rarely recognized for their specific significance. Rather, they are most often lumped with dismemberment marks as "butchering marks." An exception in this regard was Henri Martin (1907–1910), who called them "marks of evisceration" and correctly identified some of the short oblique "chevrons" occurring on long-bone fragments as being produced during the removal of the meat (see Martin 1907–1910: Plate LX, No. 2, and Plate Nos. 2, 3, and 6). No such filleting marks were reported from the Eschelman site (Guilday *et al.* 1962). Similarly, Frison (1970) cited no such marks from the Glenrock site, nor were they noted (Frison 1971) at the Eden–Farson site. I observed only one example in the Mousterian materials from Combe Grenal—a distal radio-cubitus from Level K (Quina Mousterian). On the other hand, these marks have been described from the Jurgens site (Wheat 1979: Figures 31, 32), where they were lumped with marks of dismemberment, breakage, and other characteristics that the author thought related to butchering.

Marks Produced during the Preparation of Parts for Consumption

One of the most common methods of preparing parts for consumption is in fact filleting, particularly when the meat is to be boiled or used in a stew. This introduces some ambiguity into the significance of filleting marks. This ambiguity is further exacerbated by the fact that when leg bones are prepared for marrow cracking they are typically "cleaned," which consists of cutting off adhering sections of meat or tendon that would modify the way the bone would break when impacted with a percussion blow. Figure 4.40 shown Jean Rulland cleaning marrow bones prior to their being cracked for marrow, with a woman's knife or *ulu*. This cleaning of marrow bones produces short parallel cut marks essentially indistinguishable from the filleting marks illustrated in Figures 4.37–4.39. However, much attention is given to the metapodials, distal radio-cubitus, and distal tibia, all bones that yield either no meat or very little. This means that "filleting marks" may well be abundant on the metapodials, bones not easily conceived of as having been filleted! It is interesting that Joe Ben Wheat observed this at the Jurgens site, and commented as follows:

> Light diagonal cuts were found on a number of bones including metapodials. It is not likely that the cuts on metapodials were associated with filleting meat, since there is very little meat on such bones, so perhaps some other reason must be found for these cut marks [Wheat 1979:71].

Cleaning the bones of segments of meat and tendon is not all that is done prior to cracking. The secret to controlled breakage of marrow bones is the removal of the periosteum in the area to be impacted. The Nunamiut invariably do this by scraping it back with the edge of a knife, a rough surface on a hammerstone, or almost any handy crude scraping tool. This means that longitudinal scratches and striations (Figure 4.41) along the shafts of long bones are commonly produced when bones are prepared for cracking during marrow processing. Such marks are noted in Mousterian assemblages. My notes record 18 examples (striated bones) from the site of Combe Grenal, but the fragments of long-bone shaft were not saved at this site, so the parts where most such striations would be expected were in fact never systematically examined. Henri Martin (1907–1910: Figure LXII, Nos. 1, 3, and 5) reports examples from La Quina. Most analysts do not examine all the long-bone splinters for such marks, hence there is little mention of this type of mark in the literature.

Modifications other than those just described may occur as a result of various forms of food preparation. How a part is treated (filleted or not), what size it is reduced to before consumption, and how it is reduced, (chopped, cut, broken, etc.) are all variable strategies that relate to the methods of food prepara-

FIGURE 4.40. *Jean Rulland cleaning marrow bones for cracking.*

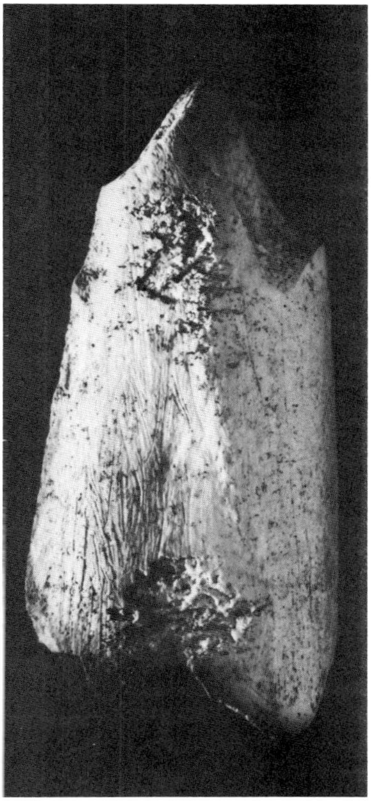

FIGURE 4.41. *Striations on bone believed to have been produced during removal of the periosteum. On top of the striations is pitting believed to have been produced by gnawing animals. (Found in the Mousterian site of Combe Grenal by F. Bordes.)*

tion. For instance, whether something is boiled, roasted, or eaten raw, and the types of containers used in food preparation, as well as the methods used (stone boiling versus direct boiling, etc.) *can be expected to leave subtle traces diagnostic of each.* Researching these differences is not the subject of this discussion; rather, I have been concerned with presenting descriptive information regarding the traces that tool-using man might be expected to have left on bones when he hunted and killed. This information might serve to distinguish the actions of tool-using man from the carnivorous actions of other animals. Traces referable to skinning, dismemberment, filleting for transport, and marrow consumption are all perhaps relevant. It is hoped that enough has been demonstrated regarding patterning in cut marks to encourage others to describe their material, so as to begin the task of developing a larger corpus of comparative material for study and use in the further specification of diagnostic characteristics reliably referable to specific actions carried out in the past. (See Table 4.04.)

TABLE 4.04
Inventory of Described Skinning and Butchering Marks[a]

Code number[b]	Part and description	Activity producing mark	Figure number	References[c]
	Skull			
S-1	Transverse cut on occipital condyles	Dismembering	4.11	Guilday *et al.* 1962; Figure 2
S-2	Skull split into left and right halves	Food preparation	4.18	
S-3	Transverse chops across cranium above and below the antlers or horns	Dismembering, consumption	4.17	Guilday *et al.* 1962; Figure 7
S-4	Cuts circling the base of antlers	Skinning	4.10, 4.11	Martin 1907–1910: Figure XLIII, Nos. 9, 10
S-5	Hole pounded into the frontal area	Killing, food preparation	Not shown	Frison 1970; 24
S-6	Cut on maxilla just above third molar	Dismembering	4.19	Frison 1971: 263; Guilday *et al.* 1962: Figure 2; Parmalee 1965: Figure 1
S-7	Nose cut off	Consumption	4.19	
	Mandible			
M-1	Transverse cut on the inferior surface of the mandibular symphasis	Skinning	Not shown	Guilday *et al.* 1962: 65
M-2	Marks across the masseteric fossa	Dismembering	Not shown	Guilday *et al.* 1962: Figure 2 Parmalee 1965: Figure 1; David 1972: 317
M-3	Diagonal cut on medial surface below P-3 and P-4	Dismembering (tongue)	Not shown	Wheat 1979: Figure 33b
M-4	Diagonal cut on the lateral face behind the third molar	Dismembering	4.07	Wheat 1979: Figure 3c
M-5	Transverse cut on the inferior surface of the mandibular condyle	Dismembering	Not shown	Martin 1907–1910: Figure XLVIII, Nos. 1–5
M-6	Longitudinal cut across masseteric fossa	Dismembering	Not shown	Wheat 1979: Figure 33c

(continued)

TABLE 4.04—*Continued*

Code number[b]	Part and description	Activity producing mark	Figure number	References[c]
	Cervical vertebrae			
CV-1	Transverse cuts across the proximal ventral surface of atlas	Dismembering	4.20	Guilday *et al.* 1962: Figure 2; Martin 1907–1910: Plate XLVII, Nos. 3, 4, 5
CV-2	Transverse cuts across posterior ventral surface of atlas	Dismembering (stiff body)	4.20	Martin 1907–1910: Plate XLVII, No. 6; Parmalee 1965: Figure 1
CV-3	Transverse cut across anterior ventral surface of axis	Dismembering (stiff body)	Not shown	Frison 1971: Figure 3aa–dd Parmalee 1965: Figure 1
CV-4	Longitudinal splitting of vertebrae	?	Not shown	
CV-5	Transverse chopping of axis	Dismembering	Not shown	
CV-6	Transverse cutting or chopping of sixth vertebrae	Dismembering	Not shown	Martin 1907–1910: Plate XLVI, No. 7, 64
	Thoracic vertebrae			
TV-1	Transverse chop or cut marks along dorsal spine into or between centra of second and third vertebrae	Dismembering	Not shown	
TV-2	Longitudinal cut along base and lower part of the dorsal spine	Filleting tenderloin	4.21	Frison 1970: 20–21, Figure 14; Gilbert 1969: 290; Wheat 1979: 67
TV-3	Transverse cut across the inferior surface of the centrum	Dismembering (ribs)	Not shown	Guilday *et al.* 1962; Wheat 1979: Figure 33g
TV-4	Transverse chop or cut marks along dorsal spine into or between centra of thirteenth to fourteenth vertebrae	Dismembering	Not shown	
TV-5	Longitudinal marks just below the articulator processes	Dismembering (ribs)	4.21	Wheat 1979: Figure 33g
	Ribs and sternum			
RS-1	Marks along superior surface just lateral to the rib head	Filleting (tenderloin)	Not shown	Martin 1907–1910: Plate XLII, No. 5
RS-2	Distal ends of the ribs cut off	Dismembering (ribs from brisket)	Not shown	
RS-3	Transverse cut on ventral rib surface just to the side of the rib head	Dismembering (ribs)	Not shown	Frison 1971: 265
RS-4	Marks on the ventral surface of sternum plates	Filleting	Not shown	
	Pelvis and sacrum			
PS-1	Sacrum trimmed down lateral edges	Secondary butchering	Not shown	

(continued)

TABLE 4.04—*Continued*

Code number[b]	Part and description	Activity producing mark	Figure number	References[c]
PS-2	Marks inside the acetabulum	Dismembering	Not shown	Guilday *et al.* 1962: Figure 2
PS-3	Marks across the lateral face of the pubis	Filleting	4.36	
PS-4	Ischiatic spine trimmed off	?	Not shown	David 1972: 317
PS-5	Cut or chop through the pubic symphasis	Dismembering	4.36	Guilday *et al.* 1962: Figure 7
PS-6	Cut across deep fossa in front of acetabulum	Filleting	4.36	Guilay *et al.* 1962: Figure 2
PS-7	Marks above the acetabulum on arm of ilium	Dismembering	4.22	Wheat 1979: Figure 32
PS-8	Marks below acetabulum on arm of ischium	Dismembering	4.22 4.24	
PS-9	Marks circling the rim of the acetabulum	Dismembering	4.22	Martin 1907–1910: Plate L, No. 12; Frison 1970: Figure 8a–b
PS-10	Marks below acetabulum on arm of pelvis	Dismembering	4.20, 4.22, 4.24	
	Femur			
Fp-1	Marks on the neck of the femur	Dismembering	4.25	Martin 1907–1910: Plate L, Nos. 8, 9, 10, 11; Guilday *et al.* 1962: Figure 2
Fp-2	Marks on the ball of femur head	Dismembering	4.25	
Fp-3	Marks circling the margin of the femur head	Dismembering	4.25	
Fp-4	Transverse marks on lesser trochanter	Filleting	4.25	
Fp-5	Transverse marks on lateral surface of greater trochanter	Dismembering	4.25	
Fp-6	Short marks in the fossa along the neck of the femur	Filleting	4.37	Wheat 1979; Figure 33
Fp-7	Short marks on the neck of greater trochanter, anterior face	Filleting	4.37	Wheat 1979: Figure 33
Fp-8	Short "chevron" marks, both transverse and oblique, below the neck of the femur both lateral and medial faces	Filleting	4.37	David 1972: 319
Fp-9	Longitudinal mark on upper shaft of femur	Filleting	4.37	Wheat 1979: Figure 33
Fd-1	Transverse across posterior surface just above condyles	Dismembering	4.25	Guilday *et al.* 1962: Figure 2
Fd-2	Nick marks on upper margins of the patellar surface—trochlea	Dismembering	4.25	
Fd-3	Longitudinal mark along the lateral face of the medial condyle	Dismembering	4.25	

(continued)

TABLE 4.04—*Continued*

Code number[b]	Part and description	Activity producing mark	Figure number	References[c]
Fd-4	Short "chevron" marks obliquely grouped	Filleting	4.38	Wheat 1979: Figure 33
Fd-5	Short oblique marks on the anterior face above patellar surface	Filleting	4.38	
	Tibia			
Tp-1	Marks on or around intercondylar tubercles—particularly the lateral one	Dismembering	4.26	Guilday *et al.* 1962: Figure 2
Tp-2	Mark across the posterior face of the lateral and medial condyles	Dismembering	4.26	
Tp-3	Short oblique marks on the lateral face of the tibial crest	Filleting	4.37	Wheat 1979: Figure 33
Tp-4	Oblique marks on medial face just below the articulator surface	Filleting	4.37	
Tp-5	Longitudinal marks on the upper tibial shaft	Filleting	4.37	
Td-1	Transverse marks across medial malleolus and just above on distal tibia	Dismembering	4.26	Guilday *et al.* 1962: Figures 2, 7
Td-2	Transverse mark across inferior articulator surface	Dismembering	4.26	
Td-3	Transverse cut across anterior face between the dorsal projection and the medial malleolus	Dismembering	4.26	
Td-4	Short "chevrons" on the anterior face of the distal tibia	Filleting	4.38	Wheat 1979: Figure 33
	Tarsals			Martin 1907–1910: 231–273
TA-1	(Astragalus) transverse mark at margins of the anterior face, midway on the bone	Dismembering	4.27	Guilday *et al.* 1962: Figure 7 Frison 1970: Figure 8c, d, e
TA-2	(Astragalus) marks across medial face midway on the bone	Dismembering	4.27	
TC-1	(Calcaneus) marks along the proximal margin of the lateral face	Dismembering	4.27	Guilday *et al.* 1962: Figure 7; Frison 1970: Figure 8
TC-2	Marks across distal end (tuber calcis)	Dismembering, filleting	Not shown	
TC-3	Marks on dorsal chest midway between tuber calcis and the articulator surface	Filleting, hanging carcass	4.27	
TE-1	(Ectocuneiform) transverse mark along margin of anterior face	Dismembering	4.28	
TNC-1	(Navicular–cuboid) transverse mark across lateral and or anterior face	Dismembering	4.28	

(*continued*)

TABLE 4.04—*Continued*

Code number[b]	Part and description	Activity producing mark	Figure number	References[c]
	Metatarsal			
MTp-1	Transverse mark across anterior face of proximal condyle	Dismembering	4.28	Wheat 1979: Figure 33
MTp-2	Transverse mark on the margin of medial face	Dismembering	Not shown	
MTp-3	Transverse mark on the margin of lateral face	Dismembering	4.28	
MTd-1	Transverse mark across posterior face of both condyles	Dismembering	4.27	Wheat 1979: Figure 33
MTd-2	Circular mark around the distal shaft	Skinning	4.16	Martin 1970–1910: Plate LIX, No. 6; Guilday *et al.* 1962: Figure 7
MTd-3	Nicks along margins of intercondylar channel	Dismembering	4.27	
MTd-4	Short "chevron" marks obliquely clustered on lateral and medial crests, anterior aspect	Filleting	4.38	Wheat 1979: Figure 33
	Scapula			
S-1	Marks along inferior border of condyle and/or at origin of triceps brachia	Dismembering	4.29	Martin 1907–1910: Plate XLII, No. 1, Plate XLIX, Nos. 1, 2, 3; Guilday *et al.* 1962: Figure 2
S-2	Marks along the neck of the scapula	Dismembering	4.29	Guilday *et al.* 1962: Figure 7
S-3	Longitudinal marks along base of spine in both the supra- and infraspinous fossae	Filleting	4.06	Frison 1971: 265; Wheat 1979: Figure 31
S-4	Longitudinal marks up and down the medial face of the scapula	Filleting	Not shown	Wheat 1979: Figure 31
	Humerus			
Hp-1	Marks along the border of the "lip" of ball, concentrated on the posterior edge	Dismembering	4.30	Martin 1907–1910: Plate XLIX, No. 4; Guilday *et al.* 1962: Figure 2
Hp-2	Marks on the apex of the lateral tuberosity	Dismembering	4.30	Parmalee 1965: Figure 1
Hp-3	Marks on the lateral face of the neck just below the lateral tuberosity	Dismembering	4.30	
Hp-4	Short "chevron" marks obliquely oriented along crest below the external tuberosity at insertion of teres minor	Filleting	4.39	
Hp-5	Short "chevron" marks obliquely oriented on medial face below the head	Filleting	4.39	
Hd-1	Transverse marks across anterior articulator face	Dismembering	4.30	Guilday *et al.* 1962: Figure 7; David 1972: 317
Hd-2	Transverse marks across medial surface	Dismembering	4.15, 4.30	

(continued)

TABLE 4.04—*Continued*

Code number[b]	Part and description	Activity producing mark	Figure number	References[c]
Hd-3	Marks on the upper margins of the olecranon fossa	Dismembering (stiff body)	4.30, 4.31	Wheat 1979; Figure 31; David 1972; 317
Hd-4	Marks on the ventral edge of the lateral condyle	Dismembering	4.30	Wheat 1979: Figure 31; David 1972: 317
Hd-5	Transverse mark across anterior face above articulator end	Dismembering	4.30	Martin 1907–1910: Plate XLIX, Nos. 5, 6
Hd-6	Oblique short "chevron" marks clustered on neck of distal end on the anterior face	Filleting	4.39	Wheat 1979: Figure 31
Hd-7	Longitudinal mark along medial crest of shaft	Filleting	4.39	
	Radio-cubitus			
RCp-1	Transverse mark inside semilunar notch	Dismembering	4.32	
RCp-2	Diagonal mark across lateral surface of olecranon	Dismembering	4.32	David 1972: 317
RCp-3	Diagonal mark across medial face of olecranon	Dismembering	4.32	Guilday *et al.* 1962: Figure 7; David 1972: 317
RCp-4	Marks on upper margin of medial semilunar notch	Dismembering	4.32	Guilday *et al.* 1962: Figure 2
RCp-5	Transverse marks on anterior margin of radial tuberosities	Dismembering	4.31, 4.32	Frison, 1970 Figure 8,f.9; David 1972: Figure 199–2; Wheat 1979: Figure 31; Parmalee 1965: Figure 1
RCp-6	Clustered oblique "chevron" marks below lateral and medial tuberosities on anterior face, of radius	Filleting	4.39	Wheat 1979: Figure 31
RCp-7	Short longitudinal marks along posterior crest of the cubitus	Filleting	4.39	Wheat 1979: Figure 31
RCd-1	Transverse mark across the anterior face along the articulator circumference	Dismembering	4.32	
RCd-2	Transverse marks across the styloid process	Dismembering	4.32	Martin 1907–1910: Plate L, Nos. 2, 3, 4; Guilday *et al.* 1962: Figure 2
RCd-3	Transverse marks on the inferior articulator surface	Dismembering	4.39	
RCd-4	Short diagonal "chevron" marks on anterior face	Filleting	4.39	Wheat 1979: Figure 31
	Carpals			
C-1	Transverse cut along articulator margin of "ulna carpal" (os pyramidal)	Dismembering	Not shown	Wheat 1979: Figure 31

(*continued*)

TABLE 4.04—*Continued*

Code number[b]	Part and description	Activity producing mark	Figure number	References[c]
	Metacarpal			
MCp-1	Transverse mark along anterior articulator margin	Dismembering	Not shown	
MCd-1	Circular cut around distal shaft	Skinning	4.16	Guilday *et al.* 1962: Figure 7
MCd-2	Transverse mark across posterior face of both condyles	Dismembering	Not shown	
MCd-3	Transverse mark across anterior face of both condyles	Dismembering	Not shown	Guilday *et al.* 1962: Figure 7
MCd-4	Short "chevron" marks on lateral and medial crests of the anterior face	Filleting	Not shown	Wheat 1979: Fgiure 31

[a] After this table was compiled I learned of an excellent study of cutmarks by Von Den Driesch and Boessneck (1975). This work should be consulted.
[b] p = proximal; d = distal.
[c] References are only to illustrations of mark in question.

Chopping and Bone Breakage as Butchering Techniques

The role of "chopping" and the use of heavy tools in butchering are points of "covert" controversy among many interested in the tools prehistoric man used for butchering. Early in the archaeological literature it was not uncommonly assumed that large "core tools" such as hand axes and cleavers were used in butchery. A most provocative study by J. D. Clark and C. V. Haynes (1970) illustrated that, in terms of actual patterns of association between tools and dismembered animal carcasses, Paleolithic butchering and meat-processing equipment consisted predominantly of small numbers of light-duty tools, cutting flakes, and small scraping tools, plus only a few of the larger elements (J. D. Clark and C. V. Haynes 1970:409). These findings as well as other experiences have led me to expect little if any major "chopping" in most butchery. Certainly, this is a bias on my part from having observed so many episodes of primary butchering among what might be described as the masters of small-tool butchery, the Nunamiut. Observations among the Navajo (see L. R. Binford and J. B. Bertram 1977:91–93) and reports by colleagues who have worked among contemporary African groups affirm the use of the ax, adz, or *panga* (a big "bush knife"), indicating that chopping and the use of heavy tools is a distinct butchering style, which may well have had counterparts in the Paleolithic. This view has been recently set forth anew (Jones, P. 1980). The following quote by an analyst of an African Iron Age faunal assemblage may provide the reader with some feeling for the contrast between butchering by chopping versus the butchering by cutting as described from groups like the Nunamiut (see L. R. Binford 1978b:47–64):

> Definite patterns of chopping, cutting and battering were visible on many bones. These are most clearly evident on bones of cattle and zebras. Study of these bones suggested the pattern of dismembering of carcasses, outlined below.
>
> Horns of cattle were chopped from the cranium, skulls were chopped open for the brains, mandibles were snapped off to remove the tongue, hind limbs were removed by means of a clean blow through the femur head, joints of meat were separated out by means of chopping blows concentrated at limb articulations. Raw meat was usually stripped off by means of chopping blows which left shallow chop marks along the length of bones, often concentrated at muscle articulations due to the difficulty of chopping through tough tendon. Bones were commonly cleft for marrow by means of chopping blows delivered at the distal end although a few were cracked, probably with stones [Welbourne 1975:12–13].

Metal tools leave rather obvious marks, as Welbourne has suggested; however, George Frison (1970, 1971) has suggested that bones were commonly used as chopping tools in the butchering of North American bison. Such implements leave less obvious marks and, as he has suggested, chopping results in more distinctive breakage of bones than any obvious chop marks as referred to in the description of African Iron Age materials.

Figure 4.42 (Frison *et al.* 1976:52) nicely summarizes the butchering-related damage inferred by the authors from the condition of the bison bones at the Hawken site in Wyoming. Twenty-nine different modifications of bone are indicated, and only 5 are cut marks; the remaining 24 are some form of breakage believed to have been produced during butchering.

The most extensive and detailed discussion of Frison's muscle-stripping argument is perhaps contained in his Glenrock (Frison 1970) report:

> Butchering was accomplished by processes of chopping, crushing, breaking, and cutting muscle origins and insertions and then stripping the muscles . . . the process as described here is a reconstruction from the writer's interpretation of the evidence of butchering activity found on the bones and then applying this to the anatomy of the buffalo along with a consideration of the tool assemblage used [Frison 1970:12].

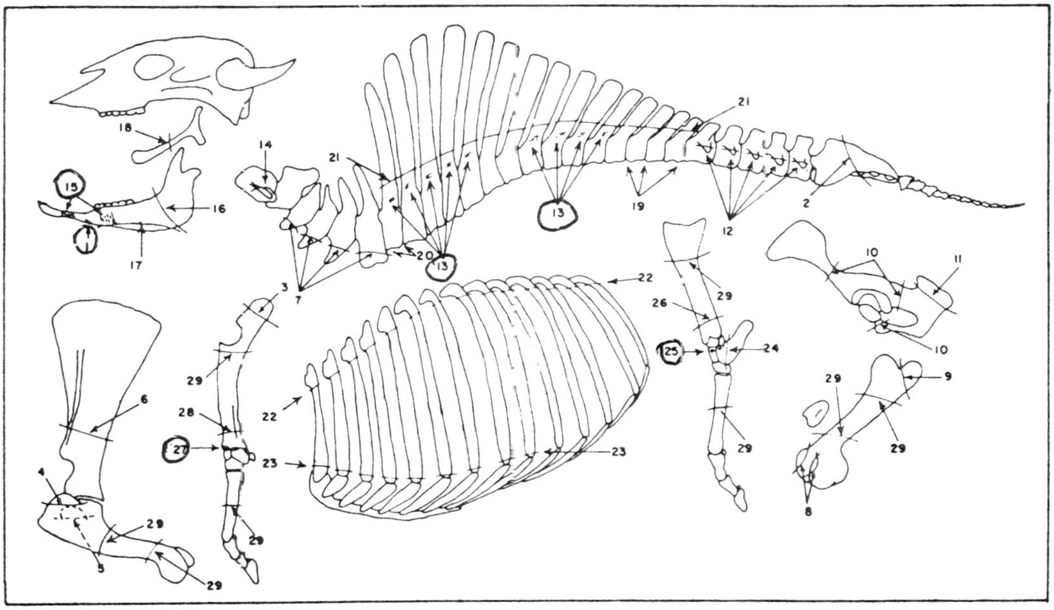

FIGURE 4.42. ***Placement of cut marks on bison bone from the Hawken site: (1) sawing marks on ventral border of mandible; (2) crushing and breaking of sacrum; (3) crushing and breaking of olecranon; (4–5) chopping off lateral and medial tuberosities of humerus; (6) breaking of scapula; (7) breaking of ventral branches of cervicals; (8–9) chopping into trochlea and breaking off major trochanter of femur; (10) breaking of the pelvis; (11) chopping off tuber coxae; (12) breaking of transverse processes of lumbars; (13) knife marks from cutting back muscles; (14) breaking wings of the atlas; (15) knife marks from cutting loose tongue; (16–17) crushing and breaking mandible; (18) breaking hyoid bone; (19–20) chopping and crushing to separate vertebral column; (21) breaking and chopping loose dorsal spines; (22–23) chopping and breaking of ribs; (24) breaking calcaneus; (25) knife marks on calcaneus and astragalus; (26) breaking distal end of tibia; (27) knife marks on distal radius-ulna; (28) breaking off distal radius-ulna; (29) long bone breakage for marrow. (Reproduced with permission from Frison*** **et al.**, ***1976:52.)***

I have difficulty with this argument. First, many of the bones illustrated in the Glenrock report (Figure 9a, b, e–h, j–o; Figure 10e, h, i, l; Figure 11a, b, e, f, i, k; Figure 12) are in my experience identical with animal-gnawed bones. Second, the patterns of assemblage composition given in Tables 1 and 2 are consistent with a faunal assemblage that has suffered from attrition; for instance, the proximal humerus and the proximal tibia are consistently represented by fewer examples than are the distal ends. (See Brain [1969] and L. R. Binford and J. B. Bertram [1977] for a discussion of survival probabilities.) I find it extremely difficult to imagine a mass kill butchering site being avoided by predator–scavengers after man left, so that all bone modifications at such a site were exclusively referable to human action. Yet Frison has *interpreted* all attrition and breakage as a result of human butchering, and has reconstructed a form of butchering to accommodate all the obvious attrition. Figures 3.34 and 3.35 illustrate a sheep pelvis, from a Navajo winter site (L. R. Binford and J. B. Bertram 1977), that has been gnawed by dogs. Illustrated in Figure 3.39 are caribou innominates that have been gnawed by wolves. It is certainly difficult to distinguish these bones from those cited by Frison as evidence for muscle-stripping butchering techniques. (See Frison 1970:17–18, Figures 11 and 12). Finally, the assemblage composition at Glenrock is consistent with a kill butchering site where anatomical elements were being removed, rather than a site where meat was being processed off the bones for transport (see L. R. Binford 1978b:475–476). Extensive muscle stripping would only make sense if one was attempting to reduce weight by removing meat from the bones and in turn abandoning the bones. As pointed out, the bones from the most meat-bearing parts were not abandoned at Glenrock! I am left with the conclusion that I agree with Frison that muscle stripping is one way of butchering that could well have been practiced, and we must attempt to find diagnostic criteria for its recognition. I must remain skeptical that the facts from Glenrock and other sites support Frison's interpretation.

In later reports, Frison acknowledges that "carnivores can and do chew off muscle attachments on a carcass and it is necessary to be able to recognize the difference between ones that were removed as the result of butchering process and those removed by carnivores [Frison 1974:28]." His reconstruction of butchering events at the Casper site (Frison 1974) is much more consistent with my experiences, and many of the types of breakage of which I was skeptical at the Glenrock site are either not present at Casper or they are not featured in Frison's reconstruction. All in all, it is my judgment that great care must be exercised in using Frison's butchering model.

Breakage patterns coupled with anatomical segmentation and cut marks have served as the basis for Frison's reconstruction. Less satisfactory data have served as the basis for reconstructing a butchering procedure exclusively from breakage patterns (see Lyman 1978). This interesting paper begins analysis against a backdrop of previous research in which it was largely postulated that butchering had been accomplished in the southeastern Washington State area almost exclusively by breaking bones. "Apparently there is little discernible evidence that butchering the carcass included a cutting process [Lyman 1978:2]." Guiding Lyman's analysis were the following principles:

> Brumley (1973, p. 24) believes that the ways and places in which a bone is broken reflect the methods and purposes of the breakage. Granting this the osteo-archaeologist examines the archaeological bone assemblage to define butchering units. Butchering units are determined by examination of bone units, breakage patterns and bone groups. A bone unit is defined as any fragment of bone. Breakage patterns are defined as the common points of fracture. A bone group is defined as the assemblage of bone units representing a particular skeletal element, such as a right femur or a left radius. Bone units, breakage patterns, and bone groups are considered to be attributes of butchering [Lyman 1978:4].

Lyman's procedure was to describe each skeletal element (e.g., scapula, femur, metacarpal) in terms of the "bone units" into which it had been broken and then present the frequency of each unit in different archaeological populations. No attention was given to cut or chop marks, with the implication that none were present. After this information was summarized, an interpretation was offered:

> butchering consisted of smashing through the limbs just proximal and distal to the joints of the long bones. The

> resultant (secondary) butchering units included articulating proximal and distal ends of long bones and associated ligaments and tendons, and the long bone shaft and its associated musculature (meat) . . . the joints were laid aside, if not discarded, and later boiled to make soup or processed for any remaining marrow. The bone shaft and its meat was cooked as a unit, the meat overlapping the bone ends such that the bone was insulated from the fire if roasted over an open fire. These butchering and cooking processes would result in the types of bone units, bone groups, and breakage patterns described above [Lyman 1978:22].

This novel interpretation assumes that cutting is used to go through the meat exposing the bone. However, the butcher then shifts to a hammer and bludgeons his way through the bone, achieving a disarticulation. If correct, this would certainly suggest that the aboriginal peoples of the southeastern part of Washington State were truly "culturally unique."

I have observed the Nunamiut use bludgeons to break leg bones, particularly frozen bones, but such a technique was only used on lower legs, where little muscle sheathed the bone. More specifically, bludgeons were used in breaking the distal tibia just above the articulator end, and the distal radiocubitus; less commonly they were used just below the proximal ends of the metapodials. Among the Navajo I have observed ribs and transverse processes of the lumbar vertebrae broken off during butchering with an ax. Among Australian aborigines, I have observed a bludgeon—specifically, a throwing stick—used to break the ribs loose from the vertebrae (Figure 4.43). In both cases of butchering with bludgeons

FIGURE 4.43. ***Alyawara Australian using a throwing stick as a bludgeon in butchering.***

or using a chopping technique, the size of the animal was relatively small (e.g., sheep and kangaroo). Among the Nunamiut, when a large animal is butchered, slightly different techniques are used, which employ more leverage and more dislocation of joints coupled with cutting rather than chopping. For instance, Figure 4.44 shows a female moose partially butchered by a Nunamiut hunter and his sister. The animal weighed approximately 780 lb and was completely butchered with a penknife. Some butchering differences, which were obvious accommodations to the size of the animal, relate to the removal of the front leg, the removal of the ribs, and the segmenting of the vertebrae.

The "large-animal" strategy of front leg removal was to strip the muscles from the humerus by cutting insertions at the humerus–radio-cubitus joint away from the humeral shaft. As the proximal humerus–scapula articulation was approached, two rather quick cuts were made "inside" the stripped muscles and around both the lateral and medial sides of the articulation. Once these seemingly superficial cuts were made, the butchers foot was placed on the dorsal surface of the scapula–humerus joint. The foreleg was then given a robust twist away from and then back toward the body, easily dislocating the joint. Once this was done, several minor cuts of connective tissue freed the forelimb from the shoulder. The scapula was then removed from the thorax by cutting between the lateral surface of the ribs and the medial surface of the scapula, which freed it handily as is the normal case when the entire front leg is removed from smaller animals.

It will be recalled that with caribou and sheep the

FIGURE 4.44. *Partially butchered female moose.*

technique of rib removal is to cut the distal ends of the ribs from the brisket, separate the main rib slab from the first two ribs, which normally remain attached to the first two thoracic vertebrae, and then grasp the rib slab along its distal margin and pull up smartly, breaking the rib heads at their point of articulation with the vertebrae. (See L. R. Binford [1978b:48–60, 94–97] for further butchering descriptions.) With the moose and with large bear that I have observed, the foregoing technique was altered. The first five ribs of the slab (ribs 3–8) were broken off at the proximal ends *one at a time*, which resulted in a very oblique break yielding a very pointed broken proximal end of the rib. Once the smaller ribs were approached, the butcher grasped the distal margin of the rib slab and broke it back all at once, as is the normal method with smaller animals. Once the breaks were produced, the knife was run between the breaks as the slab was pulled up and it was removed as a unit.

The final accommodation of the Nunamiut butchering technique to body size was in the segmentation of the vertebrae. The knife was used to cut into the articulation between the sacrum and the last lumbar vertebra. Once some of the connective tissue was severed, the entire spinal column with attached pelvis was placed upright, with the posterior end of the pelvis resting on the ground. The hunter then leaned down on the vertebral column while pushing forward, placing a strain on the articulation between sacrum and lumbar vertebra. This normally resulted in cracking the articulation, permitting adhering connective tissue to be cut readily with a small knife. This same technique was used in segmenting the column between the thoracic and lumbar vertebrae.

These and other accommodations to body size make it appear unlikely that "chopping" is a strategy that is necessarily related to large size, as is frequently suggested. Conversely, as has been mentioned, in my few ethnographic experiences with the technique it was used on relatively small animals—sheep and kangaroo.

The assumption that the use of chopping techniques was widespread has been largely fostered by functional inferences from tools that subsequent research, such as that mentioned earlier by J. D. Clark and C. V. Haynes (1970), has certainly cast in doubt. Finally, the assumption of chopping was a post hoc means of accounting for missing bone elements by the "father" of butchering analysis (White 1952, 1953a, 1953b, 1954, 1955). The continuation of this interpretative tradition by Frison (1970, 1971, 1974, 1978) and other Plains archaeologists (see Johnson 1977, 1978) fosters the view that chopping was a major butchering strategy.

As has been illustrated in Chapter 3, many of the missing bones that White explained away by a "destruction through butchering with a chopper model" were probably not preserved archaeologically because of consumption by both wild and domestic canids. Similarly many of the facts of destruction cited by Frison and others as evidence for chopping may well be gnawing destruction by canids. As will be shown, many of the patterns of bone breakage cited by White (1953b, 1954, 1955) and Frison *et al.* (1976) as well as Lyman (1978) as evidence for chopping as a butchering tactic could easily be seen as the consequence of breaking bones for marrow, or, in the case of Lyman, pounding bones for bone grease.

It is clear that chopping is documented to have been used in butchering animals, but there is *no* methodology adequate for recognizing the consequences of this tactic, in either forms of breakage or bone destruction, that is not ambiguous and subject to alternative interpretation. This is an area in need of strong methodological development.

Gnawing by Humans

Wolves, dogs, hyenas, and other nonhuman carnivores do not have a monopoly on the act of gnawing. Humans can, and most certainly do, gnaw bones. Although bone gnawing by modern man is relatively rare in the intensive sense documented earlier for the canids, we might expect some of our earlier ancestors to have engaged in somewhat more enthusiastic gnawing, particularly during the very early time ranges when many different food-preparation techniques were not yet developed or used and particularly when our ancestors were taking their first steps toward regular hunting and meat consumption.

Are there ways of distinguishing between bones

gnawed by humans and those gnawed by animals? As I and my students continue to analyze the control data collected among the Nunamiut, particularly faunal materials recovered from houses and from undoubted human use contexts, I feel confident that I will be able to provide at least some descriptive data relative to the patterns of bone modification and destruction generated by gnawing Eskimos. I have made some ethnographic observations that may aid in this work. For instance, it was noted previously (L. R. Binford 1978b 151–52) that gnawing and "sucking" of short sections of ribs was a fairly common act where fresh meat was being consumed, but when dry and/or frozen meat was being eaten steps were generally taken to prevent breakage of such parts as ribs, since even the small amount of blood and "marrow" in such bones is generally not very appetizing. The gnawing and sucking on rib "tablets" (sections of rib ranging from 2 to 6 inches in length) is almost exclusively restricted to the ends of these tablets. Scoring, punctures, or pitting along the lateral margins of ribs were never observed. This is certainly different from what canids do to ribs where they regularly crenulate the edges, as is shown in Figure 3.33. In addition to ribs, I have regularly observed human gnawing on dorsal spines of thoracic vertebrae, proximal margins of the scapula, and various very "soft" bones of young individuals. In the few incidences where I have observed the consequences of this gnawing in any detail, the pitting and the "mashing back" of edges is fairly common. Punctures were rare and *scoring was never observed.*

This is an area where diagnostic properties might well be developed through direct experimentation with modern subjects. However, the gnawing would have to be directed and controlled for the strength of the subjects and they could not be allowed to select what they wanted to gnaw. Instead, the experimenter would have to instruct his subjects to gnaw assigned anatomical parts in specified ways so as to obtain as complete a picture as possible as to what is possible and likely under different conditions. I feel it is highly unlikely that a normal hominid pattern of consumption would include gnawing to the extent that it would mimic the consequences of the classic carnivore's behavior, as is documented in Chapter 3. Nevertheless, we need data on this problem before tooth modifications on bones can be assigned to nonhuman agents in a totally reliable manner.

Marrow-Bone Breakage

As indicated in Chapter 3, patterns of bone breakage have served as focal points of concern for many who have attempted to recognize human occupations when convincing evidence in the form of stone tools was lacking. I have thus far argued that the crack-and-twist method is not essential to spiral fracture. This appears to be a simple product of green-bone breakage. Similarly, the position that spiral fracture itself was a characteristic diagnostic of man has been dispelled by my discussions of bone fracture by animals and other agents. Nevertheless, man does characteristically break bone for marrow, and we would like to know if there are properties that would permit us to recognize bones broken by man as opposed to other agents, or bones broken by man in the context of butchering as opposed to marrow extraction or perhaps other contexts such as tool manufacture.

As mentioned earlier, a study of marrow-bone breakage by Zierhut (1967) with follow-up observations by Bonnichsen (1973) is to my knowledge the first specific behavioral observation made on this subject. The observations by Zierhut were among the Calling Lake Cree Indians of Canada. The observations were made in the "yard" of a permanent residence where bones of moose had been introduced from hunting successes in the local area.

> The meat is first stripped from the bones from which marrow is to be taken. Once this is accomplished the bones are thrown on or next to the coals of an open fire. They are heated in this manner for a very short time, being turned over occasionally to prevent burning. The Indians relate that this procedure makes the bones easier to break. Likewise, fresh bones break more easily than do dry bones. Next the bones are taken from the coals and allowed to cool. Meanwhile two large oval stones are placed on the ground about 8–10 inches apart. When the bone is cool enough to handle, it is placed so that the proximal and distal ends rest upon

> the stones; the midshaft is thus suspended between the stones. The bone is then struck in the middle, using the blunt end of a small ax. With only one or two vertical blows the bone is broken into two major halves; during this operation a few small fragments or chips of bone are generally detached from the point of impact [Zierhut 1967:34–35].

These ethnographic experiences impressed Bonnichsen to the point that he seemingly accepted the Cree behavior as being *the* characteristic method of marrow-bone breakage. Bonnichsen (1978:36) refers to this as the "mid-diaphysis smash technique of spiral fracture." It should be pointed out that although no other ethnographic descriptions were apparently available, the actualistic studies of Sadek-Kooros (1972) clearly indicated ways of breaking bones other than that observed among the Cree.

Sadek-Kooros (1972, 1975) conducted some fairly extensive experiments in bone breaking; however, the overall thrust of her research was a "replicative experiment" designed to argue in favor of a bone tool assemblage at Jaguar Cave, Idaho. She states,

> During the identification of the Artiodactyla, a consistency in the shape and fracture patterns of many fragments, particularly among the long bones, was noted. While it is not inconceivable that bone would tend to fracture along lines of least resistance, that is, following inherent stresses and strains . . . yet the possibility existed that such patterning reflected purposeful fracturing aimed at obtaining specific shapes [Sadek-Kooros 1972:369].

The particular pattern that attracted the Jaguar Cave research was what my field crew in Alaska referred to as "bayonet" breaks, described as follows:

> a metatarsal bone retaining one epiphysis, usually the distal epiphysis, fractured regularly across the entire posterior or anterior surface at specific distances from the epiphysis retained, and having a shaft at least 3 times as long as it was wide on the surface opposed to the surface fractured [Sadek-Kooros 1972:371].

This type of break was obtained by placing the metatarsal on an anvil (see Sadek-Kooros 1972: Figures 1, 2, 3) and hitting it with

> one or several controlled blows (with obtuse point of the hammerstone) administered on the medial and/or lateral surface close to the midpoint of the shaft, followed by twisting the distal epiphysis. The use of leather or cloth wrapped around the hammerstone or around the bone being fractured helped reduce the shattering effect of the impact, control the force of the blow and regulate the extent of the fracture [Sadek-Kooros 1972:371].

Although this technique was of primary interest to Sadek-Kooros since it permitted her to approximate the form of breakage she had characteristically seen in her archaeological remains, it is of some interest that she recognized two other types of fracture in addition to spiral fracture: *green-stick fracture,* produced when the bone was struck across a stone anvil or another bone, and *transverse fracture,* produced when the bone was broken by bending down both ends across a stone anvil, another bone, the experimenter's knee, and so on.

Now, although Sadek-Kooros does not report the details regarding the fracture procedures that were followed, it is clear that they were successful ways of breaking a bone, even though they did not yield the forms in which Sadek-Kooros was interested.

Could I argue about the way men of the past broke bone by matching patterning in an archaeological site with the debris produced by Sadek-Kooros using "green-stick" and "transverse" bone-breaking procedures? Would this be an argument from analogy? If we are attempting to develop methods for distinguishing bone broken by humans from that broken by animals, are we not interested in understanding the range of variability that human bone-breaking might take? Is not Sadek-Kooros human? Is it not reasonable that the tactics she used to yield patterns of bone breakage different from a classic spiral fracture are also likely to have been employed by other humans at other times? I think we must answer yes to all these questions! Workers like Bonnichsen would apparently answer no, since the only technique discussed or considered as characteristic of human fracture by him was that performed by the Calling Lake Cree, the "mid-diaphysis smash technique of spiral fracture [Bonnichsen 1979:36]." Is the only form of valid argument from analogy ethnographic?

I will attempt to describe in some detail the *vari-*

ous ways I have observed the Nunamiut Eskimo break marrow bones. I will place these descriptions in the situational or systemic context in which they were most often observed.

I might comment at this point that in my previous descriptions I mentioned that the Nunamiut broke bones differently in different contexts (L. R. Binford 1978b:153), but I made no systematic attempt either to survey my notes and extract the details of these differences or to mention the total range of bone-breaking behavior that I had in fact witnessed. I was very intent on treating butchering in detail but at the time it did not occur to me to treat bone breaking with equal attention. I was attempting to understand interassemblage variability in faunal remains considered from an anatomical perspective. I was simply not asking the questions that I am now posing regarding the relationships between the morphology of broken bone and the tactics used in bone breaking. In many of the following descriptions I do not have the bear and the footprint together; most of the time I observed the bear—the behavior of the Eskimo—but I did not think to record its footprint, the resulting patterns of bone modification. In short, much of what is to follow is ethnography, not ethnoarchaeology.

Marrow Cracking during Butchering Episodes

It is not uncommon for a butcher to crack leg bones open while engaged in butchering. A comment that is frequently made is that you can tell the nutritional state of the animal by examining the bone marrow. In spring and during early summer, when nutritional conditions of caribou are generally poor, a butcher may break the front leg just above the distal articulator end of the radio-cubitus. This is done *before* skinning and while the leg is fully articulated to the animal. The procedure is to grasp the front leg, bringing the foot and metapodial to rest on one's knee, while standing behind the front leg facing the head of the animal. With the left hand the foot is pulled upward and forward; the butcher is still bearing down on the leg against the knee. In the right hand the butcher has a bludgeon (frequently the lower leg of another caribou), which is used to deliver a heavy blow down and on the posterior edge of the lateral side of the distal radius. One blow is generally all that is needed. A skinning knife is then used to cut through the skin and tendon at the break, freeing the articulated lower leg (phalanges, metacarpal, carpals, and distal radio-cubitus), which is tossed to the side after the marrow is removed from the distal end of the radius. On the other hand, if the break was too high, resulting in a considerable length of diaphysis remaining on the distal end, the butcher might use the handle of his skinning knife and strike at a protruding section of diaphysis, with the intent of driving off a flake longitudinally down along the bone, thereby exposing more of the marrow cavity.

Under normal conditions, the break remaining on the diaphysis of the radius and arm of the cubitus is very sharp and dagger-like, with a classic spiral moving up and over the posterior of the face of the radius. The lower leg would be left at the butchering site and the upper part of the front quarter would be transported to the residential location or to a cache from which it would later be introduced to a residential location. The results of breaking through the front leg during butchering are well recorded in the inventories of anatomical parts remaining on kill butchering locations (see L. R. Binford 1978b:76), where without exception the discrepancy between the frequency of proximal and distal ends of a bone in the front leg is greatest for the radio-cubitus.

Less commonly, this same behavior may be directed toward the lower third of the distal tibia with the intent of removing the lower rear leg (phalanges, metatarsal, tarsals, and distal tibia). At least among the Nunamiut, cracking the tibia is considered more of an indulgence and is only done when the butcher desires to eat good marrow. It is not a "testing" technique like breaking the radius. The only difference in the way the tibia is treated is that I have observed it broken after skinning; in fact my notes indicate in 79% of the cases (11 out of 14) the rear leg had been skinned prior to breaking the tibia for marrow. Otherwise the procedure of bone fracture was essentially the same as for the radio-cubitus. On the other hand, removal of marrow was more extensive and more attention was given to modifying the shaft remaining attached to the animal. For instance, in 3 out of 14 cases, after the lower leg had been removed by cutting the tendons and tissue around the break, the butcher walked to a nearby rock and hit the protruding fractured end of the distal tibia on the rock. The compact blob of marrow from inside the distal tibia fell out neatly onto the rock. The butcher picked it up, putting it into his mouth as he tossed the lower leg to the side. In 5 out of 14 cases the butcher partially filleted the meat from the tibia, leaving it attached to the mass of unfilleted meat on the femur. This had the effect of exposing the main part of the tibial shaft up to the point at which the shaft widens, just below the proximal end of the tibia. The butcher then scraped the proximal surface of the tibial shaft and struck a blow directed at the proximal face of the tibia about halfway between the previous break and the proximal end. In all 5 cases a classic spiral fracture developed, running in both directions so that several splinters could be pried up and the "hot dog" shaped section of marrow lifted from the bone as a single unit. The fracture tended to disappear into the uncleaned proximal end of the tibia still surrounded by substantial quantities of tissue at the point of articulation with the distal femur. In all but one case the butcher then picked up a tabular stone or a lower leg section to use as a bludgeon. The stump of the broken tibia was then held and pounded with the bludgeon on the fracture. Several substantial blows were delivered, the bludgeon was tossed to one side, and the fracture was examined to determine if loose splinters had been produced. The knife blade was inserted in the end of the bone and used as a lever until a section of bone gave way. Then the blade was run down the outside of the bone along the indicated crack removing the periosteum along the crack. The bone splinter was then bent outward, generally leaving the proximal end of the splinter still attached to the proximal end of the tibia by the periosteum and other tissue. The bent-away splinter "door" then permitted the easy removal of the marrow in the proximal end of the tibia. The proximal articulator end with bent-away splinters still adhering was left with the heavy muscled "ham," which was either later filleted or carried away as a unit to a residential location.

If any further marrow was eaten during butchering it was almost invariably done during a rest and the metatarsal was the target bone. Fracturing of the metatarsal under these conditions was essentially done in the same way that it is normally done in hunting camps and stands and will be described in that context.

Marrow Cracking in Hunting Stands and Camps

As I have deomonstrated, consumption of bone marrow is a common snacking strategy in hunting camps and stands (L. R. Binford 1978b; 179–191). The most consistent target of this consumption is the metatarsal. I observed 61 cases of breakage for marrow of the metatarsal in this context.

The first point to be made is that bone-breaking strategies in the field are variable, unlike the situation in residential camps where a redundant set of bone-breaking patterns was evident. The variability in field sites is primarily conditioned by the context. Most consumption in hunting camps and stands is snacking and is done quite expediently and as a partial function of the sociality of the moment, when several people are snacking and talking together. Under such conditions, one looks about both for a source of snacks as well as tools for processing bone. The occupants of these field camps, are usually men, and the knives they regularly carry expediently serve to break marrow bones; the blade is held and the bone is struck with the knife handle. Figure 4.45 shows the cleaned marrow bones of a rear leg together with an anvil, a stone hammer, and the skinning knife commonly carried by today's hunters. When a skinning knife is not used for breaking marrow bones, an elongated stone such as the one illustrated (Figure 4.45) is used. The anvil, although sometimes used as such, more commonly serves as a working surface on which marrow recovered from a bone may be placed or accumulated prior to consumption. It may also serve for tapping the marrow out of some articulator ends. Typically the marrow bone is cleaned to the state illustrated in Figure 4.45 before it is broken. This will vary, however, with whether snacking is a leisure activity carried out between work episodes (such as butchering, hunting, or transporting field-butchered parts to storage or to the residence) or an activity done while a worker is

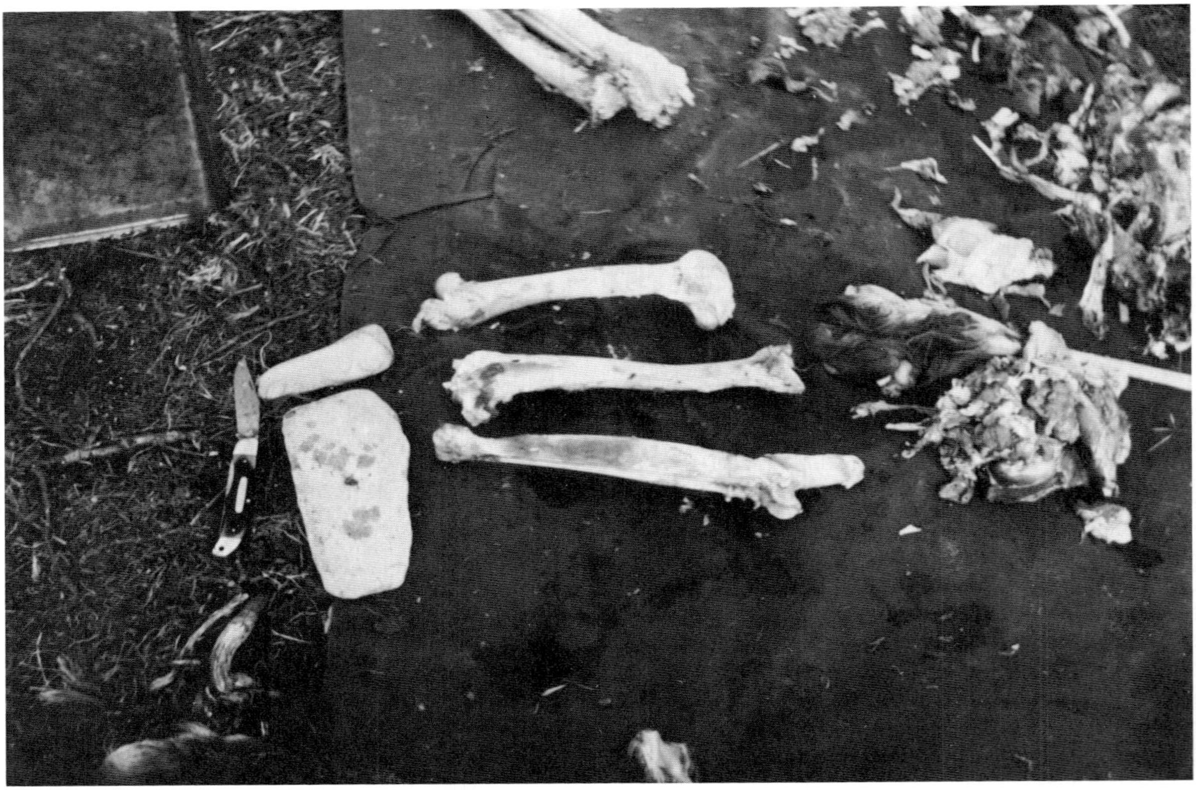

FIGURE 4.45. ***Bones and tools used in Nunamiut marrow cracking.***

engaged in other tasks. Under the latter condition the bone may be broken while articulated to others, or it may be even broken while it is only partially skinned.

Under normal conditions of snacking, when other activities are not being conducted, the bones will be cleaned as shown in Figure 4.45 and placed up close to the coals of the fire to be warmed before being cracked. A metatarsal is typically grasped by a right-handed man in the left hand, while the hammer is held in the right. The bone would be held slightly turned outward so that if the metatarsal is from the left the wide ridge along the junction of the posterior and medial faces of the bone would be uppermost. If a right metatarsal is being cracked the posterior and lateral faces of the bone would be uppermost and aligned to receive a blow. The hammer is held in the right hand and a blow is struck down on the aligned "ridge" about 2½–3 inches back from the distal end of the bone. Figure 4.46 shows Johnny Rulland holding the stone hammer and metatarsal in the proper position for striking the bone. Figure 4.47 shows the spiral fracture resulting from the neck-directed blow; the distal end is twisted off, since the periosteum held the bones together after the fractures from the hammer impact developed. Typically what happens is that the fracture runs down into the articulator end on the surface exposed to the blow and runs away and up the bone on the face opposite the impact area. This results in a pointed or short "bayonet" projection of diaphysis along one side on the distal end of the bone. The marrow cavity in the distal end of the metatarsal is smooth and not webbed cancellous tissue. Any marrow inside the distal end is picked out with a small stick or, if the knife handle is being used as a hammer, with the blade of the knife. Sometimes the broken distal end may be tapped on the anvil so the marrow will pop out. This may result in "re-touch" on the apex of the bayonet break.

The distal end with attached large section of diaphysis, or bone shaft, is then normally reversed in the hand so the proximal end of the bone is most distant from the body if it is a left metatarsal; a right metatarsal would be positioned with the distal end away. The reason for this is that the lateral side of a caribou metatarsal is much wider and slopes obliquely up to the fluted crest on the anterior face of the bone. It is along the flat face at its juncture with the anterior channel that a series of quick blows are directed. Usually three hits about 2 inches apart are sufficient to crack the "crest" of the metatarsal into a series of bone plates, which can then be picked off since they normally adhere to the bone by virtue of small segments of tissue. I use the term *plates* since the thickness of a caribou metatarsal is not great and the thickness is uniform over much of the shaft. If the blows are delivered with the flat side of a hammerstone or with a knife handle wrapped with a small tuft of skin, the blows are diffuse and have the effects of "depressing" out bone plates much like the segments of bone composing a "depressed fracture" of the skull or other large bone. Most of the time there are few if any impact notches or classic properties of a percussion point impact with this method of cracking the metatarsal. The remains of the shaft are also quite distinctive, generally exhibiting very angular profile of fracture, as shown in Figure 4.48. If the metatarsal is disarticulated from the tarsals, the

FIGURE 4.46. ***Demonstration of impact alignment for bone and hammer when a metatarsal is being fractured.***

FIGURE 4.47. ***Twisting off spirally fractured distal end of a metatarsal.***

proximal end may be split with segments of the proximal articulator surface adhering to various bone plates and splinters. Once the fractured bone splinters and plates are picked off the bone, the marrow is removed, normally by digging it out with a stick or with the knife blade as is shown in Figure 4.49.

Another bone sometimes broken for marrow in hunting camps and stands is the metacarpal. In caribou this is structurally a very different type of bone in that the diaphysis is much thicker than in the metatarsal, and the marrow cavity is much smaller. The bone is shorter than the metatarsal and it is much more symmetrical in cross section. It is prepared in the same manner as the metatarsal but it is broken somewhat differently. The position of the bone relative to the hand-held hammer is the same as illustrated in Figure 4.46. However, the blow is generally delivered with a more rounded portion of the hammer and with more force, so a point of impact about three-quarters of an inch in diameter is produced. The blow is directed somewhat lower on the side of the bone. Generally, before the distal end is twisted off a second blow is directed along the same side of the bone about 2 inches in from the distal end of the bone. Both of these blows develop classic impact scars ringed with small, incompletely fractured tiny impact chips (see Figure 4.48). When twisted after the second blow, the bone may well fall into basically four pieces: a splintered section of the anterior crest of the diaphysis, the distal end, the proximal end, and a section of the posterior face of the diaphysis. This is a hard bone, so that most of the diagnostic properties of point impact such as dis-

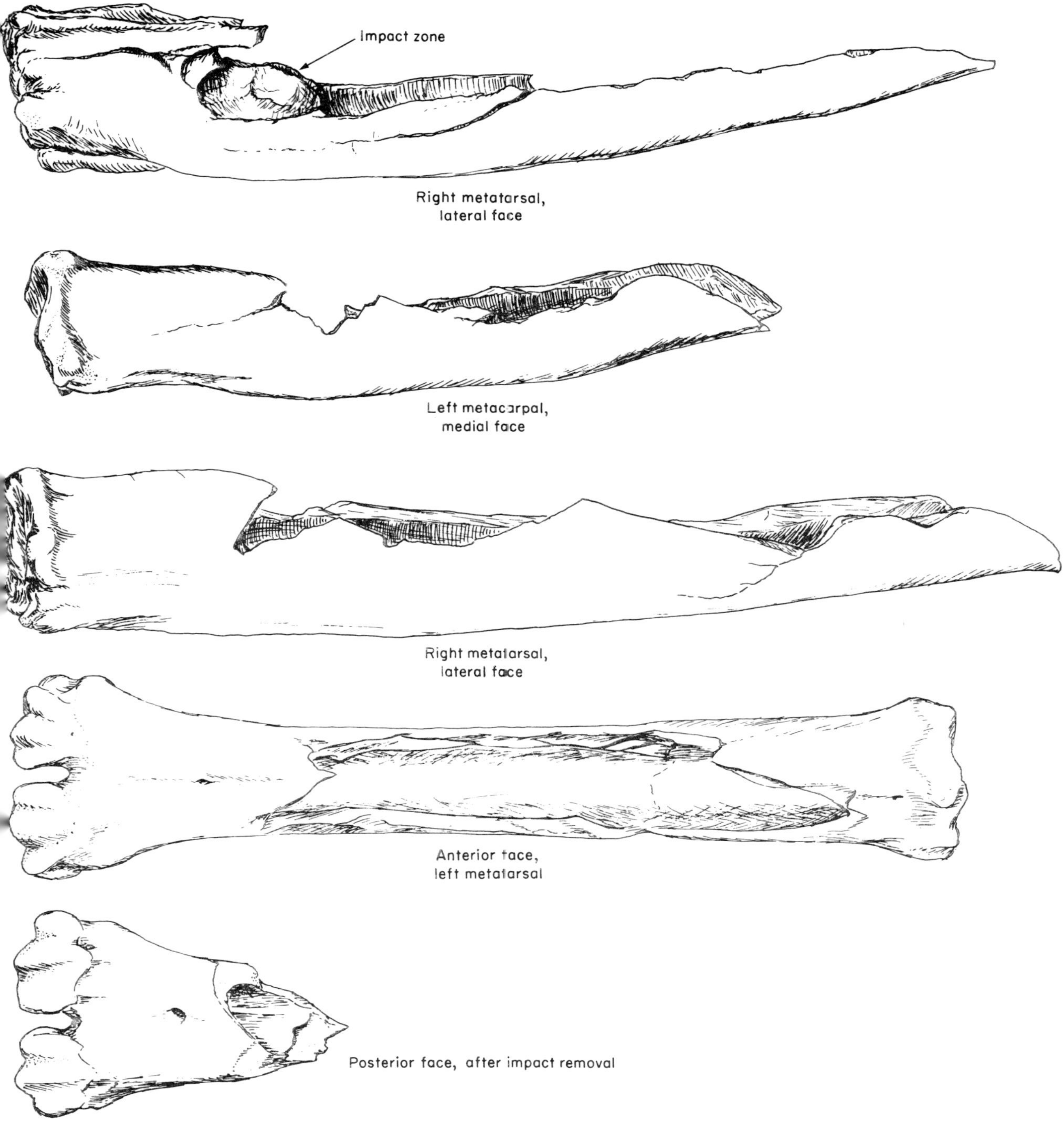

FIGURE 4.48. ***Impact and fracture patterns typical of metapodials broken for marrow.***

FIGURE 4.49. ***Removing marrow from the proximal section of a broken metatarsal.***

tinct impact notches remain on the fragments of the diaphysis. As in the case of the metatarsal, the proximal end may sometimes split out with the sections of diaphysis.

Variability arises in the degree of cleaning and the degree to which bones are disarticulated from adjoining bones. It is not uncommon to see the lower rear leg of a caribou placed up close to the fire for warming prior to skinning, sometimes with the foot and tarsal still articulated. Once the bone is warmed, the skin of the central part of the anterior surface may be removed and a series of blows directed along the juncture of the lateral face and the anterior flute, resulting in the plates and thin bone splinters described for the other method of cracking bones. These will be pulled up and commonly allowed to remain attached by tissue at the proximal or distal ends of the bone. A window is thereby exposed and warm marrow may be dug out. The result is a metatarsal with a typically angular fracture profile along the central part of the diaphysis but with proximal and distal ends remaining articulated with the foot and tarsals respectively. Somewhat more commonly the foot will be disarticulated and the bundle of tendon that runs along the posterior surface of the metatarsal will be pulled up to its attachment at the tarsals. The exposed distal end of the metatarsal will be treated in the same manner as a completely cleaned bone. The only difference is that the proximal end remains articulated with the tarsals and associated with the foot bones, which are attached at this time to the tendon bundle (see Figure 4.03).

The men explained that they used the hammer-on-bone technique in the field because it is easily done standing up, it is easily done so that articulator ends and the desired marrow do not fall to the ground, and they do not have to clean the bone perfectly or have it totally disarticulated. The last two characteristics are important relative to the techniques normally used by women for processing marrow bones in residential camp situations.

Men in the field may sometimes crack bones other than the metapodials. When this is done they normally impact the bones on the sides, either lateral or medial, with hammers as was done for the metapodials (an exception is the proximal femur, which I will describe later when I discuss strategies employed in the residential camps). They always impact the bones on the neck area just above the articulator end. With bones such as the distal femur and the distal tibia, they may impact the bone up to 4 or 5 inches from the articulator surface, aiming for a long bayonet break, which exposes the medullary cavity but does not crack the end off the bone. The latter must be twisted apart given the attachments of the periosteum. This ensures that the articulator end does not fall to the ground or otherwise get "out of hand." Bones with dense diaphyses such as the tibia tend to break with clean "percussion" related properties, yielding splinters with clean impact notches (Figure 4.50).

When meat is being cut from the bones in the field for purposes of either field drying or weight reduction for transport, complete legs or major articulated segments may be lying around, with the meat removed. Not uncommonly a butcher or person engaged in filleting the meat may grab one of these articulated but filleted legs and proceed to crack bones—generally those of the upper leg such as the humerus, or the tibia and femur—by striking them obliquely in center shaft while the bone rests on an anvil. The bone is then twisted apart and the marrow is dug out from the remaining two ends. This results in frequent articulations of distal tibia and proximal femur or proximal metatarsal with tarsals and distal tibia, clearly indicating that the bones were not disarticulated prior to having been broken in midshaft for marrow. As stated earlier, bones impacted in midshaft tend to show developed impact notches and classic spiral fractures. When this strategy is used, the impact is nearly *always* on the lateral or medial sides of the bone, since it is difficult to

FIGURE 4.50. ***Percussion impact notch on long-bone splinter.***

straighten out the joints and place a bone on an anvil with anterior or posterior faces uppermost.

Marrow Cracking during Meals in Residential Sites

Most of the time marrow bones are filleted by the women while they are preparing meals, and the bones are stored for later processing in large groups. On some occasions, however, marrow bones may be served at a meal (see L. R. Binford 1978b:145, 146). When this is done they are almost always eaten by men or at least broken by men, although women may receive choice morsels from their male relatives. Sometimes an old woman will break her own marrow bones around a household hearth, and when this is done it is almost always done in "male style," that is, using a hammer to impact the bone held in the hand. It was explained that this ensured that the ends did not fall to the ground or in the hearth, since this is generally done in front of diners inside a winter house. An alternative method is sometimes used inside houses when the marrow bones have been warmed in a stew and are very warm and slippery. This is the "on-anvil" technique, where the bone rests on an anvil—commonly one of the large stones surrounding the hearth—and is impacted with a hammer. Bones broken in this manner are seated on the anvil in a stable fashion and do not slip around. Bones commonly seated on the posterior or anterior faces are distal metapodials, distal radio-cubitus, and the distal tibia. These are almost always impacted on the posterior or anterior faces rather than on the lateral or medial faces, as is more common when the bone is held in the hand and impacted with a hammer at an oblique angle to the bone. The latter orientation biases the break in favor of a classic spiral fracture, which if done well will run up the cavity exposing a considerable portion. Impacting on the posterior or anterior faces tends to shatter the bone or produce a complex depressed fracture (common with the radio-cubitus and metapodials) or a complex bipolar fracture more likely with the distal tibia. Removal of the articulator ends in this fashion commonly results in an unbroken section of the shaft of the bone in the form of a cylinder (particularly with the femur and tibia), which is then freed of its marrow by pushing the "hot dog" out with a small stick or sometimes a sheep rib bone. Many women are more comfortable with the on-anvil technique and I have seen it employed by them at large migration hunting stands (see my descriptions of behavior at Anavik and Anaktiqtauk [L. R. Binford 1978b: 171–178]). Regardless of whether the hammer-on-bone technique or the on-anvil technique is used, the impacts are still directed at the neck area of the bone just below the articulator ends.

Processing of Marrow Bones by Women in Residential Camps

Most marrow-bone breaking in residential camps is done by women during intense processing sessions (see Figure 4.51). Commonly the marrow bones are collected for several days and then processed all at once. Occasionally when large quantities of bones are available, as during migration hunting activities, massive quantities may be processed at once.

Typically a skin is placed on the floor or ground, depending where the processing occurs (normally outside). On the skin is placed a container for the marrow and an anvil stone, and the bones to be cracked are piled nearby. The woman is seated at the

FIGURE 4.51. ***Mass of bones awaiting cracking by Anaktuvuk women during "gang" processing for marrow.***

edge of the skin, which serves as a drop cloth. Most of the time the bones have been previously cleaned but some minor cleaning may take place prior to cracking. *Actual breaking of the bones is done by striking the bone down on a handheld "hammer."* The hammer tool is called a *kaotah* and has been illustrated previously (see L. R. Binford 1979). It is typically a natural tabular section of local limestone, relatively square or thicker on one side than the other. This means that by a shift in orientation, the *kaotah* can impact a bone as a point, a very thin ridge, or a very broad, flat face. It may also be shifted from the left to the right hand (if the woman is right-handed) and used as a hammer in the male fashion if needed. When women break marrow bones in residential locations, they are concerned with breaking the bone so that little compact and dense diaphysis remains attached to the articulator ends, which will later be used in the making of bone grease on bone juice. There is a second concern: namely, that the bones be broken so that the marrow remains "clean," that is, free of impact chips commonly driven into the marrow when the dense bone of the shaft is impacted. Both of these concerns are served by striking some articulator ends as close as possible to the point at which cancellous tissue marks the transition from the medullary cavity to the articulator end. Such cancellous tissue is typically found in both ends of the femur, the proximal tibia, and the proximal humerus. On the other hand, relatively smooth inner surfaces are characteristic of the distal humerus, distal tibia, proximal and distal radius, and proximal and distal metapodial. In caribou, the metatarsal and the distal radius are relatively thin-walled and may be broken by impact crushing the area just above the distal articulator end. This results in few impact chips and only depressed plates, which can be picked off the marrow Only the distal tibia, proximal radius, distal humerus, and metacarpal are characterized by smooth inner surfaces of the medullary cavity, and thick and dense adjacent diaphyses. These bones are of less utility as grease bones because of their density. The degree to which they will be broken close to the end is therefore related to the quantity of bones available for rendering grease. For instance, if many bones are available, as in the case of Figure 4.51, these bones will be struck obliquely in the section of shaft judged to be most perfectly round, since obliquely impacting these with either the hammer-on-bone or bone-on-hammer technique, as illustrated in Figure 4.46, tends to produce clean spiral fractures running both directions with little impact notching and hence the driving of chips into the marrow. Marrow contained in the longer unborken section of the distal tibia, for instance, may be dug out; occasionally the spirally fractured end may be tapped on an anvil, knocking the marrow free. If this is the strategy employed, the dense bones, distal tibia, distal humerus, proximal radius, and ends of the metacarpal would not be processed for grease without further reduction of the diaphysis. The latter would be done as part of processing articulator ends for grease rendering, rather than as part of the marrow-processing activities. The technique of bone-on-hammer cracking very close to the articulator end leaves very distinctive types of breaks as well as impact areas.

FEMUR

Figure 4.52 illustrates an Eskimo woman about to strike the proximal end of the femur on the hand-held hammer–anvil. Figure 4.53 shows the type of break that results from a blow of this kind. The proximal end is broken off, with little attached diaphysis. Hints of slight depressed fracturing of the thin-walled bone are still visible at the upper end of the break. Impact is invariably on the anterior face of the proximal femur, just between the femoral head and the greater trochanter. The fractures are generally clean and have some of the properties of having been snapped off rather than spirally fractured.

FIGURE 4.52. ***Proximal femur about to be hit across a hand-held anvil.***

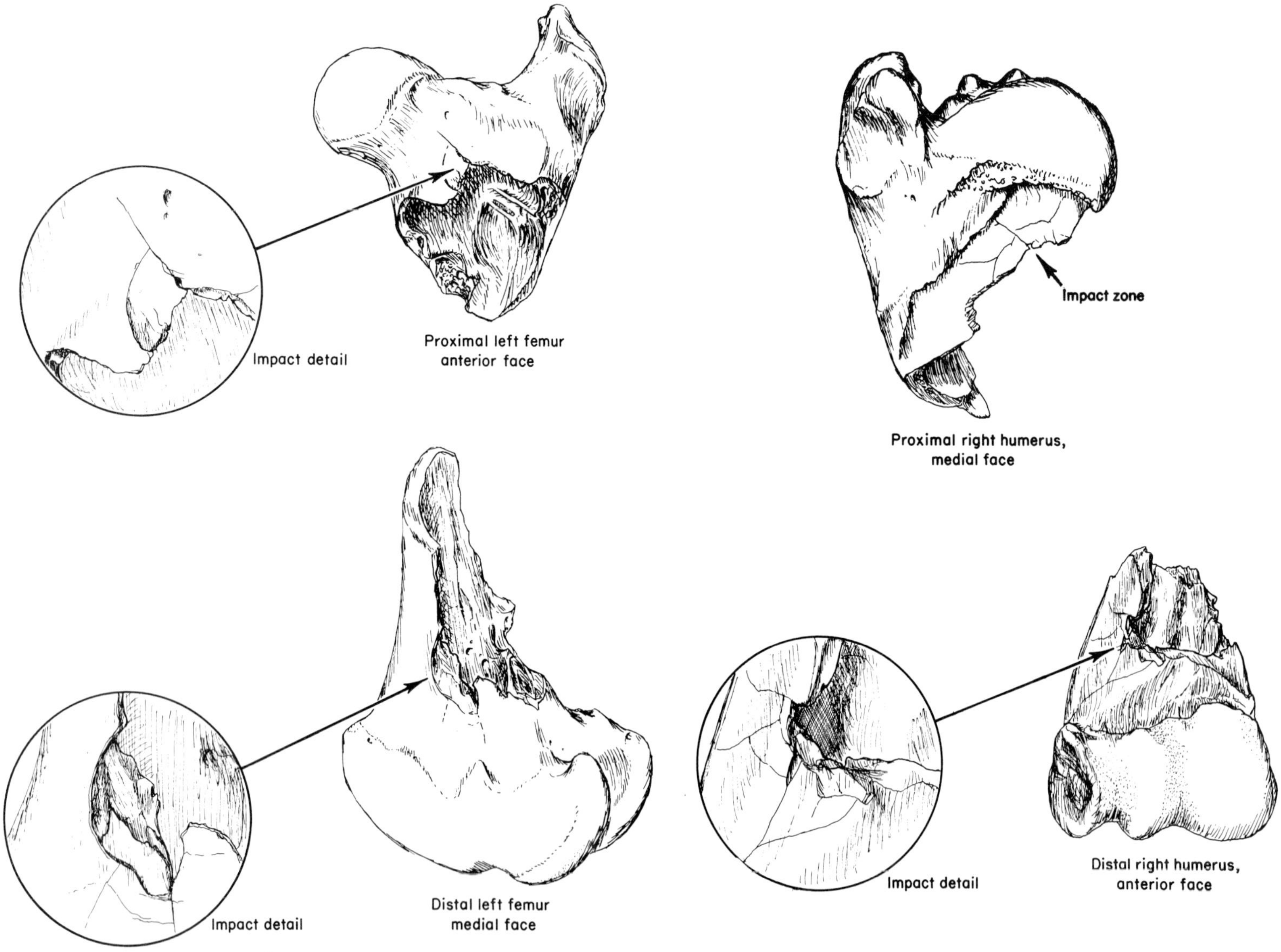

FIGURE 4.53. ***Typical impact scars and patterns of fracture for selected long-bone articulator ends.***

The distal end is impacted as shown in Figure 4.54; that is, obliquely across the hand-held anvil just behind the distal articulator end. Impact is normally on the lateral surface. The curvature is relatively slight on this surface and a massive impact zone is developed with some of the properties of a depressed fracture, with numerous small chips arranged in radial "tiers" around the zone of impact (see Figure 4.53). A short spiral fracture normally comes up from the impact zone around the shaft on the side away from the person. The person cracking the bone is then left holding the cylindrical segment of the femoral shaft if the spiral fracture does not intercept a fracture produced when the proximal end was bro-

FIGURE 4.54. ***Distal femur aligned over hand-held anvil just before impact.***

FIGURE 4.55. ***Pushing marrow from a femoral cylinder with a marrow pusher.***

ken. In the latter case the femoral shaft would essentially break open in the hand. If this does not happen, marrow from the cylinder is pushed out (see Figure 4.55) and the marrow from the cavity of the distal end is dug out both with a stick or sometimes a more substantial marrow pusher, such as a lower sheep rib

TIBIA

The tibia is normally held by the distal end and struck across the hand-held hammer–anvil so the impact is received on the posterior face just below the condyles. As is the case with the femur, this bone is relatively thin in this area and depressed fractures remain along the margins of the actual fracture. Typically all the bone on the posterior face is broken away, leaving only a pointed projection of the tibial crest. The bone is then reversed in the hand and the distal end of the bone is struck down across the hammer–anvil approximately 2 inches back from the distal end of the bone. If the impact is done well, the distal end is snapped off and falls to the floor below the anvil. Classic spiral fractures are not so clearly represented with this method. If the bone is particularly robust it may be impacted across the corner of a stationary anvil resting on the ground. In any event, more or less straight across but jagged breaks are normally the result. The tibia is generally more irregular in section than the femur, so it is not uncommon to impact an unfractured cylinder using the *kaotah* as a hammer, splitting the cylinder and thereby exposing the marrow.

HUMERUS

The humerus is a shorter bone than either the tibia or the femur, yet the distal end is very hard and dense. For this reason the distal end is normally broken off first. The strategy used is similar to that used for the distal tibia, except that the dital end is almost always seated on a stationary anvil, medial face down, and impacted with the *kaotah* used as a hammer on the lateral face directly on the ridge marking the anterior from the lateral face of the bone (see Figure 4.53). A strong bipolar pattern of scarring is commonly visible on the broken end. The bone treated in this manner does not normally have a very distinctive spiral fracture. Instead, a jagged transverse fracture is common.

The bone is reversed in the hand and the proximal end is then impacted obliquely down on a hand-held hammer–anvil so the blow is to the medial face of the bone just below the lip of the large proximal articulator condyle. Sometimes this is cracked off, using the same strategy as for the distal end, since the butcher already has the hammer in the right hand and the anvil is in use. When this is the case, the bone is seated on the anvil with the edge resting across the lateral face and the impact is delivered to the medial face just below the articulator surface.

RADIO-CUBITUS

Fracturing of the radio-cubitus is more directly related to the degree of previous cleaning of tissue from the bone than holds true for any other marrow bone. I have already mentioned that it may be frequently broken during butchering. This bone is rarely served in residential sites, since it is considered "awkward" to break inside the house.

If the bone is cleaned the technique is to strike the distal end of the bone down on the hand-held anvil oriented obliquely, as was shown for the metapodials (see Figure 4.46). Impact is generally slightly on one side, although it may sometimes be on the anterior face back about 1½ inches from the distal end. The distal end usually breaks off rather transversely when this strategy is used, although the break may be somewhat jagged. The bone will then be reversed in the hand and impacted on the hand-held hammer--anvil all along the anterior face of the radius, beginning very close to the articulator surface, and then hit at about 1-inch intervals down the shaft of the radius. Blows are generally just slightly off center but still tend to be on the anterior face of the bone. This mode of fracturing has the effect of breaking the proximal end of the radius off very short. With mature animals the posterior face of the radius remains attached to the cubitus, and impact scarring may be overlapping down short sections of the margins of the fractures on the body of the bone remnants.

Marrow is primarily contained in the medullary cavity of the body or shaft of long bones. This shaft is shaped like a cylinder, so access to the medullary cavity and hence the marrow is facilitated by collapsing or fracturing the cylinder longitudinally. Transversel fractures in the center of long-bone shafts do *not* provide ready access to the marrow.

If one simply breaks a long bone through the center shaft, then one must dig out the marrow, which is very time-consuming and not very satisfactory, or proceed to break the bone again, longitudinally, to expose the contained marrow. This results in curved fractures on short splinters. The most likely bone form discarded is an articulator end with very little attached diaphysis. This is not the form of bone breakage that is generally cited as deriving from marrow-bone breakage and the crack-and-twist method. Breaking bones for marrow results in certain very regular patterns as observed among the Nunamiut Eskimo. That what the Nunamiut do is indeed a *general* strategy has been confirmed for me by the examination of many faunal assemblages from many different time periods and places where intentional breakage by man is not in question. The breakage strategy generally involves percussion impact, whether it be (*a*) with a hammer on a hand-held bone, (*b*) with the bone on a hand-held or rigid anvil, or (*c*) with a hammer while the bone is resting on an anvil. Regardless of the loading tactics used, the target impact zone on a bone is generally the same. The marrow bone to be cracked is generally held by the most robust articulator end and initial impact is just below the neck of the more compact articulator end. This initial heavy blow normally results in considerable fractures developed through the shaft of the bone, and the resulting splinters are peeled back, as they may be held in place by the periosteum. This peeling back exposes the cylinder of bone marrow, which, if the animal is in good nutritional condition, can be picked up and either eaten or added to an accumulating pile of marrow. If the initial impact did not produce fractures extending the length of the bone shaft, secondary impacts are made along the base of the neck of the opposite end. The force of these impacts are generally less than the initial impact that fractured off the opposite articulator end. These more modest impacts fracture the bone so splinters may once again be peeled or struck off the cylindrical shaft. This strategy of directing percussion impacts to the neck of the long bone (the area directly below the articulator end) results in many long-bone splinters and isolated articulator ends with little attached diaphysis.

Control Collections

As mentioned earlier, these descriptions are culled from behavioral observations made while Eskimo were breaking marrow bones, and I did not make a practice of gathering the bones I observed being cracked. Part of this failure on my part was from not knowing at the time what patterning had been observed in the archaeological record to which my observations might be of relevance. For instance, at the time of my fieldwork, I was unaware of Frison's argument about bone choppers. It therefore never occurred to me to inspect in detail the bones I saw cracked for marrow during butchering when the butchers hit the apex of spiral fractures on a rock to dislodge the marrow. In fact, the potential importance of this act did not really register until I became aware of Frison's observations. I took it for granted, since I can remember discussing marrow cracking with Kent Flannery and he reported that this was a common act in the Near East.

> In order to shake the marrow out of the shaft of the bone the villagers seem to have tapped the broken end against a hard surface. The broken edges of some limb bones show a resultant chipping which (although irregular) almost always resembles percussion techniques on flint. Dr. Junius Bird of the American Museum of Natural History reports that he observed the Indians of Patagonia using the technique of marrow removal when he visited them in the 1930's. At that time he noted the scale-like "retouch" produced on the broken edges of the bone. This "retouch" also occurs on gazelle bones from Ali Kosh (see Fig. 123c, e) [Hole *et al.* 1969:291].

This illustrates that when one is studying living peoples one simply does not know what is relevant without a detailed knowledge of the patterning that has been recognized in the archaeological record and, even more important, patterning for which interpretative claims have been made.

I did not ignore marrow cracking, however, since I do have a number of collections of bone known to have been cracked for marrow and about which I have considerable informant information, although I did not witness the bone breaking. From a series of such informant-documented assemblages, some of the properties of bone fragments generated by the tactics described can be demonstrated. Impact scars from hitting the bone during marrow cracking are quite distinctive. First, they are almost always at a *single* impact point, which results in driving off short but rapidly expanding flakes inside the bone cylinder. At the point of impact the bone may be notched, in that a crescent-shaped notch is produced in the fracture edge of the bone. (See Figure 4.50.) These have the appearance of Clactonian notches in the Bordean taxonomy of lithic fracture but are somewhat smaller given the cylindrical nature of the bone. I have *never* observed two such impacts directly side by side. This makes a great deal of sense, since the notches are only produced when the loading successfully breaks the bone. If that happens there is no further need for fracturing at that exact spot. Any additional hitting of the bone will be along an unfractured surface. The frequency of such impact notches (Figure 4.50) in a marrow-cracked assemblage is indicated in Table 4.05, which presents the observed frequencies of splinters (diaphysis fragments exceeding 2.9 cm in length) with impact notches in populations of bone splinters from six control samples collected at Anaktuvuk.

It should be clear that the production of percussion impact scars on the diaphysis is a regular phenomenon and can be expected to exhibit distinctive numerical patterning depending on the impact tactics used and the size and morphology of the bones being broken. For instance, in samples 1, 3, and 4, where there is a bias in favor of metapodials, there is a slight reduction in the percentage of bone splinters exhibiting impact notching. This is consistent with the earlier observations that fracturing, particularly of the metatarsal, was more like producing a series of depressed fractures from yielding bone plates lacking impact notches. On the other hand, assemblages known to have contained more upper limb bones such as the femur, tibia, and humerus clearly exhibit a higher frequency of impact-notched bone splinters. Comparative study of the frequencies of impact notches is believed to be a fruitful approach to distinguishing between fractured bone derived from animal versus human actions.

Obviously the impact scars are not exclusive to bone splinters. They regularly occur on articulator ends and in distinctive places, since the placement of blows is a regular phenomenon varying with the morphology of bones and only minimally with the loading strategies. Impact notches as illustrated in Figure 4.50 are the scars remaining after a section of bone has been removed by virtue of impact. There is another type of impact scar that commonly occurs when impact is over a wide surface, when impact is on a surface "backed" by cancellous tissue, or when the bone is exceedingly strong or dense and thick. This is what I have termed a *splintered and depressed* margin (see Figure 4.53). Morphologically speaking, this is commonly seen along a fracture where there may be some small segments of depressed surface surrounded by a kind of rosette of very small "chips" that have fractured on the surface of the bone; the chips remain attached because the fractures are not carried through the bone. This is well illustrated in Figure 4.53. (This pattern was first described by H. Martin [1910].) Table 4.06 summarizes the frequencies of both impact notches and depressed margins observed in three populations of marrow-cracked bones collected among the Nuna-

TABLE 4.05
Bone Splinters Recovered from Six Marrow-Cracking Experiments

	Unscarred splinters		Impact-notched splinters		
	No.	%	No.	%	Total splinters
Sample 1	59	.86	10	.14	69
Sample 2	63	.84	12	.16	75
Sample 3	50	.85	9	.15	59
Sample 4	114	.84	22	.16	136
Sample 5	144	.83	30	.17	174
Sample 6	312	.83	64	.17	376
Totals	742	.83	147	.17	889

Sample descriptions

Sample 1	From the Mask site (see L. R. Binford 1978b: 178–182); represents the activities of several hunters during a 2-day period; biased in favor of metapodials
Sample 2	From marrow-cracking demonstrations (reported in L. R. Binford 1978b: 155)
Sample 3	From the Kongumuvuk fall hunting stand (L. R. Binford 1978b: 357); represents the activities of two hunters during one evening
Sample 4	Bones saved for study by the women of an Eskimo household during April 1971; freshly killed animals introduced during the last two days of the collection; biased in favor of metapodials; relative to samples 5 and 6
Samples 5 and 6	Bones saved by the head of an Eskimo household over a considerable period of time during late fall and early winter; represent household consumption of marrow bones largely from stored meat; biased in favor of upper limb parts (femur, tibia, humerus)

TABLE 4.06

Frequency Distribution of Impact Scars on Articulator Ends of Long Bones[a,b]

Anatomical part	Rulland site[c]			Kakinya site[d]			Sample[e]			Total		
	No.	N	D	No.	N	D	No.	N	D	No.	N	D
Cancellous												
Proximal femur	6	1	4	17	0	6	4	0	2	27	1	12
Distal femur	8	1	2	16	0	5	4	1	1	28	2	8
Proximal tibia	5	0	3	14	2	3	4	0	2	23	2	8
Proximal humerus	4	1	2	7	1	3	4	0	1	15	2	6
Subtotal	23	3	11	54	3	17	16	1	6	93	7 (8%)	34 (37%)
Thin, smooth												
Distal radio-cubitus	5	0	2	10	1	3	5	1	0	20	2	5
Distal metatarsal	0	—	—	7	0	1	4	0	1	11	0	2
Proximal metatarsal	2	1	0	2	2	0	5	1	0	28	3	0
Proximal metacarpal	1	1	0	1	0	1	5	1	1	7	2	1
Subtotal	8	2	2	20	3	5	18	2	2	47	7 (15%)	8 (17%)
Thick, smooth												
Distal humerus	2	1	0	12	2	2	5	2	0	19	5	2
Distal tibia	6	0	0	10	1	0	4	1	0	20	2	0
Distal metacarpal	0	0	0	1	0	0	5	1	0	6	1	0
Proximal radius	6	2	1	11	2	2	5	0	1	22	4	4
Subtotal	14	3	1	34	5	4	19	4	1	67	12 (18%)	6 (9%)
Grand total	45	8 (18%)	14 (31%)	108	11 (10%)	26 (24%)	53	7 (13%)	9 (17%)	206	26 (13%)	48 (23%)

[a]No. = number of articulator ends observed; N = number of impact notches observed; D = number of marginal depressions observed.
[b]Values in parenthesis are percentages of sample totals.
[c]See Binford 1978b: 374.
[d]See Binford 1978b: 374–391.
[e]See Table 4.05 for description.

miut Eskimo. I have arranged the articulator ends into three basic groups, ranging from those with very cancellous interiors through those with smooth but thin-walled marrow cavities to those with thick and dense but smooth interior marrow cavities. It will be recalled that the Nunamuit were aware of these anatomical differences and broke bones slightly differently depending upon the morphology.

Several things are of interest in this table. Reviewing the totals it is clear that moving down the table from cancellous to smooth and thick interiors there is a correlated increase in the percentage of articulator ends exhibiting distinct and clean impact notches. Only 8% were observed among the ends judged cancellous, whereas 18% of those judged to have thick and smooth marrow cavities exhibited impact notches. Conversely, only 9% of the thick and smooth ends exhibited marginal depressions from points of impact, whereas 37% of the cancellous ends exhibited such scars. There are several implications of these data. Clearly the more cancellous tissue there is forming an "undersurface" to the bone, the more evidence of direct impact there is remaining on the bone. Similarly, the more dense and compact the bone, the cleaner the fracture and the less direct evidence of impact there is. Perhaps this is not as indicative a correlation as it might appear, since good examples of marginal depressions and "chip rosettes" are in evidence on some long-bone sections from bison bone illustrated by Frison (1974: 301, Figure 1.13b, 1.24; 1978: 307, Figure 8.2a). Similarly, the bovid bone impacted by Bonnichsen (1979: Plate IV-2, No. 10) experimentally with the "mid-diaphysis smash technique" clearly shows the chip rosettes, and the impact depression characteristic of my more cancellous bones. This effect may be more a function of thickness than of bone density. In any event, the evidence from impacting bone is common on articulator ends, occurring on over 35% of the total number examined. As indicated earlier, the placement of these impact traces is *regular*, occurring in the same locations relative to the morphology of the different bones.

I think it is clear that the combination of the disposition and frequency of impact scars, on both articulator ends and diaphysis fragments, coupled with the breakage morphology of the articulator ends, should correlate well with the overall marrow-bone-breakage tactics.

Breakage Related to Other Forms of Bone Processing

Previous descriptions exist regarding the use of bones in rendering bone grease (L. R. Binford 1978b:157–163; Leechman 1951) and bone juice (L. R. Binford 1978b:163–165). In both processes the bones carrying grease and fat in the cells of cancellous tissue are pounded as to increase the surface area of the bone exposed. These fragments are then boiled, thereby rendering off the grease. Splitting and otherwise pulverizing articulator ends is an integral part of this processing, but as we have seen it is essentially never a consequence of marrow-bone breakage (with the occasional exception of the proximal metapodials). I have previously discussed at some length how one can identify the consequences of this behavior (see L. R. Binford 1978b:157–165).

Man-to-Man Comparisons, or Alternative Human Behaviors

Given the central problem of diagnosing variability in bones, one of the greatest sources of variability is most likely to be the different ways man has developed for accomplishing similar things. Clearly, if we are going to compare man to animals, consideration of this potential variability is crucial.

I have by no means described all the alternative tactics which the Nunamiut are capable of performing. The slight accommodations of technique to the different anatomical properties of the marrow bones, their degree of articulation, and the degree of previous cleaning, all lead us to expect that breaking the bones of larger animals, or bones from animals with different anatomical properties, would be accommodated by modifications in technique. In addition to the responsiveness of the Eskimo to properties of the bones, there is a correlated response to the situation in which the activity is carried out. For instance, one informant pointed out that using the hand-held anvil inside a winter house during a meal might result in bone fragments or articulator ends going into the fire or someone's lap, or it would just be considered messy and "impolite." On the other hand, using a hammer or an on-anvil technique inside a house did not carry these risks. Similarly, men in hunting stands repeatedly told me that using the hammer-on-bone strategy was possible standing up and one did not run the risk of having ends of bones, marrow, and splinters fall to the ground and get dirty. Men also laughed and pointed out that finding an anvil and spreading a drop cloth was not what you did while sitting around a fire with friends! Finally, it should be pointed out that there is an interaction between one set of conditions of tactics and another. The amount of disarticulation and previous cleaning of bones is apt to be much less in a hunting camp or stand. This is also where stand-up techniques of bone breakage are more appropriate. This means that there will be some correlations between technique used and fracture patterning that are not the direct result of the character of the technique, but instead derive from the characteristic coincidences between degrees of prior bone preparation and the conditions influencing the type of technique used. This type of interaction should encourage archaeologists greatly, since the strength of such interactions for conditioning the resulting morphological properties of broken bone should be strong and primarily informative of the role of different types of sites in the overall settlement–logistical system. This, if true, has strong potential for methodological development. These descriptions will, I hope, prompt the imagination of archaeologists to consider the nature of the variables man is responding to when behaving in such a way as to produce different morphological patterns in broken bone. Doing the latter is theory building and is the basis for building a reliable set of methods for giving meaning to the archaeological record. It is beginning the interesting task of *explaining* cultural differences and similarities.

By way of contrast, the ethnographic descriptions by Zierhut (1967) and Bonnichsen (1973) present us with a classic normative description of behavior. One gets the impression that the Calling Lake Cree break bones by the "mid-diaphysis smash technique" everywhere and at all times, in hunting camps, in the house, snacking on marrow bone while walking along a forest trail, while butchering animals, and so on. They do it that way not because it helps to accomplish a given task in a given setting, but because they are Calling Lake Cree! My guess is that this impression was structured by the investigators and not by the Calling Lake Cree. My guess is also that the Cree rarely eat marrow with meals in the house, that they assumed the context from the place where the questions were asked, and their demonstrations were more reminiscences of things from the past than actions daily performed. What was demonstrated by the Cree was probably roughly equivalent to the Eskimo's "bone on hand-held anvil" strategy, generally performed in residential camps. I further guess that the marrow was intended to be added to a stew in small quantities and not to yield a large quantity, as was most Eskimo women's bone processing. Part of my reasons for thinking these things relates to my experiences as a young boy in Appalachia. I was taught to break marrow bones by my grandmother in a manner similar to what Zierhut (1967) describes for the Calling Lake Cree, except we invariably used an ax. You might call our technique the "kindling wood technique" although it meets all the criteria for the "mid-diaphysis smash technique" that has so enamored Bonnichsen (1979:54, Plate IV-1). Stanley South and I once discussed how we were taught to break marrow bones up in the eastern mountains. Stanley said the first time he was told to take a big beef bone out and "break it up" because "Uncle so-and-so was coming who really liked bone marrow" was a real learning experience. Stan said he went out to the woodshed and put the big bone down with each end resting on a split section of log and gave a mighty swing with the butt end of the ax. The bone went flying off into the woodpile. He tried again and the same thing happened. He then turned the ax around and hit the bone midshaft with the bit of the ax and the bone broke nicely in half, but one end flew up and hit the roof of the woodshed. Although my experiences were largely the same, I had been told by my grandmother always to hit the bone in the middle with the bit of the "old dull ax out by the stump harness." This was an ax used for cutting roots when stumps were being pulled by a mule team. It was always well polished from having been driven into the soil and fairly dull from having hit so many stones. I had also been told not to hit it as hard as a piece of firewood "because it might fly up and hit you in the head." Clearly I had no knowledge of bone anatomy; any bone given us was treated in the same fashion. Similarly we certainly had no knowledge of fracture mechanics or any understanding of the relationships between bone anatomy, impact location, or strength of loading, as is illustrated by Eskimo behavior. We just learned that if we hit it too hard it might "fly up and hit us in the head." In short, we used our knowledge of chopping kindling wood to fracture bones. Although I have argued against the analogy with lithics as a guide to differentiating between animal and humanly broken bones, clearly the knowledge used by the Eskimo was more akin to the knowledge needed in lithics reduction and tool manufacture than to that needed in kindling cutting with an ax!

The kindling-wood strategy that I learned as a young boy, which Zierhut (1967) and Bonnichsen (1973) observed among the Calling Lake Cree, and which I observed among the Navajo (see L. R. Binford and J. B. Bertram 1977:94), appears to be a preparation technique associated with relatively sedentary peoples using boiling as an almost exclusive technique of food preparation. It appears to be modeled on the mechanics of cutting kindling wood with an ax rather than on an understanding of the fracture mechanics of the bone itself and the relationships of impact to bone morphology. Both the apparent model for the technique and the contextual features common to the Cree, Navajo, and Appalachia suggest that this is a very poor ethnographic analogy for modeling the behavior of big-game hunters, men of the Pleistocene, or for distinguishing the products of men from the products of animals.

I am not suggesting that we substitute the almost certainly incomplete descriptions from the Nunamiut Eskimo as our generalized configurational model of marrow cracking. Instead I am suggesting that we attempt to understand *why* one group of people does things one way and another does similar things in different ways. In short, explaining cultural differences and similarities is our job. If we cannot do this we cannot develop means for reliably giving meaning to our observations. For instance, in a very important seminal study of marrow cracking, the fragment morphology at the famous sites of Star Carr and Kongenmosen was compared with those from two other well known Mesolithic sites from northern Europe (Noe-Nygaard 1977). The author clearly demonstrates that the morphology of fragments remaining on the Star Carr site is similar to that at Kongenmosen, and both differ from the fragment morphology characteristic of the Praestelyngen and Muldbjerg I sites. At Star Carr long bones tended to be broken in midshaft with most of the diaphysis remaining attached to the articulator ends. By way of contrast, at the sites of Praestelyngen and Muldbjerg I the bones seem to have been cracked for marrow using a technique that directs the impact in the area of the articulator ends, similar to that generally described here for the Eskimo. In fact, comparison of the illustrations (Noe-Nygaard 1977: Figure 8) and the specimens illustrated here (see Figure 4.53) strongly suggests that similar techniques were used by the Eskimo in residential camps and the occupants of Praestelyngen and Muldbjerg I. On the other hand, the materials illustrated from Star Carr evidence breaks similar to those produced in the butchering of frozen carcasses rather than those produced in marrow recovery. The low frequencies of femurs (Noe-Nygaard 1977:222) are consistent with a kill butchering location or a field camp (see L. R. Binford 1978b). Similarly, the low frequencies of the proximal humerus relative to the distal humerus plus the reported presence of femoral cylinders strongly suggest that the remains from Star Carr were ravaged by predator–scavengers. The totality of the differences noted between the four sites strongly suggests a major contrast between residential locations and special-purpose sites related to hunting logistics. These approaches to understanding the differences contrast with the more traditional view:

> Differences in marrow fracturing technique between various cultures reflect different methods of preparing food and may therefore reflect differences in cultural level, such as aceramic as opposed to ceramic cultures. It may thus be possible to use the character of bone splitting as an ecostratigraphical tool [Noe-Nygaard 1977:235].

Clearly we need far more detailed pattern-recognition studies, designed to elucidate fracture patterning, of archaeological bone assemblages. In some cases we can obtain "controls," in that the species can be reliably identified and any correlations with body size or with bone density can be investigated. Similarly, where we have good reason to place some faith in our functional interpretations of sites, such as bison jumps and kill sites versus residential locations, or processing areas associated with bison kills, we may begin to determine how sensitive fracture patterning is to site functional differences. Similarly, if we can gain control over "season of occupancy," we can begin to investigate the degree to which butchering strategy may be altered to "bone breakage" versus "joint cutting" during seasons of extreme cold. We have not even begun to investigate the potential information about the past being carried in the archaeological record by such mundane and unspectacular things as broken bones.

As interesting as such implications for future study may be, our real interest here is with criteria that might distinguish bones broken by men from those fractured by nonhuman predator–scavengers.

Comparing Man and Beast

I have thus far developed a number of contrasts, such as cut marks from tools versus tooth-scoring marks. Impact fracturing from the use of tools in bone breaking has been contrasted to the gradual destruction of soft parts by the progressive weakening of bones and their collapse under animal persistence. Standing behind these contrasts is the recognition that animals and man may well dismember a skeleton differently because each acts under different mechanical constraints.

There is little chance than an observer of modified bone would confuse cut marks inflicted during dismembering or filleting by man using tools with the action of animals. It was noted that cut marks are concentrated on articulator surfaces and are relatively rare as transverse marks on long bones except where encircling marks may betray skinning activities. It was suggested that cut marks from stone tools are most commonly made with a sawing motion resulting in short and frequently multiple but roughly parallel marks. Such marks are generally characterized by an open cross section. Another characteristic of cut marks derived from the use of stone tools is that they rarely follow the contours of the bone on which they appear. That is, the cut does not show equal pressure in depressions and along prominent ridges or across the arc of a cylinder. In almost all ways, the marks of animals' teeth are somewhat different. They commonly follow the contours of the bone's surface, as is well illustrated in the specimen shown in beautiful color by *National Geographic* (Canby 1979:356) cited as evidence of man's presence at the site of Tagua Tagua in Chile. The caption reads, "Deliberately fractured and engraved with parallel strokes, a horse leg bone was found in a natural trap. . . ." This is almost a type specimen of tooth scoring by gnawing animals! Tooth marks may frequently take the form of a depressed or mashed lines, as appears to be the case on the bone from Chile. On many of the wolf specimens, the tooth mark under magnification appears as a "cracked" surface scar rather than as a cut or incision in the bone as is the case particularly with metal tools.

Claims for the use of tools should be supported by the citation of marks produced by tools.[1] Admittedly, distinguishing between tooth scoring and cut marks is sometimes difficult, but the attempt must be made. For instance, increases in the amount and density of scoring and pitting are noted around areas that have been partially destroyed or furrowed, particularly around long-bone shafts when the articulator end has been partially destroyed by gnawing. This association of scoring with patterns of destruction is not to be expected when man dismembers an animal with tools. More commonly the modifications produced by animals have been inter-

[1]I must emphasize that I am speaking of a substantial population of animals and not a single animal. I am also "hedging" on the size of the population, for if one excludes the mandible the highest proportion of marked bones is 30–40% of those present in the Mousterian sample (see Table 4.03). On the other hand, around 50% were marked in some classes represented in the modern Nunamiut material (see Table 4.02) Obviously in small samples we could expect some examples of humanly butchered animals exhibiting few if any marked bones.

preted as either intentional modification of bone for use as tools or attritional modification from the bone's having been used as a tool. The use and production of bone tools have not been discussed here, for several reasons. Most important is the fact that quite literally this is frequently the "bone of contention," the interpretation of uncertainty. If I accepted everything that has been cited as a tool and proceeded to describe it, then almost all the documented modifications produced by animals would have been described. On the other hand, if I had demanded the same actualistic controls, the bear and the footprint seen together, as the basis for describing bone tools "characteristically produced" by human agents, then I would face the unlikely uniformitarian assumption that contemporary patterns of manufacture and use were the same as in the past! We are not talking about products generated because of "intrinsic" mechanical and behavioral constraints or limitations on the way a bone can be modified, as is likely to be the case of animal modification. Human tool design and use are responsive to situation, mechanical needs, and the character of knowledge, equipment, and raw materials available. Tool production is responsive to tool need and recognized options, among other things.

We know that the range of tool needs, the character of knowledge, available equipment, and accessible raw materials are certainly different and representative of a different range of variability today than would have been the case at different periods in man's past. Without knowing the things we seek to know about the past, we could never guess what contemporary experiences might be directly relevant to either a specific archaeological case or all the past taken in a generic sense. *We cannot assert that we understand the very phenomenon we seek to understand in order to achieve a method for its study!* Arguments about bone tool use and manufacture must be arguments from elimination, where alternative explanations are systematically eliminated through the use of a robust and informed methodology. Even after we have narrowed the identity of the causal agent to man, the question then remains as to what man did with the item and why he made it in the manner observed. For instance, I have made frequent reference to the fact that animals produced modified bones that were to me indistinguishable from what have been called fleshers by Frison (1978) and others. In this case I am questioning the identity of the causal agent. On the other hand, Frison has repeatedly noted a pattern of modification on the fractured ends of long bones at several bison kill butchering sites. He has interpreted these as choppers used expediently in the butchering process (see Frison 1974: 51–56, 1978: 301–316). I agree that many of the items illustrated by Frison have no direct analogues in animal-modified assemblages. Vague similarities can sometimes be found for some bones with dense shafts such as the distal tibia, but I have been unable to find anything similar to the proximal tibia and proximal femur choppers illustrated by Frison.

My experiences with butchering and expedient marrow eating by butchers while engaged in work, as well as Junius Bird's experiences in Patagonia as reported by Flannery, (Hole *et al.* 1969:291) suggest that the same patterns of modification Frison observed on his choppers might derive from striking the broken ends of fractured bones down on hard surfaces during marrow removal.

This is the type of problem to which experiments can be directed. For instance, in the case of Frison's post hoc interpretative model regarding chopper use, a bone could be used in the manner envisioned by Frison. Similarly, an identical bone could be treated in the manner described for marrow cracking. The consequences of the two separate acts could be compared, both could be compared to the archaeological bone, and one of the alternatives could, it is hoped, be eliminated.

There is still another research approach, the approach that stresses the "integration of knowledge" as a criteria of plausibility. For instance, as long as there was reason to believe that ancient men on the Plains muscle stripped animals as a normal butchering strategy, then the use and presence of choppers were internally consistent with the facts cited as evidence for muscle stripping. I have seriously questioned the degree to which the facts cited in support of the muscle-stripping argument are referable to human action. Most of the forms of bone destruction interpreted as evidence of muscle stripping are seemingly indistinguishable from animal-ravaged faunal material. If this is the case, then there should be no necessary correlation between choppers and the alleged evidence of muscle stripping. Carrying out such comparative studies would almost certainly result in the recognition of previously unrecognized patterning in the archaeological record. These patterns then would present us with new challenges and potential sources of new information regarding the past. Knowledge gained through such prompted research[2] then would increase the strength of our methodology and in turn our science would grow in both knowledge and sophistication.

The problem of surficial modification, although complicated, is relatively minor compared to the problem of arguments that cite structural modifications and breakage patterns as causally referable to human agents. Most claims for human causality have reference to overall morphology or the presence of obvious postmortem modifications on bones, such as polish or abrasions considered indicative of use. It seems to me that before interpretations are offered for such modifications one must attempt to assess the relative role of different agents as general contributors to the populations of bones from which such "interesting" items might be singled out. Arguing for human modification of a particularly interesting item abstracted from a population of otherwise exclusively animal-modified bone is not very convincing. This is criteria of the "integration of knowledge" somewhat reversed and used as a procedural suggestion for research. Are there not properties of populations that might be evaluated to provide the context for looking at the characteristics of particular items? The following section presents a compilation of some properties of humanly modified faunal assemblages that contrast in a generic sense with properties of assemblages produced by animals.

[2]What I have called here *prompted research* is similar to what Richard Gould calls the *argument by anomaly*. I recognize that when we are surprised or observe unexpected things, we should investigate (see L. R. Binford 1972b:115). In this I concur with Gould. That this in any way is a more epistomologically "convincing approach than the use of analogy" (R. A. Gould 1980:138) is sheer nonsense. An argument from analogy is a form of inductive argument. Gould's argument by anomaly is not a form of argument but a guide to asking interesting questions. The answers to such questions may be given through arguments from analogy, which is exactly what Gould does in his example of "anomalous" raw material usage at Puntutjarpa (see R. A. Gould 1980: 153–156), despite his disclaimers.

Morphology of Bone Breakage

The degree to which bones have been chewed and fed upon conditions the character of the surviving faunal inventory. However, the pattern of bone gnawing is redundant and results in a characteristic pattern of bone destruction. This pattern was first described during my Alaska fieldwork by Dan Witter, who was given the task of studying the bones remaining in several well-documented dog yards in Anaktuvuk village. Later, when I worked over my documentation of wolf kills, it became clear that the same types of destruction were present.

There is a fairly regular sequence to the destruction of bones by both wolves and dogs. The redundancy in the chewing strategy results in there being very distinctive patterns to surviving long-bone parts. A general classification of these parts was worked out by Dan Witter, and his terminology will be adopted here.

1. *Articulator end* (Figure 4.56, specimens on left side): the articulator end when there is a broken diaphysis just below the epiphysis of a bone. That is, the articulator end has been broken off immediately below the "head" or in the neck of the bone, with little attached diaphysis.
2. *End plus shank* (Figure 4.56, middle specimen): the articulator end with attached diaphysis complete to approximately halfway between the two ends.
3. *End plus shaft* (Figure 4.56, specimens on right side): an articulator end with the attached complete diaphysis ranging from half the length of the bone as far as the neck at the opposite end. The bone is essentially complete except for the opposite end.
4. *Cylinder* (Figure 4.57): the central part of long bones or segments thereof, *minus the articulator ends*. The behavior that produces this type of part is simply the alternate gnawing away of the articulator ends leaving segments of the long-bone shaft remaining as a cylinder.

It should be noted that as the bone is reduced by gnawing there is a correlated destruction of anatomy. A bone preserved as a distal articulator end plus shaft (Figure 4.56, right side) is rarely as "de-

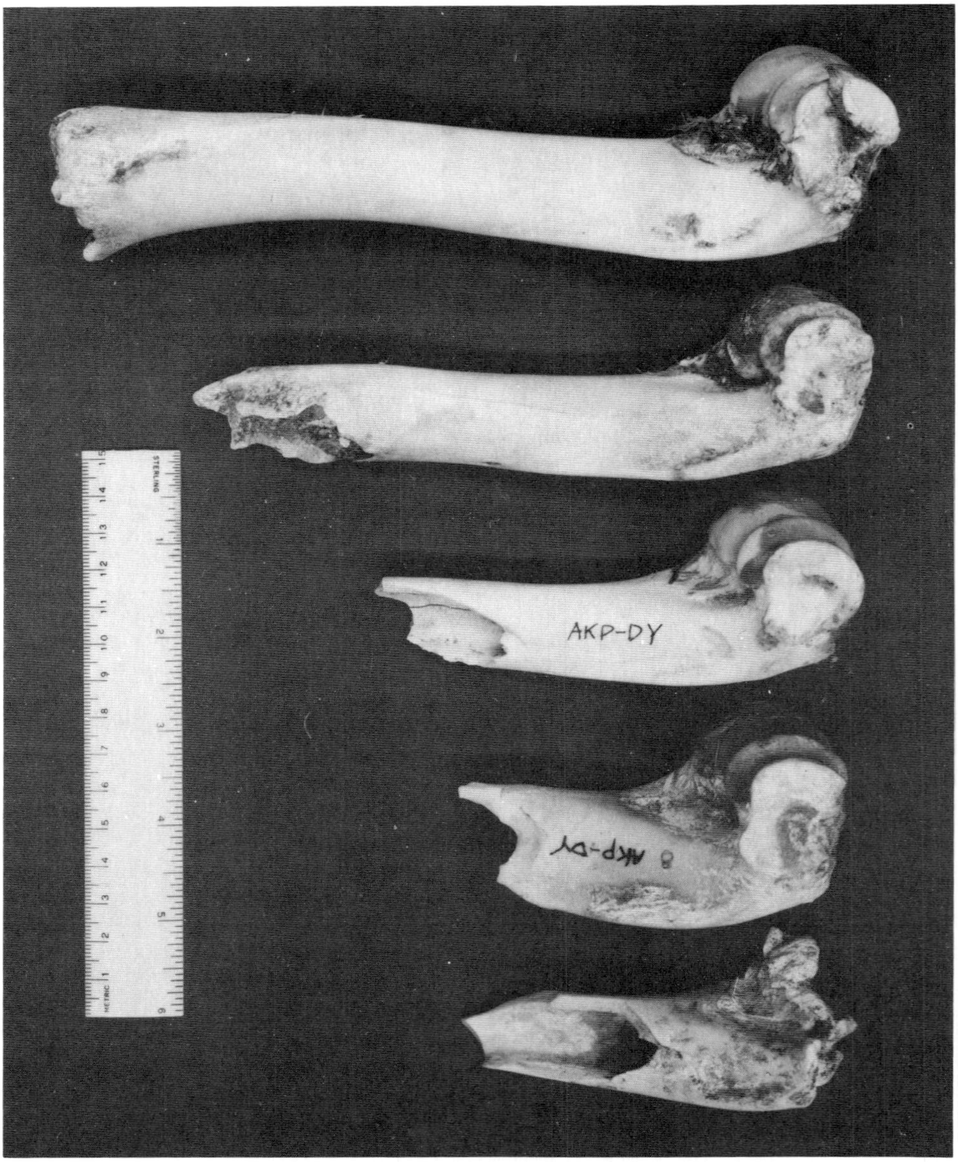

FIGURE 4.56. *Examples of different degress of breakage for animal-gnawed long bones.*

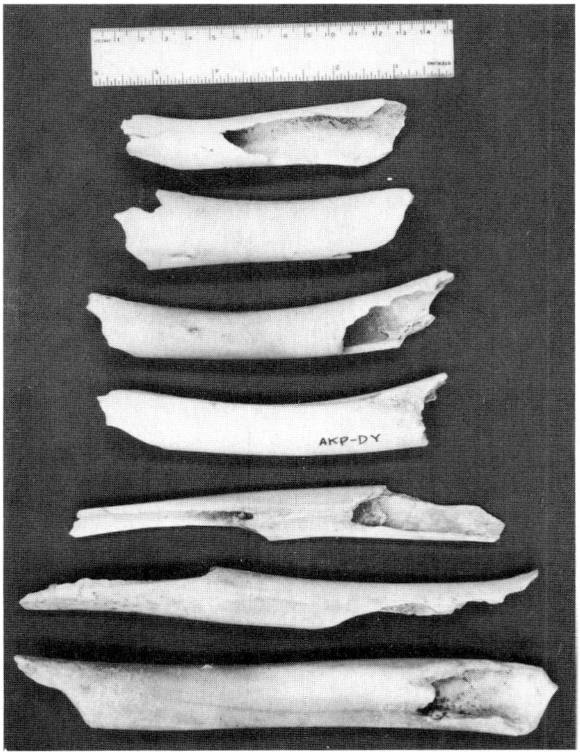

FIGURE 4.57. ***Cylinders—a diagnostic product of animal gnawing.***

stroyed" on the distal end as is a distal articulator end when both the shaft and shank have been removed. *The relative frequency of these progressively reduced parts is a measure of the level of exploitation by gnawing animals.* Table 4.07 summarizes data from the wolf kills, three dog yard samples, and two assemblages processed by man as to the relative frequencies of identifiable long-bone parts representing different stages of breakage or attrition. The patterns documented here will be presented as diagnostic of gnawing animals as opposed to the patterns associated with tool-related breakage of bones by man for marrow extraction.

1. Animal gnawing is an attritional process resulting in the progressive deletion of certain parts from the assemblage. Breakage of bone for the recovery of marrow by man is a process that results in morphological changes but does not necessarily result in attrition.

2. Animal gnawing is a sequential process resulting in a progressive development of destructive patterning that has a signature quality correlated with the pattern of breakage. Exploitation of bones for marrow by man is a direct process such that there is no necessary correlation between the morphology of breakage and the pattern of survivorship in bone parts.

Given tools, man is able to go directly to the target of exploitation. If skilled, he first cleans the bone so control of percussion is maintained (see L. R. Binford 1978b:152–157), and then using a hammer or anvil breaks the bone generally just below the head or articulator end of the bone. Regardless of the varying skills involved and the fact that there are different ways of addressing the bone with percussion tools, the point being made is not altered: breakage is accomplished *by tools directly,* and is not dependent on weakening the bone through the successive destruction of parts, as is the case with gnawing.

Figure 4.58 illustrates the relationship between percentage frequencies of broken bone parts as tabulated in Table 4.07, columns 2 and 14, for the wolf kills displayed against the sample from the Kakinya site, columns 12 and 24. There are many examples of high percentage values for articulator ends in the Kakinya sample and corresponding low values in the wolf kill sample. Conversely, there are high percentage values for articulator ends with attached shafts and shanks in the wolf kill population and corresponding low values in the Kakinya sample. The basic distribution is of an indirect Poisson form with a separate clump of articulator ends that show moderate to high values in the wolf kill sample. These are both ends of the femur, the proximal metatarsal, the proximal tibia, the distal humerus, and the distal radio-cubitus. To state the patterning somewhat differently, high frequencies of articulator ends with attached shafts and shanks are characteristic of animal-gnawed assemblages. Isolated articulator ends may occur in both types of assemblages but are largely the exclusive or dominant forms in assemblages cracked for marrow by man.

TABLE 4.07
Variation in Bone Breakage for Animals and Human-Derived Assemblages

	Dog yards								Human-derived assemblages			
	Wolf kills		Sod 2		Sod 3		Sod 1		Anavik stand		Kakinya site	
Anatomical part	MNI (1)	% (2)	MNI (3)	% (4)	MNI (5)	% (6)	MNI (7)	% (8)	MNI (9)	% (10)	MNI (11)	% (12)
Humerus												
Proximal only	1.0	.40					1.5	.50	.5	1.00	4.0	1.00
Proximal + shank	.5	.20										
Proximal + shaft												
Complete	1.0	.40	2.0	1.00			1.5	.50				
Total	7.5		2.0		0		3.0		.5		4.0	
Distal only	.5	.05	.5	.10	2.0	.57	4.0	.19	.5	1.00	5.5	.83
Distal + shank	2.5	.26	1.0	.20	.5	.14	9.0	.43			.5	.08
Distal + shaft	5.5	.58	1.5	.30	1.0	.29	6.5	.31				
Complete	1.0	.11	2.0	.40			1.0	.07				
Total	9.5		5.0		3.5		21.0		.5		6.0	
Radio-cubitus												
Proximal only	.5	.06	.5	.11	.5	.14	1.5	.09	1.0	.66	1.5	.25
Proximal + shank	1.5	.19	1.0	.22	.5	.14	5.0	.29			1.5	.25
Proximal + shaft	4.0	.50	.5	.11	.5	.14	5.5	.32				
Complete	2.0	.25	2.5	.56	2.0	.57	5.0	.29	.5	.33		
Total	8.0		4.5		3.5		17.0		1.5		6.0	
Distal only	1.5	.33							4.0	.89	5.0	1.00
Distal + shank	.5	.11	1.0	.25			1.0	.14				
Distal + shaft	1.0	.22	.5	.13			1.0	.14				
Complete	2.0	.44	2.5	.63	2.0	1.00	5.0	.71	.5	.11		
Total	4.5		4.0		2.0		7.0		4.5		5.0	
Metacarpal												
Proximal only	.5	.09	.5	.20							1.0	.50
Proximal + shank	1.0	.18			.5	.50	1.5	.30				
Proximal + shaft	1.5	.27	.5	.20			1.5	.30				
Complete	7.5	.45	1.5	.60	.5	.50	2.0	.40	2.5	1.00	1.0	.50
Total	5.5		2.5		1.0		5.0		2.5		2.0	
Distal only	1.5	.27					1.0	.33	1.5	.38	.5	.33
Distal + shank	1.0	.18										
Distal + shaft	.5	.09										
Complete	2.5	.45	1.5	1.00	.5	1.00	2.0	.66	2.5	.63	1.0	.67
Total	5.5		1.5		.5		3.0		4.5		1.5	

(continued)

One of the final forms of reduction by animals is the production of long-bone cylinders (Figure 4.57); data on cylinders and splinters are given in Chapter 5, Table 5.01. Figure 4.59 illustrates the relationship between the number of bone splinters per MNI from the wolf kills (discussed in Chapter 5) and the percentage of the maximum number of articulator ends that could have been present but were in fact missing (24 in the case of kills of single animals minus the number present divided by 24). There is clearly a positive linear relationship between the number of splinters and the articulator ends that were *missing*.

TABLE 4.07—*Continued*

	Dog yards								Human-derived assemblages			
	Wolf kills		Sod 2		Sod 3		Sod 3		Anavik stand		Kakinya site	
Anatomical part	MNI (13)	% (14)	MNI (15)	% (16)	MNI (17)	% (18)	MNI (19)	% (20)	MNI (21)	% (22)	MNI (23)	% (24)
Femur												
Proximal only	1.5	.43	1.0	.67	1.5	1.00	1.0	1.00	1.0	1.00	8.5	.94
Proximal + shank	.5	.14									.5	.06
Proximal + shaft	.5	.14										
Complete	1.0	.29	.5	.33								
Total	3.5		1.5		1.5		1.0		1.0		9.0	
Distal only	5.0	.71	.5	.14	.5	1.00	1.0	.67	1.0	1.00	7.5	.79
Distal + shank	1.0	.14	1.5	.43							2.0	.21
Distal + shaft	1.0	.14	.5	.14			.5	.33				
Complete	1.0	.14	.5	.14								
Total	7.0		3.5		.5		1.5		1.0		9.5	
Tibia												
Proximal only	1.5	.30			.5	1.00	2.5	.71	3.0	1.00	7.0	1.00
Proximal + shank	.5	.10										
Proximal + shaft												
Complete	3.0	.60	.5	1.00			1.0	.29				
Total	5.0		.5		.5		3.5		3.0		7.0	
Distal only			1.5	.25			1.5	.11	3.0	.86	4.5	.90
Distal + shank	.5	.05	1.5	.25			3.0	.22	.5	.14	.5	.10
Distal + shaft	7.5	.68	2.5	.42	1.0	1.00	8.0	.59				
Complete	3.0	.27	.5	.08			1.0	.07				
Total	11.0		6.0		1.0		13.5		3.5		5.0	
Metatarsal												
Proximal only	5.0	.59	3.0	.67	.5	.50	3.5	.70	3.0	1.00	2.5	.63
Proximal + shank												
Proximal + shaft	.5	.06			.5	.50						
Complete	3.0	.35	1.0	.33			1.5	.30			1.5	.38
Total	8.5		3.0		1.0		5.0		3.0		4.0	
Distal only									5.0	1.00	1.5	.38
Distal + shank	1.0	.22					1.0	.33			1.0	.24
Distal + shaft	.5	.11					.5	.17				
Complete	3.0	.67	1.0	1.00			1.5	.50			1.5	.38
Total	4.5		1.0		0		3.0		5.0		4.0	

A totally different situation holds for sites where humans were responsible for the faunal assemblage. Except in cases where there was manufacture of bone grease or bone juice (intentional destruction of articulator ends), there was a fairly constant relationship between the number of bone splinters and the number of articulator ends *present* on the site when corrections were made for the number of complete bones and the number of shafts and cylinders (see L. R. Binford 1978b:157, 464, Table 9.1).

Another interesting relationship is demonstrated in Figure 4.60, where the number of splinters is dis-

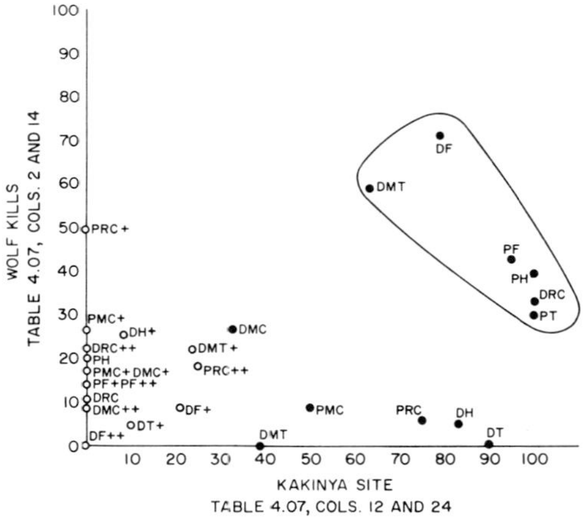

FIGURE 4.58. ***Relationship between breakage frequencies at the Kakinya site (human) and on the Alaskan wolf kills (animal).***

played against the numbers of cylinders per MNI (taken from Table 5.01). It is clear that a strong positive linear relationship is indicated up to an approximate value of 55 splinters per MNI. At that point the trend is reversed and the number of cylinders is reduced in an apparent linear fashion with further increases in the numbers of splinters. This simply reflects that cylinders are one of the "last stage" forms

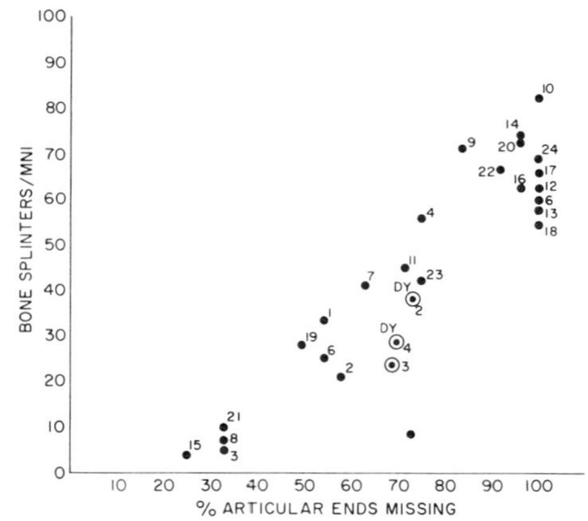

FIGURE 4.59. ***Relationship between number of bone splinters per MNI and the percentage of articulator ends potentially present that were in fact missing. (Numbers relate to columns in Table 5.01.)***

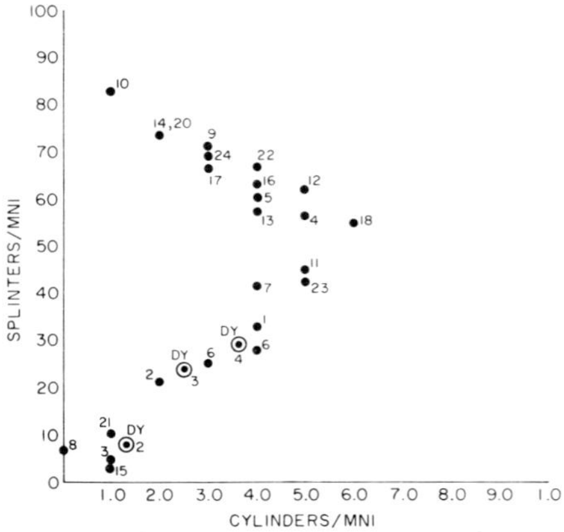

FIGURE 4.60. ***Relationship between number of bone splinters and bone cylinders per MNI. (Numbers relate to columns in Table 5.01.)***

to be produced by gnawing animals; hence breakage increases as the animal progresses in the reduction of a bone to a cylinder. Figure 4.61 illustrates the converse of this relationship, namely that the number of cylinders per MNI decreases as the proportion of the articulator ends represented by complete bones increases. The more the destruction, the more cylinders and the fewer complete bones there are. In turn, the more the destruction of bones per se,

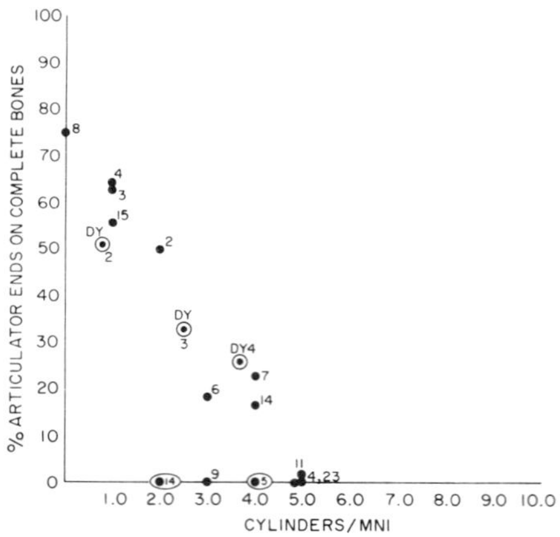

FIGURE 4.61. ***Relationship between the percentage of total articulator ends represented by complete bones and the number of cylinders per MNI. (Numbers relate to columns in Table 5.01.)***

the more splinters remain. For instance, using the methods found so successful for estimating the expected number of bone splinters present on a site given a count of the articulator ends from humanly broken bones, we obtain some surprising results. The total number of articulator ends observed in the combined sample of wolf kills was 145, of which 50 were articulator ends on complete bones. There were thus 95 broken articulator ends. Using this figure in the formula worked out previously (L. R. Binford 1978b:157),

$$x = [A - (2E + G)](y)$$

where

x = expected number of bone splinters
A = number of broken articulator ends
E = number of cylinders (Table 5.01)
G = number of ends plus shaft (Table 4.07)
y = constant of relevance, in this case 2.85 or the mean number of splinters produced per articulator end when bones are broken for marrow

Substituting in this equation the foregoing data, we obtain the following results:

$$\begin{aligned} x &= [95 - (152 - 43)]\,(2.85) \\ &= (-100)\,(2.85) \\ &= -285.0 \text{ splinters} \end{aligned}$$

In short, no splinters are expected given the surviving population, yet we observed 1119! This illustrates nicely that the splinters are related primarily to what is *no longer present* in the animal-gnawed assemblage, whereas in the humanly borken assemblage the splinters are related to what *remains* as undestroyed articulator ends.

I feel confident that I have demonstrated that in an animal-gnawed assemblage we may partition the assemblage into basically two components—splinters and cylinders—which vary in their frequencies primarily with regard with what is no *longer present* in the form of articulator ends or complete bones. Complete bones vary positively with shafts and articulator ends representing the elements or survivors not attacked by the gnawing animals.

Summary

In Chapter 1 I suggested that in the earliest days of archaeological research the paradigmatic view considered there to be two "creators"—God, the molder of nature, and man, the molder of culture and civilization. Early arguments regarding evidence for man's antiquity developed the position that the patterned modifications observed on some stones could not have been produced "in nature" and had therefore to be a consequence of the hand of man. Incorrect judgments were made in the early days of archaeological research as to what forms of patterning nature was incapable of producing. This ignorance of nature led to the spurious assignment of many things to the hand of man. In most cases this same situation can be viewed as responsible for inaccurate interpretations of modified bones as tools, spiral fractures as only produced by man, and many other such inaccurate or at best ambiguous assignments of meaning to bone modifications. In Chapter 2 I suggested that as far as middle-range research is concerned, we are seeking to isolate causal relationships between energy sources and modifications in material things. If we can gain a causal understanding of certain relationships, then there is established a *relationship of necessity* between the cause and the effect. When such a necessary relationship can be established then the inference of the cause when effects are observed is particularly secure.

When the early researchers into man's antiquity addressed the subject of stone tools, the comparisons were between one agent—hominids—and other inanimate natural forces that might alter stones. The difference between patterned goal-directed behavior of man and the accumulation of modifications that were the result of non-goal-directed sequencing rendered the recognition of stone tools largely unambiguous. The criterion of "manufactured to standards" or repetitively "produced in the same way" served for recognizing human products in stone.

A relationship of *necessity* was implied between complex sequential modification of matter to a pattern and man as the actor. This was largely correct for stone tools, since man is essentially unique in his use of stone as a raw material for the production of tools. Since this relationship between bear and footprint was essentially *one of necessity* or uniqueness in nature, we have been largely successful inferring actions by men, (the bear), from his stone tool products (the footprint).

A tactical analogy with lithic studies is probably the basis for most of our inaccurate interpretations of bone tools. As illustrated in Chapter 3 nonhominid animals regularly modify bones; they manipulate both bones and prey animals in ways that result in regular, sequentially generated, and patterned modifications. These modifications extend from individual bones through units of larger anatomical organization to complete faunal populations. Man is not related to patterned sequential modification of faunal material in a unique or necessary way, as is the case with lithic materials. Failure to take into consideration alternative causal sources of patterned modifications noted on bones further exacerbated the problem rooted in the assumption that all things found in association with stone tools could also be attributed to the hand of man. Such an argument seemed particularly secure when those associated things exhibited pattern modifications, which were taken to indicate "manufactured to standards" or "produced in a standard way."

In Chapter 3 I spent some time illustrating how many patterned properties of modified bones, cited by others as evidence for human modification or use, were in fact *regularly* produced by animals. I attempted to argue that many of the modifications were necessary consequences of the anatomy of an animal's jaw, such that if modifications were to be made they would inevitably be made in certain ways. I sought to illustrate that an animal tearing apart a skeleton was *limited* to a certain sequence and to a given strategy by virtue of the constraints on the use of *jaws* as the animal's basic mechanical device for carrying out work on bones and carcasses. That is, I have attempted to at least suggest some of the unique, or necessary, linkages between gnawing animals and their products.

In this chapter, I have explored some of our knowledge regarding the ways man manipulates and modifies bones. Although our knowledge is surprisingly sparse, several generalizations seem in order.

Perhaps the most important point to be made is that a bone may take on properties from a number of sequentially independent acts. Treating all the morphological properties as if they were referable to a single historical episode or behavior is a major error. We need informed and detailed diagnostic criteria for recognizing properties that derived from independent acts in the past. This should be clear with regard to separating consequences of animal behavior from that of humans but it is equally important to be able to recognize derivatives of different human actions.

Of equal importance to the recognition of mulitple causes is the realization that similar properties such as patterns of bone breakage may be produced in a number of different contexts. Breakage may occur in the context of butchering, in the context of field snacking for marrow, in the context of residential processing of marrow bones, and so forth. Treating all breakage as referable to butchering, or marrow cracking, or any other single behavioral context by convention is a major error. In the absence of truely diagnostic means for recognizing the consequences of one context of breakage from another, interpretation by convention will certainly lead to myth making.

In terms of distinguishing properties generated by animals versus characteristics referable to hominid actions, a common strategy has been to seek characteristics that might be generically referable to "human" behavior. We have seen this with the claims of uniqueness for the crack-and-twist method of bone breakage, the longitudinal breakage of long bones, spiral fracture, pressure "flaking" of bones, polishing, striations, and the like, which have all been considered by some to be diagnostic of hominid behavior. Can we cite such diagnostic properties based on the studies and knowledge reported here as *necessary consequences* of hominids acting upon bones?

We might ask initially whether there are any general points of contrast between what we know of human behavior and what we know of animal behavior. In this we must answer that there are provocative differences.

Man is not limited to the physical manipulation of material things with a single mechanical device, jaws. Man conceived in very simple terms as a tool user can manipulate physical things with a variety of mechanical devices and in terms of a number of mechanical principles. Although man is capable of rendering a wide variety of patterned modifications on bones it is suggested that his primary strategy is cutting, not tearing and pulling, and hammering, not vising down on objects. Animals manipulate bones and carcasses by literally tearing them down structurally. This pattern is one of gradual weakening and with its attendant sequential modification of weakened locations results in accumulated patterns of both surficial and structural modification of bones. Man's strategy could be characterized as direct. Using cutting tools he can achieve dismemberment or filleting directly without first weakening the joints through an attritional or destructive preliminary process. Animals cannot do this, except perhaps where the prey or bone being manipulated is much smaller than the size of the jaws of the active animal. Using percussion tools, man can break bones directly without first weakening them through the gradual destruction of their morphology prior to breaking. In a very general sense, the more an animal acts on a bone the more it is modified, and structural destruction is an inevitable consequence of this action. Man may dismember a carcass, prepare the meat for consumption, and break the bones for marrow without necessarily destroying bone material. He may break it and dissipate the anatomical structure but he does not destroy it, only renders it into smaller pieces. *As a normative statement the contrast of animals as bone destroyers and meat "tearers," with men as bone breakers and meat cutters, is probably statistically accurate.* This is what I have referred to as an empirical generalization or an inductive argument from enumeration (see L. R. Binford 1978a). It is quite probable that this is an accurate generalization and could be restated to meet all of Schiffer's (1976) criteria as a "law" to be used in archaeological inference. It would not lead us to a secure knowledge of the past, simply because there is no condition of necessity shown to be determining the customary actions of man as generalized.

Man can produce bone tools by destroying sequentially and to a pattern segments of bone morphology—hence the ambiguity between animal-gnawed bone and pseudotools. Man can butcher animals using bludgeoning techniques and techniques designed to weaken the anatomical structure of an animal gradually so that the results may look very much like the destruction caused by animals gnawing apart a carcass—this is certainly illustrated by Frison's work. It is also proved by Frison's replicative experiments for there at least man did mimic the products of scavenging animals. Man's flexible mechanical capabilities made possible through the use of tools render the specification of any relationship of necessity between man conceived of as biological species and the products of his work almost impossible. Man could under certain conditions replicate any mechanical consequence of either other animals' behavior or action by physical forces. We can imagine situations in which tool-using man might be forced to such destructive chopping techniques for disarticulating the humerus from the scapula resulting in the destruction of the proximal humerus (see Frison 1970; White 1953a; and many others). We can imagine situations in which tool-using man might even pick up a bone previously modified by an animal, for instance, a so-called humerus "flesher" (see Frison 1974) and use it as a flesher! Certainly replicative experimenters have done this or in some cases they have made facsimiles of such bones and used them successfully. Why not early man? The flexibility of our own researchers to invent accommodative models for all the forms of destruction I have shown to be regularly produced by animals is sufficient in itself to show that man could do these things; there is no linkage of necessity between man and a particular footprint in this case. Man seems to be capable of, mechanically speaking, producing traces on bones that could mimic any other "natural" footprint thus far known.

I would therefore suggest that the empirical generalization offered, although not particularly useful for interpreting the products of human behavior, can serve as a useful guide in the research design. For instance, if one observes a pattern of bone destruction and knows that destruction is the normal consequences of animal behavior, one should view as one's task the disproof of the proposition that animals were responsible for the observed pattern. One approach might be to demonstrate the pattern to exist independently of recognized evidence of animal behavior (such as tooth scoring; pitting; chipping back; production of chipped-end, longitudinally fractured diaphysis fragments; cylinders; scooped out soft cancellous ends of long bones; and bitten off bone protrusions such as the greater trochanter on the femur). Failure to follow this strategy is probably basic to the errors made by Frison (1970) as well as that of Stanford (1979a) at the Selby and Dutton sites and most of the workers at Old Crow Flats (Bonnichsen 1979). One might suspect that the reverse strategy might prove helpful when a pattern of bone breakage or modification by percussion is noted. Namely, knowing that breakage is a normal consequence of human behavior, one should view as one's task the disproof of the proposition that man was responsible.

This is a much more difficult task, since we have already suggested that man is capable of producing modifications that can mimic most of the effects of any natural process. In order to do this, we would have to have exhaustive knowledge of all the possible modifications on bone of which "nature" was capable, and then have performed sufficient middle-range research to permit us to differentiate all these possible modifications from their natural analogues.

For instance, it can be shown that animals regularly produce spiral fractures. Spiral fractures seem to be the characteristic result of at least some types of impact loading of bones. The very presence of spiral fractures suggests that animals may also modify bones in ways that produced at least some formal consequences similar to impact loading. For instance, I have seen punctures in bone shafts that if they had been seen in isolation from the remainder of the assemblage, would compare favorably with impact notches known to result from the use of hammer in breaking marrow bones. Similarly, the patterned modification I described in Chapter 3 as "mashed edges" bears many morphological similarities to the impact rosettes or "depressed fractures" produced by hammers on thin-walled bones. We have a few studies of the breakage morphology and descriptions of the statistical distribution of forms of scarring both from impact and from cutting on different segments of animal anatomy. We do not yet have similar statistical descriptions (although Hill [1980] has made a start in this direction) regarding the placement and patterns of association between scarring and differing anatomical markers on prey animals. These detailed descriptive materials are needed and may eventually serve to permit rather specific identifications of agents. We do not have this capability at the present time because the actualistic middle-range studies have not been conducted.

This problem is perhaps well illustrated by the current work of Richard Morlan (1981), mentioned in Chapter 3. Morlan has correctly dismissed many of the claims that man was the causal agent of modified bones recovered from the Old Crow Flats locality. However, as a consequence of his research he further documented a pattern of breakage previously noted by Bonnichsen (1978, 1979), the patterned flaking and fracturing on segments of bones from extinct elephants. Morlan has correctly sought the recognition of patterning manifest in items from deposits remaining from the past. This is fundamental and an essential part of archaeological research. Having established that a pattern exists, we must face the question of what it means. Morlan has reasoned that the bones of elephants are too large to have been modified in this relatively robust manner by carnivores, and that the patterning is too redundant to be the result of accidents of nature. Therefore, he has interpreted these bones as evidence for man's participation in the events represented in the admitted paleontological deposits at Old Crow.

The reader will undoubtedly recognize this as a classic argument from elimination. It might also be recalled that for such an argument to be correct the assumption must be met that all the possible causes have been listed and all but one eliminated. To what degree do you think we are capable of listing all the possible cause of broken and flaked elephant bones? Given our current ignorance of taphonomic conditions surrounding the burial of elephant bones in differing circumstances, it is impossible to meet the criteria for a valid argument from elimination. What is needed is middle-range research aimed at learning about elephant bone taphonomy. This would include actualistic studies of dead elephant carcasses and the behavior of other animals in the presence of elephant bones (the effects of trampling, tossing bones around as is recorded for elephants themselves, and so on). Research might also look at known paleontological assemblages where the presence of man can be ruled out on historical grounds to see if such items occur in assemblages known to be unrelated to human actions. All such studies must be done by us. As stated earlier I know of no other science concerned with explaining properties of the archaeological record.

I discussed in some detail Bonnichsen's (1973, 1979) use of ethnographic experience among the Calling Lake Cree as the basis for identifying a "human" pattern of marrow cracking. The weakness of this approach is that one cannot argue for there being a necessary connection between man considered as a biological species and cracking bones by the "mid-diaphysis smash technique" or butchering bison by "muscle stripping" etc. Equally clear is the fact that men could and did break bones by other techniques. Men of the past may or may not have butchered animals using Frison's techniques, yet man is clearly capable of doing so, as demonstrated by Frison. All forms of arguments from analogy must attempt to justify an assumption of necessity. If research sufficient to establish a necessary connection had been conducted, then an argument from analogy would not be needed. This means that all arguments from analogy are subject to error, as are any other purely indicative arguments. I suggest that as long as such arguments serve as the basis for a method of inference as was developed around Frison and Bonnichsen's work, then we operate with a set of conventions for "knowing" the past that have not been subjected to scientific investigation. This procedure results most commonly in modern myths.

The traditional archaeological procedure was to investigate the archaeological record and seek to document patterning (pattern-recognition studies). Once patterns had been recognized, then one invented a post hoc accommodative argument, which if true would account for the patterning observed. If it appeared plausible and consistent with what was considered to be our general knowledge of man and nature, then the post hoc argument would gradually be accepted and become part of archaeological theory. In short, a convention would grow up such that when one observed the patterning originally isolated by the initial researcher one would interpret it (i.e., give meaning to it) by restating the original post hoc argument. "Reliability testing" normally consisted, under such circumstances, of pointing out how well the facts fit the theory! This is unacceptable; we must have middle-range research.

Implied in the materials presented thus far is a principle that has not generally characterized traditional approaches: the implication that a pattern that may be recognized by an archaeologist can frequently be expected to be the accumulative consequence of a variety of causal conditions. Presenting a single scenario of the past as an accommodative argument for the patterning observed is likely to be wrong, since the constituents of the pattern are likely to have arisen from different "causes" and to be combined in characteristic ways in terms of still additional determinant conditions. Given such expectations, middle-range research must provide the diagnostic tools for analysis of patterning, thereby providing information about a wide variety of past conditions, rather than the "one pattern, one interpretation" form of most traditional archaeological inference.

One might reasonably ask at this point that if we cannot establish a pattern of bone modification unambiguously referable to man, why study the faunal products of man and seek greater understanding of his highly variable behavior? The answer to this is simply that the basic task of anthropology—of which archaeology is a part—is to seek an understanding of man's variable cultural behavior, to explain cultural similarities and differences.

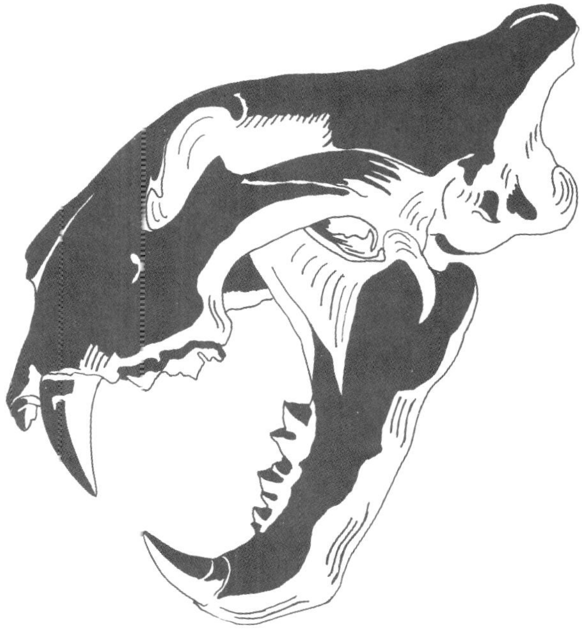

Chapter 5

Assemblage composition: Patterns of association stemming from the behavior of man versus that of beast

Assemblage composition is a relatively new domain of research. Within lithic studies the most common approach in the past has been characterized by Sackett (1968) as "the attempt to equate artifact fossils, cultural stages, and geological horizons as narrowly as possible into mechanical systems whose parts articulated with almost robotic precision [Sackett 1968:66]." As Sackett points out, the comparative study of the full range of associated lithic materials, an assemblage, has been an innovation largely referable to François Bordes and his colleagues in France. I suggest that there was a secondary center of development in Africa stemming largely from the influence of J. D. Clark, F. C. Howell, and Maxine Kleindienst (see Vincent [1978] for an insightful survey of Africanists' research).

The earliest description of a faunal assemblage presented as anatomical part frequencies, of which I am aware, was the truly seminal work of Alfred Rust (1937, 1943, 1958) and his collaborators Walter Krause (1937) and Walter Kollau (Krause and Kollau 1943). In these remarkable archaeological studies the fauna was described in detail and the relative frequencies of anatomical parts were summarized for large reindeer populations. These studies have been influential in a number of fields, for instance in the study of site structure as acknowledged by André Leroi-Gourhan and Michel Brézillon (1972:246) in their pioneering work at Pincevent. Despite the innovative priority that the works of Rust, Krause, and Kollau certainly enjoy in the history of archaeology, it was White's work [1952, 1953a, 1953b, etc.] that had an impact on the field so significant that studying faunal assemblages in terms of anatomical parts became a common approach. As will be described later Dart's work had a similar impact on old world studies.

New World Beginnings—Man as the Agent

In North America the earliest work on assemblage composition viewed from the perspective of anatomical parts was done by Theodore White (1952, 1953a, 1953b, 1954, 1955). The principles for interpreting differential frequencies among anatomical parts were stated in one of White's earliest papers:

> The accompanying table shows discrepancies which cannot be accounted for by accident of preservation. It is difficult to escape the inference that the parts either were not brought into camp, as with the thoracic vertabrae, or that they were mutilated beyond recognition while cutting up the carcass, i.e., the proximal end of the humerus [White 1952:337].

White considered variability in anatomical part frequencies to be referable primarily to differences in butchering practices and secondarily to the differential abandonment of anatomical parts in the context of hunting logistics. He also mentioned the selective use of bones as tools. As discussed earlier, most of White's arguments were post hoc accommodations to account for the patterns of anatomical variability noted in his samples. His model of prehistoric butchering included an assumption of the regular use of large chopping and hacking tools and a simple assumption of site functions—villages versus kill sites.

After White's initial work, there were several comparative studies of faunal assemblages. One of the earliest ones was concerned primarily with recognizing "ethnic markers" in "butchering practices" (Wood 1962). Following this early comparison there was a considerable body of information comparatively summarized in the report on the Boarding School Bison Drive site (Kehoe 1967) and Bonfire Shelter (Dibble and Lorrain 1968). Both reports provided reviews of ethnohistorical and ethnographic sources on bison procurement, descriptions of the sites including summaries of anatomical part frequencies by provenience unit, and fairly complete comparisons of the authors' materials with previously published assemblages. In both cases most of the assemblages available for comparison were from the "villages" studied by White. This meant that there was concern for the mirror-image arguments that were implied by White's inferences regarding how kill sites should look. As one might imagine, there were some reasons for confirming and some for rejecting White's arguments. Nevertheless, in both cases there was a large body of variability, the causes of which were not apparent. There was some indulging in post hoc argumentation, but in general there was a kind of glossing over the variability; a "what can you say?" attitude was expressed. Although it was not discussed at the time, the demonstrated variability signaled that things were not as simple as a patterned contrast between kill sites on the one hand and villages on the other. Roughly contemporary with these comparative studies on the American Plains, which were largely using White's methods, the first "exports" of White's procedures to other areas of the world appeared. By 1969 White's procedures had been applied in the Near East and, as we will see, in Africa.

Perhaps the most bizarre argument to surface was published by D. Perkins and P. Daly (1968) regarding a "Neolithic" site in Turkey (see also Daly 1969). Oddly enough this is perhaps the best-known argument regarding fauna. These authors have been called the discovers of the "schlepp effect." The schlepp effect has been called a "law" by Mike Schiffer (1976:21). It is, however, not a law but a post hoc accommodative argument that seems to have very little to recommend it. It is not based on ethnographic analogy; it is sheer accommodative fantasy, yet it has served as the interpretative basis for a number of arguments about fauna.

> Archaeologists working at Paleo-Indian sites in the New World have shown that, if only a few parts of a bison skeleton are found, the site is probably a more or less

permanent camp to which the hunters brought back meat from animals they had killed and butchered elsewhere. If on the contrary most of the bison's bones are present, the site is probably a "kill" site to which the successful hunters summoned their families [Perkins and Daly 1968:104].

The first thing wrong with this proposition is its assumption that all sites are either kills or residences. It also assumes that decisions to transport bones and process meat are always made in the same way (e.g., we always fillet meat, we always transport only filleted meat), and that the decision to transport animal parts is not conditioned by either distance to be covered or the quantity of meat to be transported.

A more detailed interpretative model was invented by Perkins and Daly to accommodate the specific facts they observed in a Turkish site of the sixty-fifth century B.C. They observed that bones from the feet and lower legs of cattle were more common in the site than were the bones of the upper leg.

> When a Suberde hunting party killed a wild ox, they apparently butchered it on the spot and used the animal's own hide as a container for carrying the meat home. They evidently stripped the forequarters and hindquarters of meat and threw the leg bones away. They apparently left the feet attached to the hide, perhaps because the feet made convenient handles for dragging the meat-filled hide. Perhaps they also valued the feet; this part of the animal contains useful sinews and has been called "the hunter's sewing kit." We have named the disparity between the number of cattle foot bones and leg bones that resulted from this treatment of the prey the "schlepp effect," after the German verb meaning to drag [Perkins and Daly 1968:104].

Now it is very clear that the feet and lower legs yield little if any meat. A "principle" that had also guided Plains faunal interpretation from the very beginning was that man acted differently toward parts that yielded low food returns per unit weight. For instance, White states, "Because of their greater weight the skull, scapula, pelvis, and metapodials of the bison were not transported as far as the corresponding elements in the deer and antelope and were left at the kill [White 1954:256]."

If Perkins and Daly had treated the Plains literature more fully they would have realized that the two basic propositions were in conflict in their case: The number of bones was relatively small but those bones present were from low-utility parts—yet they believed they were excavating a residential site. They ignored the more basic proposition regarding transport, namely that one preferentially moves to points of consumption parts with high meat yield per unit weight, and they invented a "just-so story" to accommodate the facts and still preserve their beliefs that (*a*) these were Neolithic hunters and (*b*) it was a residential site or village. *This is myth making of the first order.*

The amazing thing is that people adopted the schlepp effect as a methodological principle. For instance, the small cave of Palegawra (approximately 5 m deep and 6 m across), some 70 m above the valley floor, was interpreted as a "residential site" by virtue of the implications of the schlepp effect.

> The relatively high proportion of numbers of individuals which would have been counted on the basis of mandibles and distal bones of the limbs, as compared with maxillae and more proximal bones of the limbs, is apparent . . . the disproportion of these bones we attribute to "schlepping" . . . the favorite butchering techniques with a dead ungulate seems to have been to skin it where it fell, leaving the feet and mandibles (less often the maxillae also) in the skin. The meat was then cut off the carcass and piled on the spread skin, after which the feet and jaw provided handles for sometimes dragging, and sometimes carrying the load back to the cave. This explanation fits the facts as we find them at Palegawra [Turnbull and Reed 1974:136–137].

Again to imagine the past in this manner is sheer myth. The picture of the ancient Zarzian hunters dragging and pulling a skin full of meat 70 m up to their "residence" is probably as realistic as imagining the meat being transported to the cave by ancient astronauts. The facts at Palegawra had been "fit into" the assumed site typology of residences and kill sites. The frequencies given for Palegawra, although not well described, appear much more comparable to those reported by Raymond Wood (1968) from Vista Shelter. In a very important paper, Wood explored the idea of organizationally more complex subsistence–settlement systems.

Wood argued that the faunal remains of deer and bison found in Vista Shelter had to be viewed as derived from the activities of hunting parties originating in large permanent agricultural settlements miles away from the hunting grounds in which the shelter was located. In addition to suggesting that there could well be other types of sites with other characteristic faunal assemblages, Wood's study pointed to facts that had frequently been ignored in earlier studies: "All deer and bison humeri and radii are fragmentary, the shaft having been shattered by heavy blows, leaving only the articular ends nearly intact. The stubs of the shafts often show pitting resulting from these blows [Wood 1968:175]." In terms of the descriptions presented here, the pits mentioned by Wood were most certainly impact depressions. Wood also noted butchering marks and provides some descriptions of these marks. He was arguing for a hunting camp; at least one more type of site was acknowledged. Although Wood's work represents a major step forward, his line of investigation was not developed further for some years.

In African studies, Brian Fagan (1967; Fagan *et al.* 1969) reports the assemblage composition for both wild and domestic animals from a number of Iron Age sites in Zambia. Little interpretation is offered and differences noted between sites are generally referred to differences in butchering practice as manifestations of minor cultural differences. Of prophetic importance however, were two things: (*a*) the recognition that there were major differences in the relative frequencies of anatomical parts at the same site among species of differing sizes (see Fagan 1967:83), and (*b*) the recognition of the problem of preservation: "Another unknown factor is that of differential preservation. The relative resistance of different body parts to destruction by natural means is unknown, and it may be that our distributions are a reflection of preservation rather than habit [Fagan 1967:82]." This type of skepticism, in the context of the clear inability of White's models to offer guidelines for interpreting the wide variety of patterns being revealed as more descriptive studies appeared, surfaced in North America as well. For instance, Pillaert (1969) analyzed the whitetail deer bone from the Millville site in Wisconsin and concluded, "the percentages do not produce a pattern that would be conducive to determining the butchering process [1969:101]."

Perhaps the most crucial objection that began to be voiced regarding White's methods had reference to his assumption about preservation. White had acknowledged that the counts of bone parts might be subject to variation as a result of differences in preservation, but suggested that each part had an equal chance of being preserved, so the MNI ratios would be an accurate reflection of the parts originally present on the site (White 1952:337). John Guilday (n.d. and personal communication) voiced the pessimistic view that all the variability in anatomical part frequencies was related to the differential destruction of anatomical parts by "village dogs."

This was essentially the reseaıch environment during which important work by Joe Ben Wheat was initiated. Wheat was finding the unmistakable remains of bison that had been driven into a small arroyo. Unlike the site dug by Kehoe, Olsen–Chubbuck appeared to represent a single episode. The resolution of the deposit was high, in that only a few sequentially related events were represented, uncomplicated by the buildup of remains from many events, almost certainly the case at the Boarding School site (Kehoe 1967). Wheat's excavation strategy was sufficient to permit the recognition of some very important associations in the deposit.

> As we excavated the bone deposit we first uncovered the upper layer containing the single bones and articulated segments of skeleton. It was soon apparent that these bones were the end result of a *standardized Paleo-Indian butchering procedure.* We came to recognize certain "butchering units" such as forelegs, pelvic girdles, hind legs, spinal columns and skulls. Units of the same kind were usually found together in groups numbering from two or three to as many as 27. Similar units also formed distinct vertical sequences. . . . Where the order of deposition was clear, the bones at the bottom of each pile were foreleg units. Above these bones were those of pelvic-girdle units. Sometimes one or both hind legs were attached to the pelvic girdle, but by and large the hind-leg units lay separately among or above the pelvic units. The next level was usually composed of spinal-column units . . . at the top of nearly every pile were skulls. The jawbones had been removed from most of them, but some still retained a few of the neck vertebrae [Wheat 1967:149–50].

Wheat was recognizing an associated pattern, both in the character of bones remaining articulated and in the anatomical associations between the recognized units and their vertical disposition within the narrow arroyo. Wheat's interpretation of this pattern was justified by citing generalizations advanced after an extensive survey of the ethnohistorical sources describing Plains butchering procedures. (See Wheat [1972] for a review of this literature.) In short, Wheat's inference as to the behavioral meaning of his findings and in turn his reconstruction of the butchering procedure used over 8500 years ago at Olsen–Chubbuck was a classic argument from ethnographic analogy: "it seems reasonable to assume that the Paleo-Indians followed the same initial steps in butchering that the Plains Indians did in recent times [Wheat 1967:50]."

One of the steps in the procedure as described ethnohistorically was the removal of the "blanket of flesh."

> Directly under the skin of the back was a layer of tender meat, the "blanket of flesh"; when this was stripped away, the bison's forelegs and shoulder blades could be cut free exposing the highly prized "hump" meat, the rib cage and the body cavity [Wheat 1967:50].

This model of the past was strikingly different from the model of butchering presented by White, where there was an assumption of regular use of choppers and cleavers with the consequent destruction of anatomical parts during butchering operations. No such picture was presented by Wheat's work. One obtained the impression that the butchering was primarily done by cutting through the joints, and chopping and smashing were not so obviously practiced.

The next most important development in Plains research was the work of George Frison (1970), previously cited. His interest in butchering practices was stated from the very onset of his analysis of the Glenrock Buffalo Jump, but his approach was different from that of either White and his "followers" such as Dibble and Lorrain (1968) and Kehoe (1967), or Joe Ben Wheat (1967) who was working with ethnohistory and units of articulation. Frison (1970:8–10) made basically two arguments: (*a*) that all the marks or modifications on the bones at the Glenrock jump had been produced by the "tools" remaining at the site, and (*b*) that tools were recognized by virtue of a patterned form of breakage, a patterned form of structural modification, and/or a patterned form of surficial scarring. This leads directly to the correlated argument of there being many bone tools at Glenrock.

I have already treated Frison's work at Glenrock in some detail. It is clear to me that the bison bones at Glenrock had been ravaged by predator–scavengers. Many of the modifications and much of the distruction manifest on the bones was a consequence of nonhuman animal behavior. Frison, however, operating on the assumption that all modifications were human, invented a post hoc accommodative argument—muscle stripping executed largely with bone tools. Frison's knowledge of both tool mechanics and animal anatomy served to ensure that his model was in fact feasible. When he butchered bison in a replicative experiment, he could produce many of the modifications observed on bones at Glenrock. This demonstration was convincing to both Frison and many others. Among a group of largely Plains and early man enthusiasts the replicative experiment became an epistemological device. A classic statement heard at meetings in reply to someone's suggestion as to how things were done in the past was the quip, "How many times have you done that?" If the answer was "never" then a knowing glance circulated among the in-group and the person and his idea were dismissed as being not worthy of further discussion (see the exchange between Stanford and Johnson in Davis and Wilson [1978:310]). This "going native" approach is illustrated by the following statement from Frison: "hunting can never be fully understood in an anthropological sense until the investigator has had to fill his empty stomach by means of his own expertise and ability as a hunter [Frison 1978:364]."

The American public has been exposed to photographs of Dennis Stanford using an atlatl (Stanford 1979b), butchering an elephant (Park 1978), and eating bison meat "Eskimo style" (Odyssey film, *Seeking the First Americans*, 1980)—presumably under the assumption that all these things establish his credentials as an expert on hunting and gathering. We may also assume that all these same achievements provide the basis for "understanding" the "first Americans." Why should the way I do something today in the context of a kind of "boy scout" romanticism or even while "trying to fill my belly" ensure that the people of the past did things in similar ways? To argue that they did is to argue from *egographic analogy.*[1]

For instance, George Frison discussed his reconstruction of butchering, which was based on his imagination and replicative experiment, relative to the reconstructions that Joe Ben Wheat had offered for Olsen–Chubbuck based on ethnographic analogy. Reasoning from egographic analogy, Frison (1970:40) discussed Wheat's "blanket of flesh":

> It is difficult to understand what this so called "blanket of flesh" might refer to unless it is the thin layer of flesh that adheres to the hide during skinning. Actual butchering of buffalo fails to reveal anything that satisfied this description. It was not possible to find any way to "work up under" any blanket of flesh to remove the front quarters [Frison 1970:40].

I do not doubt that Frison could not find the blanket of flesh but it was not because such a blanket did not exist; it was simply because Frison did not know how to butcher a bison in the appropriate manner. See, for instance, my illustrations of the "blanket of flesh" removed by the Nunamiut Eskimo when butchering for drying (Binford 1978b:95, Figure 3, 2). Certainly a bison could be butchered in an analogous fashion. If we take knowledge of our "culture" as the basis for knowledge in general, or as the standard for judging what is plausible or possible, then we commit the worse kind of ethnocentric error.

Unfortunately, the approach of Frison has tended to somewhat dominate faunal studies in the Americanist literature in recent years. Assemblage variability treated in terms of anatomical part frequencies has not been studied recently on the Plains even at the pattern-recognition level of research. Those who have tended to be unimpressed by Frison have quietly sided with Joe Ben Wheat. For instance, a student at a Plains university only last year announced to me that "we don't study anatomical part frequencies or breakage, we study the articulated segments of the anatomy as a clue to butchering practices." Another student at the same university confided that "we think Joe Ben Wheat was right about Plains butchering."

I must admit that I too "think Joe Ben Wheat was right" and that muscle stripping probably never was performed on the Plains, at least not in the manner invented by Frison. In this case ethnographic analogy is likely to be much more relaible than egographic analogy. The sad fact is, however, that anatomical part frequencies can reveal much more than just butchering practices. In fact, they probably reveal very little directly about butchering practices per se, since cut marks, patterns of articulation, and patterns of context association are all apt to yield more direct information about butchering than are relative frequencies of anatomical parts. White's original assumption that the absence of certain parts was referable to their destruction during butchering now no longer appears plausbile as a general condition. It became clear that there are many patterns; in fact, almost every new assemblage described and compared—for example, Wood (1962), Dibble and Lorrain (1968), and Kehoe (1967)—had unique properties. This wide range of variability was hard to accommodate to the simple model of settlement that White and other early students of faunal assemblages tended to imagine.

The major methodological implication of the study of anatomical part frequencies for New World studies was simply that one could tell the difference between kill and residential sites. In fact there was little difficulty in accomplishing this task on the American Plains when bison was the major animal involved. It may be that this banal characteristic of the methodological implications of studying faunal assemblages contributed to the lack of interest that seems to have characterized the work of the 1970s.

[1] I am in no way implying that practical experience, familiarity with one's subject matter, and replicative experiments are not of value to a researcher in an educational sense. They are very important and should be strongly encouraged. I am only suggesting that it would be a very rare thing indeed if any of us could become "completely educated" by such self-teaching tactics. I would be very skeptical of the claim that any one expert was knowledgeable about the full range of behavior that mankind in general had perfected and executed in differing circumstances, many of which are certain to be unappreciated by the single individual working from even a very complete knowledge of his particular experiences.

African Problems and Assemblage Composition

Only several years after the works of White had turned the attention of some New World archaeologists toward the potential of anatomical part frequencies as sources of information, Raymond Dart (1957) published a very important study in which he described properties of the Makapansgat fauna from the perspective of anatomical part frequencies. Dart had previously argued from patterns of breakage, particularly of baboon skulls, that *Australopithecus* was a hunter, and that the faunal remains in the deposits from which the early fossils had been obtained were "kitchen-middens" (Dart 1949:2). The 1957 report on his studies of the fauna from Makapansgat included a tabulation of relative frequencies of anatomical parts. The variability in part frequencies noted by Dart was interpreted as deriving from the selective removal of bones from kills for use as tools by *Australopithecus*.

> The disappearance of tails was probably due to their use as signals and whips in hunting outside the cavern. Caudal and other vertebrae may also have disappeared because of the potential value of their bodies as projectiles and their processes as levers and points . . . femora and tibiae would be the heaviest clubs to use outside the cavern, that is probably why these bones are the least common. Humeri are the commonest of the long bones; probably because they would be the most convenient clubs for the women-folk and children to use at home [Dart 1957:85].

Although these interpretative arguments appear myth-like today, they were central in prompting a great deal of important research. As pointed out earlier, it was largely the claims of Raymond Dart regarding the nature of the *behavior* of the early hominids that demanded that behavioral questions be answered using archaeological remains. Actualistic research was prompted by alternative arguments such as Washburn's (1957) that the early hominids had not accumulated the deposits but were instead present by virtue of having been prey to whatever predator–scavenger was responsible for the deposits. Arguments over the agent responsible for the deposits, the criteria for recognizing tools, and what would be acceptable evidence of early hominid hunting were all questions that stemmed from the controversy surrounding Dart's claims. This is a very different intellectual environment from one that seeks to answer questions such as these: Was the Mousterian the cultural ancestor of the Aurignacian? Out of which "tradition," Clactonian or Acheulian, did the Quina Mousterian develop? Instead of traditional historical questions, the researchers of early man were faced with controversy regarding the character of the early hominids' behavior, and they had to seek answers through means that were *outside* the domain of traditional archaeological research procedure. I can remember teaching classes on Lower Pleistocene materials in the early 1960s and pointing out that although I thought Dart's post hoc arguments, such as the one just quoted, were sheer fantasy, the patterning he demonstrated at Makapansgat was interesting and essentially unanswered. That is, we did not have reliable methods for interpreting such patterning. Why were the anatomical part frequencies at Makapansgat so different from the frequencies in a living animal? Here was a classic pattern-recognition study. What we needed was middle-range research to investigate the causes of such patterning. If we isolated the causes we might have the basis for a powerful set of archaeological methods. The Makapansgat fauna seemed to me to be a critically challenging phenomenon. For instance, I had become accustomed to archaeologists "discovering" new patterns in the archaeological record but in most cases such patterns were cited as distinctive of the people who produced them. There was no reason to view them as "strange" or unexpected since, after all, cultures varied—we all know this. On the other hand, the claims of Dart were different. He was comparing the frequencies of anatomical parts observed in the archaeological record to "expected" frequencies derived from a knowledge of the anatomy of animals. When the observed differed from the expected he demanded an explanation—what behavior, events, or causal conditions could have "happened in history" to produce the differences between the expected condition and the observed facts? He was able to make a uniformitarian assumption regarding the anatomy of animals. He was able to say with some confidence what the frequencies of anatomical parts had been all those thousands of years before. He was therefore able to ask a very specific question about some events that we knew had to have taken place, namely those events that resulted in the modification of the anatomical part frequencies away from the known prior condition.

It is true that essentially the same structural situation characterized the earlier arguments by Theodore White. But there was a difference. White phrased the comparison as a basis for recognizing cultural differences. It was viewed as a way of monitoring "ethnic history," and specific explanations were not demanded as to the behaviors responsible for the differences. It is true that a number of post hoc arguments were offered, but one got the impression that the truth of such arguments was not as important as the demonstration of cultural differences, as was suggested in the comparative studies by Wood (1962) and Dibble and Lorrain (1968). This was of course the procedure of traditional archaeology; patterning varied because it reflected cultural differences. The culture varied as a function of the accidents of history. The relationships between behavior and the archaeological record were not a really serious concern. *Dart's argument demanded a behavioral evaluation.* He was saying that the anatomical parts varied from the expected frequencies because of specific behaviors carried out by specific agents, and that these behaviors were distinctive to those agents, the early hominids. Controversy here had to be with behavioral models, not vague "historical" models, as to which archaeologically defined unit influenced another or where certain traits were invented and where they diffused. In the Dart case one had to argue over (*a*) how hominids behaved and (*b*) what archaeological facts were unambiguous indicators of such behaviors. This was the intellectual climate of New Archaeology, and hardly anyone recognized it!

Nevertheless, research shifted rapidly to actualistic studies. There was a major set of research pushes with contemporary primate studies, and Washburn and DeVore were instrumental in starting what might be called "modern" hunter–gatherer research consciously considered of relevance to understanding something about the character of early hominid lifeways. (See the introductory statement to *Man the Hunter* [Lee and DeVore 1968].)

One of the earliest research "spin-offs" from Dart's claims focused on the behavior of the African porcupine. Based on analogy with Dart's claims for bone tools at Makapansgat, similar claims were made for items from the Hopefield site: Ronald Singer (1956) reported that those "tools" could be understood as products of gnawing by the African porcupine. Dart (1958) countered by first illustrating ethnographically documented bone tools, showing that man did use bone for tools, and then illustrating alleged bone tools from a site excavated by R. J. Mason called Kalkbank (R. J. Mason *et al.* 1958). This site was dated by carbon-14 and by the style of the few stone tools found there to around 15,000 years ago. Dart and Kitching claimed there to be a very large number of bone tools present, arguing that the location was a living site in an area where there was little available stone, therefore the occupants had used bone as a raw material for tool production. They recognized that the African porcupine has been active in modifying the bones at Kalkbank (over 25% of the total bones were considered porcupine gnawed) but they insisted that this gnawing took place after the bones had been broken or otherwise modified by man. In many cases modification prior to porcupine alteration does seem to have been the case. The question remained, who broke or modified the bones before the porcupine did its work? Of course Dart and Kitching argued that man did. They illustrated spiral fractures and longitudinally split bones, which may well have been broken by man or by predator–scavengers for marrow. Dart and his companion assumed that man broke the bones in the observed manner in order to obtain tools (see Dart 1958). In addition, they interpreted the observed differential frequency of anatomical parts as arising from selectivity on the part of man in returning parts of potential use as tools to his living site. Later Hendey and Singer (1965) reported on a porcupine lair as well as the contents of an adjacent site occupied periodically by man. It was noted that there was a major difference in the degree of fragmentation characteristic of the bones from the human site versus those from the procupine lair. The other point of contrast was in the frequency of identifiable porcupine gnawing—.3% in the human site, but an impressive 60.0% of the bones in the procupine site. This interest in the African porcupine as an agent of both bone modification and bone accumulation continues today (see Brain 1980). This, however, is getting somewhat ahead of the story.

Although the porcupine controversy led to the study of contemporary behavioral dynamics of porcupines, the methodology that I suggested as the only one really appropriate to beginning middle-range research, the studies of more direct relevance to assemblage composition viewed from the perspective of anatomical part frequencies, were carried out by C. K. Brain. Brain's research directly addressed two important questions: What caused the differential frequencies of anatomical parts observed by Dart (Brain 1967b, 1969, 1976b)? What agent was responsible for the accumulation of the deposits at some of the sites yielding early hominid bones (Brain 1968, 1970, 1976a, 1980)?

Brain carried out actualistic studies of modern leopard behavior and observed a number of distinctive characteristics that convinced him that, at least at the site of Swartkrans, leopards, not the early hominids, had been the agents responsible for accumulating the faunal deposits. The study that did more to change the direction of much African research than almost any other was his ethnoarchaeological study of the faunal assemblages occurring at contemporary Hottentot villages (Brain 1967b, 1969, 1976b). In this actualistic setting Brain was able to obtain a number of *controls:*

1. the Hottentot were goat herders; all animals killed for food were initially present on the site as complete animals.
2. Once the Hottentot were finished with the bones they were discarded about the camp and free roaming dogs scavenged the garbage.
3. Collections from these sites yielded populations of bones that were extremely variable in the relative frequencies of anatomical parts represented.

Brain reasoned that since the complete skeleton had been present to begin with, the differential frequency could not be due to selective introduction to the site, as Dart's arguments would demand. Instead they had to be due to differential deletion from the population. Brain noted that the bones that were underrepresented were soft and characterized by cancellous tissue. He argued that the sequence of epiphysial union would differentially affect the susceptibility of a part to destruction given a variable aged population. He then proposed that the differential destruction of anatomical parts as a function of their intrinsic strength by the free roaming dogs of the Hottentot camp was responsible for the differential anatomical part frequencies. Brain next compared his Hottentot goat samples to the relative frequencies reported by Dart from Makapansgat, demonstrating an impressive similarity:

> The evidence of the Kuiseb River goat bones strongly suggests that the disproportions which Dart encountered do, in fact, not require any special explanation. Artificial selection of certain skeletal parts need not be postulated. If for instance, antelope were hunted as they came to drink at a waterhole in the entrance of the cave and were then consumed by australopithecines and scavenging carnivores, a considerable bone accumulation could have been built up in the lower parts of the cavern. The bones preserved would have been those best able to survive the destructive treatment to which they had been subjected [1969:22].

This is a classic piece of middle-range research. Brain was able to gain informational control over the dynamic conditions in terms of which the bone assemblage was modified. No selective introduction of parts characterized the situation. Yet the result was a highly variable set of differential frequencies for the anatomical parts remaining on the sites. This had to derive from a series of *deletions* from the assemblage. Brain then went on to seek an understanding of the factors conditioning the removal processes.

One could almost detect a great sign of relief coming from early man research teams in Africa. Isaac (1971) considered the problem of Dart's claims for Makapansgat solved and in no need of further study.

Brain's findings were gladly incorporated into the discussion of Olduvai Gorge (M. D. Leakey 1971:275–281). Mary Leakey begins her discussion of fauna by pointing out that the disproportionate frequencies of parts of the skeleton had long been a "puzzling feature" that had never "been satisfactorily explained in the past." She then summarizes Brain's Hottentot work and concludes, "There is thus a combination of human and scavenger activity similar to conditions at early living sites where hyaenas and other scavengers are known to have gnawed the bones discarded by the hominid population [M. D. Leakey 1971:277]."

This conclusion incorporates an interesting assumption about Olduvai Gorge, namely that the destruction of anatomical parts at her "living sites" by hyenas and others occurred after the hominid population had discarded the bones. Presumably this would be after the site was abandoned—how does Mary Leakey know this? A table of comparison is presented illustrating a combined sample from Olduvai Gorge, the Makapansgat bovids, and the Hottentot goat sample. It is concluded that they are all very similar, and although minor contrasts are evident Leakey is comfortable with the conclusion that all are the result of essentially the same causal conditions: human utilization, with subsequent scavenging by predator–scavengers. Leakey does not consider the added assumption that the animals present were originally complete skeletons, or that Brain's work illustrates no faunal characteristics referable to human utilization in so far as differential anatomical part frequencies are concerned. How can one distinguish an assemblage of bones ravaged exclusively by animals from one where man had played a role? No one had addressed this question. Man's role was being assumed in all the interpretations, both of Makapansgat by Brain (1969), and Olduvai Gorge by Mary Leakey (1971). We had no procedure for recognizing the consequences of man's behavior, particularly if one discounted spiral fractures, as was done by Mary Leakey (1971:277–278). Was man bringing whole animals back to his living sites after successful hunts? Of course this is what one would have to imagine if Mary Leakey's arguments were accepted. At least Brain had the animals coming to Makapansgat and being killed on the site; no such convenient reasons seemed available to account for complete animals on the "living sites" of the early hominids at Olduvai Gorge! This implication of adopting Brain's argument seems to have been conveniently overlooked.

The next major development in the ongoing study of assemblage variability from an African perspective came from the work of Cathy Read-Martin and Dwight Read (1975). They argued that some insight could be gained on the Makapansgat situation by comparing the profile of anatomical part frequencies from the North German sites originally reported by Rust's collaborators Krause (1937) and Krause and Kollau (1943) with the Makapansgat profile. Preservation was reported to be excellent at the German sites of Stellmoor and Meiendorf. I would agree. There appears to be no suggestion that any post-depositional destructive process was at work in the German sites. The faunal assemblages seem to represent unmodified primary depositional assemblages. The character of the German fauna was interpreted as representing primary butchering near the kill location as well as the food consumed in support of the hunting group while they engaged in the hunting–butchering. The variable anatomical part frequencies on the German sites were interpreted in a way similar to the models argued in the American literature: Differential frequencies of anatomical parts reflected differential transport or abandonment of anatomical units in the context of hunting logistics. Comparison of the German sites with Makapansgat revealed that in the latter site scapulae, pelves, some vertebrae, and ribs were relatively underrepresented. On the other hand, heads and some parts of the front limbs were relatively more abundant. It also appeared that limb sections were introduced to the Makapansgat location after having been disarticulated from the pelvis or the scapula. These comparisons were made despite Brain's demonstration of a strong set of correspondences between the profile of part frequencies at Makapansgat and the *ravaged* assemblage from the Hottentot villages. There was strong support for the recognition of the Makapansgat assemblages as having suffered considerable destructive attrition, and no such evidence of destruction was characteristic of the assemblages from the German sites. That there was a difference between Makapansgat and the German sites was clear. That the differences were referable to the original composition of the assemblages was not clear. The ravaged character of the Makapansgat assemblage had been convincingly demonstrated by Brain. The unravaged character of the German assemblages seemed not in dispute. Thus the difference in composition between the assemblages could derive from destructive agents operative on one and not operative on the other assemblage. This was not considered. Another difference not considered by Read-Martin and Read was

role of differential preservation through his study of contemporary Hottentot peoples and their faunal debris. Simultaneously he studied the behavior and the faunal consequences of African animals, initially leopards (Brain 1968) and later porcupines (Brain 1980). Brain not only worked to obtain recognition criteria for animal-generated assemblages, but also moved in the direction of investigating the *causes* of differential bone preservation. Because there was a large segment of the Africanist community that generally disagreed with Dart, there was an emphasis on the study of animal-produced assemblages, and this bias continues although not so marked as perhaps during the early 1970s.

You might say that my work and the work of some Near Eastern researchers (Perkins and Daly 1968) has been largely "spin-off research" in that the original problems that prompted the research on the American Plains, and in Africa, were not the major motivators for either my work or that of others who began to explore faunal assemblage variability in other areas. Certainly my own work was motivated by the implications of the Americanist research. For instance, when I went to France in 1968 to study the fauna of Combe Grenal, I had a vague idea that site functions should be manifest in faunal assemblage variability. This was a projection of the Plains arguments regarding kill versus residential sites. I arrived in France with a copy of the then very new study by Kehoe (1967). I hoped that I could use faunal variability to inform me about the causes of lithic assemblage variability. It took me some time to realize that we had insufficient knowledge as to the causes of variability in faunal assemblages, and could not use the patterning seen at Combe Grenal as a methodological device for learning about the behavioral contexts in which the stone tool assemblages were generated. It was at that juncture that I came to adopt many of the positions presented in Chapter 2, namely that middle-range research was crucial and that we had to do it ourselves, since no other field sought to explain facts of the archaeological record. In addition, it was clear to me that explaining the facts of the archaeological record consisted of understanding the dynamic conditions that produced those facts. In short, we had to do ethnoarchaeological and experimental studies. I went to study the Nunamiut Eskimo in 1969. We now have a vast body of behaviorally controlled material (L. R. Binford 1978b) that should leave little doubt that the model of the past that admitted only residential and kill sites was extremely naive. I am of the opinion that the beginnings of a methodology have been worked out for identifying site functions, at least with regard to hunting logistics. We need at least as much research investment in the behavioral consequences of animal-generated faunal assemblages.

Other Types of Assemblage Variability

I have focused on assemblage variability considered exclusively in terms of anatomical part frequencies. Clearly this is not the only property of assemblages. We could study assemblages from the perspective of their frequency composition in terms of different species, different ages of the different species, or a number of other zoological properties. Studies of this type have been conducted for a long time. The composition of assemblages in terms of species frequencies is most commonly interpreted as a reflection of environmental composition. Much of the current work in taphonomy is concerned with the degree to which paleontological assemblages, even assemblages referable to the actions of a single species, are representative of the frequencies of species in the surrounding habitat (see Behrensmeyer and Dechant-Boaz 1980, Brain 1980, and particularly Gifford 1981). There is, however, another possible conditioner of species frequencies, and that is the bias of predator–scavengers for exploiting prey niches rather than taking a random sample of prey in the habitat. Important work of direct relevance to the problem of identifying different predator–scavengers products has been done by Vrba (1975, 1980). Of very great importance is a provocative observation by Richard Klein (1976a, 1980) that carnivore species, in terms of both number of species and number of individuals, are much more common in animal den assemblages than in other types of either human or paleontological assemblage. This supports my observations that animal-produced pseudotools claimed from European sites are more common from levels and sites where carnivore remains are extraordinarily common. This appears to be true of the famous site of Choukoutien, and probably applies to other such locations as well. For instance, the famous burial cave of Teshik-Tash (Movius 1953) appears to have been an animal den for a considerable period of time. To what degree are the evidences of ritual cited at this site referable to the behavior of animals? The widespread use of European caves by nonhominid predator–scavengers is well illustrated by the following generalization regarding European sites: "The nature of the site controls to some extent the nature of the fauna. Cave sites, for example, are often rich in carnivore remains out of all proportion to the abundance of the creatures in the wild [Evans 1978:36]."

In fact, two discussions of hunter–gatherer subsistence settlement strategies during the Paleolithic of Germany were almost certainly "led astray" by the old assumption of the monument and relic era that things found in association with tools showing the hand of man were also referable to the actions of men. Gamble (1979:39) accepts the numerous predators (sometimes representing 45% of the faunal remains exclusive of cave bear) as largely the prey of human hunters. He noted that sites with high frequencies of predators tend to be small caves and rockshelters that also generally yield relatively small artifact assemblages. He contrasts such sites with large archaeological sites in the region that are numerically dominated by herbivore remains, in an attempt to obtain a "regional" perspective on the hunting strategies of ancient man. It is very likely that he is primarily contrasting the behavior of carnivores with the behavior of men.

In another study Gamble (1978) is concerned with the changes in both resource utilization and settlement through a time sequence. He clearly demonstrated that there is a major change between the Middle Paleolithic and early Upper Paleolithic seen as a unit and the subsequent Magdalenian in patterns of faunal association. The sites of the earlier time period have large numbers of carnivores, whereas the Magdalenian has very few. Gamble considered this to be a measure of human behavior. I strongly suspect that it refers to the patterns of interaction between competing predators—man and others—in the region. Our inability, or in many cases our failure to try, to recognize the consequences of hominid behavior as opposed to that of other animals continues to ensure that our views of man's past are largely distorted myths.

Faunal assemblages have been studied in terms of the relative frequencies of animals of different ages. This has been a common strategy among researchers seeking to recognize domesticated as opposed to wild animals exploited by ancient men. These studies (see Chaplin 1969, 1971; Ducos 1969; Higgs and Jarman 1972) all make a basic assumption regarding the exploitation strategies of hunters:

> The basic hypothesis is that pre-agricultural man will have exploited resources in a "natural," i.e., random, fashion and that the increased control over his resources given by domestication allows planned cropping, which will be seen in the form of exploitation patterns which deviate from the natural random one [Higgs and Jarman 1972:7].

Despite the fact that the proposition is phrased as a hypothesis, no attempt has been really made to test it. It is like the hypothesis of Bonnichsen (1979) regarding spiral fracture; it is in fact a convention used to ascribe meaning to patterned observations made on the archaeological record. The use of this convention led the late Eric Higgs to recognize evidence for domestication in the Paleolithic of Europe! Such conclusions were reached because there was no understanding regarding the factors that conditioned the differential frequencies of animals of variable age appearing as debris from hunter–gatherer adaptations. In the absence of understanding of causal processes, many myths are generated. I have made no attempt to survey this literature, although it is extensive and in need of critical review.

I will restrict further discussion to assemblage variability conceived in terms of anatomical part frequencies.

Studies of Assemblage Composition Caused by Beasts

Interest in assemblages generated by animals certainly grew out of the controversy surrounding Raymond Dart's work. An early argument about the Makapansgat fauna was that it had been produced by hyenas. One of the earliest "recent" studies (see Dart

1956 for review) treating the degree to which animals, specifically hyenas, accumulate bones at a den was done by Hughes (1954). He observed a den excavated in a soft substrate in which no bones or feces were deposited. This research was really done in the context of answering the question as to whether animals actually accumulated bones in their lairs and dens. As late as 1970 the accuracy of the claim that animals actually accumulated bones was still in doubt. Sutcliffe's work (1970) seemed to put this question to rest with an unequivocal yes. Subsequent research has been more oriented toward obtaining samples of animal-generated faunal populations. Andrew Hill collected faunal remains from two African locales (Hill 1975). Shipman and Phillips (1976, 1977) made collections from still another area. Diane Gifford has initiated time-series studies of carcasses to determine what the sequence and interrelationships between disarticulation and bone modification might be. These studies have not yet been completed. A number of hyena den studies have been conducted (Henschel *et al.* 1979; Owens and Owens 1979) and we look forward to more that are currently under way by Hill and by Bunn (Glynn Isaac, personal communication). Faunal assemblages have been collected by Behrensmeyer and Dechant-Boaz (1980) but have not yet been reported in terms of anatomical part frequencies. Although we can certainly look forward to a large body of materials in the near future, the only study published to date that addresses the problem of recognizing the behavior of animals as opposed to that of humans is the work of Andrew Hill.

In his thesis, Hill (1975) sought to identify assemblages as typical of human behavior and contrast them with assemblages assignable to animal behavior. In this he chose what was at the time one of the few widely available descriptions of a faunal assemblage clearly referable to human behavior, that from Olsen-Chubbuck reported by Wheat (1972). Without understanding what made it distinctive, archaeologists accepted it as ontologically "human" in its composition and other characteristics.

Ironically, the Olsen-Chubbuck Site is one where (*a*) there was a mass overkill and (*b*) the animals were processed for transport by removing the meat from the bones. These conditions ensured that the meat was removed from the parts of highest utility and lowest weight-to-surface-area ratios, resulting in the biased abandonment at the site of bones from parts most desirable for consumption. The assemblage therefore looked much like an animal kill assemblage where consumption had been on the spot. Hill's comparisons made it appear that man behaved much like animals.

The "search" for typical or ontologically characteristic assemblages when there is little understanding of what conditions their form, other than a vague suspicion that intrinsic properties of the agent are responsible is just a slightly modified version of the traditional archaeological assumption that variability is referable to differences in the persons or agents producing the assemblages. That is, assemblage differences are referred to ethnocultural differences among men. Things are not that simple. We must understand the processes that condition differences and similarities before we may give meaning to them reliably. This is another way of saying what was stressed in Chapter 2—middle-range research is of necessity rooted in actualistic studies, studies of dynamic phenomena. This normally means studies of contemporary living systems.

I have already studied in detail a tremendous range of variability as it was generated by the Nunamiut Eskimo (L. R. Binford 1978b). Although the variability produced by man in this case appears almost overwhelming, there was patterning and there was some redundancy in the types of patterning produced correlated with the types of setting, the structure of the site, and a wide variety of other conditions. In that study, I was interested in the dynamics of the behavior that stood behind the assemblages. In a similar way it is important to begin the task of understanding, from a behavioral perspective, the structure of bone populations produced by animals. As a step in this direction, I will present summary information on the behavior and treatment of prey by wolves.

Observations of Wolves and Their Behavior

The observations presented here were made during a field investigation among the Nunamiut Eskimo, inland hunters of caribou. The Nunamiut live at the drainage divide of the Brooks Range of north central Alaska. Major segments of the Arctic caribou herd biannually move through this area from their winter range in the forests to the south, to the summer feeding areas on the tundra to the north (see L. R. Binford 1978b:13, Figure 1.1). Most of the wolves of the central Brooks Range are dark brown to black; a lesser number are gray, and I have observed an albino wolf on several occasions. Of those killed by the Nunamiut during the winter of 1970–1971 (five individuals, three males and two females), the average weight was 87 lbs. During the fieldwork in and around the Anaktuvuk Pass area, I spotted 23 single wolves and 8 packs. Of the packs observed the mean size was 6.38 animals with the smallest group being 2 animals and the largest 11; a total of 51 animals with the smallest group being 2 animals and the largest 11; a total of 51 animals were observed in packs. I observed only one kill by a pack, but came upon feeding wolves on two other occasions. I had the opportunity to examine in some detail two wolf dens and made observations on 36 wolf kill and feeding sites.

Dens and Lairs

In earlier studies almost any repeatedly used location was called a den. For instance, what Hughes (1954) called a den was actually a nest area where young pups were born and kept while their eyes were still closed. I will continue to use *den* to refer to this type of site. However, I must distinguish between a den and a *lair;* the latter is the central place where young localize their activities and where adult wolves spend most of their time adjacent to a den. Lairs generally consist of two areas, a traffic area and a defecation area; the older animals tend to move away from the core activity zone of the lair for defecation. In the Brooks Range of Alaska most dens are dug into south-facing cut banks and slopes; however, there are a few rockshelters and caves in the limestone that occasionally serve as dens. In the immediate area of Anaktuvuk, most caves or rockshelters are quite high along the mountain slopes and the Eskimo report that wolves prefer a lower place where they can "run" out into the valley better."

Robert O. Stephenson, who has conducted extensive studies of wolves in the Anaktuvuk area, reports, and my direct observations as well as those of informants confirm, that wolves dig burrows (or enlarge other animal burrows) in sand and clay and make use of fissures in stone substrates for their dens (Stephenson and Johnson 1972:29). Den collapse is common in sand, from overly dry conditions in the area, and in clay from excessive moisture (Stephenson and Johnson 1972:29), resulting in only limited reuse of a given den. This is obviously not the case for rockshelter, cave, and crevice dens, which may be reoccupied almost endlessly: "Naturally occurring rock formations are the most stable substrate in which wolf dens occur and, in some cases, have provided essentially unchanging den sites for decades and perhaps even centuries [Stephenson and Johnson 1972:29]."

The actual pattern of reuse of a given den was in the opinion of the Eskimo not continuous; that is, dens were believed to be reused on an every-other-year basis by the wolves. Informants told both Stephenson (Stephenson and Johnson: 1972:21) and me that this was because of the lingering smell from the earlier year; "they wait" till the smell goes away.

The females are considered by the Eskimo to have "denned up" by the first week in May. Each den is considered to be "like a family" with a single mature female, a mature male, and a variable number of immature animals. Activities of the family are believed to be centered on the lair throughout the summer. Only with the fall migration of the caribou are the families believed to come together into larger packs that move toward the timberline behind the caribou. Wolf lairs are, then, highly seasonal and specific to spring and early summer. As the pups become larger, the adults may increase their ranging in search of prey and may take the pups "out," as the Eskimo say, leaving them in a thicket, ditch, or other protected location while the adults travel over the territory hunting, returning to the pups either to move them again or to direct them to a kill. Such locations may be used over and over again and are referred to in the following discussions as *territorial centers* or *rendezvous* points (see Figure 5.01). These will be discussed again after some details on lairs have been presented.

ROCKSHELTER DEN ALONG THE ANAKTIQTAUK

A rockshelter den was visited during a helicopter survey of sites on June 17, 1972. According to infor-

FIGURE 5.01. ***Wolves relaxing at a rendezvous point.***

mants this shelter (Figure 5.02) had been used as a den many years ago, with the comment that "only crazy wolves den this high." Located along the north face of mountains bordering the Anaktiqtauk Valley, the rockshelter is approximately 235 m above the valley floor and faces south. It is relatively large, 12.6 m wide at the mouth and 5.4 m deep. When visited, there was a very small chamber extending up under the back wall, which branched into several fissure-like small chambers. These extended back into the rock at least an additional 3 m. Along the drip line was a substantial "cone" deposit standing about 1 m above the base of several very large blocks of roof fall, immediately inside the shelter. There was an irregular "floor" among these blocks, and this was the level on which the solution chambers in the back passed into the rock. Along the west wall of the shelter was a substantial accumulation of sheep dung, and between a large block of roof fall and the east wall was a stone cairn, probably built as a temporary meat cache. It contained no artifacts or bones. Bone resulting from the wolf occupation of the shelter was scattered along the entrance and down the talus deposit (see Figure 5.02). In addition, there were a few scattered bones on top and at the base of a large block of roof fall directly in the entrance. Most of the

FIGURE 5.02. ***View inside the rockshelter den along the Anaktiqtauk.***

bone along the talus deposit had the appearance of the remains of scat. Considerable washing of this surface had gone on since the site was used, so the telltale "bone meal" was not evident. Bone inside and among the boulders at the entrance was more heavily gnawed and appears to be the remains of parts introduced and gnawed but not ingested. The inventory of bone recovered from this den is as follows:

1. *Talus deposit:* Two astragalus, three tarsals, five carpals, one distal tibia, five halves of the distal articulator condyles of metapodials, and seven phalanges (two ungual, two second, and three first), all of Dall sheep, were recovered. In addition there were first phalanges, seven sesamoids, three mandibular teeth, and one distal end of a calcaneus of caribou. Four maxillary teeth of sheep were also noted. No attempt was made to collect bone splinters since there had been obvious washing and vegetation was fairly developed at the mouth of the shelter.

2. *Entrance deposit:* Five half-ribs, one tibia with the proximal articulator end chewed away, one proximal metatarsal plus shaft with attached calcaneus, with heavy chewing on the distal end of the calcaneus, were on the rocks at the entrance. One cranium of a female caribou with attached but heavily gnawed antlers was partially buried under sheep dung. All these bones were heavily weathered; there were developed longitudinal cracks and superficial checking. Inside the shelter were considerable accumulations of relatively recent sheep dung and evidence of recent washing. Since excavation was

impossible no attempt was made to collect bone splinters or to search for small bones.

This collection is not very informative as a picture of the actual population of bones that must have been present in this site. Only excavation could produce such an assemblage. Nevertheless, the differences between the character of superficial bones remaining along the talus versus those inside the entrance are probably indicative of real differences between a "latrine" sample and one from the gnawing–resting area of a lair.

WOLF DEN NEAR THE MOUTH OF ITIKMALAIYAK CREEK

In a south-facing sandbank that is heavily windswept is a den entrance just behind a slumped pile of sand representing a collapse of the bank in the not too distant past. Animals enter the den by going between the bank and the slumped earth. There is a slight overhand to the bank at this point, and an excavated hole back into the sandbank. The entrance is approximately 20 inches in diameter and the chamber curves away to the east and down. Probing the curved inner chamber with a fishing pole indicated that the chamber was then about 10 feet long and varied in width in the nest area from about 2 to over 4 feet in width. There were no scats, bones, or other organic debris inside the chamber. All bone debris and scats were around the entrance. Scats were generally concentrated in front of the slumped bank; the bones were localized in the open, in the area just before the path moved into the narrow walk between the bank and the slump.

Collected in the bone area were the articulated horn cores of two Dall sheep (*Ovis dalli*) with attached segments of the frontal. A fragment of the occiput and 27 additional fragments of parietal and frontal bone were noted. One maxillary arc with both rows of upper teeth in place and 17 isolated teeth, 3 half-mandibles, 1 nearly complete radio-cubitus with the olecranon largely gnawed away, 1 proximal radio-cubitus with gnawed and radially scarred broken midshaft, and the olecranon essentially gnawed away, and 1 metacarpal with gnawed away dorsal ridge just below the proximal articulator end were collected. Two additional proximal metacarpals with short attached areas of shaft, 1 distal tibia with attached shaft, 1 distal tibia heavily gnawed and with only short segments of attached diaphysis, 1 complete metatarsal, 1 distal metatarsal with attached shaft, and 2 proximal metatarsal fragments rounded out the collection of sheep bones. Caribou (*Rangifer tarandus*) were represented by the occiput of a large bull, 7 gnawed sections of what appeared to be a cow's antler (one very polished), 5 very small rib fragments, 11 isolated maxillary teeth, and 7 mandibular teeth. In addition, there were 6 first phalanges, 4 second phalanges, 2 third phalanges, 2 complete metacarpal with gnawed distal end, 1 proximal metacarpal with attached shaft, 1 distal tibia with attached shaft, 3 calcanei with gnawed distal ends, 1 astragalus, and 1 proximal metatarsal with attached shaft.

Other animals represented at the den included moose (*Alces alces*): one heavily gnawed scapula, and one distal tibia with attached shaft; marmot (*Marmota caligata*): one complete skull with attached mandible; ground squirrel (*Citellus undulatus*): five complete skulls, three with attached mandibles and no other remains; ptarmigan (*Lagopus* sp.): five complete wings (proximal humerus missing on all examples—otherwise complete); domestic dog (*Canis familiaris*): one heavily gnawed mandible.

In the area where defecation had occurred regularly, 46 scats were still complete. These were collected and taken apart in order to obtain size information on the included splinters. These data are summarized in Table 3.02. Collected in the area of the scats and remaining from weathered scats were the following recognizable anatomical parts: 16 isolated teeth (12 mandibular and 4 maxillary), 13 "roots" or broken bases of mandibular teeth, all of caribou; 7 ungual phalanges of sheep, 4 third phalanges, 2 second phalanges, 1 first phalange, 8 carpals, 3 fragments of distal condyle from metapodials, 2 femur heads, and 3 astragalis. In addition, recognizable concentrations of fish bone and small rodent bone, considerable quantities of both caribou and sheep hair, and some bird bone occurred in the area. There were large quantities of very small bone fragments, almost the consistency of bone meal, bone chips and splinters, as well as cancellous bone chunks in substantial numbers.

The den may be described as exhibiting an assemblage dominated by skulls, teeth, feet, and lower limbs for the large to moderate-size mammals; small rodents were represented by skulls only, and birds by wing bones only. There were two rather distinct spatial localizations, a feeding area outside the den entrance, and a defecation area just to the east of the entrance and feeding area. The latter location was characterized by large quantities of small chips, splinters, bone-meal-like concentrations of small bone "nuggets," and relatively large numbers of chunks of cancellous tissue. In addition, there were occasional phalanges, roots of broken teeth, rodent bones, fish scales, and much associated hair.

It should be pointed out that this assemblage is an accumulation of debris from a number of years of use and the location may have been occupied first by foxes. Informants knew of this den having been used for at least seven separate spring births of pups.

Information on Lair Behavior

I was unable to observe directly the behavior of wolves in lairs. However, I did interview Eskimo hunters who had much experience in both seeking out wolf dens and in observing wolf behavior. Most Eskimos were of the opinion that most (whole) bones in dens were introduced by the pups and juvenile animals. They were also quick to comment that many small bones would be introduced to a lair assemblage in adult feces. Informants recognized that there would be a diversity of faunal remains at a lair with a fair representation of small game. This was considered to be a function of both season and the fact that the predator was "tethered" during the time that the pups were very young. "When the mother had to stay close she could only hunt around the den and then she would mostly only find little animals." Some variety in the small mammals and rodents could also be introduced by the pups themselves, who begin killing small mammals and rodents nearly as soon as they can move around. In addition, the Eskimo mentioned "play hunting" by the young. A pup may pick up rocks, sticks, and bones from kills and death sites that may be in the area of the lair, and others will chase them and "compete" over such items much as adults compete over parts at a kill. Such items are transported to the lair and may be played over by the pups and subsequently gnawed by both pups and adults.

Almost all informants mentioned that wolves in lairs, particularly females during the first weeks of the pups' lives, compulsively chew bones. The taphonomic potential of wolf lairs seems best estimated simply by the stability of suitable denning areas in caves, rockshelters, and fissures. Stated another way, reuse of locations as lairs appears to be a function of the degree to which dens do not collapse. Therefore, substantial lair accumulations can be expected in caves and rockshelters.

Although the accumulation of bones at lairs by predators was recognized early in the history of cave research (Buckland 1823), most discussion has centered on the hyena:

> The hyena plays an important role as a bone collector, to the benefit of the paleontologist. Although it will mutilate the bones, it always leaves the teeth intact and systematically they are the most important part of the skeleton. As a result the activities of hyenas in the European caves have been of great importance for our knowledge of the Ice Age and its fauna. The hyena itself is one of the most commonly found fossil mammals and the evolutionary history of the family Hyaenidae is thus comparatively well known [Kurten 1968:63]. . . .
> The spotted hyena or cave hyena (*Crocuta crocuta* . . .) is one of the best-known fossil mammals of the Ice Age. The species is represented in most European bone caves and in some, used as dens for thousands of years, bones of the cave hyena occur in profusion. A famous cave of this type is Kirkdale Cave in Yorkshire, described by William Buckland (1823). Buckland showed that the enormous accumulation of hyena bones resulted from continuous use of the cave as a living site by the hyenas over a very long period. . . .
> Here are found remains of hyenas of all ages from newborn to senile and in addition their characteristic feces and the bones of their prey. Another British hyena cave with a rich hyena stratum of F-Emmian date is Tornewton Cave in south Devon (reckoned to have yielded more than 20,000 hyena teeth), while Kent's Cavern in Torquay contains an immensely rich hyena stratum of more recent date, 4-Wurm. On the continent several hyena caves are known and at least one—the Teufelslucken at Eggenburg in Austria—is of the same class [Kurten 1968:69].

Other predators (with the exception of the Ice Age cave bear) do not seem to produce the spectacular bone accumulations attributed to the hyena nor do other predator lairs appear dominated by remains of the occupants, particularly the young and old, as is the case with both the cave bear and the hyena. Two factors may well contribute to this. First, hyenas are basically nocturnal and stay holed up much of the day. Dens and lairs may serve this function, resulting in much more continuous use of a lair by hyenas than by canids who primarily use them only when the young are born. An additional behavior or species difference between hyenas and canids is in the care of the young. Canids very quickly begin taking the young out to kills for feeding, and dens are abandoned quickly. It appears from the literature that this is not so much the case with hyenas, where the young stay around the den for up to 8 months; many suffer severe malnutrition because of the ranging activities of the females:

> During the dry season, Serengeti hyenas may have to travel more than 50 km from their den to the nearest concentrations of game, and female hyenas leave their suckling cubs for several days on end. At this time, cubs are not infrequently emaciated and have been found dead near the entrance of the den [Kruuk 1970:370].

In addition, the cubs appear to attempt to stay in the den as long as possible. When they are juveniles and young adults, they may return rapidly to dens when threatened or wounded while competing over food.

Despite the longtime recognition by paleontologists that hyenas accumulate bones at dens and lairs, no direct field studies were made until very recently. Even at the present time there are no very complete field descriptions of hyena dens and lairs.

Hans Kruuk has made the following observations regarding the introduction of bones to hyena lairs:

> Hyenas also do not carry any substantial amount of food to their cubs; they may occasionally take a bone or a head along, but this seems comparable to taking it to their lying-up place, where they can chew it in peace. Cubs may chew these bits, but only after the original owner has finished with them. Older cubs are more likely to carry bones to their den than adults. They sometimes take them down their holes but they are usually left in front of the den, or in the entrance . . . but this does not often happen; it may be more common among brown hyenas (Pienaar 1969). . . . Hyenas rarely defecate near the den; they usually go at least 20 m away to deposit their droppings and walk back to the den again [Kruuk 1972:243–244].

Sutcliffe had excavated paleontological deposits in Great Britain that were generally recognized as hyena lairs, and as in most other European occurrences from the Pleistocene these deposits were largely dominated by the remains of the hyenas themselves. These observations led Sutcliffe to seek knowledge of the bone-accumulating activities of contemporary African hyenas. The result is certainly the most provocative natural history observation available to date:

> In the Queen Elizabeth Park, Uganda, two burrows in alluvial sediment . . . were completely excavated. The first was of simple plan with a single terminal chamber, about three feet below ground level, opening into the surface by a trifurcating tunnel. A few fragments of bones of antelopes and of a juvenile hippopotamus were found in the access tunnel. The terminal chamber was empty except for the complete skeleton of a baby spotted hyena and one dropping. The second lair was more complicated, having a central chamber . . . and a series of tunnels leading to at least ten openings situated around it. The tunnels contained numerous bone remains, including skulls of baby hippopotamus, warthog and kob, bones of buffalo, a humerus of a baby elephant and a fragment of a skull of a baby hyaena. Many of these remains were splintered and gnawed. . . .
>
> I also inspected two further hyaena lairs, near Kajiado, Kenya. Both were natural horizontal cracks in lava which apparently extended some distance into the rock, but were too low for human exploration beyond about 15 feet. Both tunnels contained substantial quantities of bones (including domestic donkey, cow and dog). . . . In this lair were also ostrich bones, part of an ostrich egg, a human jaw, three human crania and a few other human remains. A great quantity of bone fragments were strewn about outside . . . local Masai villagers . . . expressed the opinion that the human skulls had been carried to the lair from Kajiado hospital cemetery, about two and a half miles away. An inspection of the cemetery showed that many of the graves had indeed been plundered by hyaenas and that there were innumerable splintered human bone fragments scattered around them [Sutcliffe 1970:1111].

I think there can be little doubt that the accumulation of bones in lairs is a natural condition and one that may play a considerable role in the formation processes of many deposits in caves and rockshelters, as well as in some open-air deposits. Several points are worthy of emphasis here. First, my data on wolf lairs, and observations on hyena lairs, all indicate that bones are introduced to such locations in low frequencies. Any substantial accumulation of bones in a lair context must have occurred over a substantial period of time. Such an accumulation is a palimpsest, or the aggregated result of numerous small independent events. In addition, there is provocative evidence in the form of Eskimo informant observations, naturalists' observations, and faunal assemblages that stable locations, such as caves and rockshelters, may well have a successional history; different animal forms may make a location attractive to other forms, and thereby set up the possibility of natural succession in its use. For instance, the Eskimos who provided information to me on wolf behavior were of the opinion that foxes originally dug the holes in soft substrates and wolves moved in later, enlarging the holes. Still later, porcupines moved in. Their presence tended to discourage further use by both foxes and wolves, since the quills commonly left in the chambers and passageways could be very painful when picked up by canids and could even cause death.

Kruuk observed that most hyena dens were not originally dug by hyenas: "I do not know whether hyenas ever start an entirely new den; in every case I watched, there was always some beginning of a hole already there, made by warthogs, springhares, jackals, or others [Kruuk 1972:242]."

It is hard to imagine the complex interaction that might well characterize a cave or rockshelter in Pleistocene Europe, where there were bears (*Ursus spelaeus* and *Ursus arctos*), wolves (*Canis lupus*), foxes (*Vulpes* sp.), and felines (*Felis leo, Felis spelaeus, Felis pandus* etc.). In addition there were hyenas (*Crocuta crocuta* and *Crocuta spelaea*), lynxes (*Lynx spelaeas*), and wolverines (*Gulos gulo*). Several different Carnivora may well have participated in complex types of successional interaction at caves and rockshelters, and there were also a large number of birds, such as owls, and rodents, such as rabbits, ground squirrels, marmots and porcupines, who would join in by adding their contributions to the palimpsest that we may see today as the cave or shelter deposit. For instance, archaeologists working in the caves and rockshelters of the Dordogne region of France have observed that during the excavation season the shelter will be abandoned by birds, particularly the owls, but when the dig closes down and the crew departs the birds return and begin adding their unmistakable contributions to the deposits below. Add to this panorama of competing occupants for caves and rockshelters that interesting animal, man, and the stage is set for some of the most complicated palimpsest deposits one can imagine. So the next step is to imagine the periodic appearance of man and his use of the same caves and rockshelters that have offered shelter to other animals. Clearly the addition of man as a potential contributor to the palimpsest deposits presents the archaeologist with formidable interpretative problems —namely, *how do we reliably relate evidence of hominid presence, in the form of either stone tools or skeletal remains, to the other debris that make up the deposit?* What methods do we use in understanding man's role in the formation processes of the deposit? Unfortunately, most often archaeologists have no methods, or have not even seriously considered the problem. Instead, as we have seen earlier, if evidence of hominids was present, archaeologists assumed that all other remains spatially associated with the hominid remains had also been introduced into the deposit by hominids.

Hunting Behavior of Wolves and Their Feeding Behavior at Kills

One might conclude from the preceding discussion that the prehistorian need be aware of animals as potential contributors only to caves and rockshelters deposits where artifacts have been recovered. This is not the case. Animals are equally capable of contributing to open site deposits in complicated and confusing ways. The most common predator–scavenger activity that may contribute to deposits is the actual killing of animals by wolves and other predators (Figure 5.03).

Repeated kills over substantial periods of time may result in a background of faunal remains dis-

FIGURE 5.03. ***Air view of wolves feeding on a young moose.***

tributed over the landscape that reaches an impressive magnitude, depending on the environmental conditions favoring burial and preservation. In many cases bones of prey may be transported by carnivores and later moved again and accumulated by natural agents, particularly water. Substantial accumulations may build up as naturally aggregated bones derived from kills by predators, or natural death populations that may have been scavenged. How do we distinguish such an accumulation from one resulting from human or hominid behavior? Finding the answer to this question must begin with behavioral observations on animals and comparisons with man to determine if there are diagnostic differences.

I was fortunate to observe several episodes of wolves killing and/or feeding on their prey. These episodes are described next.

Episode One (August 1969)

Four caribou were observed moving south along the west edge of Tulugak Lake, and some distance behind were four wolves following at a leisurely pace. (The ground had approximately 22 inches of new snow, yet the temperature was only 26°F and the sun was out.) I observed two wolves leave the group as the caribou approached the north end of the lake and I could not see where they were going. The other two continued to follow the caribou but increased their pace slightly along the edge of the lake and gradually began to overtake the caribou, who continued moving at a steady pace. As the caribou approached the south end of the lake the two wolves in pursuit speeded up and the caribou responded by breaking into a run

At the south end of the lake is a series of moraines and eskers that had remained windswept during the snow of the previous several days. As the caribou crested the first moraine at the end of the lake the other two wolves came into view running from the west at top speed toward the caribou. As the caribou shifted to the east in response, running along the snowfree margin of the moraine the two wolves who had been following began attacking the upper left rear leg of one animal. When the caribou swung its head to the left, the two wolves that were approaching from the west jumped for the neck and ear region on the right side. With all four wolves attacking, the caribou ran only approximately 50 yards before being brought down. It fell to its knees first, then rolled over away from the wolves, who had it firmly by the neck on the right side. Once the caribou went down, it rolled down the moraine out of view.

By the time I could approach from the east side of the valley moving along the talus so I could gain a view, the wolves were already feeding on the caribou, a yearling calf. Two wolves were concentrating in the "stomach" area, tearing at the walls of the belly while the other two were tearing at the wounds in the left rear leg. I was unable to stay out all day watching the wolves but I did come out to the "viewing spot" three times during the day (the kill was early in the morning). Each time the four wolves were either feeding at the carcass or lying up along the moraine in the sun, seemingly sleeping. The following morning I went out to see the kill and no wolves were in sight. However, I could not see well to the west of the kill, so I made no attempt to approach it. Around four-thirty I again went out to observe the kill, and this time all four wolves were lying in a dispersed manner along the moraine, chewing on bones that they had dragged away from the immediate kill area. I returned again around ten in the evening, and still all four wolves were near the area of the kill, lying along the moraine chewing on bones. From my position, it was clear that the skeleton had been largely disarticulated and dragged about by this time; remaining near the place of the original kill were the neck vertebrae with attached skull. The morning of the third day there were no wolves at the kill, although many ravens were around the kill feeding. Similarly at noon, no wolves were seen; but once again around four-thirty two wolves were observed near the carcass chewing on bones. This was the last time I observed wolves at the kill.

EPISODE TWO (JULY 1972)

During a helicopter survey of the mountain areas to the west of Anaktuvuk village we spotted a pack of seven wolves feeding on what appeared to be the carcass of two different animals. Later, however, when the site was investigated from the ground, it was found to be the disarticulated remains of a single very large bull caribou. Some time later we walked into the location so it could be recorded and while on the kill observed what appeared to be the same pack resting in the edge of a small willow stand approximately 1 mile from the kill. I observed the pack feeding for about 2 hours. During this time the pattern was for an animal to move leisurely into the central kill area and sniff parts that were scattered around the basic thoracic and neck section. After several turns around the kill area, the wolf would very slowly drag a part off to the periphery of the kill area, lie down, and begin gnawing the bone or partially articulated section that had been dragged to the edge. Sometimes a single wolf might work on such a part for as long as an hour. Frequently, when other wolves would move into the central area, others would get up, sniff, tail wag, and move around some; then they would return to the gnawing of "their" pieces.

EPISODE THREE (JUNE 1972)

While going between Anaktuvuk village and our camp at Tulugak Lake I decided to walk the high trail along the toe of the mountain talus between Anavik and Tulugak Lake. As I approached Akvalutak Creek (see L. R. Binford 1978b:173) I saw a moose feeding in the upper edge of the willows growing along the creek. I decided to approach as silently as possible, staying well to the east of the willows in order to get a good picture of the moose with my long lens. I was able to approach the east edge of the willows along the upper talus of the mountains, about 300 m from the moose. As I was setting up to take pictures I noticed through the viewfinder (a 350-mm lens) a dark object between me and the moose crouched in the very edge of the willows. It became clear that the object was a very large grizzly bear crouched within 50 yards of the feeding moose. With an explosive dash the bear ran at the moose from the rear, and the moose began running; however, the bear overtook the moose in short order and hurled its full weight on the rear haunches while reaching forward with its front paws for the neck of the moose. As the weight of the bear buckled the rear legs of the moose, its head jerked up and back at the same time the bear grabbed the neck and pulled backward. The moose and the bear crashed to the ground and the moose never moved again.

The bear moved around the head and neck area of the moose biting and tearing the head area as if to ensure that the moose was dead. After a few moments of this, the bear moved off, defecated, and then lay down licking its paws for approximately 20 minutes. It then got up and violently tore open the belly area with its front claws and began feeding on the viscera and chewing and tearing into the upper thigh and pelvis area. By this time ravens had arrived and were circling and sitting around the kill area. I began to get very chilly from sitting so still, and decided to go on into camp and return the following day to observe the feeding. The kill had occurred at about nine in the evening and I had gotten to bed about two in the morning. By the time I had breakfast and inspected the excavations at the Palangana site it was ten the following morning. I went south to the top of Akvalutak Creek and, much to my surprise, as I approached from the north expecting to see a bear, three wolves bolted from the moose carcass and ran west into the willows. Clearly, I had been detected. I retreated north for about a half-mile and then moved upslope and walked south again along the high talus, watching the kill with the long lens. Shortly, the wolves returned and began feeding directly on the still articulated moose body. They fed for approximately an hour and a half and then slowly moved up to the head of the creek and began climbing up the talus almost to the spot where I had originally seen the moose the previous day. They lay around on large rocks in the sun and apparently went to sleep. I then returned to camp and on subsequent days never saw any additional action at the kill except the feeding of the ubiquitous ravens.

Information on Kill Behavior and Comparisons[2]

Having experienced these episodes, I was in a better position to ask questions of the Eskimo hunters who had witnessed many wolf kills during their lives. They generally agreed that the behavior of wolves was very regular both in killing and in feeding. They suggested that the level of exploitation of a kill by wolves varied with the season, which in turn was related to game density in the area. During summer, the season when all the described episodes occurred, the Eskimo agreed that the diet of wolves was largely ptarmigan (*Lagopus* sp.), Arctic ground squirrel (*Citellus undulatus*), grayling (*Thymallus arcticus*), and occasionally sheep (*Ovis dalli*). Caribou and moose were rare elements of diet in the summer mountains. Because of the scarcity of large game in the summer, the Eskimo were of the opinion that wolves spent more time feeding on large game kills, particularly feeding much more extensively on bones after the flesh had been consumed. During spring and fall, when caribou are migrating through the passes of the Brooks Range, in the opinion of the Eskimo the wolves rarely disarticulated the skeletons of prey and fed almost exclusively on flesh, frequently leaving the animal completely articulated and only defleshed in the sternum, abdominal, and rump areas (see Figure 5.04).

The Eskimo generalized that the more extensively the wolves fed on a kill the more dispersed the bones and the more disarticulated the skeleton becomes. They agreed that the feeding behavior of wolves at a kill might be justifiably divided into a communal phase and a solitary phase. Feeding on flesh prior to the basic disarticulation of the skeleton was the communal phase, where all wolves fed directly on the body. After this, there was a sleep or rest period. *If the wolves returned*, the kill was then gradually disarticulated and the parts dragged off to a perimeter around the kill where each wolf fed on the bones individually (Figure 5.05).[3] This second phase of feeding was where most bone destruction and disarticulation occurred. The amount of this second phase behavior varied with the size of the feeding group relative to the size of the kill and the game density, which tended to vary seasonally. The Eskimo agreed that the degree of skeletal destruction was related to the size or age of the prey in addition to the number of feeding animals. Figure 5.06 shows a wolf kill as it looked two years after the kill.

A major difference between kill and lair assemblages is that a large contributor of the assemblage at a lair is bone contained in scat. Although scats are present around kills, they are dispersed and most commonly deposited after initial feeding. Another contrast with kills is that there is a diversity of faunal remains at a lair; both large game and relatively small game are typically included. As we have seen, this is a function of season and the fact that the predator's hunting range is restricted to the area around the den when the pups are very young. For these reasons a wide variety of small mammals and rodents are generally present in the lair assemblage. In midsummer the pups begin play hunting, killing small game and rodents. Remains of wolves themselves may also be found in lair assemblages. The Eskimo have observed that when they have killed pregnant wolves there are almost always more fetuses than there are pups in an average litter. It is the belief of the Eskimo that the mother wolf cleans up the birth chamber, eating the afterbirth, and any stillborn young together with the bloody soil associated with the birth. It was further observed that female wolves may at least partially eat pups that die at a later date. The Eskimo believe that this is to keep the babies from eating a sick pup and catching the "death." Another source of wolf bones at lairs is the untimely collapse of the nest area resulting in the death of both female and young. One

[2] The reader should consult Mech (1966) for additional observations on wolf behavior.

[3] Important observations on this aspect of feeding have been recently made by Magoun 1979.

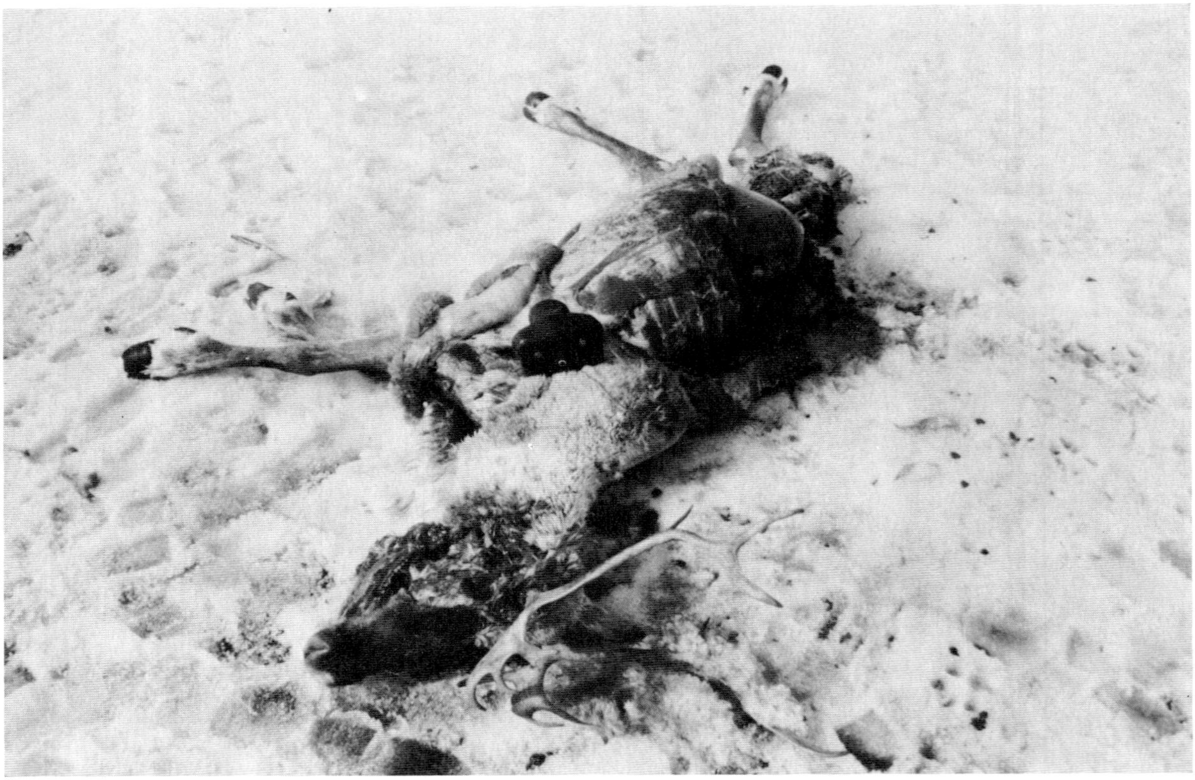

FIGURE 5.04. ***Caribou shortly after having been killed by wolves.***

informant was of the belief that old females also frequently die giving birth. When that is the case the Eskimo believe that the den site is abandoned by the family and either foxes or porcupines may then move in. "Wolves would not go there until somebody else cleans up good."

One of the characteristic actions of pups at a lair is a kind of play competition similar to the competition that goes on among adults at a kill. This play competition may focus on practically any items that a young wolf may pick up in the area around the lair. Rocks, sticks, and bones from kills and death sites that may be in the area of the lair are transported to the lair and played over by the pups; they are subsequently gnawed by both pups and adults. This behavior results in the introduction to the assemblage at the lair of bones that are most resistant and apt to remain longest in unaltered condition at kills and death sites. These include teeth of large mammals, mandibles, parts of skulls, and isolated fragments such as the acetabulum. Bones introduced as food and part of the dispersion of parts normally characteristic of a kill may well be conditioned in their frequency by the likelihood of disarticulation given some competition over the carcass.

In addition to structural contrasts between kills and lairs, there are situational factors that differentially condition the further participation of kills in subsequent natural processes.

Wolves tend to follow rather regular paths in their hunting travels and during winter these are normally determined by the distribution of ice. They travel on the frozen rivers and across lakes. In addition they have frequently been observed to follow prey long distances, waiting to attack until they are on ice. The Eskimo report that the claws of the wolf enable him

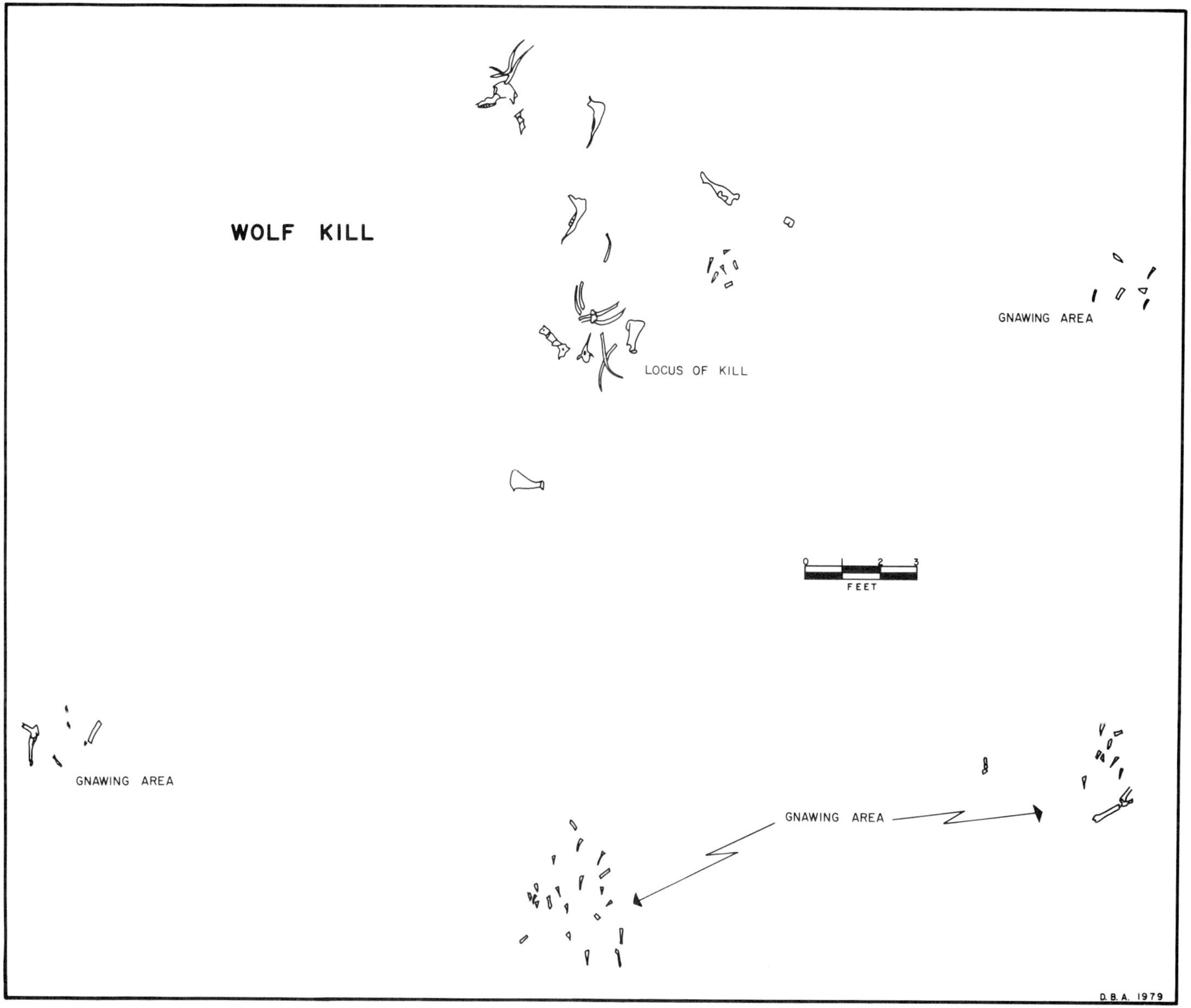

FIGURE 5.05. *Map of wolf kill showing core area (locus of kill) and peripheral gnawing areas.*

FIGURE 5.06. ***Wolf kill as it appeared at time of recording.***

quickly to outrun a caribou or moose on ice, whereas on snow of moderate depth a wolf has great difficulty keeping up and generally follows in the prey's tracks. The result is an "ice bias" in winter wolf kills. Come spring breakup of the rivers and the melting of lakes, the bones from winter kills are commonly swept away by the violent spring rivers or dropped neatly to the bottom of lakes. Ice-rafted bones are apt to be substantially modified during transport, since the flow of ice chunks is fast and violent. Melting around bones is differential since they are dark, and they are therefore frequently partially exposed while being swept along by the spring rivers. I have personally observed (*a*) bones from my previous summer camp, (*b*) a footlocker full of cooking gear, and (*c*) a tent together with poles, rafted down Kongumuvuk Creek (see L. R. Binford 1978b:173) in the spring of 1972. The bones that were recovered in a backwater along with the tent and the shattered footlocker were crushed and broken with typical spiral fractures; in fact, one large sheep horn was completely sheared off at the tip, leaving what appeared to be an intentional cut at the tip of the horn. This is not the place to enter into an extended discussion of taphonomic processes in polar and near-polar settings where there is violent melting and associated flooding. Enough has been indicated to illustrate that bones may be moved and redeposited, and that modifications may well result in "fresh bone" breaks, chipping, or crushing. This, however, is getting slightly ahead of the story. More important than considering the factors affecting the transport and modifications of bones by natural agents is the establishment of a general understanding as to the occurrence of bone deposits that have not been transported.

I think it can be reasonably concluded that predators do accumulate bones in lairs and dens. In addition it is clear that predators and scavengers do produce populations of bones as a result of their normal participation in the ecosystem. Both conditions may be coterminous with human or hominid behavior, resulting in the spatial association of items derived from both hominid–human actions and nonhuman predator–scavenger behavior.

It is to the latter task that the remainder of this chapter is directed. I address the interesting question of what a population of bones generated by animals under different behavioral contexts does in fact look like. I am concerned with the degree to which there are distinctive properties that might permit the identification of the agent and the behavioral context.

Control Collections of Animal-Structured Assemblages

Data available at the present time are from 24 individual wolf kills of caribou, two very small wolf den assemblages described earlier in this chapter, and the Bent Creek den assemblage. Informants were aware of a series of wolf dens that had been dug into a bank up Bent Creek. Several years before the collection was made the bank collapsed and the wolves abandoned the location. What remained was an accumulated scatter of bones on a small gravel bar directly in front of where the wolves had previously denned. This scatter represents an accumulation of bones from an unknown number of years, deposited

TABLE 5.01
Assemblage Composition of Alaskan Wolf Kills (Caribou MNI)[a]

	Wolf kill site inventories (MNIs)[b]									
	(1)	(2)	(3)	(4)	(5)	(6)	(7)	(8)	(9)	(10)
Skull	1.00			1.00	1.00			1.00		1.00
Mandible	1.00	.50	1.00	1 00	1.00	1.00	1.00	1.00	1.00	1.00
Atlas	1.00			1 00	1.00			1.00		1.00
Axis	1.00	1.00		1 00		1.00	1.00	1.00		
Cervical vertebrae	1.00	.80		.60		1.00	1.00	1.00	.20	
Thoracic vertebrae	1.00	.29	.79	.57	.36	1.00	1.00	1.00	.79	.64
Lumbar vertebrae	1.00	.20	1.00	1.00	1.00	1.00	1.00	1.00	1.00	.60
Pelvis	1.00	1.00	1.00	1.00	1.00	1.00	1.00	1.00	1.00	1.00
Ribs	1.00		.29	1.00		.61	1.00	1.00	.39	.57
Scapula	.50	.50	1.00	1.00		1.00	1.00	1.00	.50	.50
Proximal humerus	.50		.50					.50		
Distal humerus	.50	.50	1.00	.50		1.00	1.00	1.00	.50	
Proximal radio-cubitus	.50	.50	.50	.50		.50	.50	1.00	.50	
Distal radio-cubitus	.50		.50			.50	.50	.50		
Carpals	.50		.50			.50	.50	.50		
Proximal metacarpal	.50	.50	.50			.50	.50	.50	.50	
Distal metacarpal	.50		.50			.50	1.00	.50	.50	
Proximal femur	1.00	.50	.50			1.00	.50	.50		
Distal femur		.50	.50					.50		
Proximal tibia	.50	1.00	1.00			.50		.50		
Distal tibia	1.00	1.00	1.00	.50		1.00	.50	1.00		
Tarsals	1.00	.50	.50	.50				.50		
Astragalus	1.00	.50	.50	1.00				.50		
Calcaneus	.50	.50	1.00	.50				.50		
Proximal metatarsal	1.00	.50	1.00	1.00				1.00		
Distal metatarsal	.50		.50	.50				.50		
First phalange	.38		.50	.25				.12		
Second phalange	.25		.25	.25						
Third phalange	.12		.25	.25						
Site totals										
Splinters	33	21	5	56	60	25	41	7	71	83
Cylinders	4	2	1	5	4	3	4	0	3	1
Complete bones	0	2	5	0	0	1	1	6	0	0
Articulator ends on complete bones		4	10			2	2	12		
Articulator ends	11	8	16	6	0	11	9	16	4	0
Percentage of ends complete	0	50	63	0		18	22	75	0	
Percentage of articulator ends missing	54	58	33	75	100	54	63	33	83	100

(continued)

TABLE 5.01—*Continued*

	Wolf kill site inventories (MNIs)[b]									
	(11)	(12)	(13)	(14)	(15)	(16)	(17)	(18)	(19)	(20)
Skull					1.00	1.00	1.00	1.00		
Mandible	.50	1.00	.50		1.00	1.00	.50	.50	1.00	
Atlas					1.00	1.00	1.00	1.00		
Axis	1.00	1.00	1.00		1.00		1.00	1.00	1.00	
Cervical vertebrae	1.00	.60	1.00		1.00		.80	.60		
Thoracic vertebrae	1.00		1.00		1.00				.71	
Lumbar vertebrae	1.00	.60	1.00		1.00				1.00	
Pelvis	1.00	1.00	1.00	1.00	1.00		1.00	1.00	1.00	1.00
Ribs	.79	.28			1.00				1.00	
Scapula	1.00	.50	.50		1.00				.50	
Proximal humerus					.50					
Distal humerus	.50				1.00				.50	
Proximal radio-cubitus	1.00				.50				.50	
Distal radio-cubitus	.50				.50				.50	
Carpals					.50					
Proximal metacarpal	.50				.50				.50	
Distal metacarpal	.50				.50				.50	
Proximal femur					1.00				.50	
Distal femur					1.00				.50	
Proximal tibia					.50					
Distal tibia	.50			.50	1.00				1.00	
Tarsals					.50				.50	
Astragalus					.50				.50	
Calcaneus					1.00				.50	
Proximal metatarsal					1.00	.50			1.00	.50
Distal metatarsal					1.00				.50	
First phalange					.50					
Second phalange					.50					
Third phalange					.38					
Site totals										
Splinters	45	62	58	74	4	63	66	55	28	73
Cylinders	5	5	4	2	1	4	3	6	4	2
Complete bones	0	0	0	0	5	0	0	0	0	
Articulator ends on complete bones					10					
Articulator ends	7	0	0	1	18	1	0	0	12	1
Percentage of ends complete	0			0	56				0	
Percentage of articulator ends missing	71	100	100	96	25	96	100	100	50	96

TABLE 5.01—*Continued*

	Wolf kill site inventories (MNIs)[b]				Total wolf kills		Bent Creek den		Chandler Lake kill	
	(21)	(22)	(23)	(24)	(25)	(26)	(27)	(28)	(29)	(30)
Skull	1.00	1.00			11.0	48.0	12.0	86.0	4.0	38.0
Mandible	1.00	1.00		1.00	18.5	80.0	12.5	89.0	6.0	57.0
Atlas	1.00	1.00			12.0	52.0	9	66.0	2.0	19.0
Axis	1.00	1.00	1.00	1.00	17.0	74.0	5	36.0	2.0	19.0
Cervical vertebrae	1.00	.20	1.00	1.00	12.8	36.0	.86	6.0	.2	2.0
Thoracic vertebrae	1.00		.60		12.15	53.0	.29	2.1	.3	3.0
Lumbar vertebrae	1.00	.40		.80	15.6	68.0	1.7	12.0	.8	8.0
Pelvis	1.00	1.00		1.00	23.0	100.0	9.5	68.0	3.5	33.0
Ribs	1.00		1.00	.86	10.8	47.0	0	0	.5	5.0
Scapula	1.00	.50			12.0	52.0	8.0	57.0	4.0	38.0
Proximal humerus	.50				2.5	11.0	3.5	25.0	.5	5.0
Distal humerus	1.00	.50			9.5	41.0	14.0	100.0	3.5	33.0
Proximal radio-cubitus	1.00	.50			8.0	35.0	10.5	75.0	3.0	29.0
Distal radio-cubitus	.50				4.5	20.0	6.5	46.0	1.5	14.0
Carpals	.50				3.5	15.0	4.5	32.0	5.5	52.0
Proximal metacarpal	.50				5.5	24.0	10.0	71.0	8.5	81.0
Distal metacarpal	.50				5.5	24.0	9.5	68.0	7.5	71.0
Proximal femur	1.00		.50		7.0	30.0	2.0	14.0	.5	5.0
Distal femur	.50				3.5	15.0	3.0	21.0	.5	5.0
Proximal tibia	.50		.50		5.0	22.0	2.0	14.0	.5	5.0
Distal tibia	1.00		1.00		11.0	48.0	8.5	61.0	4.5	43.0
Tarsals	.50		.50		5.0	22.0	5.0	36.0	6.5	62.0
Astragalus	.50		.50		5.5	24.0	4.5	32.0	7.5	71.0
Calcaneus	.50		.50		5.5	24.0	5.0	36.0	6.5	62.0
Proximal metatarsal	.50		.50		8.5	37.0	7.0	50.0	10.5	100.0
Distal metatarsal	50		.50		4.5	20.0	4.0	29.0	8.5	81.0
First phalange	.38		.50		2.63	11.0	3.25	23.0	6.3	60.0
Second phalange	12		.50		1.87	8.0	2.25	16.0	5.4	51.0
Third phalange			.50		1.5	7.0	1.0	7.0	4.2	40.0
Site totals										
Splinters	10	67	43	59	1119					
Cylinders	1	4	5	3	76					
Complete bones	5	0	0	0	25					
Articulator ends on complete bones	10				50					
Articulator ends	16	2	6	0	145					
Percentage of ends complete	63	0	0		34					
Percentage of articulator ends missing	33	92	75	100	75					

[a] Average weight of adult caribou in the Arctic herd is 222 lb (100.91 kg).
[b] All MNIs are calculated as outlined in Binford (1978b: 69–72, 478–479).

in front of a den and in the defecation area to one side. I made no attempt to collect bone splinters, since I suspected some bias from washing and transport at this location. Inventory data on this sample are given in Table 5.01. Another collection was made near the south end of Chandler Lake from a surface scatter of bones that informants assured me were the remains of wolf kills. It was suggested that the remains "might" represent wolf kills made over about a 10-year period. I was not convinced that Eskimo-killed animals were not also represented and that there might not be an additional bias introduced by the nearby lake surface, which would have been ice during most of the kill episodes; any bones remaining there would have gone to the bottom of the lake on melting.

TABLE 5.02
Andrew Hill's Kill Assemblage Data from East Africa [a]

	Lake Rudolf, Kenya						
	Delta flat			Delta pond		Totals	
Anatomical part	Zebra (1)	Topi (2)	Gazelle (3)	Topi (4)	Gazelle (5)	Topi (6)	Gazelle (7)
Skull	60.0	75.0	20.0	67.0	29.0	84.0	24.0
Mandible	100.0	69.0	80.0	22.0	100.0	70.0	88.0
Atlas	60.0	67.0	20.0	67.0	29.0	76.0	24.0
Axis	40.0	88.0	60.0	89.0	29.0	100.0	47.0
Cervical vertebrae	28.0	83.0	24.0	71.0	40.0	92.0	31.0
Thoracic vertebrae	29.0	62.0	30.0	29.0	37.0	65.0	33.0
Lumbar vertebrae	40.0	77.0	40.0	63.0	81.0	85.0	57.0
Pelvis	60.0	100.0	100.0	22.0	100.0	100.0	100.0
Ribs	5.0	49.0	22.0	25.0	24.0	52.0	23.0
Scapula	30.0	60.0	40.0	67.0	43.0	70.0	41.0
Proximal humerus	20.0	35.0	30.0	44.0	14.0	42.0	24.0
Distal humerus	20.0	44.0	70.0	100.0	14.0	60.0	47.0
Proximal radio-cubitus	20.0	42.0	40.0	44.0	14.0	48.0	29.0
Distal radio-cubitus	20.0	33.0	40.0	44.0	29.0	40.0	35.0
Carpals							
Proximal metacarpal	20.0	27.0	30.0	56.0	14.0	36.0	24.0
Distal metacarpal	10.0	27.0	30.0	33.0	14.0	32.0	24.0
Proximal femur	20.0	46.0	40.0	22.0	14.0	48.0	29.0
Distal femur	20.0	42.0	50.0	22.0	0	44.0	29.0
Proximal tibia	20.0	50.0	70.0	22.0	14.0	52.0	47.0
Distal tibia	40.0	60.0	70.0	33.0	29.0	64.0	53.0
Tarsals							
Astragalus	30.0	46.0	30.0	33.0	0	50.0	18.0
Calcaneus	10.0	46.0	30.0	33.0	0	50.0	18.0
Proximal metatarsal	30.0	42.0	30.0	33.0	14.0	46.0	24.0
Distal metatarsal	30.0	40.0	30.0	33.0	0.0	44.0	18.0
First phalange	0	30.0	30.0	28.0	7.0	34.0	3.0
Second phalange	2.0	30.0	30.0	25.0	7.0	34.0	3.0
Third phalange	0	30.0	30.0	22.0	7.0	33.0	3.0
MNI in sample	5.0	24.0	5.0	4.5	3.5	28.5	8.5
Body weight (kg)	164	82	22	82	22	82	22

Continued

A valuable body of comparative data has been made available by Andrew Hill from research conducted in both Kenya and Uganda (Table 5.02). Hill's material represents six African species (approximately 62 individual animals). These assemblages are all presumably kill assemblages. Richard Klein (1975) has reported on the excavation of a bone deposit from South Africa that is believed to be the product of animal behavior, perhaps hyena (Table 5.03). At the latter site, Swartklip, Klein recorded 32 different species, of which 7 bovid species are of interest here. Among the latter there were a minimum of 130 individuals.

These samples of bone populations are quite different in their control characteristics. Some of my data represent the total remains referable to known

TABLE 5.02—*Continued*

	Murchison Falls National Park, Uganda								
	Collection A			Collection B			Totals		
Anatomical part	Waterbuck (8)	Reedbuck (9)	Oribi (10)	Waterbuck (11)	Reedbuck (12)	Oribi (13)	Waterbuck (14)	Reedbuck (15)	Oribi (16)
Skull	22.0	0	60.0	33.0	66.0	50.0	28.0	33.0	54.0
Mandible	100.0	50.0	100.0	100.0	63.0	100.0	100.0	67.0	100.0
Atlas	44.0	100.0	40.0	44.0	0	25.0	44.0	67.0	31.0
Axis	89.0	50.0	0	44.0	0	25.0	67.0	33.0	15.0
Cervical vertebrae	49.0	40.0	0	31.0	0	25.0	40.0	33.0	15.0
Thoracic vertebrae	46.0	22.0	6.0	19.0	33.0	22.0	33.0	33.0	15.0
Lumbar vertebrae	89.0	67.0	53.0	4.0	78.0	46.0	46.0	83.0	49.0
Pelvis	67.0	100.0	100.0	33.0	66.0	88.0	50.0	100.0	92.0
Ribs	48.0	38.5	5.0	26.0	0	22.0	37.0	24.0	15.0
Scapula	100.0	50.0	40.0	11.0	100.0	38.0	56.0	83.0	38.0
Proximal humerus	44.0	50.0	20.0	11.0	0	50.0	28.0	33.0	38.0
Distal humerus	44.0	25.0	40.0	33.0	33.0	75.0	50.0	33.0	62.0
Proximal radio-cubitus	56.0	25.0	60.0	11.0	33.0	38.0	33.0	33.0	46.0
Distal radio-cubitus	44.0	0	40.0	0	0	63.0	22.0	0	38.0
Carpals									
Proximal metacarpal	33.0	0	20.0	22.0	0	38.0	28.0	0	31.0
Distal metacarpal	33.0	0	20.0	11.0	0	63.0	22.0	0	46.0
Proximal femur	56.0	0	60.0	0	33.0	50.0	28.0	17.0	54.0
Distal femur	33.0	0	40.0	0	0	38.0	17.0	0	38.0
Proximal tibia	67.0	0	40.0	0	0	75.0	33.0	0	62.0
Distal tibia	50.0	25.0	40.0	22.0	0	75.0	39.0	17.0	62.0
Tarsals									
Astragalus	11.0	25.0	0	0	0	63.0	6.0	17.0	38.0
Calcaneus	11.0	0	0	0	0	38.0	6.0	0	23.0
Proximal metatarsal	44.0	50.0	40.0	11.0	0	88.0	28.0	33.0	69.0
Distal metatarsal	33.0	50.0	40.0	0	0	75.0	17.0	33.0	62.0
First phalange	0	0	0	3.0	0	35.0	1.0	0	21.0
Second phalange	0	0	0	0	0	28.0	1.0	0	17.0
Third phalange	0	0	0	0	0	19.0	1.0	0	12.0
MNI in sample	4.5	2.0	2.5	4.5	1.5	4.0	9.0	3.5	6.5
Body weight (kg)	131	37	17	131	37	17	131	37	17

[a]Data from Hill 1976.

TABLE 5.03
***Richard Klein's Den Data from South Africa*[a]**

	Swartklip den assemblages					
Anatomical part	Grysbok (1)	Springbok (2)	Reedbuck (3)	Topi roan, sable (4)	Wildebeest (5)	Kudu, eland (6)
Skull	55.0	79.0	45.0	100.0	67.0	45.0
Mandible	100.0	100.0	100.0	100.0	44.0	45.0
Atlas	10.0	43.0	8.0	50.0	26.0	27.0
Axis	17.0	50.0	21.0	10.0	33.0	9.0
Cervical vertebrae	7.0	21.0	13.0	10.0	7.0	18.0
Thoracic vertebrae	7.0	7.0	5.0	10.0	7.0	9.0
Lumbar vertebrae	10.0	14.0	18.0	20.0	14.0	18.0
Pelvis	48.0	50.0	26.0	10.0	44.0	64.0
Ribs	3.0	7.0	3.0	5.0	4.0	5.0
Scapula	10.0	50.0	21.0	20.0	19.0	9.0
Proximal humerus	0	14.0	3.0	0	4.0	9.0
Distal humerus	17.0	64.0	11.0	60.0	78.0	73.0
Proximal radio-cubitus	17.0	43.0	8.0	50.0	37.0	73.0
Distal radio-cubitus	7.0	36.0	11.0	30.0	52.0	100.0
Carpals	7.0	14.0	8.0	20.0	14.0	27.0
Proximal metacarpal	7.0	43.0	16.0	50.0	70.0	82.0
Distal metacarpal	14.0	50.0	16.0	50.0	100.0	91.0
Proximal femur	7.0	7.0	5.0	10.0	7.0	18.0
Distal femur	17.0	21.0	13.0	40.0	26.0	27.0
Proximal tibia	21.0	29.0	5.0	20.0	19.0	18.0
Distal tibia	34.0	29.0	8.0	90.0	89.0	64.0
Tarsals	10.0	7.0	8.0	10.0	7.0	27.0
Astragalus	24.0	14.0	8.0	40.0	37.0	55.0
Calcaneus	21.0	29.0	13.0	60.0	33.0	64.0
Proximal metatarsal	41.0	43.0	18.0	40.0	41.0	45.0
Distal metatarsal	31.0	43.0	13.0	60.0	41.0	64.0
First phalange	17.0	21.0	5.0	20.0	11.0	27.0
Second phalange	17.0	21.0	5.0	20.0	11.0	27.0
Third phalange	17.0	21.0	5.0	20.0	11.0	27.0
MNI in sample	29	14	38	10	27	11
Body weight (kg)	12	23	37	82	128	200

[a]Data from Klein 1975.

wolf kills; that is, individual caribou were killed and these are the bones remaining in the area after wolves had fed and abandoned the carcass. Hill's data are areal samples; that is, they are all the bones remaining on the surface in a given sample area and the inference is made that animals killing as well as the scavenging of natural deaths was the dynamic context of their having been dispersed over the land surface. This is similar to my Bent Creek and Chandler samples. Klein's data derive from the excavation of a deposit where there is no evidence of man, and the inference has been made that animals were responsible for the accumulation. This deposit has been interpreted as a den assemblage.

Before beginning the discussion of these assemblages, it is perhaps important to outline some of the analytical problems that must be faced in comparing faunal assemblages, particularly in attempting to document diagnostic patterning in the faunal remains of both animal and human activities.

When viewing an assemblage, interpretation consists of identifying the agents and the behavioral context of the agents' activities that brought into being the particular association of bones remaining. This statement of the problem *assumes* that the association is behaviorally meaningful and the result of single agents behaving relatively consistently. Dealing realistically with the latter assumption will be the subject of my later analysis of a number of well-known archaeological assemblages. For the moment I simply want to treat in realistic terms the activities of animals that differentially contribute to the composition of assemblages.

There are basically two actions that animals are capable of performing that condition the composition of assemblages remaining for us to observe: (*a*) the transport of bones and (*b*) the destruction of bones. If we are to *understand* and accurately *identify* the behavioral context of the accumulation of bone deposits, we must understand the factors conditioning the selection of anatomical parts for transport and the levels of redundancy in the patterns of bone removal from kill sites. For instance, we wish to recognize whether we are dealing with an assemblage that has been transported or with a *residual* population, that is, what remains after other parts are transported. Further complicating the matter may be the presence and/or absence of destruction coupled with transport as well as different levels of destruction in different settings.

How do we begin unraveling such a complicated set of possible conditions? I think we must begin with understanding destruction, since that is the area that has received most research up until this time (see Brain 1969 as well as L. R. Binford and J. B. Bertram 1977). It is apparent that the survival potential of a given bone subjected to destructive agents is a function of its strength or simple ability to resist destructive agents. Earlier work showed that this resistance to destruction was monitored by measures of bone density. In fact, the probability of a bone surviving attack by destructive agents was measured and a third-order equation was generated experimentally; that is, on the basis of fitting an equation to a set of data for which the actual survivorship was known, as were the bone densities of the parts subjected to destructive agents. The equation that is believed best to approximate the relationship between bone density and survival probabilities is reproduced here from my earlier work (L. R. Binford and J. B. Bertram 1977:138):

$$\text{Survival percentage} = -352.778 + 1050.4(d) - 1008.69(d^2) + 332.882(d^3)$$

where d = bone density of the element of interest.

Table 5.04 summarizes the values of the survival probabilities (percentages) calculated for each bone element using the equation and the data on bone density given previously (L. R. Binford and J. B. Bertram 1977:109) for a 6- and a 90-month-old sheep, as well as a 2½-year-old caribou. Simple inspection of the table reveals two points: (*a*) The percentage of bones surviving after being addressed by a destructive agent is much greater in old animals (columns 5 and 7) and relatively insignificant in young animals (column 1), and (*b*) some bones survive better than others regardless of age (e.g., the scapula, the distal humerus, the mandible, and the distal tibia). Two bones, the humerus and the tibia, are particularly diagnostic of the action of destructive agents. Clearly there is a consistently great difference in survival potential between the proximal and distal ends of the humerus, and a smaller but nonetheless consistent difference between the proximal and distal ends of the tibia. These conditions can be seized upon for purposes of *diagnosing* the relative role bone destruction plays in modifying an assemblage.

What seems clear from Figure 5.07 is that there is a differentiation indicated by the humerus between assemblages subjected to destruction and those not subjected to destruction. This picture is not so clear for the tibia (Figure 5.08). It is true that all assemblages suffering destruction of parts fall below the diagonal line and most nondestroyed assemblages fall along the diagonal. (See Table 5.05 for inventory of assemblages.) However, there is an overlap, with some nonravaged assemblages mixed with the ravaged assemblages below the diagonal line. This condition arose from the differential

TABLE 5.04
***Survival Percentages for Bones of Animals of Different Ages*[a]**

	Sheep					Caribou	
	6 months		Mean value 6 + 90 months		90 months	30 months	
Anatomical part	SP (1)	SP 47.7 (2)	SP (3)	SP 73.8 (4)	SP (5)	SP (6)	SP 87.57 (7)
Skull	39.0	84.0	69.6	94.0	100.0	69.62	80.0
Mandible	47.7	100.0	73.8	100.0	100.0	87.57	100.0
Atlas	11.8	31.0	42.5	58.0	73.2	62.15	71.0
Axis	17.0	41.0	32.3	44.0	47.5	47.24	54.0
Cervical vertebrae	17.0	41.0	26.1	35.0	35.1	27.53	31.0
Thoracic vertebrae	16.1	39.0	29.9	40.0	43.6	31.41	36.0
Lumbar vertebrae	16.1	39.0	37.9	51.0	59.6	43.18	49.0
Pelvis	17.7	42.0	58.8	80.0	100.0	76.95	88.0
Ribs	16.7	41.0	20.7	28.0	24.6	22.28	25.0
Scapula	21.7	50.0	60.8	82.0	100.0	47.64	54.0
Proximal humerus	0	0	7.10	10.0	19.1	14.89	17.0[b]
Distal humerus	19.8	46.0	59.9	81.0	100.0	51.21	58.0
Proximal radio-cubitus	17.7	42.0	58.8	80.0	100.0	35.99	41.0
Distal rado-cubitus	11.8	31.0	44.4	60.0	76.9	43.38	50.0
Carpals	1.9	12.0	26.6	36.0	51.2	22.32	25.0
Proximal metacarpal	9.7	27.0	22.5	30.0	35.26	31.40	36.0
Distal metacarpal	18.2	43.0	30.7	47.0	43.2	30.66	35.0
Proximal femur	9.1	26.0	43.0	58.0	76.97	35.05	40.0
Distal femur	16.75	41.0	33.98	46.0	51.21[c]	31.05	35.0
Proximal tibia	13.81	35.0	30.53	41.0	47.24[c]	22.32	25.0
Distal tibia	9.12	26.0	54.6	74.0	100.0	62.36	71.0
Tarsals	8.08	24.0	46.1	62.0	84.0	35.05	40.0
Astragalus	3.70	15.0	41.9	57.0	80.0	31.40	36.0
Calcaneus	12.1	32.0	49.9	68.0	87.6	31.40	36.0
Proximal metatarsal	12.1	32.0	27.8	38.0	43.4	35.99	41.0
Distal metatarsal	12.1	32.0	22.1	30.0	32.1	35.26	40.0
First phalange	6.0	13.0	12.5	17.0	21.0	15.0	17.0
Second phalange	6.0	13.0	10.0	14.0	15.0	11.5	13.0
Third phalange	6.0	13.0	9.0	12.0	11.0	8.75	10.0

[a]SP = survival percentage (the percentage of original number expected to survive).
[b]Density for caribou proximal humerus, remeasured since earlier study, is 1.01.
[c]Error discovered in Table 3.9 (Binford and Bertram 1977: (109). Density should read 1.41 for distal femur, 1.38 for proximal tibia.

breakage and butchering of the tibia, which is particularly common when animals are frozen. Under such conditions disarticulation of the lower leg is achieved by breaking through the tibia shaft so that the two ends of the bone are differentially transported. The result is the mix seen in Figure 5.08. I have never found empirical support for the argument that the humerus was butchered through the shaft. Claims have been made for destruction of the proximal articulator end during butchering, but this is of course destruction and what is of interest here. It is suggested that proportional plots of the standardized

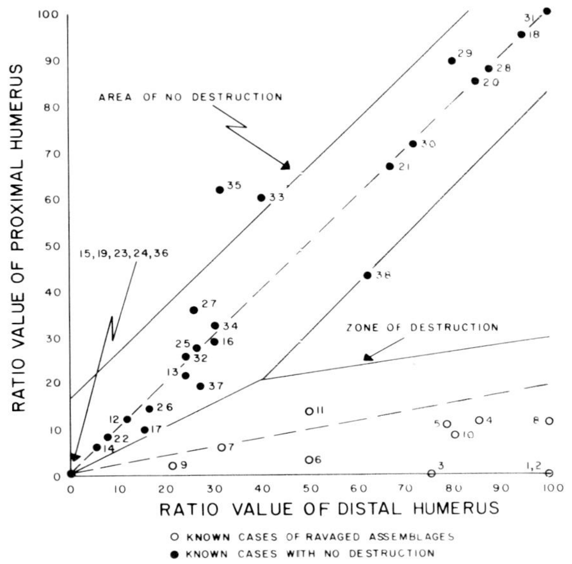

FIGURE 5.07. ***Relationship between frequencies of proximal humerus and distal humerus in a series of known assemblages (listed by number in Table 5.05).***

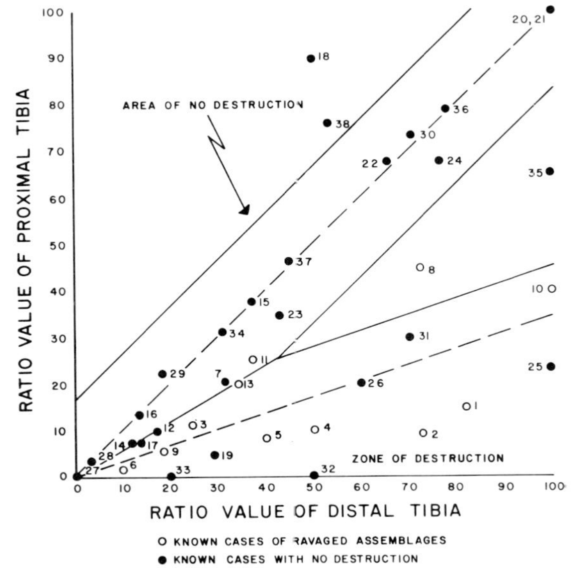

FIGURE 5.08. ***Relationship between frequencies of proximal tibia and distal tibia in a series of known assemblages (listed by number in Tale 5.05).***

frequencies for the proximal versus the distal humerus be used as the basic diagnostic technique for assessing the degree of destruction of bone parts an assemblage has suffered.

Figure 5.09 demonstrates the control assemblage from Alaska, Hill's (1976) African data, and Klein's (1975) data. It is clear that both Klein's data from the Swartklip animal den and my Alaskan assemblages have suffered attrition at the hands of destructive agents, presumably hyena and wolf, respectively. On the other hand, Hill's assemblages from animal kills all cluster in or on the margins of the graph area where "pristine" assemblages are expected. This placement suggests zero to very minimal destruction of bone in these assemblages. The difference between Hill's data and Klein's—pristine versus ravaged assemblages—might be related to differences between den and kill assemblages except for the fact that the wolf kills are clearly within the ravaged distribution.

As a further confirmation of the diagnosis of pristine versus ravaged assemblages the distribution of proximal versus distal tibia was plotted (Figure 5.10).

The pristine character of Hill's assemblages is supported by the data on the tibia, where his assemblages fall along the diagonal in a more convincing fashion than for the humerus. Similarly, the Alaskan data all appear as ravaged assemblages as before; however, Klein's data distribute somewhat differently. The large animals all unambiguously appear as ravaged assemblages but the three small bovids either appear ambiguous or clearly within the zone of pristine assemblages. A glimpse at Figure 5.09 shows that even with the humerus Klein's column 1 (grysbok) and column 3 (reedbuck) could be interpreted as ambiguous. Thus it is quite clear that the degree to which the faunal remains of a species are destroyed by carnivores is not a simple function of whether one is viewing a kill or a transported den assemblage. This not a very practical generalization, since in the materials compared the destruction by wolves at kills is most likely related to the *absence of competitors*. As noted in the behavioral descriptions, the wolves tend to feed, sleep, and then return to the kill, dragging off elements of the skeleton to peripheral "gnawing" areas where the destruction takes place. In addition to this unmolested exploita-

TABLE 5.05
Inventory of Assemblages Plotted on Figures 5.07 and 5.08

Assemblages known to have suffered destruction

1. Anaktuvuk dog yard, fall 1970 (Binford and Bertram 1977: 80, Table 3.1, column 2)
2. Anaktuvuk dog yard, late fall 1970 (Binford and Bertram 1977: 80, Table 3.1, column 4)
3. Anaktuvuk dog yard, winter 1971 (Binford and Bertram 1977: 80, Table 3.1, column 6)
4. Anaktuvuk dog yard, winter 1971 (Binford and Bertram 1977: 80, Table 3.1, column 8)
5. Anaktuvuk dog yard, spring 1971 (Binford and Bertram 1977: 80, Table 3.1, column 10)
6. Anaktuvuk dog yard, late spring 1971 (Binford and Bertram 1977: 80, Table 3.1, column 12)
7. Navajo winter sheep assemblage (Binford and Bertram 1977: 101, Table 3.5, column 2)
8. Navajo summer sheep assemblage (Binford and Bertram 1977: 101, Table 3.5, column 4)
9. Rulland dogs (Binford 1978b: 207, Table 5.9, column 8)
10. Kongumuvuk high summer hunting camp (Binford 1978b: 270, Table 6.6, column 5)
11. Amalgamation site dog yard (Binford 1978b: 322, Table 6.21, column 6)

Assemblages known not to have suffered destruction

12. Fall dispersed kills (Binford 1978b: 76, Table 2.8, column 2)
13. Winter dispersed kills (Binford 1978b: 76, Table 2.8, column 4)
14. Spring dispersed kills (Binford 1978b: 76, Table 2.8, column 6)
15. Summer dispersed kills (Binford 1978b: 76, Table 2.8, column 8)
16. Anavik spring kill site (Binford 1978b: 78, Table 2.9, column 2)
17. Anaktiqtauk spring kill site (Binford 1978b: 78, Table 2.8, column 4)
18. Ice cellar processing area (Binford 1978b: 124, Table 3.7, column 2)
19. Ice cellar spring clean-out (Binford 1978b: 124, Table 3.7, column 4)
20. Contents of ice cellars, 1971 (Binford 1978b: 126, Table 3.8, column 2)
21. Contents of ice cellars, 1972 (Binford 1978b: 126, Table 3.8, column 4)
22. Anavik spring hunting stand (Binford 1978b: 176, Table 5.1, column 2)
23. Anaktiqtauk hunting stand (Area A) (Binford 1978b: 176, Table 5.1, column 4)
24. Anaktiqtauk hunting stand (Area B) (Binford 1978b: 176, Table 5.1, column 6)
25. Mask site (Binford 1978b: 180, Table 5.3, column 4)
26. Rulland Tulugak residential site (Binford 1978b: 207, Table 5.9, column 4)
27. Big Surround kill site (Binford 1978b: 237, Table 5.16, column 2)
28. Long Rope cairn (Binford 1978b: 237, Table 5.16, column 4)
29. Site 41 cairn (Binford 1978b: 237, Table 5.16, column 10)
30. 1971 village inventory (Binford 1978b: 259, Table 6.1, column 6)
31. Site 17 hunting camp (Binford 1978b: 322, Table 6.18, column 2)
32. Schoolteacher residential site (Binford 1978b: 322, Table 6.21, column 2)
33. Amalgamation residential site (Binford 1978b: 322, Table 6.21, column 4)
34. Akmoglik residential site (Binford 1978b: 322, Table 6.21, column 12)
35. Big Happy hunting stand (Binford 1978b: 351, Table 7.1, column 4)
36. Little Happy hunting stand (Binford 1978b: 351, Table 7.1, column 4)
37. Rulland fall residential site (Binford 1978b: 376, Table 7.11, column 10)
38. Kakinya fall residential site (Binford 1978b: 381, Table 7.13, column 16)

tion of the kill by the wolves, there is an additional behavior that the Eskimo report, a behavior that would certainly contribute to the destruction of bones at the kill. This is the feeding of pups and young at the kill. It is reported that when a female wolf begins hunting after having whelped she will hunt in the immediate area of the den. If she is successful she will secure the kill from any competitors and then by howling alert the juveniles as to her location. The juveniles then either carry or guide the pups from the den to the kill location where they all feed under the watchful eye of the female. Later in

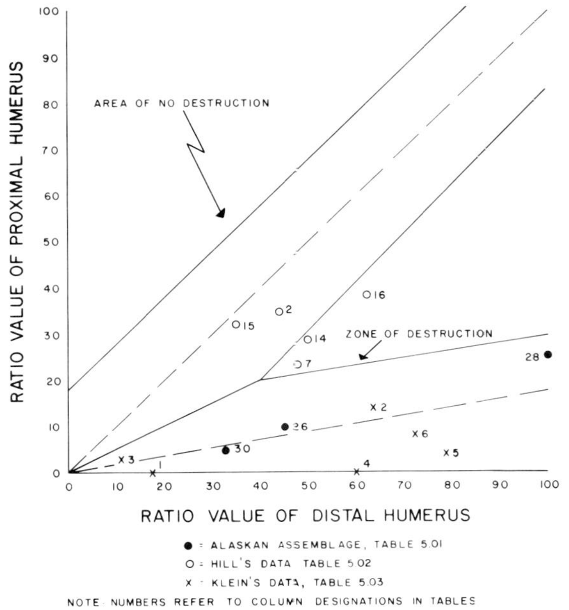

FIGURE 5.09. ***Relationship between frequencies of the proximal and distal humerus in the control assemblages from Africa and Alaska.***

the summer and early fall the mother may take the juveniles and pups away from the den to a rendezvous spot where the young remain "playing" while the female hunts. (See Figure 5.11.) Again she will howl, allowing the juveniles to find her easily and guide the pups to the kill. It is my understanding that in Africa competitors are commonplace, such that there is much greater transporting of parts away from kills so as to reduce face-to-face competition among animals fighting over a kill. Second, the field reports suggest that among hyena, and presumably Cape dogs, juveniles are not fed at the kills where they would be easy prey themselves for competing carnivores.

The result of these major contrasts in the "sociology" of predation between wolves in Alaska where competitors are few and various predators in Africa where competition is variable but generally high compared to Alaskan settings, is that meat is bolted at the kill, but later in Africa parts are systematically dragged off (transported) to "safe" spots away from the kill proper and gnawed and consumed. Dens or lairs are focal recipients of transported parts, and in Alaska there are open area rendezvous locations (as in Figure 5.11) where bone accumulations may occur. It is my guess that the degree of bone destruction at the kill will be an inverse function of the intensity of competition among predator–scavengers. In turn there may well be differences among species in the way young and juveniles are fed that are responsive to the competitive environment. It is my opinion that for African materials the "grab and run" strategy is and was most common, resulting in a pattern similar to that indicated here in the African material. Namely, there is evidence of transport of parts away from kills but little evidence of in situ destruction of bone parts remaining at kill sites per se. On the other hand, there may well be bone scatters concentrated in vegetation and other "protected" locations—deposits resulting from the accumulation of transported parts.—In additon there are lairs where ravaged assemblages are the rule as is the case with Klein's material.

It is proposed that one of the diagnostic differences between residual faunal remains of kill locations and transported remains, either at lairs or as

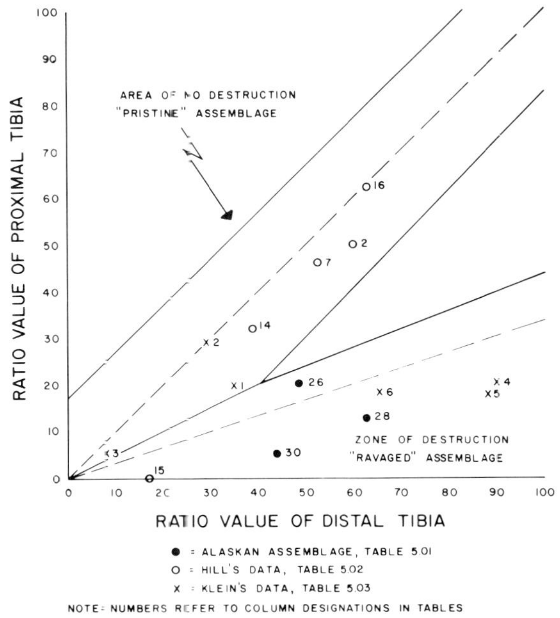

FIGURE 5.10. ***Relationship between frequencies of proximal tibia and distal tibia in the control assemblages from Africa and Alaska.***

FIGURE 5.11. ***Wolves on a rendezvous site.***

palimpsests of scattered bone, will be the relative evidence of bone destruction from gnawing. This will be manifest not only in survivorship, as demonstrated here, but also in the formal modifications made on the bone surviving. Furrowing, chipping back, channeled breakage, cylinders of long bones, tooth puncture marks, bone scoring, and the like will be greatest in den assemblages, moderate in palimpsests of transported bone, and least evident in residual remains at kill locations or scavenged death locations.

Transport of bone by carnivores as indicated here is related to competition at the source of food as well as to behavior related to the rearing of the young. In order to obtain a picture of what was transported by carnivores to the dens documented here we must reconstruct the original assemblage from the surviving bones in the ravaged assemblages. However, this requires a transformation of the data and the skeptic may object to such a radical treatment. Is there not some way we may obtain a picture of transport without resorting to a transformation of the observational data? The answer is "not really." However, because all the Alaskan assemblages are shown to be ravaged, we may hold destruction constant and compare these assemblages directly, as a way to gain a first appreciation for the character of the parts transported. This strategy may be linked with a summary of generalizations that others have offered regarding the character of parts remaining at kills. For instance, Shipman and Phillips (1976) state,

> Our data on the disarticulation of animals observed from the time of death suggest that this assumption is true of limbs, skulls, and mandibles, but not of ribs and vertebrae. With the exception of the first two cervical vertebrae, which may be transported with the skull, disarticulated vertebrae and ribs often remain at the kill site [1976:170].

This observation was made as part of a study of carcasses and bone remains occurring on the surface in Awash National Park in Ethiopia. They found that anatomical parts remaining at kills in the 80–90% level were skulls, mandibles, vertebrae, ribs, and pelves, whereas limb-bones were only represented by 69.9% and scapulae by 67.6% of the total potentially present. (See Shipman and Phillips 1976:170, Table 1.) Logically speaking, a transported assemblage should look like the "opposite" of the nontransported assemblage; that is, it should be the proportional inverse of the parts that were not transported. The Bent Creek den assemblage, when plotted against my wolf kill data, should therefore exhibit an inverse or indirect and unidimensional relationship.

Figure 5.12 illustrates the actual relationship between the kill and den assemblages, and it is clear that my "logical" expectations are not met in any particular. No clear unidimensional relationship is shown, no indirect relationship indicated. Points are scattered all over the graph, but when we begin grouping these by basic anatomical units believed of

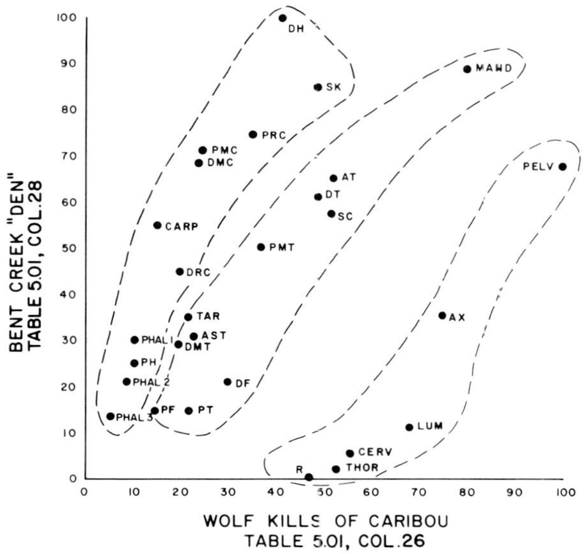

FIGURE 5.12. ***Relationship between Alaskan wolf den and kill assemblages.***

importance in animal kill transport situations, namely the axial skeleton minus the skull and upper neck and the appendicular skeleton, some regularity appears. The axis vertebra through the pelvis plus the ribs appears as a positively correlated array to the right of the graph. In the center of the distribution is the mandible, atlas vertebrae, scapula, and the complete rear leg; to the left of the scatter is the skull and the complete front leg plus phalanges. Each subset is positively and linearly distributed. What we see is that at least in this contrast certain anatomical units appear to be inversely correlated: skull, front leg, and phalanges more frequent in the den assemblage and less frequent in the kill. On the other hand, bones making up each subset are positively correlated. Similarly, the mandible, scapula, and rear leg are nearly equally represented in the two assemblages, and as in the previous situation the bones making up the subset in each population are positively correlated. Finally, the axial skeleton is very common in the kill and less well represented in the den. Once again the bones making up the subset are positively correlated. What this means is that units being acted upon are (*a*) the front leg below the scapula, (*b*) the scapula, (*c*) the complete rear leg, (*d*) the skull with occasionally attached atlas vertebra and mandible, and (*e*) the axial skeleton plus ribs.

This comparison suggests that units of transport are major anatomical segments. These do appear inversely related between kills and dens, but the individual bones making up any one anatomical segment are positively related within each set.

Further insight may be supplied by a comparison between the two kill assemblages, one a summary inventory of bones remaining at discrete kills, and the other a palimpsest population of bones distributed in a given area where bones were not assignable to particular dead animals. Figure 5.13 illustrates the relationship between these samples.

Structurally speaking, the relationship between the two assemblages appears similar to the comparison between dens and kills! However, examination of the details reveals some interesting differences. To the left is a distribution that represents parts relatively common in the Chandler sample compared with the monitored kills. These are the lower parts of both rear and front legs. In the center is an array of

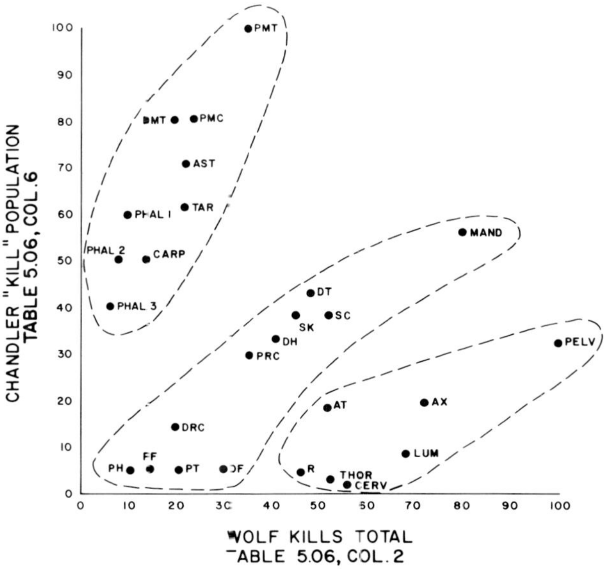

FIGURE 5.13. ***Relationship between discrete and palimpsest wolf kill assemblages.***

the mandible, the skull, and the upper parts of both front and rear legs. The axial skeleton including pelvis and ribs distributes as an array to the right as in the previous comparison, illustrating that as an anatomical unit the axial skeleton is more common in the population of discrete animal carcasses. Thus, unlike the segmentation illustrated in the kill–den comparison (Figure 5.12), here the segmentation is into three subsets: (*a*) lower legs, (*b*) upper legs including scapula plus skull or mandible, and (*c*) axial skeleton. This patterning suggests that there is a scattering effect such that bones of the extremities become more widely dispersed than other anatomical units. What we appear to be seeing here are the bones that remain in the gnawing areas around the typical wolf kill (see Figure 5.05); bones remaining at the kill proper are generally missing in the Chandler assemblage. The pattern of transport among the wolves is not different with regard to parts. It simply differs in scale or distances moved away from the kill location. In some kills the parts of the lower leg may be moved less than 15 m, and when a known situation is present such "short transport" may be seen as part of the residual bone population at a kill. On the other hand, under some conditions (e.g., high winds on the lake ice), bones may be transported into protected environmental settings (dissected terrain of low willow growth at the south end of the lake) and therefore separated from the kill location proper.

In African settings where competition is high it will be interesting to determine if a similar pattern of transport is recognizable, and to what degree parts introduced into dens are different from parts scattered through attempts of individual animals to feed unmolested. In order to accomplish this comparison we must reconstruct the African assemblages from the Swartklip den so direct comparisons may be made to the unravaged kill data. This was successfully accomplished previously (Binford 1978b) by dividing the percentage ratio of each bone in an assemblage array by its survival percentage, providing an estimate of the original frequency in the pristine assemblage. This sounds relatively straightforward; however, the real survival percentage for a given bone element in a given assemblage is a function of the species and the age structure of the original population. (See L. R. Binford and J. B. Bertram [1977] for a demonstration of this point.) In the case of the control assemblages, we lack analysis of the age structure and we do not have measured bone densities for any of the African species being compared. On the assumption that there is some relationship between body size and skeletal structure, I will use the survival percentages for caribou for all animals of moderate to large size. These include topi, waterbuck, zebra, wildebeest, kudu, and eland (for simplicity identified in tables and figures as "large prey"). For the small bovids—grysbok, springbok, reedbuck, gazelle, and oribi—I will use survival percentages derived from studies of sheep (Table 5.04). I do not have age-controlled data for caribou, so it will be impossible to adjust for age variability with the larger animals. However, for the small bovids some "fitting" to accommodate possible differences of age structure among the populations is possible.

Estimates of the composition of the original assemblage before the differential destruction of parts by animals are given in Tables 5.06 for the Alaskan data and Table 5.07 for the Swartklip African data. These estimates were obtained as previously indicated by dividing the observed frequencies (expressed in percentage ratios of the most common part) by the fractional survival percentage as summarized in Table 5.04.

Comparison between the original assemblage estimates for the Alaskan sample proves very instructive. Comparisons analogous to those made earlier (Figures 5.12 and 5.13) are given in Figures 5.14 and 5.15. Several points of contrast are obvious. First, what appeared as a series of linear positive relationships in the raw data now appear as extreme Poisson relationships. What is frequent in a den or transported assemblage is *infrequent* in the kill assemblage. Our earlier expectations for mirror-image relationships between kills and transported assemblages (as at lairs and in "scattered" or derivative palimpsest assemblages) are now realized. Why did the raw data comparisons fail to exhibit these properties? The answer is that, when ravaged assemblages are compared, patterning that derives from the differential destruction of parts dominates the frequencies. Regardless of the form of the original population, the surviving populations are similar in their pattern of differential survival. This condition accounts for the fact that positive linear correlations were observed among elements of anatomical segments that seem to have been treated as indepen-

TABLE 5.06
Reconstruction of Original Assemblage Composition for Alaska Samples

Anatomical part	Wolf kills: Table 5.01, col. 26 × Table 5.04, col. 7 (1)	Wolf kills: Col. 1 / 188.0 (2)	Bent Creek: Table 5.01, col. 28 × Table 5.04, col. 7 (3)	Bent Creek: Col. 3 / 197.22 (4)	Chandler: Table 5.01, col. 30 × Table 5.04, col. 7 (5)	Chandler: Col. 5 / 210.0 (6)
Skull	60.0	32.0	107.5	55.0	30.4	8.0
Mandible	80.0	42.0	89.0	45.0	57.0	14.0
Atlas	73.24	39.0	92.96	47.0	26.7	7.0
Axis	137.04	73.0	66.67	34.0	35.2	9.0
Cervical vertebrae	170.97	91.0	19.35	10.0	6.5	2.0
Thoracic vertebrae	188.89	100.0	5.83	3.0	8.3	2.0
Lumbar vertebrae	113.64	60.0	24.49	12.0	16.3	4.0
Pelvis	188.0	100.0	77.27	39.0	37.15	9.0
Ribs	96.3	51.0	?		20.0	5.0
Scapula	96.3	51.0	105.56	54.0	70.4	18.0
Proximal humerus	64.7	34.0	147.06	75.0	29.4	7.0
Distal humerus	70.69	37.0	172.41	87.0	56.9	14.0
Proximal radio-cubitus	85.37	45.0	182.93	93.0	70.73	18.0
Distal radio-cubitus	40.0	21.0	92.0	47.0	28.0	7.0
Carpals	60.0	32.0	128.0	65.0	208.0	52.0
Proximal metacarpal	66.67	35.0	197.22	100.0	225.0	56.0
Distal metacarpal	68.58	36.0	194.29	99.0	203.0	51.0
Proximal femur	37.5	20.0	35.0	18.0	12.5	3.0
Distal femur	85.71	45.0	60.0	30.0	14.3	4.0
Proximal tibia	88.0	47.0	56.0	28.0	20.0	5.0
Distal tibia	67.61	36.0	85.92	44.0	60.6	15.0
Tarsals	55.0	29.0	90.0	46.0	155.0	39.0
Astragalus	66.67	35.0	88.89	45.0	197.0	49.0
Calcaneus	66.67	35.0	100.0	51.0	172.0	43.0
Proximal metatarsal	90.24	48.0	121.95	62.0	243.9	61.0
Distal metatarsal	50.0	26.0	72.5	37.0	202.5	51.0
First phalange	64.71	34.0	135.29	69.0	352.0	88.0
Second phalange	61.54	33.0	123.08	62.0	392.0	98.0
Third phalange	70.0	37.0	70.0	35.0	400.0	100.0

dent units regardless of the relationships among such segments (see Figures 5.12 and 5.13). The effect of differential survival is removed in the comparisons shown in Figures 5.14 and 5.15 and the structural differences and similarities among the original assemblages are revealed.

This is a very important observation and one that essentially renders the interpretations Jack Bertram and I offered regarding the Makapansgat fauna open to being seriously wrong (L. R. Binford and J. B. Bertram 1977). This also "explains" why the comparisons between Hottentot goats, Makapansgat, and Olduvai Gorge (M. D. Leakey 1971: 277) appeared so similar. If two assemblages that are essentially opposites of one another can be rendered similar by attritional agents, raw data comparison among assemblages uncorrected for attrition, when it can be reasonably suspected, clearly becomes a meaningless exercise.

The nature of the relationships between a transported and a kill assemblage is clear. In kills the axial skeleton is most variable, with some elements

TABLE 5.07
Reconstructed Den Assemblages from Swartklip

	Grysbok		Springbok		Reedbuck	
Anatomical part	Table 5.03, col. 1 × Table 5.04, col. 4 (1)	Col. 1 / 147.48 (2)	Table 5.03, col. 2 × Table 5.04, col. 4 (3)	Col. 3 / 194.57 (4)	Table 5.03, col. 3 × Table 5.04, col. 4 (5)	Col. 5 / 135.5 (6)
Skull	79.02	54.0	113.51	58.0	64.69	48.0
Mandible	135.5	92.0	135.5	70.0	135.5	100.0
Atlas	23.53	16.0	101.1	52.0	18.82	14.0
Axis	52.63	36.0	154.08	79.0	65.05	48.0
Cervical vertebrae	26.82	18.0	80.46	41.0	49.81	37.0
Thoracic vertebrae	23.41	16.0	23.41	12.0	16.72	12.0
Lumbar vertebrae	26.0	18.0	37.0	19.0	47.0	35.0
Pelvis	81.63	55.0	85.03	44.0	44.22	33.0
Ribs	14.49	10.0	33.82	17.0	14.49	11.0
Scapula	16.45	11.0	82.24	42.0	34.54	25.0
Proximal humerus	?	–	140.0	72.0	30.0	22.0
Distal humerus	28.38	19.0	106.84	55.0	18.36	14.0
Proximal radio-cubitus	28.91	20.0	73.13	38.0	13.61	10.0
Distal rado-cubitus	15.77	11.0	81.08	42.0	24.77	18.0
Carpals	26.32	18.0	52.63	27.0	30.08	22.0
Proximal metacarpal	31.11	21.0	191.11	98.0	71.11	52.0
Distal metacarpal	45.6	31.0	162.87	84.0	52.12	38.0
Proximal femur	20.6	14.0	20.6	11.0	14.71	11.0
Distal femur	55.68	38.0	68.78	35.0	42.58	31.0
Proximal tibia	68.78	47.0	94.99	49.0	16.38	12.0
Distal tibia	62.27	42.0	53.11	27.0	14.65	11.0
Tarsals	21.69	15.0	15.18	8.0	17.35	13.0
Astragalus	57.28	39.0	33.41	17.0	19.09	14.0
Calcaneus	42.08	29.0	58.12	30.0	26.05	19.0
Proximal metatarsal	147.48	100.0	154.68	79.0	64.75	48.0
Distal metatarsal	140.27	95.0	194.57	100.0	58.82	43.0
First phalange	100.0	68.0	123.5	63.0	29.41	22.0
Second phalange	100.0	68.0	123.5	63.0	29.41	22.0
Third phalange	100.0	68.0	123.5	63.0	29.41	22.0

being common such as the thoracic and cervical vertebrae and the pelvis. In transported assemblages components of the appendicular skeleton are most variable with elements of the extremities being most common. We can think of a structural dichotomy, with residual kill populations dominated by axial skeletal parts and derivative (transported) assemblages dominated by appendicular parts. In the Alaskan data there are at least two types of derived assemblages: (*a*) the scatter, which is dominated by extremities such as lower limb bones and phalanges, and (*b*) dens, which are characterized by leg bones, including upper limb parts, and more skulls.

The degree to which observations represent general conditions is perhaps best indicated by a comparison of the African data on kills and dens. Table 5.07 summarizes the reconstructed assemblages, or our estimates as to the original composition of the faunal assemblages introduced to the animal den at

TABLE 5.07—*Continued*

Anatomical part	Topi + roan + sable: Table 5.03, col. 4 × Table 5.04, col. 7 (7)	Col. 7 / 166.67 (8)	Wildebeest: Table 5.03, col. 5 × Table 5.04, col. 7 (9)	Col. 9 / 285.71 (10)	Kudu + eland: Table 5.03, col. 6 × Table 5.04, col. 7 (11)	Col. 11 / 260.0 (12)
Skull	125.0	75.0	83.75	29.0	56.25	22.0
Mandible	100.0	60.0	44.0	15.0	45.0	17.0
Atlas	70.42	42.0	36.62	13.0	38.03	15.0
Axis	18.52	11.0	61.11	21.0	16.67	6.0
Cervical vertebrae	32.26	19.0	22.58	8.0	58.06	22.0
Thoracic vertebrae	27.78	17.0	19.44	7.0	25.0	10.0
Lumbar vertebrae	40.82	24.0	28.57	10.0	36.73	14.0
Pelvis	11.36	7.0	50.0	18.0	72.73	28.0
Ribs	20.0	12.0	16.0	6.0	20.00	8.0
Scapula	37.04	22.0	35.19	12.0	16.67	6.0
Proximal humerus	—	—	23.53	8.0	52.92	20.0
Distal humerus	103.45	62.0	134.48	47.0	125.86	48.0
Proximal radio-cubitus	121.95	73.0	90.24	32.0	178.05	68.0
Distal radio-cubitus	60.0	36.0	104.0	36.0	200.00	77.0
Carpals	80.0	48.0	56.0	20.0	108.0	42.0
Proximal metacarpal	138.89	83.0	194.44	68.0	227.78	88.0
Distal metacarpal	142.86	86.0	285.71	100.0	260.00	100.0
Proximal femur	25.0	15.0	17.5	6.0	45.0	17.0
Distal femur	114.29	69.0	74.29	26.0	77.14	30.0
Proximal tibia	80.0	48.0	76.0	27.0	72.0	28.0
Distal tibia	126.76	76.0	125.35	44.0	90.14	35.0
Tarsals	25.0	15.0	17.5	6.0	67.5	26.0
Astragalus	111.11	67.0	102.78	36.0	152.78	59.0
Calcaneus	166.67	100.0	91.67	32.0	177.78	68.0
Proximal metatarsal	97.56	59.0	100.0	35.0	109.76	42.0
Distal metatarsal	150.0	90.0	102.5	36.0	160.0	62.0
First phalange	117.65	71.0	64.7	23.0	158.82	61.0
Second phalange	117.65	71.0	64.7	23.0	158.82	61.0
Third phalange	117.65	71.0	64.7	23.0	158.82	61.0

Swartklip. Figure 5.16 compares the profiles for the reconstructed assemblages of small versus medium and large ungulates. Important differences in the way animals transport parts of large versus small animals are illustrated. The large-animal graphs support the generalization made from the Alaskan data with the exception that the head was less commonly transported. That is, parts of the axial skeleton are nearly nonexistent in the den assemblage. As is the case with the Alaskan data, lower extremities of the front leg as well as substantial numbers of the upper front leg parts minus the scapula are consistently represented and are absolutely dominant. The extremities as well as the upper limb elements of the rear leg are highly variable, occasionally approaching the levels achieved by the front leg. This is the same condition observed in the wolf den assemblage at Bent Creek (Figures 5.12 and 5.14). The front leg (which, it will be recalled, is the first to be disarticulated from the skeleton) is the most consistently rep-

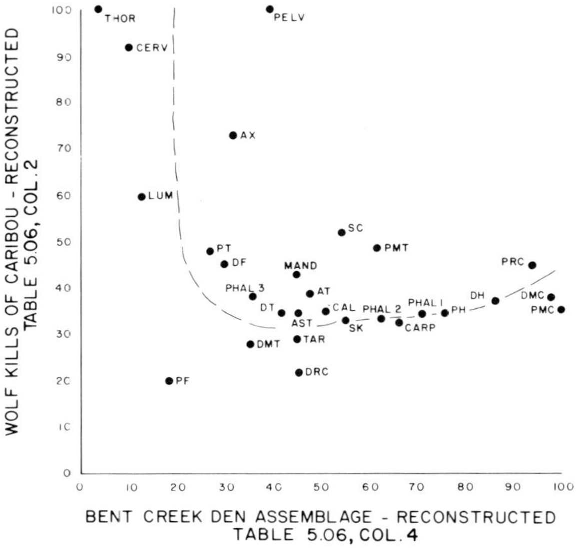

FIGURE 5.14. ***Relationship between reconstructed Alaskan wolf den and kill assemblages.***

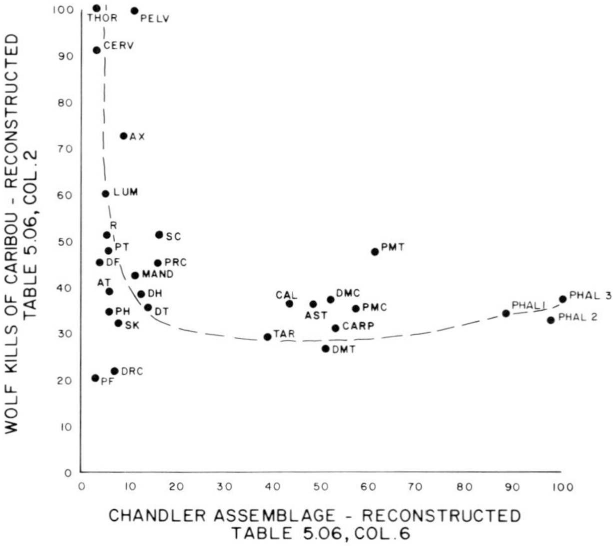

FIGURE 5.15. ***Relationship between reconstructed discrete and palimpsest wolf kill assemblages.***

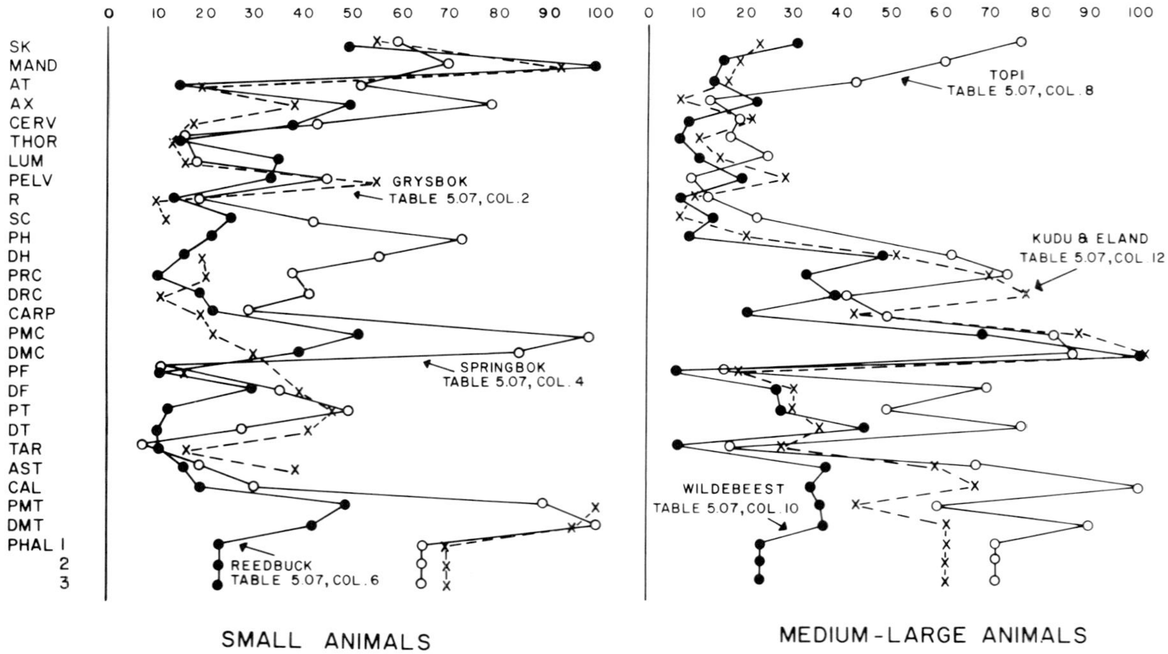

FIGURE 5.16. ***Comparison between medium-to-large and small ungulates introduced to animal dens—Swartklip reconstructed fauna.***

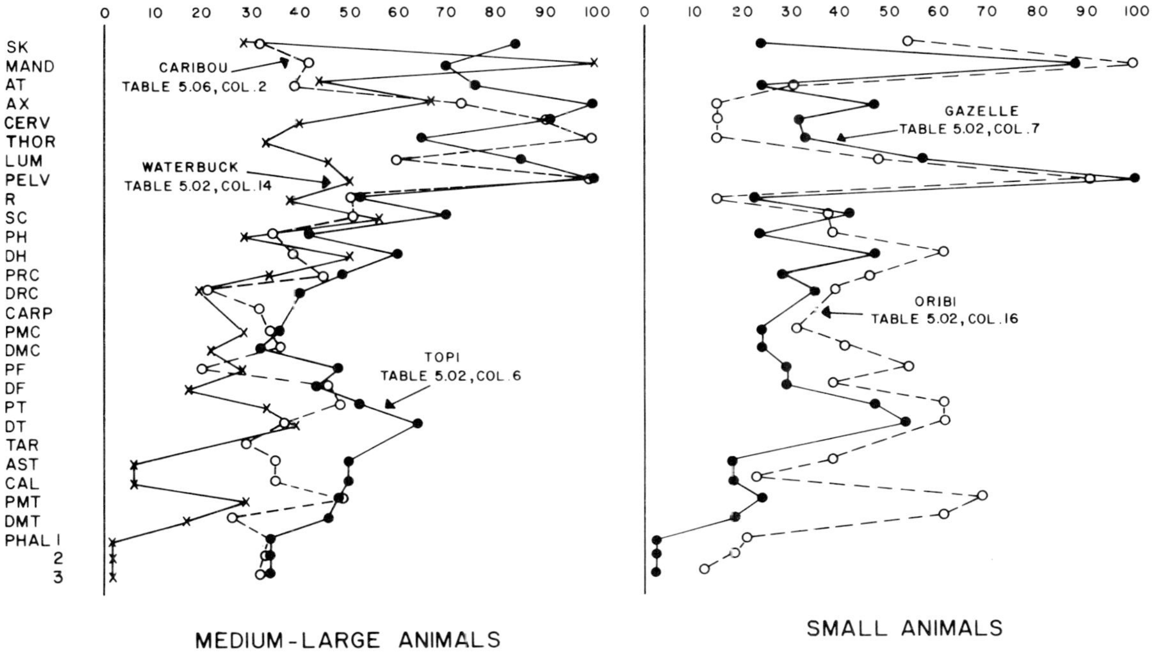

FIGURE 5.17. ***Comparison between medium-to-large and small ungulates, parts remaining at kill locations.***

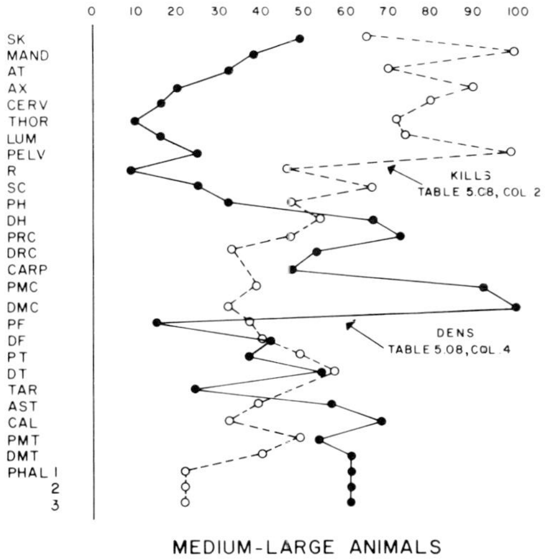

FIGURE 5.18. ***Comparison between kills and dens for medium-to-large animals.***

resented in the den assemblage. A very different pattern is evident for the small animals. There are greater numbers of mandibular and neck parts. As for large animals, thoracic and lumbar vertebrae as well as ribs are rare. Pelves and scapulae are moderately represented, unlike the situation for large animals. The front and rear legs are about equally variable, with the lower limbs dominating the parts introduced and the upper limb bones moderately represented. Ignoring the leg parts, which are structurally similar, the difference appears to be in the transportability of dismembered elements of the axial skeleton, mandible, neck, pelvis, and scapula, all parts that regularly remain at the kill sites of large animals.

The best way to appreciate the mirror-image patterning between transported assemblages and residual kill site assemblages is to compare the Swartklip patterning with that from documented kill site assemblages. Figure 5.17 summarizes the *observational* data of Andrew Hill for small and medium/large animals and the inverse of the patterning demonstrated by the *reconstructed* Swartklip as-

TABLE 5.08
Summary Data for Human and Nonhuman Kills and Transported Bone Assemblages

	Animal-produced assemblages							
	Large prey[a]				Small prey[b]			
	Kills		Dens		Kills		Dens	
Anatomical part	$\bar{X}$[c] (1)	Col. 1 / 78.0 (2)	$\bar{X}$ (3)	Col. 3 / 92.25 (4)	$\bar{X}$ (5)	Col. 5 / 97.33 (6)	$\bar{X}$ (7)	Col. 7 / 87.33 (8)
Skull	51.0	65.0	45.25	49.0	37.0	38.0	53.33	61
Mandible	78.0	100.0	34.25	37.0	85.01	87.0	87.33	100.0
Atlas	54.75	70.0	29.25	32.0	40.67	42.0	27.33	31.0
Axis	70.0	90.0	18.0	20.0	31.67	33.0	54.33	62.0
Cervical vertebrae	62.75	80.0	14.75	16.0	26.33	27.0	32.00	37.0
Thoracic vertebrae	56.75	73.0	9.25	10.0	27.0	28.0	13.33	15.0
Lumbar vertebrae	57.75	74.0	15.0	16.0	63.0	65.0	24.0	27.0
Pelvis	77.5	99.0	23.0	25.0	97.33	100.0	44.0	50.0
Ribs	36.25	46.0	8.0	9.0	20.67	21.0	27.33	31.0
Scapula	51.75	66.0	23.5	25.0	54.0	55.0	26.00	30.0
Proximal humerus	36.5	47.0	29.25	32.0	31.67	33.0	38.68	44.0
Distal humerus	41.75	54.0	61.0	66.0	47.33	49.0	29.33	34.0
Proximal radio-cubitus	36.5	47.0	66.5	72.0	36.0	37.0	22.67	26.0
Distal radio-cubitus	25.75	33.0	49.0	53.0	24.33	25.0	23.67	27.0
Carpals	—	—	43.75	47.0	—	—	22.33	26.0
Proximal metacarpal	29.75	38.0	84.75	92.0	18.33	19.0	57.0	65.0
Distal metacarpal	25.0	32.0	92.25	100.0	23.33	24.0	51.0	58.0
Proximal femur	29.0	37.0	14.0	15.0	33.3	34.0	12.0	14.0
Distal femur	31.5	40.0	38.75	42.0	22.33	23.0	34.67	40.0
Proximal tibia	38.0	49.0	32.75	36.0	36.33	37.0	36.0	41.0
Distal tibia	44.75	57.0	49.75	54.0	44.0	45.0	26.67	31.0
Tarsals	—	—	23.25	25.0	—	—	12.0	14.0
Astragalus	30.75	39.0	51.75	56.0	24.33	25.0	47.2	54.0
Calcaneus	25.25	32.0	62.75	68.0	13.67	14.0	26.0	30.0
Proximal metatarsal	38.0	49.0	49.5	54.0	42.0	43.0	75.67	87.0
Distal metatarsal	31.25	40.0	56.25	61.0	37.67	39.0	79.33	91.0
First phalange	17.25	22.0	56.0	61.0	8.0	8.0	51.0	58.0
Second phalange	17.25	22.0	54.25	59.0	6.5	7.0	51.0	58.0
Third phalange	17.25	22.0	47.5	51.0	5.0	5.0	51.0	58.0

[a]Large prey = caribou, zebra, topi, sable, roan, wildebeest, kudu, eland, and waterbuck.
[b]Small prey = gazelle, reedbuck, oribi, grysbok, and springbok.

TABLE 5.08—*Continued*

	Human-produced assemblages (caribou)					Anatomically based models			
	Kills		Residential	Transported		GUI[c]		IGUI[d]	
Anatomical part	[f] (9)	Col. 9 / 85.0 (10)	[g] (11)	[h] (12)	Col. 12 / 15 (13)	Sheep (14)	Caribou (15)	Sheep (16)	Caribou (17)
Skull	85.0	100.0	3.5	3.0	20.0	25.74	17.49	90.98	93.12
Mandible	68.7	81.0	3.5	4.0	27.0	11.65	13.89	91.94	87.88
Atlas	40.0	47.0	24.2	3.0	20.0	35.6	23.56	66.95	77.77
Axis	38.6	45.0	24.2	3.0	20.0	45.5	29.64	57.68	71.71
Cervical vertebrae	36.4	43.0	26.0	3.0	20.0	55.33	35.71	46.47	65.61
Thoracic vertebrae	19.4	23.0	87.9	5.0	33.0	46.49	45.53	55.69	55.59
Lumbar vertebrae	23.4	28.0	82.8	4.0	27.0	38.9	32.05	63.59	69.35
Pelvis	22.2	26.0	86.9	4.0	27.0	81.51	47.89	19.25	53.18
Ribs	16.9	20.0	100.0	6.0	40.0	100.0	49.77	0	51.26
Scapula	16.4	19.0	81.0	7.0	47.0	45.06	43.47	57.18	57.69
Proximal humerus	14.9	18.0	71.5	7.0	47.0	29.50	30.23	73.37	71.21
Distal humerus	16.7	20.0	71.5	7.0	47.0	29.31	29.58	73.56	71.86
Proximal radio-cubitus	19.4	23.0	69.7	7.0	47.0	15.18	16.77	88.27	84.94
Distal radio-cubitus	26.5	31.0	69.7	7.0	47.0	15.18	17.82	87.68	83.87
Carpals	35.8	47.0	60.9	7.0	47.0	5.00	5.51	98.86	96.42
Proximal metacarpal	40.2	47.0	51.9	7.0	47.0	6.37	8.24	97.44	93.64
Distal metacarpal	42.9	50.0	51.9	7.0	47.0	6.79	8.83	97.01	93.04
Proximal femur	9.7	11.0	72.8	15.0	100.0	79.38	98.32	21.56	1.72
Distal femur	10.8	13.0	72.8	14.5	97.0	80.58	100.0	20.21	0
Proximal tibia	15.7	18.0	72.2	11.0	73.0	22.71	27.57	80.43	73.91
Distal tibia	20.3	24.0	71.0	9.0	60.0	23.41	29.46	79.71	71.98
Tarsals	25.0	29.0	67.4	8.5	57.0	6.64	11.20	97.16	90.62
Astragalus	25.6	30.0	67.4	8.5	57.0	6.64	11.23	97.16	90.59
Calcaneus	25.6	30.0	67.4	8.5	57.0	7.27	12.40	96.69	89.4
Proximal metatarsal	28.1	33.0	54.4	11.0	73.0	8.02	15.03	95.73	86.71
Distal metatarsal	28.8	34.0	54.4	11.0	73.0	8.46	16.24	95.73	85.48
First phalange	41.6	49.0	38.7	9.0	60.0	4.33	3.52	99.57	98.45
Second phalange	41.6	49.0	38.7	9.0	60.0	4.10	3.03	99.81	98.96
Third phalange	41.6	49.0	38.7	9.0	60.0	3.49	1.85	100.00	100.00

[c]L. R. Binford 1978b: 73.
[d]Obtained by subtracting columns 14 and 15 from 100 and standardizing.
[e]Value obtained from class sums of Tables 5.01, 5.02, and 5.03.
[f]Data from L. R. Binford 1978b: 75, column 10.
[g]Data from L. R. Binford 1978b: 259, column 6.
[h]Data from L. R. Binford 1978b: 301, column 9.

semblages. (For a clearer demonstration see Figure 5.18.) For the small animals we note assemblages dominated by mandibles and pelves, with neck parts poorly represented (these were well represented in the den assemblages). It is also interesting that the front leg, the part most consistently represented in transported assemblages, is only moderately represented, and then parts of the upper leg rather than the extremities commonly appear. Extremities are most variably represented by the lower rear leg, which was observed to be most variable in the den assemblage. For medium or large animals, the mirror-image effect is most striking in that when axial skeleton parts are common, appendicular parts are rare. When leg parts are found, they are most commonly front leg parts. The reader should keep in mind the remarkable circumstances of these comparisons. The Swartklip assemblage is an ancient excavated assemblage from South Africa, which has been "corrected" using anatomical information obtained from domestic sheep and Arctic caribou. This is ancient corrected data compared to modern observational data independently collected in East Africa. The structural fit between the two is remarkable. I take this fit as strong support for the position that inductive generalizations may be confidently made from these data.

Since generalization appears justified, I have calculated the mean values (Table 5.08) for large and small animal assemblages as a statement of central tendency for kills and transported assemblages as classes. These summary data provide a very easy way of comparing the residual and transported components of skeletons.

For instance, the mirror-image character of kill and den assemblages is clearly indicated in the summary data for medium to large animals illustrated in Figure 5.18. The overall similarity of den versus kill assemblages is shown when such assemblages are compared by body size class, as in Figure 5.19. The summarized contrasts can be taken as first approximations for the diagnostic recognition of residual versus transported assemblages produced by animals.

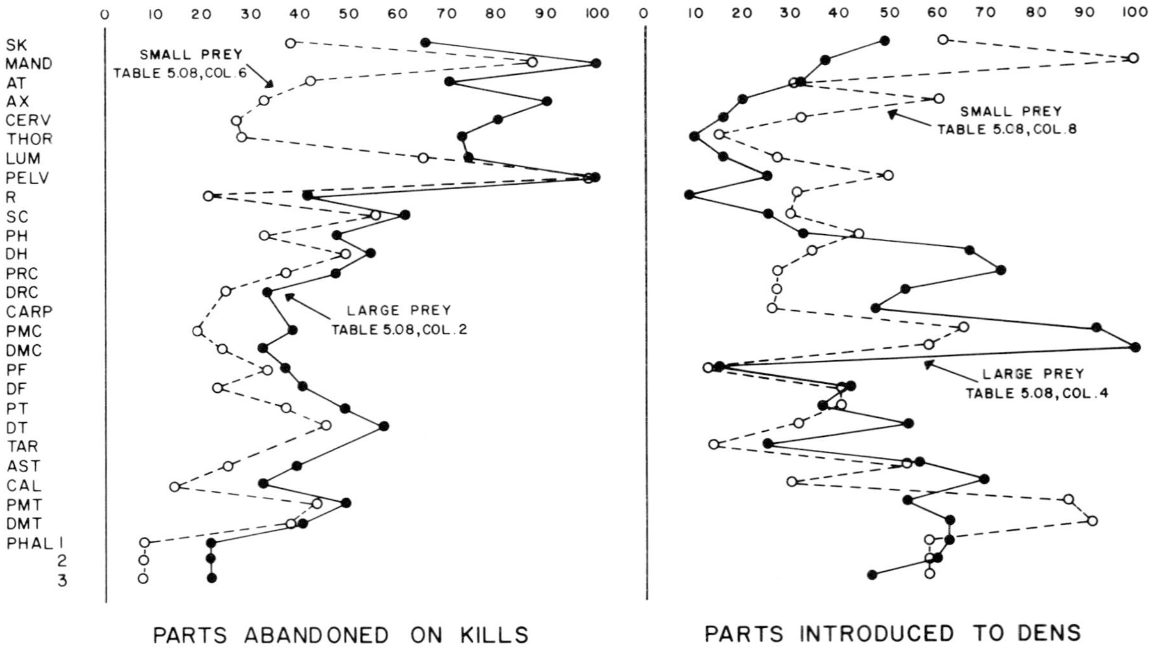

FIGURE 5.19. ***Comparison of mean values for remains of large and small animals abandoned by predators at kills and transported by them to dens.***

The important question must now be considered. Are such assemblage types distinctive to animals, and can we distinguish assemblages generated by man from those generated by nonhuman predators? Answering this question must rest initially with an *opinion*, namely the degree to which there are in fact assemblage types distinctive of animals and then in turn the degree to which there are assemblage types distinctive of human behavior. Using such an opinion, we may then address the question as to the diagnostic properties of each that would permit unambiguous and accurate recognition. It should be clear by this time that, based on the data available from widely different areas and dealing with different predators (wolf and hyena) as well as prey (a large list), it appears that there is in fact a regular and redundant pattern to the consumption and transport of anatomical parts by nonhuman predator–scavengers, resulting in regular and relatively unambiguous assemblage types.

My earlier studies of faunal assemblages (L. R. Binford 1978b) from a single hunting society yielded very different types of conclusions. It was observed that there were many assemblages forms produced by a single society. Certainly there was nothing that could be recognized as a distinctively human assemblage. The factors that conditioned this variability were found to be numerous; men responded to many situational variables that conditioned the character of assemblages produced in differing logistical, social, and systems state conditions. Upon reflection, this very responsiveness may well be characteristic of humans, since in the animal assemblages such responsiveness appears largely restricted to (*a*) the level of competition among predator–scavengers and (*b*) the bulk or size of the parts available for transport. For instance, front leg parts are more commonly present in dens and scatters, and Hill's observations (1976:143–145) show that the front leg becomes dismembered earliest in the dismemberment sequence. In addition, the differential frequencies of parts from large versus small animals illustrate nicely that the selection of parts transported is a simple function of what is accessible, and what is feasible given the size of the predator–scavenger versus the prey. It would also appear that transported parts are the consequences of either marginal competition by subdominant predator–scavengers or the sequential scavenging of abandoned kills. This suggestion derives from the fact that (*a*) parts represented in transported assemblages have been considerably dismembered prior to transport, and (*b*) they are consistently parts of relatively low food value. In fact, this point is so strikingly a characteristic of the animal assemblages that their transported or den assemblages are most like human kill assemblages where rational choices dictated the differential abandonment of parts of marginal utility. This is well illustrated in Figure 5.20, where the accumulated total of parts abandoned on approximately 277 individual caribou kills by Nunamiut Eskimo is displayed against the mean values for medium to large prey found in predator dens, as given in Table 5.08, column 4. The overall structural similarity is clear. The major difference between the two is that the den assemblage is more radically variable than is the Nunamiut kills graph, where relatively smooth trend lines are characteristic. More particularly, the den has higher frequencies of more marginal elements than do the human kill sites, with the exception of the head and neck; as we have seen this is variable with predator–prey body size relationships such that the head and neck of smaller prey are much more common than those of larger prey.

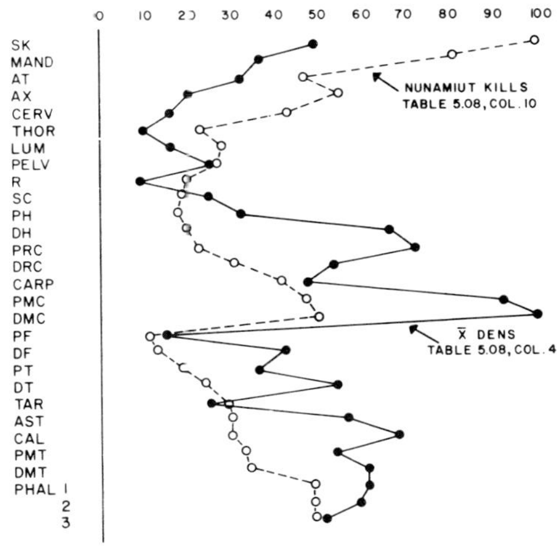

FIGURE 5.20. ***Comparison of mean values for large animals found in predator dens and found abandoned on Nunamiut Eskimo kill sites.***

In my earlier studies of the economic anatomy of sheep and caribou and the way Nunamiut Eskimo employ knowledge of this anatomy, one interesting characteristic was noted and models based on the anatomy had to be modified to account for it. This was the phenomenon of riding, mentioned in Chapter 3: if an anatomical element of high utility was selected for transport, attached or articulated elements of lesser utility would be transported with the selected element as long as the package size remained within a given transport tolerance (see discussion in L. R. Binford 1978b:271–273). This condition implies a number of facts. The degree of dismemberment at the kill is a direct function of the anticipated transport problems and schedule. Therefore, transport choices were made among butchering segments of the anatomy rather than among basic anatomical elements *as if* the animal were completely dismembered prior to making transport decisions. Given these "facts" the strict anatomical scales of utility (general utility index, L. R. Binford 1978b:73) for different elements of the anatomy segmented primarily in terms of individual bones or groups thereof (the lumbar, thoracic, and lower cervical vertebrae were treated as sets rather than as individual bones), had to be *modified* to accommodate the reality that decisions were made in terms of sets of bones and that elements of low to minor utility frequently rode along with elements of high utility. (See L. R. Binford 1978b:72–75.) It is just this characteristic that is absent from the animal den assemblages. This fundamental difference is perhaps best illustrated by comparing the distributions of the mean den assemblages and the Eskimo kill population with the general utility index (GUI) developed earlier from the analytical study of the anatomy of sheep and caribou. Figure 5.21 illustrates the interesting phenomenon of riding characteristic of the bones of the leg remaining on Eskimo kills. It is clear that the femur and scapula distribute as components of an inverse curvilinear distribution—namely, the higher the food utility of the part, the lower its frequency in the kill site population. On the other hand, the remaining bones of both legs form an underrepresented subdistribution relative to the utility index such that fewer elements of low utility are present as a curvilinear function of the *anatomical distance* away from the high utility part (the femur or the

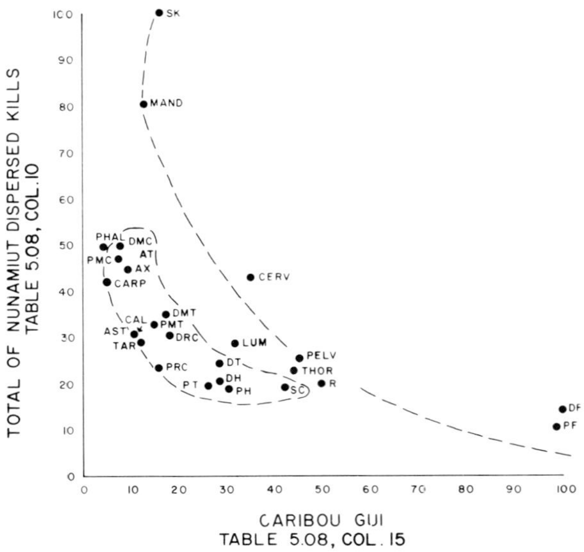

FIGURE 5.21. ***Relationship between total Nunamiut dispersed kills and the GUI for caribou.***

scapula–humerus). These were simply parts that "rode" with parts of high utility selected for transport to another location for consumption. Another way of viewing this situation is in terms of dismemberment. It was found among the Eskimo that the degree of anatomical disorder as measured by dismemberment ratios (see L. R. Binford 1978b:64–69, 251, 420) was a direct function of the degree to which an animal had been used or consumed. Typically hunters kill an animal and field butcher it into several parts. These are then carried ultimately to a consumer location, where the parts are further dismembered into units appropriate to differing food-preparation techniques. The act of eating further disarticulates and dismembers the anatomical parts. Finally, processing of parts for bone grease or marrow may further dissipate the anatomical organization present in the living animal. Thus, the greatest disorganization occurs at the sites of consumption, not at kill locations or locations of procurement. This of course assumes the very human characteristic of food sharing, sleeping, and eating in base camps. It appears that the general principles involved in understanding the degrees of anatomical organization relative to consumption also apply to predator–

scavengers: Major anatomical disorganization can be expected at the primary place of consumption. For nonhuman predators, the primary place of consumption is the kill site. Any parts removed from the kill after the dominant predator completes his feeding will be (*a*) anatomical parts of minor food value, and (*b*) parts dismembered or disarticulated as a consequence of prior consumption that was centered on the kill after a successful hunt by the nonhuman predator. We therefore do not expect much evidence of riding, since low-utility parts have been separated from high-utility parts during consumption. An assemblage transported by scavengers away from a predator kill site should therefore be anticipated by the inverse general utility index (L. R. Binford 1978b:73), whereas parts abandoned by human hunters at a kill site should be anticipated by the *inverse modified general utility index* (L. R. Binford 1978b:72). The difference is referable to the fact that nonhuman predators primarily consume prey at the kill site, whereas human predators primarily consume prey at a base camp where "dependents" tend to be localized. Also, the selection of parts for transport to human consumers from a butchered but unconsumed carcass is made from sets of bones remaining anatomically organized as a simple function of decisions as to the size of the units that can be transported. This is in contrast to the scavenger, who picks up parts remaining from acts of consumption, residual elements that are generally present in inverse anatomical frequencies to high-food-value parts. In short, they are removed as discrete elements rather than as components of a set in which a high-food-value part is the focus of transport.

Figure 5.22 illustrates the relationship between the mean medium–large prey den assemblage and the general utility index. Clearly there is an inverse curvilinear distribution, with some "noise" contributed by the phalanges, carpals, and tarsals, as well as the atlas and axis vertebrae. The phalanges, carpals, and tarsals are all little bones and in my experience they are commonly swallowed whole. They may be passed complete in the feces of predator–scavengers, but they may also be dissolved beyond recognition in the digestive tract (particularly the small tarsal and carpal bones, which are consistently underrepresented in Klein's Swartklip assemblages but relatively well represented in the Bent Creek wolf den

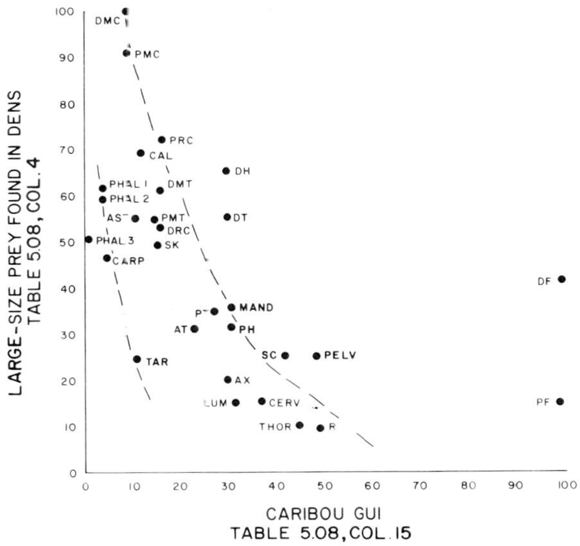

FIGURE 5.22. ***Relationship between mean frequencies for bones of large prey found in dens and the GUI for caribou.***

assemblages). It is my guess that Klein's sample does not include bone from the defecation area around the original Swartklip den, which is where most of these small bones might be expected to occur. The "noisy" behavior of the atlas and axis stems from the fact that I treated these vertebrae individually in my anatomical studies and in my experience with animal dismemberment they actually ride with the skull and cervical vertebrae, respectively. That is, the atlas more commonly rides with the skull and the axis more commonly rides with the cervical vertebrae. Given this knowledge, I have altered the general utility index for the atlas by multiplying the value of the skull by 2 and the value of the cervical vertebrae by 1 and dividing by 3. The same procedure was used for the axis, with the exception that the double weighting was in favor of the cervical vertebrae. Table 5.08 gives the general utility index as altered for the atlas and axis in both direct and inverse forms.

I turn now to the interesting question of the relationship between what men carry away from a kill for consumption elsewhere and what animals leave at a kill after consumption. Does the interesting opposition continue to hold between animal and human assemblages? In short, is what men carry away similar to what animals abandon at kills? A partial answer is supplied by Figure 5.23, where the mean for the documented animal kill site assemblages (Table 5.08, column 2) is plotted against the frequencies of parts actually transported by the Eskimo to Anaktuvuk village and placed in storage during the spring of 1972 (Table 5.08, column 13). It is clear that the overall structure of the two populations is similar. Nevertheless, there are differences that may well be important for distinguishing between the two contexts:

1. An underrepresentation of the femur relative to the tibia at animal kills, presumably representing destruction during dismemberment or biased removal during competitive feeding (see also Figure 5.21).
2. An underrepresentation of the proximal humerus and the distal radio-cubitus at animal kills, presumably deriving from their destruction during dismemberment by biting and gnawing
3. More pelves and fewer ribs at animal kills, with the reverse pattern for human-transported materials.
4. Relatively high frequencies of head and neck parts at animal kills, with greatly reduced numbers of these parts in human-transported assemblages.

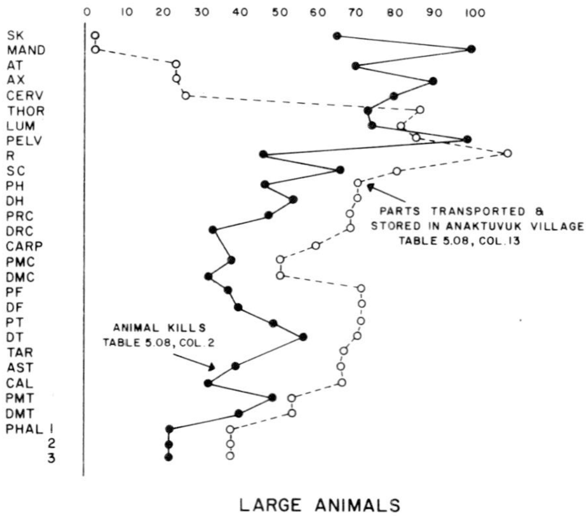

FIGURE 5.23. ***Comparison between parts transported for storage at Anaktuvuk village and mean values for parts abandoned on predator kills.***

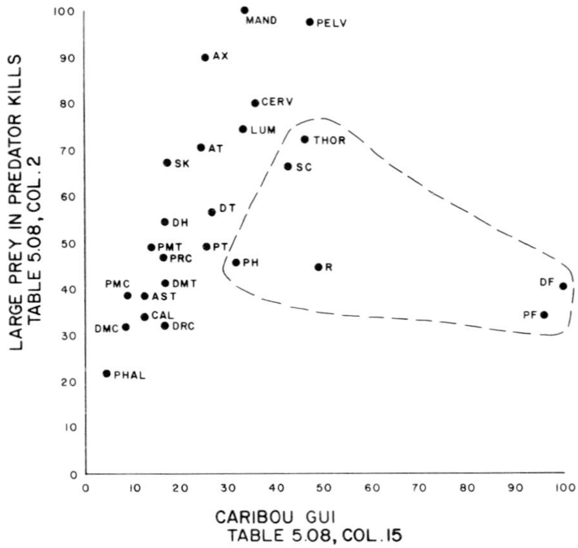

FIGURE 5.24. ***Relationship between parts remaining at the "average" kill site for large-sized prey and the GUI for caribou.***

These differences are believed to be the consequence of two conditions:

1. Animal kills are largely dispersed as a function of the relative strength of articulations. Necks, for example, remain attached to thoracic vertebrae on animal kills but may be removed with cutting tools in human contexts.
2. There is a size-related probability to parts being dispersed or transported from an animal kill, whereas man may generate appropriate-size units through controlled dismemberment using tools.

These contrasting properties are perhaps seen more clearly when the predator kill assemblage is arrayed against the general utility index (Figure 5.24). In this display a generalized positive and linear relationship is indicated, with some interesting "noisy" parts: the femur, ribs, scapula, thoracic vertebrae, and proximal humerus, which are underrepresented relative to other parts as arranged relative to the general utility index. These are high-food-value parts and we might view their moderate frequencies at kills in several ways. First, the femur is demonstrably overrepresented in the den assemblages, suggesting that if a predator is harrassed he may well secure the choice part and withdraw for consumption under calm, protected, noncompetitive conditions. This postulated behavior might well account for the minor underrepresentation of the thoracic vertebrae and ribs, parts apt to be articulated early in a sequence of consumption as would be the scapula and proximal humerus, also marginally underrepresented.

Whatever the behavioral basis for the moderate occurrence of the highest utility parts in the animal kill assemblages, it is clear that, as in the earlier comparison, *what man transports animals generally leave behind and what animals leave at a kill man generally transports*. This behavior is understandable in terms of the locus of consumption. Animals do the major part of their consumption at the kill, whereas man transports the consumable parts away from the kill to a camp or residence. Since dismemberment and differential destruction of parts associated with consumption occur prior to the transport of parts by animals, and man makes transport decisions *before* consumption and related dismemberment and destruction, it has been argued that diagnostic differences between the faunal assemblages generated by animals versus men can be expected. Transported assemblages such as those found in animal dens are anticipated quite well by the inverse general utility index, whereas the untransported human analogy—the kill site—is anticipated by the inverse *modified* general utility index. The latter was developed to accommodate the particularly human characteristic of riding, that is parts of lower utility moving with parts of higher utility, since consumption-related or complete dismemberment does not take place at a human kill site.

There is a similar opposition between analogous assemblages of parts remaining at predator kills versus the parts introduced by men into base camps.

Summary

Diagnosing formation processes operative in the past as well as the agents responsible for the bones included in deposits appears quite feasible. On the other hand, our current knowledge of variability as generated by man appears almost overwhelming in its diversity (see Binford 1978b). More confusing is the fact that man is capable of generating faunal assemblages very similar to those generated by other animals. As more studies are conducted, the diversity in assemblage composition as documented for man will almost certainly grow. Despite this increase in knowledge of diversity there is no reason to believe that any of it *necessarily* represents the form of patterning our earliest hominid ancestors generated when they took their first exploratory steps into the world of meat eating, scavenging, and predation with the attendant generation of faunal assemblages. What has been forcefully demonstrated, once again, is that man's behavior is extremely variable and responsively flexible to external conditions. With regard to behaviorally generated phenomena such as faunal assemblages, one cannot expect to understand the variability in terms of Aristotelian ideas of "formal causation." These assemblages are not the way they are because of the humanness inherent in them. You might say the diversity itself may be referable to the humanness of their production, but this is of little aid in diagnosing a given case. It should be clear by now that there appears little chance of recognizing generalized formal causes for diagnosing man as the exclusive agent of faunal remains perseved to us from all times and places.

In my earlier study of Nunamiut behavior I demonstrated over and over again that an assemblage was the vectoral sum of a number of strategy decisions and their linked behaviors. Redundancy in patterning was a function of redundancy in the coterminus of a number of similar conditioning and or causal determinants. Assemblages were not similar because each had a common cause, but instead because of having been conditioned by a similar configuration of numerous causal factors. Many of the same determinants frequently conditioned very different looking assemblages because of playing differing "roles" in the overall causal configuration. Our task in facing such a situation is the recognition of the dimensions of determinacy and ways of measuring their various "strengths" in the overall configurations of determinant conditions. We should never invent a particular scenario to accommodate each unique combination. Our job is the *analysis* of patterning, not the accommodation of each unique form to a unique situational picture of the past. A very important research step must be added to the old procedure of pattern recognition followed by post hoc accommodation or interpretation. We must seek ways of analyzing the patterning for purposes of isolating the number and identity of determinants that acted to generate the observed pattern. Only after we can identify the determinant can we even hope to begin the difficult task of "reconstructing" the character of the past from the remnant effects of past dynamics. In the case of faunal assemblages, part of the analytical strategy must be the ability to recognize unambiguously the effects of action by nonhominid agents.

As we saw in the last chapter, there does appear to be an impressive degree of redundancy among faunal assemblages produced by various nonhominid predator–scavengers operating on a variety of prey forms. It also appears quite possible to specify diagnostic formal causes or essential characteristics referable to nonhuman predator–scavengers as the agents responsible for certain faunal assemblages. This possibility is strongly suggested by the comparative studies presented in this chapter. It would appear that we can make strong uniformitarian assumptions about nonhominid predator–scavenger behavior as a cause of redundant patterning in faunal assemblages. The necessary relationship between the animal—the bear—and its footprint, particular formal characteristics of faunal assemblages, seems to be referable to the anatomical limitations of these animals for performing certain mechanical tasks. In short, the necessary connection arises as a result of biological conditioners or limitations on the animals' behavior. No such limitation seems operative with respect to tool-using man viewed in a generic sense.

Our knowledge of human behavior indicates much greater variability, and even some behavior that might appear very similar to nonhuman predators. I think we would be foolish to attempt to make uniformitarian assumptions to all times and places about the documented patterns of human behavior as manifest in animal bone assemblages. It is very difficult indeed to imagine what the behavior of the early hominids might have been, or how it might compare with that of modern humans, or ancient nonhominids for that matter. One thing that we might well anticipate, particularly since there could be some ambiguity between animal assemblages and modern human assemblages, is that isolating the characteristics of faunal assemblages, which are referable to very early hominids, represents a methodological challenge not apt to be as simple as developing contrasts between animal behavior and human behavior.

The histories of research as summarized here provide an interesting set of contrasts relative to my earlier arguments that actualistic studies are central to furthering our abilities as archaeologists to know with confidence certain things about the past. American research clearly exhibited great skill in pattern-recognition studies. White's (1953a, 1953b) study of anatomical parts clearly was a major advance in recognizing patterning preserved in archaeological materials. White offered provocative post hoc arguments to accommodate the facts as he had observed them. These arguments cited several different causes of variability:

1. Differential destruction of anatomical parts during butchering
2. Differential abandonment of anatomical parts during transport of meat between kill locations and residential locations
3. Biased introduction of some parts to residential sites for use as tools

White's pattern-recognition procedures were followed and assemblages of fauna described in a number of later studies. The result was the description of increasing variety in assemblage composition. There was also some increasing uneasiness that all the causes of variability had not been specified, particularly regarding differential preservation. Then, there was a seeming lack of interest expressed and no investigation of the "causes" for observed patterning was started; little comparative work continued. Instead, new pattern-recognition studies were conducted by Wheat on anatomical segments, and by Frison on bone destruction and modification. In turn new interpretative arguments were offered and each researcher tended to guide "followers" in the procedures worked out by themselves. Earlier pattern-recognition procedures, for instance, those developed by White and expanded by Wood (1968), Dibble and Lorrain (1968), and Kehoe (1967), were largely ignored. This seems to be a classic example of the type of "one-sided" or incomplete archaeological research that I (L. R. Binford 1968a, 1968b) argued against in the late 1960s. I suggested that archaeology largely operated as if it did not need to evalute the arguments advanced about the past.

> In stressing induction and the drawing of sound inference, then, the stress falls on the psychological issue, . . . of how to make meaningful statements about archaeological remains and what they represent from the past. What is argued here is that the generation of inferences regarding awareness of as great a range of variability in sociocultural phenomena as possible and the citation of analogy to living peoples are not belittled here, the main point . . . is that independent means for testing propositions about the past must be developed. Such means must be considerably more rigorous than evaluating an author's . . . professional competence or intellectual honesty [L. R. Binford 1968a:17].

This failure even to recognize the role of evaluating ideas was derived directly from the canons of traditional archaeology so well summarized by Raymond Thompson: "The final judgment of any archaeologist's cultural reconstruction must therefore be based on an appraisal of his professional competence [Thompson 1956:331]."

Given this viewpoint, the traditional archaeologist would not be concerned with evaluating his ideas per se, only with performing in such a way as to enhance his professional status. Given a growth in professional competence, then, one's ideas would be both accepted and adopted by appeals to one's authority.

The entire epistemology of traditional archaeology was a complicated set of role conventions considered appropriate to the status of the archaeologist. If an archaeologist performed well in these roles, competence was gradually established and his ideas would be taken seriously. Eventually they might be accepted by virtue of "established" competence. Epistemology was linked to the sociology of the field. Truth was the natural result of a person performing well in a sociological sense! What were some of the guidelines for establishing one's competence?

1. "The importance of the archaeologist's particular skills and views in the recognition of the indicative quality has been particularly emphasized [Thompson 1956:331]."

This is simply the ability to recognize patterning in the archaeological record. Patterning has "indicative potential" and the ability to recognize this is important in the eyes of traditional archaeologists. There was an additional emphasis on the skills of a traditional archaeologist as a typologist, one who had the ability to specify patterning in material things. "Competence" or "status" was certainly enhanced when a typology was adopted by others or when new forms of patterning were discovered, as was the case with White's recognition of patterning in faunal assemblage variability. The number and variety of new patterns recognized was considered a measure of one's professional competence:

2. "These same faculties are employed . . . first in the search for analogous data, and then in the demonstration of the nature, of the relationship between this material and the archaeological evidence [Thompson 1956:331]."

Here Thompson gives us another criterion for assessing professional competence—scholarship. Scholarship is the ability to recover from the literature of history, ethnohistory, and archaeology evidence for similarities and differences. Similarities between archaeologically recovered material and things described in the literature are related as formal analogies:

> What actually happens is that he compares an artifact type which is derived from archaeological data with a similar type in a known life situation. If the resemblance in the form of the two artifact types is reasonably close, he infers that the archaeological type shares the tech-

nique, behavior, or other cultural activity which is usually associated with the ethnographic type [Thompson 1956:329].

Prudence was the watchword for interpreting archaeological remains. Since the basic criterion for judging whether an interpretation was warranted rested with the degree to which one could document an ethnographic analogy, it was clear that views of the past would be directly linked and limited to the information or knowledge that was by chance preserved ethnographically and/or historically. One could not demonstrate professional competence through speculation or by demonstrating imaginative insight. The latter was what was cited by the frequent derogatory remark that someone's argument was "only theory." Professional reputations could only suffer from indulging in "theoretical" discussion:

3. "The sensitivity for the indicative quality is not an uncanny and inexplicable ability which only certain gifted individuals possess. Rather, it is the combination of the*investigator's anthropological background or training, his archaeological experience,* which is often called familiarity with the material, and *his intellectual capacity* [Thompson 1956:328; emphasis mine]."

This final criterion for judging one's professional competence may be translated into practical terms in a variety of ways. In the field of archaeology as I was introduced to it, the criterion of "the investigator's anthropological background" was best seen as the prejudged quality of the school from which the investigator was graduated. If one was from a "good" school, then one's background was assumed to be also "good." The criterion of "archaeological experience" is best translated into several separate criteria. Clearly the amount of experience was of utmost importance. This ensured that the older one was the greater one's "competence" and hence the more sage one's words. Young people should be "seen and not heard."

The character of an investigator's experience was also of importance. To speak with authority, one had to have had extensive field experience in the area of relevance to the discussion. For instance, I can remember frequent comments after I wrote "Post-Pleistocene Adaptations" (Binford 1968c) to the effect that such a paper should not be taken seriously since "after all he has never even been to Jarmo." One should not attempt statements about things not seen, touched, counted, and otherwise "communicated with" quite directly. This meant that for all practical purposs careers were forced into more and more provincial molds. One really gained very little competence capital by moving from one area to another. This also meant that the only literature of useful relevance to one's competence rating was the ethnographic and historical literature of one's own local area. "Why should I read about Eskimos when I study Plains archaeology? After all Eskimos are not analogous to the Arikara or Mandan!"

The final implication was that one should not really talk about the past, since it could not be experienced. All one could do "safely" was to offer generalizations about the archaeological record itself. Speculation as to what it all meant was left to barroom conversation.

Truth viewed as a natural consequence of career success dictated that middle-range research played no role whatsoever in archaeology. We were assured of endless digging, taxonomic redefinition, and very "safe" interpretations from ethnographic analogy.

Against this backdrop Joe Ben Wheat's work was prudent, based as it was on ethnographic analogy and clearly representing a demonstration of his ability to dig and to recognize patterning. What is important is that this patterning was different and had not been previously discussed. George Frison attempted a real innovation. What better authority for warranting an inference regarding bison hunting, or butchering, or rustic living than to say that you had shared the experiences, and done the things being inferred. Egographic analogy became the final appeal to authority. Traditional archaeology moved along, never evaluating ideas, only people. Change proceeds at roughly the rate of generational replacement, other things being equal, and the character of the field during any one generation is largely stylistically dominated by several "authorities." Their approaches to the field are like any other fad, being dropped as new "mentors" establish their "professional competence" and their particular styles of archaeological discourse. This state of the art in American archaeology is certainly the reason that no middle-range research was conducted to evalute the arguments of White, to seek an understanding of the patterning observed, or to build constructively from the contributions of colleagues.

In Africa, the situation was somewhat more rough and ready. How does one argue from ethnographic analogy to the behavior of *Australopithecus*? How does one cope with controversial or even "reckless" arguments when they are advanced by the recognized authorities in the field?

This book started with a dedication to David Clarke and a discussion of archaeological metaphysics. In attempting to characterize the "state of the art" in Africa, for instance, relative to North America it occurred to me that they differed significantly in what David Clarke (1973) referred to as "innocence." In several places I have suggested that the paradigm shift to a behavioral concern for the causes of the archaological record was a shift that occurred in Africanists' studies in the late 1950s. I see this largely as a response to the behavioral claims of Raymond Dart that was consistent with interest in learning something about the behavior of our earliest ancestors. This shift was an "innocent" shift in that the questions flowed naturally from the research and did not follow a major call for "new directions" or any philosophical discussions of what archaeology was and where it should be going. Actualistic studies were initiated as the "obvious" things to be done.

On the other hand, in North America there was a clear "loss of innocence" regarding the overall character of the field of archaeology beginning in the late 1950s and continuing through the early 1970s. This loss was so fundamental and basic that there was a seemingly endless debate over procedure, definitions, concepts, and goals. Nothing seemed to flow naturally from the questions being investigated, since the questions were largely self-conscious questions regarding the field itself, not its subject matter of interest—the past.

Both of these states are considered counterproductive. Africanist studies must lose their innocence. From an American perspective, regardless of one's disciplinary bias, many of the arguments offered by leading Africanists seem painfully innocent. At the same time these researchers seem to appreciate the role of middle-range research and the importance of actualistic studies. On the other hand many Americanists' studies, while spouting forth anything but "innocence," show an amazing lack of methodological insight and rarely exhibit any appreciation for the role of middle-range research. Egographic analogy is offered as preferable to ethnographic analogy. Others may claim to avoid analogy completely and seek "anomaly" as a means to understanding! Still others claim to avoid inferential methods and to seek direct empirical knowledge of the past. Finally, others claim that science is improper in archaeology since man behaved with free will; therefore explanation in any scientific sense is fruitless. Africa is long on innocence, and methods; America is short on innocence and short on methods.

There is an implied set of conclusions embedded in this study. Under traditional ideas of archaeology the field was thought to be advanced by increased exploration in the archaeological record. Digging more sites, increasing the documentation of field observations, piece plotting finds, saving bones, and so forth, all these advances in observational strategies were taken as indisputably good. It is true, as I have frequently emphasized, that until we have recognized a pattern in the archaeological record, we cannot be expected to carry out middle-range research seeking to clarify its meaning. I have also pointed out that when we do carry out middle-range research, we most often find that we have not recorded the "facts" of the archaeological record in usable ways. Many times this is not just a matter of format in reporting. It frequently implicates the more important area of field recording. We cannot go back and record facts that were destroyed through excavation, nor can we go back and recover things that were discarded as irrelevant at the time of excavation (e.g., bone splinters flint flakes, carbon samples). "Who would discard carbon samples?" I would have to reply almost all the archaeologists working before the perfection of carbon-14 methods of analysis. The object lesson is the same for any methodology that might be developed through middle-range research. Until we have methods, we really do not know what facts are useful. Archaeology must progress by virtue of methodological development alternating with its application. The latter exposes to us new forms of patterning, which then prompts more middle-range research, and so on. At any one time the field might be considered pattern rich and method poor; and at other times it appears method rich and short on new knowledge of the archaeological record. *We are method poor.* Traditional archaeology did not recognize middle-range research as part of the field, or as even necessary to the field. We have a backlog of investigations of the archaeological record that are almost certain to prove inadequate once we have developed methods for reliably giving meaning to archaeological phenomena. The facts demanded by the methods are almost certain to have been overlooked or recorded in ambiguous ways, rendering the "data" from even our best excavations relatively useless. *We need to discourage continuous exploratory excavation at the present time.* It will almost certainly prove a waste of the resources in the archaeological record. We need a concentrated effort on middle-range research. We need to develop whole new methodologies. If successful, new excavations will yield information never dreamed of by contemporary excavation-oriented archaeologists.

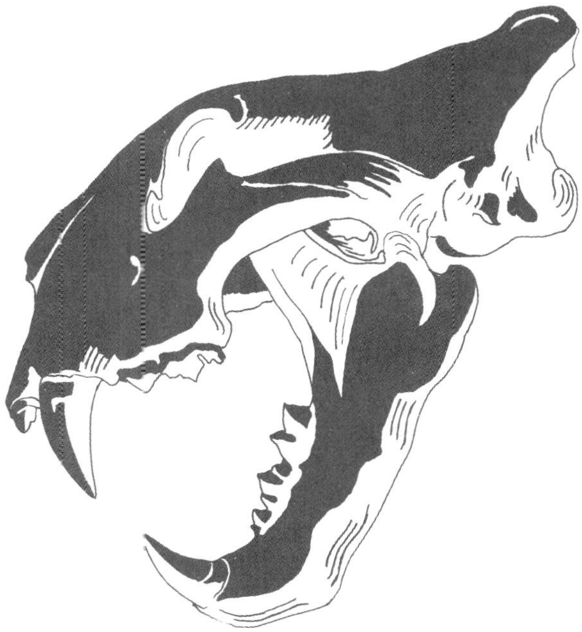

Part III
Putting our knowledge to work: Seeking to know the past

In Part I of this book I discussed the limitations of the use of archaeological observations for the development of methodology. In Part II, I examined in some detail the characteristic patterns of bone modification and distribution that might render recognizable the actions of nonhuman predator–scavengers as opposed to hominids as contributors to assemblages yielding archaeological remains. I discussed the logic of middle-range research, and then proceeded to execute, through actualistic study, research designed to yield relevant information. In this part of the book we face a very different problem, a question that was asked earlier but never answered: How do we carve out knowledge from ignorance? Having discussed what constitutes relevant information for the development of method, and having engaged in such research, I now wish to address the important question of how we use secure knowledge to elucidate conditions or events in the past about which we desire knowledge. In short, how do we develop research methods? We now face squarely the procedural challenge of how to use contemporary knowledge to investigate situations and conditions of which we have no direct experience. How do we use our observational language, and our secure intellectual "anchors" to probe a past the character of which is unknown to us?

It has been pointed out many times that there were likely to have been forms of cultural systems extant in the past, as well as states of systems and conditions of humanity, that are not even vaguely represented in the modern world (see, e.g., S. R. Binford 1968a; Freeman 1968; R. A. Gould 1980). How do we proceed, using only a secure knowledge of actualistic experience to recognize and diagnose accurately events and conditions that are no longer represented in the world of our experience?

This is a challenge in the area generally referred to as the "context of discovery." Until quite recently philosophers of science tended to concern themselves with the context of justification or confirmation (see Suppe 1977:716). Ideas were treated as "received" and one concerned oneself with evaluating them after they had been generated (see Hempel 1965:6). One good reason for this avoidance was that it was not clear how generalizations could be offered regarding the various ways scientists proceeded in their attempts at discovery. The apparent lack of susceptibility to formal systemization led to the suggestion that "archaeologists should be concerned more with how to justify their conclusions and less with how to generate them [Sullivan 1978:185]."

I consider this position to be shortsighted and self-defeating for practicing archaeologists. We must make decisions and design research strategies. We must therefore rationally consider and discuss ways that we can use our secure knowledge to tease our ignorance for new insights and ideas.

One very important strategy is conducting middle-range research itself. This is well illustrated in the ideas generated by archaeologists about the character of the past, such as Frison's (1970) muscle stripping with bone tools model, or Perkins and Daly's (1968) schlepp effect model, or Brain's whole-animal model of the Makapansgat past. Each of these was an innovative post hoc accommodative argument that was offered to explain specific archaeological observations. As such they were fantasy, generally unsupported by middle-range research. If such arguments were adopted, as methods, as secure procedures for inference, as has been done by many—Stanford (1979a) and Johnson (1978) on the Plains, Klein (1976a) with his use of the schlepp effect to explain his African data, and my own acceptance of Brain's whole-animal model for Makapansgat in the face of at least 25% unexplained variance—we are on the road to *myth making.*[1] *The interaction between post hoc arguments as to what the past was like and scientific investigations to evaluate components of such arguments is at least one valuable context in which knowledge has the chance of being refined and accumulated. It is this interaction that has ensured that research treating the problem of the character of early hominid behavior has been some of the most progressive in archaeology. I term it progressive because the investment in middle-range research through actualistic field studies has been high.*

Conducting middle-range research itself, however, is not enough. One must be able to use the knowledge gained through such studies to design research procedures that will be likely to yield more provocative observations of relevance to the past. As an example, I will take the research previously reviewed regarding differential bone preservation. Brain's middle-range work established the role of differential preservation and began to explore something of its causes. He did not provide us with a set of methods. My work in conjunction with Jack Bertram (L. R. Binford and J. B. Bertram 1977) carried further the middle-range research started by Brain. We perhaps refined the arguments about the properties of bone that conditioned its susceptibility to destruction, and isolated some of the factors that conditioned the distribution of this property in a population of animals (e.g., the age structure of the population acted upon by destructive agents). We did not develop practical methods for using this information to isolate other sources of variability or causes that may have also conditioned the character of archaeologically important faunal assemblages. We were in the position of knowing rather securely one source of potential variability in faunal assemblages. We did not know how to use this information as an aid for isolating other causes that may well have also operated on an assemblage in the past. This situation was seen in the tendency of Africanists to note the similarities among assemblages, presumably deriving from their all having been subjected to various destructive agents, and therefore dismiss the "unpatterned" variability remaining as "noise" (as in the case of my work with Bertram [L. R. Binford and J. B. Bertram 1977] or as in the case of Mary Leakey's treatment of the Olduvai fauna discussed in Chapter 6). The point is that assemblages generally have multiple causes. We must use our knowledge of some causes linked to their diagnostic effects much as is done in chemical qualitative analysis. Given an unknown we perform various tests with each, eliminating known elements or compounds from the unknown material. That is, we continually break the unknown down into recognizable constituents. By this eliminative procedure we increasingly isolate our

[1]It is only when post hoc argument remains unevaluated and its prior existence is cited as a justification for subsequent use that conventions arise and myths become established.

ignorance. The residue from our use of knowledge that should receive our research attention eventually appears uncomplicated by the constituents about which we already have some understanding. This is essentially John S. Mill's canon of residues: "Subduct from any phenomenon such part as is known by previous inductions to be the effect of certain antecedents, and the residue of the phenomenon is the effects of the remaining antecedents [Mill 1850:233]."

This seems to me to be particularly appropriate to a situation in which we suspect that hominids contributed to an assemblage, for instance by virtue of there being stone tools in a deposit. We are also aware that other agents probably contributed to the deposit, as in Mary Leakey's recognition (1971:277) that other predator–scavengers acted on the assemblages at Olduvai Gorge: "hyaenas and other scavengers are known to have gnawed the bones." If we can use our knowledge of animal-generated assemblages as well as animal-modified bones to eliminate their contribution to the Olduvai assemblages as documented in the field, we may gain thereby a clearer, uncluttered picture of the characteristics of the fauna that are referable to the behavior of hominids. This is the overall methodology I will attempt to develop using the knowledge generated through middle-range research into animal behavior and its consequences.

I do not wish to imply that the methodology being explored here is the only one or even the best one. It is, however, a methodology that has the merit of decreasing the chance that we will make George Frison's and Raymond Dart's error and develop post hoc accommodative arguments to "explain" in human terms facts that were generated by gnawing animals. The procedure does not ensure that we will not err in our reconstructions of the past, but it does tend to offer some insurance against reconstructions based on facts irrelevant to hominid behavior.

The reader may see some similarities between my adoption of Mill's method of residues, and the "method of multiple working hypotheses" and "arguments from elimination" mentioned elsewhere in this book. It is true that they all have some elements in common, particularly Mill's approach and that of Chamberlin's (1965) method of multiple working hypotheses:

> an adequate explanation often involves the coordination of several agencies, which enter into the combined result in varying proportions. The true explanation is therefore necessarily complex . . . we are so prone to attribute a phenomenon to a single cause, that, when we find an agency present, we are liable to rest satisfied therewith, and fail to recognize that it is but one factor, and perchance a minor factor, in the accomplishment of the total result [Chamberlin 1965:756].

In this sense, this entire book is an exercise in the method of multiple working hypotheses—that is, in the sense of recognizing that multiple causal agents and conditions are likely to influence the forms in which we see both things and associations of things derived from the archaeological record. I am seeking methods that will permit the assessment of the relative roles of different agents and/or causes. In short, I am seeking ways of measuring the contributions of different agents and causes. This is very different from the way that the term multiple working hypotheses *is used by some. Not uncommonly we hear the claim that the method of multiple working hypotheses is being used when an investigator cites several possible causes—"it could be the work of lions, it could be the work of hominids, it could be the work of space men, it could be an accidental occurrence of nature," and so on. The investigator then employs an argument from elimination to warrant his opinion as to which cause is correct. This is a kind of quick and dirty probabilistic evaluation for purposes of offering an opinion based on the general "knowledge" available. This is not a conclusion arrived at through the application of scientifically founded methods designed to measure the effects of specified causes. We need the latter type of capability; anyone can offer opinions based on almost any level of knowledge and/or ignorance. The method of multiple working hypotheses is a powerful scientific procedure, and it is not to be confused with the method of opinionated elimination of biased guesses.*

> Disproof is a hard doctrine. If you have a hypothesis and I have another hypothesis, evidently one of them must be eliminated. The scientist seems to have no choice but to be either soft-headed or disputatious [Platt 1964:350].

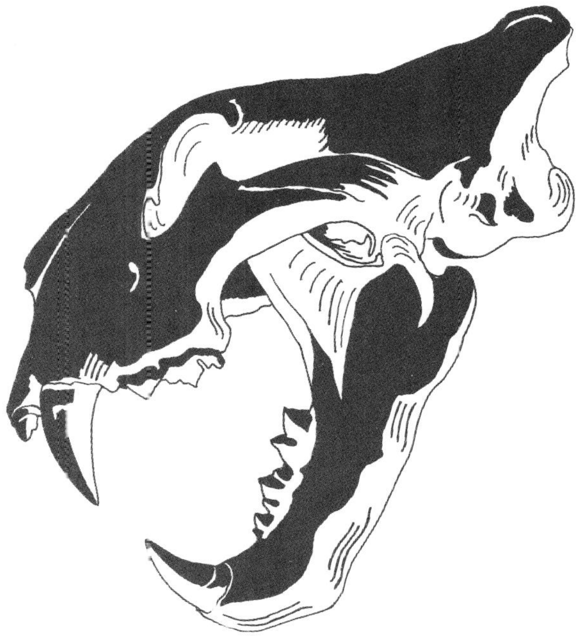

Chapter 6

Application: A new look at Olduvai Gorge

Contemporary Views of the Early Hominids

In the beginning of this book I pointed out that early discussion regarding the archaeological significance of bones largely centered on the question of the "cultural" status of early hominids and the behavior in which they engaged. As it was then, so it is today. What was early man like? Did he hunt his food? Did he live in bands? Did he share food, did he . . ., did he . . . We want to know what our ancestors were like and how they lived. We also want to know the forces at work that molded our ancestors so that we "modern men" were the evolutionary outcome.

It has become rather fashionable to engage in speculations as to the answers to these basic behavioral questions. It is equally interesting that the speculations about our past championed by Glynn Isaac (1971, 1975a, 1975b, 1976, 1978a and b) have increasingly been accepted as the "orthodox" view of hominid behavioral evolution (see, for instance, the acceptance of Isaac's position by Richard Leakey and Roger Lewin [1977] and Grahame Clark [1977]). Isaac's view has been stated many times and rests with the specification of points of behavioral contrast between modern man and modern great apes. After some of the behavioral contrasts that render man unique are specified, this question is asked: How and when did such behaviors arise? Here is one of Isaac's lists of behavioral contrasts[1] (Isaac and Isaac 1975:31):

– Speech communication
* Manufacture and use of tools, equipment, and structures
\+ Food sharing and collective responsibility for subsistence
* Propensity to consume animal protein (importance of hunting and fishing)
– Division of labor among age, sex, and other classes
* Organization of varied daily movement around a temporarily fixed location (home base or campsite)
– Existence of family units consisting of pair-bonded male(s) and female(s) plus children
* Involvement in symbols, ornament, and art
\+ Observance of detailed rule systems and codes that regulate the mode of execution of almost all individual and social functions

Having phrased the question of hominid behavioral evolution in terms of the behavioral contrasts between man and great apes, Isaac and Isaac (1975a:32) then asks themselves the interesting question, "In what order and with what dynamic interrelationships have these differences between the human and the chimp phyletic lines arisen?" In answer to this question, Isaac proposes a "model" of hominid behavioral evolution with two basic steps. Those behaviors that we would generally recognize as "culture" (e.g., those dependent upon or consistent with speech–language, the use of symbols and the behaviors such capacities facilitate, "insight and cunning," "technical ingenuity") are viewed as having arisen during the Middle Pleistocene, between 1 million and 100,000 years ago.

Much earlier, during the Miocene–Pliocene *from 8.10 to 3.0 million years ago*, basic and important changes are imagined:

> The model entails the concept that the present condition of mankind incorporates two tiers of evolutionary change; the first consists of an adaptive complex with the following set of mutually reinforcing ingredients—bipedalism, the carrying of food and implements, varied tool making and using, hunting, food collecting, food sharing, division of labor, organization of movements around a home base [Isaac and Isaac 1975:33].

This model is a speculation as to what the past was like. It is conjectural history and involves no theoretical arguments that I can recognize. As has been pointed out many times, archaeological observations are *contemporary observations*. They are observations made by the archaeologist during the excavation or recording of a location yielding archaeological remains. To move from these contemporary descriptive statements about the archaeological record to statements about the past we must give meaning to the observations made on the archaeological record. We do this through arguments regarding the processes that brought the archaeological record into being. That is, if we can accurately relate the facts of the archaeological record to the conditions in the past that would have generated these facts, then we have succeeded in giving historical meaning to the contemporary facts. When we have a conjectural model of the past as has been presented by Isaac, we must have some robust archaeological methods (see L. R. Binford 1968a) if we are to succeed in evaluating it. That is, we must observe archaeological remains of relevance to the events postulated and then employ a reliable methodology for giving meanings to these facts as a way of gaining some idea if the conjectured past is anything like that past as in fact it occurred. For Isaac, the keys to the archaeological remains of relevance are locations yielding the earliest stone tools, since tool use and manufacture make up one component of the postulated behavioral syndrome. Now we would expect some development of method and at least some strong arguments regarding recognition criteria for assessing the degree to which all the other postulated behaviors in the "model" were in fact characteristic of the hominids that left the early

[1]Behaviors on the list that leave clear-cut archaeological traces are identified with an asterisk; those that can be reasonably securely inferred, albeit indirectly, are marked with a plus; the others, which are not yet definitely detectable in the Pleistocene, are marked with a minus.

stone tools. My search of the literature has turned up only one associational fact, which Isaac and others cite in support of their *belief* that the archaeological record supports their conjectured history of human evolution. This fact is the association of stone tools with bones of animals. Given this simple association, the following arguments are generally made:

1. Meat eating was regularly practiced and persistent hunting almost certainly took place (Isaac and Isaac 1975:17).
2. Localities at which both discarded tools and bone refuse accumulated are more readily explained as home bases in the distinctive sense (Isaac and Isaac 1975:17).
3. The consumption of food at a home base involves transporting food from the place where it was obtained. The quantities that can be estimated suggest that far more food was transported than was needed for feeding infants; thus extensive food sharing seems an inevitable conclusion (Isaac and Isaac 1975:17).

The Isaac's arguments are dependent on the accuracy of the *assumption* that the integrity of the deposits is great—more important, that hominids were equally responsible for the presence of tools and animal bones in these *geological* deposits. In addition, his arguments are dependent on the *implicit assumption* that, given the accuracy of the integrity assumption, the resolution of the deposits was high; namely, that the "assemblages" derived from the actions of hominids at a given location were high-resolution assemblages rather than a palimpsest of many different event episodes in which hominids participated. Basically *no* methods beyond wishful thinking have been applied to the Early Pleistocene sites for evaluating the degree to which the foregoing assumptions are warranted or the degree to which the models are realistic.

Mary Leakey has at least shown an awareness of the problem and offered some minimal criteria for accepting or rejecting a given deposit as an "occupational floor." She assigned the sites at Olduvai to four basic classes (see M. D. Leakey 1971:258):

1. *Living floors,* the occupational debris found on a paleosol or old land surface with a vertical distribution of only a few inches
2. *Butchering or kill sites,* artifacts associated with the skeleton of a large mammal or with a group of smaller mammals
3. *Sites with diffused material,* artifacts and faunal remains distributed through a considerable depth of a clay or fine-grained tuff
4. *River and stream channel sites,* where hominid debris has become incorporated in the filling of a former river or stream channel

What is interesting is the mixture of criteria and assumptions. River and stream channel sites are recognized by geological criteria for identifying the formation processes of the deposit. Sites with diffused material are simply locations with a certain formal set of properties characteristic of the deposits, with no identification of either the geological or behavioral context of formation. Nevertheless, butchering or kill sites and living floors are recognized by virtue of *inferences about* the formation processes responsible for the content and associational patterning within the deposit.

First of all, there have been no attempts to treat or even seriously discuss the problems of potential variability in the *integrity and/or resolution* of the so called living sites. All the facts gleaned from the deposits *interpreted* as living sites have served as the basis for making up "just-so stories" about our hominid past. No attention has been given to the possibility that many of the facts may well be referable to the behavior of nonhominids. Second, the complex problem of resolution and the character of the behavioral regime in terms of which hominids might generate archaeological remians has not been addressed. For instance, the "just-so stories" produced to date by Richard Leakey and Roger Lewin (1977:116–117, 1978:8–12), as well as Glynn Isaac (1976:484–485), have all adopted an abstracted idea of a *gathering and hunting way of life* largely based on generalizations offered by Lee and DeVore (1968). (See R. Leakey and R. Lewin [1978:94–125] regarding their experiences with the Dobe !Kung Bushmen.) Ethnographic generalizations have simply been projected into the past. A student of mine, after reading Leakey and Lewin (1978), commented that the only thing hominids had not yet developed at 2 million years ago was the stock market! Very simply, the researchers have a generalized idea as to what the past was like and they have then *accommodated* all the archaeological–geological facts to this idea. This is not exactly science. There are no methods for giving meaning to the archaeological observations that are independent of interpretative arguments offered by the investigators. In short, they have no middle-range theory and, in turn, no way of evaluating their ideas about the past other than tests of plausibility and the fit between ideas and *interpreted observations.* We should all know that observations may be "interpreted" in almost any way that suits our biases.

The development of approaches whose purpose is the identification of the *agents* responsible for the presence and condition of the bones found in association with remains of hominid behavior is the first step in the necessary assessment of the *integrity* of an ancient deposit yielding traces of hominid behavior. When we focus on isolating the conditions under which a deposit containing archaeological remains of interest was produced, we similarly must face squarely the problems of the character of the behavioral regime of the hominids themselves and the possible behavioral contexts in terms of which archaeological traces might be generated. For instance, the very idea of a site or living floor assumes conditions in the past for which there is no demonstration. In fact, it assumes the very "knowledge" we would like to obtain from the archaeological remains. Did early man sleep on the ground and consume food at a "base," or did he simply range, consuming food where it was found and sleeping in the trees as do most other primates? Site and living floor identifications presuppose that concentrations and aggregations of archaeological and other materials are only produced by man. Are there not other conditions of deposition that could result in aggregations of considerable density found on old land surfaces? The answer must be a resounding yes. Paleontological locations commonly exhibit vast degrees of variability in the relative density of bones and in their patterns of aggregation in geological deposits. Similarly, extreme patterns of variability with regard to size and heterogeneity of inclusions in geological deposits occur by virtue of normal depositional processes. Given such conditions we must have *diagnostic criteria for recognizing the derivatives of hominid behavior of interest as it may be distinctive from that of other animals.*

In this study I have developed arguments as to the significance of formal variability, both in the survival conditions of bones and in assemblage composition. It is proposed that these arguments may be used for diagnosing two aspects of problems of archaeological remains found in geological deposits.

1. I have shown that there is a basic difference in the manner of destruction and modification seen for bones damaged by animal consumption versus those resulting from the human recovery of marrow through the use of tools, knives, and hammers.
2. I have shown that the behavior of human hunters and nonhuman predator–scavengers leaves a distinctive pattern of assemblage composition related to the *location of consumption,* somewhat independent of the species involved. That is, where consumption takes place is a more powerful conditioner of assemblage composition than who the consumer might be.

The second point was manifest by the rather remarkable similarity between the population of anatomical parts transported for consumption by Eskimo hunters and that remaining at kill sites where nonhuman predators consumed prey. Similarly, the parts transported under conditions of competition by nonhuman predators were strongly analogous to the marginal parts abandoned by human hunters at the kill. These findings illustrate that *locus of primary consumption* is the basic conditioner of assemblage composition for anatomical parts of prey *even when* the predators are as different as man, wolves, hyenas, and lions. This means that *the analysis of the condition of the bones in an assemblage may permit the identification of the agents responsible for it,* but *the assemblage composition in terms of relative frequencies of anatomical parts betrays the context of food consumption by the agents.*

These approaches carry the following implications:

1. *If man was living in base camps and hunting animals,* we would expect him to transport parts of high utility, resulting in the introduction to the base camp of an assemblage analogous to the parts consumed by other predators at kill sites and introduced

by modern human hunters to residences—in short, an assemblage that is anticipated by the GUI or the MGUI (as developed in L. R. Binford 1978b) presented here in Table 5.08.

2. *If man was living in base camps and scavenging animal food from the kills of other predators,* we would expect him to transport parts of low utility as do other competitive predator–scavengers, resulting in the introduction to base camps of assemblages of parts anticipated basically by the inverse GUI.

3. *If tool-using man was hunting animals but not consuming in base camps,* we would expect "human kills" to look like other nonhuman kills of prey with the possible exception of the traces of dismemberment strategies and bone-breakage techniques that would betray the use of tools.

4. *If tool-using man was scavenging but not consuming in base camps,* we would expect such "transported" faunal assemblages to look like those of other predator–scavengers but to exhibit distinctive bone-breakage patterns deriving from the basic differences between a hammer and a vise as the technical means of gaining access to the medullary cavity of long bones.

In the context of the foregoing reasoning, I will address the important assemblages reported from Olduvai Gorge.

Olduvai Gorge, a Challenge to Our Methods

I have chosen to attempt to use the method of residues on the Olduvai Gorge faunal materials. The analysis will be forced to depend almost exclusively on facts of assemblage composition, since no published descriptions of bone breakage, modification, or surficial scarring exist.[2] The bones have been saved and perhaps future studies of characteristics of bone modification can be used to evaluate some of the arguments to be developed here. My analysis is further weakened by the need to make certain assumptions regarding the anatomical properties actually referred to by some of the "categories" used in tabulating the assemblages reported by Mary Leakey (1971:276, Table 8).[3] Despite these weaknesses, my analysis will represent the application of a methodology developed from control information obtained in actualistic studies. I will follow a procedure designed to minimize ambiguity. The earlier analysis of assemblage composition (Chapter 5) clearly indicated that the differences between animal-produced assemblages and some humanly generated assemblages were by no means great, when viewed strictly from the perspective of assemblage composition, so I have chosen to follow a conservative set of research tactics. The analysis is set up so that ravaged assemblages serve as the models for analysis. It has already been shown that destruction tends to obscure original differences between assemblages. Most of the assemblages in Olduvai Gorge are demonstrably ravaged. If we could reconstruct these assemblages using good data on the survival potential of bones from African antelopes, we might be able to isolate additional facts potentially referable to the behavior of the early hominids. We do not have such reliable middle-range research information, so I have analyzed raw information against normative models of animal behavior to determine if there were any facts not anticipated by the known patterns of animal behavior.

Mary Leakey, in her report on the excavations within Olduvai Gorge conducted between 1960 and 1963, recognized that all the locations were not geologically similar and classified the investigated sites into the four basic classes as discussed earlier (see M. D. Leakey 1971:258). Table 6.01 summarizes the sites in Olduvai Gorge as evaluated by Mary Leakey as to their "character."

BK is considered to be a "derived" deposit where the materials have been redeposited from their former location. Clearly it is not likely to exhibit anything but gross *resolution* and its *integrity* is not apt to be great. Similarly, deposits described as diffuse material represented items scattered through considerable depths of deposit, 4½ feet at MNK Main (M. D. Leakey 1971:137) and 2 feet at HWK East

[2]See postscript at the end of this chapter.

[3]I attempted to obtain clarification regarding this table through correspondence sent to the International Louis Leakey Memorial Institute for African Prehistory, but never received answers to my letters.

TABLE 6.01
Mary Leakey's Classification of Olduvai Sites [a]

Living floors	Kill sites	Diffuse material	Channel fill
DK Level 3 (DK 3)	FLK North Level 6 (FLK 6)	DK Level 1 (DK 1)	BK II (BK II)
FLK NN Level 1 (NN 1)	FLK North *Deinotherium* Level	DK Level 2 (DK 2)	
FLK NN Level 3 (NN 3)		FLK NN Level 2 (NN 2)	
FLK 22 *Zinjanthropus* Level (ZINJ)		FLK Level 13 (FLK 13)	
HWK East Level 1 (HWK 1)		FLK Level 15 (FLK 15)	
EF–HR (EF–HR)		FLK North Levels ½ (FLK ½)	
FC West, Floor (FC FL)		FLK North Level 3 (FLK 3)	
TK Upper Floor (TK UF)		FLK North Level 4 (FLK 4)	
TK Lower Floor (TK LF)		FLK North Level 5 (FLK 5)	
		HWK East Level 2 (HWK 2)	
		HWK East Levels 3–5 (HWK 3–5)	
		MNK H. 13 (MNK H13)	
		MNK Main (MNK MAIN)	
		FC West Tuff (FC TUFF)	

[a] Abbreviations used in figures and some of the tables are given in parentheses.

Levels 3, 4, and 5 (p. 96). Such diffuse scatters *must* represent numbers of different event episodes, *and the longer a deposit is accumulating the less integrity it is apt to exhibit.* Both conditions of integrity and resolution are dependent on the rate of accumulation, in that the longer it takes the deposit to form the greater the number of events apt to be recorded in the deposit and, in turn, the greater the number of agents apt to be represented by events. Sites classified as living floors and kill sites are those where integrity is expected to be great and the resolution relatively unambiguous. Let us examine the excavators' observations regarding these important issues.

DK LEVEL 3

Materials found in DK Level 3 are reported in a context of the margins of a permanent body of water. Natural deaths and "water's edge" behavior of nonhominids such as water birds, fish, and crocodiles are clearly present and are so interpreted by the excavators (see M. D. Leakey 1971:24). Also found were artifacts, an enigmatic stone circle, and "other" faunal materials. To what degree is the association between stone tools, a stone circle, and "other" fauna—largely bovids—indicative of hominid behavior? Is it likely that one would find "other" fauna in deposits formed at the margins of bodies of water? The answer must be yes. In fact, it is one of the most likely places to find preserved faunal material derived from both natural deaths and predators' prey. Such deposits are common paleontological contexts and are recorded today (see Hill 1976). The behavioral significance of the stone circle derives from the belief that it occurs in a living site. The stone artifacts offer no direct clues to the events in which hominids participated. *We are left with what is most certainly a deposit of dubious integrity, and we have no idea as to the character of the resolution.* However, for there to have been 4600 teeth shed by juvenile crocodiles (M. D. Leakey 1971:249) in an area of approximately 255 m^2, the deposit certainly did not accumulate overnight. In fact, a considerable period of time appears to be represented. *In a deposit of dubious integrity, how do we decide what is a meaningful association?* This question has not been addressed. Nevertheless, the terrestrial faunal assemblage is believed to be the food of the hominids who made the tools and who lived in the "shelter" indicated by the stone circle. *This is all unsupported.*

FLK NN LEVEL 3

This is frequently referred to as the "tortoise floor," since a number of tortoises (17 in all) occurred in largely undispersed individual clusters on the old land surface. This site is additionally important since it yielded remains of hominids themselves. As in the case of DK Level 3, this site is a deposit accumulated along the margins of a major body of water. Fish bones were observed in considerable concentrations, reminiscent of fish trapped in small pools as the waters of a lake recede during the dry season. Amphibians were common, as were birds and rodents. Numerous coprolites "together with tooth marks on some of the bones—including the hominid right parietal and astragalus—indicate that scavengers were active [M. D. Leakey 1971:43]." Finally, the excavators observe that

> with the exception of one small specimen found intact with the various parts in position, the remains of the more complete tortoises were scattered over areas 2–3 feet in diameter. Careful scrutiny of the carapace and plastron fragments has been carried out, but apart from a very few small indentations, such as could have been caused by a pointed stone, no signs of damage are visible. It seems likely that the shells were not broken open but fell apart after they had lain in the open for a time [M. D. Leakey 1971:251].

Fitting these pieces together, we note a location along the margins of a major body of water (which presumably fluctuates in level), where a normal "marsh" fauna is indicated, where some tortoises died, where carnivores were active, and where man was present. In short, the *integrity of the deposit is minimal.* Regarding resolution, the accumulation of 17 tortoises and their remains in isolated clusters suggests that the deposit was not disturbed greatly by natural agents such as water but represents sufficient time for 17 tortoises to die natural deaths in a relatively small area.

One gains the impression that not nearly as much time passed during the development of the accumulation at FLK NN Level 3 than was the case at DK Level 3. However, this is simply my impression and there is no clear measure of relative time represented by the deposit. It should be pointed out that the bovids occurring in the deposit are reported to be largely of the genus *Kobus,* which includes waterbuck and reedbuck, both animals that are almost constantly near permanent water (M. D. Leakey 1971:251). These are, of course, the animals most likely either to die or to be taken as prey in such a setting, adding further support to the assessment that the deposit lacks integrity and the level of resolution for the included material is clearly suspect. There is, therefore, no justification for the interpretation of associations in this deposit as deriving exclusively from the action of hominids.

FLK NN LEVEL 1

This is described as a scatter of artifacts and bones where there is noted to be poor preservation of the animal bones (see M. D. Leakey 1971:44). No further information is given to substantiate the classification as a living floor.

FLK 22 ZINJANTHROPUS LEVEL

This is the famous *Zinjanthropus* "floor" where the first associations between hominid bones, artifacts, and animal remains were found in East Africa. As on the other living floors described, amphibians, birds, and rodents are reported to be common. On the other hand fish are relatively rare but present, as are crocodile teeth. A species of slug recovered from the floor is similar to forms living in evergreen forests with greater than 35 inches of rain per year and where mists are common (M. D. Leakey 1971:252). All in all, the deposit, as the others thus far described, appears to have formed near a body of water and in this case where vegetation was relatively dense. Thus, as in the other cases, a natural background of remains of animals inhabiting the margins of a lake are present, with no implication of hominids being responsible for their presence in the deposit. As for evidence of other agents contributing to the character of the deposit, Leakey states, "A number of indented tooth marks on some of the bone fragments, such as might be caused by the larger scavenging animals, suggests that the bones on the floor may have been broken by scavengers as well as by the hominids [M. D. Leakey 1971:50]."

Thus, we have at least a natural background—contributions by scavenging animals in addition to contributions by hominids. This is not exactly a high-integrity assemblage. As far as resolution goes, there are no clues published by the excavators. Therefore, as in most of the other cases, there is no justification for the interpretation of associations within this deposit as necessarily deriving from the behavior of hominids.

HWK EAST LEVEL 1

This is described as a paleosol that has been eroded into a sculptured surface (see M. D. Leakey 1971; 89). Nevertheless, the materials scattered on the surface of this eroded paleosol are considered in place. Little is actually said about this location except that the "proportions of bones belonging to the various faunal groups followed the pattern typical of Oldowan living floors (p. 254)." Assessments of integrity and resolution are impossible given the information provided.

FC WEST, FLOOR

This deposit is described as a dense aggregation of material with a low frequency of faunal material. In addition, it was observed that

> no heavily rolled artifacts or bones [were observed] yet a proportion of both is either weathered or abraded to some extent and is not in the mint-fresh condition usually seen in material on Bed I living floors. It seems possible that here, in common with the majority of Bed II sites, there has been some displacement of the material by water action [M. D. Leakey 1971:157].

Thus, the excavators point to a derived composition for remains on this floor. Such deposits are not likely to exhibit a high degree of integrity or resolution.

TK UPPER FLOOR

This deposit is described as resting on the eroded surface of a clay deposit overlain by a tuff. As in many of the assemblages from Bed II, faunal remains relative to numbers of artifacts were low. No basic consideration of the problems of integrity or resolution are touched upon in the published descriptive material. As regards resolution, it was noted that there was an unusually dense concentration of quartz debitage, suggesting an in-place distribution of lithic debris.

TK LOWER FLOOR

The only comment here regarding the status of the assemblage was that everything was found on a single horizon (M. D. Leakey 1971:174). In commenting on both TK locations, Mary Leakey states,

> it is perhaps significant that the faunal remains and a proportion of the artifacts from this site are weathered, indicating that they were exposed on the surface longer than usual before being buried and so were possibly displaced from their original positions [M. D. Leakey 1971:261].

Thus, as was the case for the FC location, the living floors at both TK locations cannot be taken as exhibiting high resolution or integrity and, therefore, the meaning of associations is unclear.

This review of the information supplied by the excavators is not encouraging, regarding the interpretation of associations observed within the living floor deposits as necessarily having anything to do with the hominids. Nevertheless, such observations as have been made leave the situation ambiguous, since it is conceivable that hominids could have been the agents responsible for at least some of the fauna found in association with the tools. Perhaps some of the ambiguity may be reduced by applying some of the knowledge of agent-related breakage described in Chapter 3 to the materials recovered from the Olduvai locations.

Analysis of Olduvai Fauna

It was found in our earlier analysis of wolf kills and dog yards that when animal consumption was responsible for the absence of articulator ends, there was a positive and linear relationship between the number of long-bone splinters observed in the assemblage and the percentage of articulator ends potentially present that were in fact missing—in this case the maximum MNI for each sample multiplied by 24 (the number in a single animal), with the total number of articulator ends observed subtracted. The remainder is then divided by the total number potentially present.

To determine if this is the case using the data from the Olduvai is not possible, since the numbers of long-bone splinters were not counted as such. Nevertheless, there is a category of "indeterminate fragments" (M. D. Leakey 1971:256, Table 3), which presumably included most of the long-bone splinters but is likely to include skull, pelvis, scapula, and rib fragments that are too small for referral to either anatomical part or taxa. Data converted from Leakey's Tables 3 and 8 (M. D. Leakey 1971) are summarized in Table 6.02 for the assemblages from Olduvai Gorge. Figure 6.01 illustrates the relationship between the percentage of articulator ends missing from the total potentially present plotted against the number of indeterminate fragments per MNI for each sample.

We note immediately that instead of a single linear positive correlation, as was observed among the control data from the wolf kills, there is a complicated segregation of cases into essentially three groups. The first is a cluster corresponding to the positive pattern observed among the wolf cases. This includes FLK North Level 5, MNK Main, and MNK H. 13 as well as FC West, Tuff and Floor. In these cases deletion of articulator ends is correlated with increases in unidentified bone, suggesting consumption on the spot by agents destroying articulator ends. As we have seen, this is most common when animals are the destructive agents. The second grouping isolates cases exhibiting negative correlations between the numbers of articulator ends missing and the numbers of unidentified bones. That is, as there is an increase in the numbers of missing ends there is a proportional decrease in the numbers of unidentified bone fragments. This condition could only arise if there was differential transportation of the parts in question. Either the articulator ends were destroyed before being introduced to the sites, or they were removed from the location together with attached shafts. This group of sites appears split into two subgroups consisting of FLK Level 2 HWK East Level 1, and FLK NN Level 2, and then a larger set consisting of FLK North Levels 1/2, FLK North Level 4 FLK North Level 3, FLK Level 13, FLK Level 15, HWK East Levels 3–5, HWK East Level 1, FLK NN Level 1, and FLK NN Level 3. The third cluster includes the relatively unique sites BK II and the *Zinj* floor; both of these are characterized by a considerable percentage of missing articulator ends but a truly impressive quantity of indeterminate fragments. Some extraordinary conditions of bone destruction are indicated at these two locations.

In the earlier studies of bone breakage produced by animals it was found that the production of shaft cylinders was a characteristic by-product of animal bone gnawing. It was also found that the number of cylinders produced by gnawing wolves per MNI increased with the number of long-bone splinters up to a point of about 6 cylinders and 55 splinters per MNI. With further increases in splinters there was a reversal and the number of cylinders was thereafter regularly reduced (see Figure 4.60). This reversal simply represented a threshold in the sequence of bone destruction typical of animals: first the destruction of articulator ends, then the chipping back and channeling of long-bone shafts with the production of cylinders, and, finally, the collapse of cylinders, which represents the most advanced stage of destruction. Splinters as by-products of destruction increased with cylinders up until the stage of cylinder collapse and then cylinders were reduced into splinters. Figure 6.02 illustrates the relationship between indeterminate fragments (M. D. Leakey 1971:256) and items listed in the faunal tables as limb-bone shafts (p. 276).

For such shafts to be identifiable as to taxa they are most certainly relatively complete, and if they are relatively complete and lacking articulator ends, they are most likely to be cylinders. What is clear in Figure 6.02 is that the threshold destruction observed among the wolf kill data is essentially duplicated in the Olduvai data for the following assemblages: MNK H. 13, FC West, Tuff, MNK Main, FLK North Levels 1/2, 3, 4, 5, 6, HWK East Levels 3–5, FLK NN Level 1, FLK Level 15. What appears to be a parallel distribution with the number of cylinders inflated is represented by BK II, FC West, Floor, HWK East Level 1, FLK NN Levels 2 3, as well as FLK Level 13. This situation could arise if the MNI for the maximum part underestimates the MNI actually represented, therefore giving the appearance of an in-

TABLE 6.02
Ancillary Facts: Olduvai Faunal Assemblages

		TK		FC West		MNK		HWK East Level		
	BK II (1)	Upper Floor (2)	Lower Floor (3)	Tuff (4)	Floor (5)	Main (6)	H. 13 (7)	3–5 (8)	2 (9)	1 (1
1. MNI of most common part[a]	15.76	2.97	1.53	1.91	.74	13.06	1.50	4.02	6.45	7
2. Number of articulator ends (o)	190.46	13.0	9.84	6.0	5.0	104.24	7.90	30.36	54.80	96
3. Number of articulator ends on complete bones	44.68	6.08	1.9	0	0	32.42	1.96	1.78	27.04	39.
4. Percentage of total articulator ends on complete bones	.23	.47	.19			.31	.25	.06	.49	
5. Articulator ends expected	1378.24	71.28	36.72	45.84	17.76	313.44	36.0	96.48	154.8	189
6. Percentage expected articulator ends missing	.50	.82	.73	.87	.72	.67	.78	.69	.65	
7. Number of expected articulator ends missing	187.78	58.28	26.88	39.84	12.76	209.2	28.1	66.12	100.0	92
8. Number of cylinders[b]	97.16	5.94	13.02	3.03	5.0	35.01	0	4.91	24.44	25
9. Number of unidentified fragments[c]	1594		(161)	151	55	869	135	43	23	48
Ratios										
10. Articulator ends/MNI	12.06	4.38	6.43	3.14	6.76	7.98	5.27	7.55	8.5	12
11. Missing ends/MNI	11.91	19.62	17.57	20.86	17.24	16.02	18.73	16.45	15.5	11
12. Cylinders/MNI	6.16	2.0	8.51	1.59	6.76	2.68	0	1.22	3.79	3
13. Unidentified fragments /MNI	101.14	(35.78)		79.06	174.32	66.54	90.0	10.7	3.57	6

[a]Information from M. D. Leakey 1971: 276, Table 8, as summarized in Table 6.03.
[b]Information from M. D. Leakey 1971: 276, Table 8, (limb-bone shafts).

flated cylinder count per MNI. As in the previous comparison, the *Zinj* floor stands out uniquely with an extreme count of indeterminate fragments.

The final breakage comparison concerns the relationship between destruction as measured by the percentage of the total articulator ends that occur as components of complete bones plotted against the number of cylinders per MNI. This comparison is shown in Figure 6.03.

What is immediately clear (compare to Figure 4.61) is that the same general relationship is observed among the Olduvai assemblages as was noted among the animal-gnawed assemblages. As the percentage of the total articulator ends present that occur on complete bones goes up, the number of cylinders per MNI decreases. This relationship supports the argument that destruction is related to cylinder production and, as noted in our earlier studies, this is only true for assemblages destroyed and broken by gnawing animals. The actual distribution shown in Figure 6.03 is suggestive of the operation of additional conditioners among the Olduvai assemblages than was noted among the wolf kill materials. There are at least two, and probably three, subpopulations indicated in Figure 6.03; one subpopulation exhibits a higher cylinder-to-complete-bone relationship, which appears to be slightly curvilinear. This could happen if there was some body size or bone density related bias manifest in the indicated assemblages. That is, broken bones with detached articulator ends might be either destroyed or sorted relative to both complete bones and cylinders, since the shafts of bones are very dense. This would have the effect of increasing the number of cylinders

BLE 6.02—*Continued*

FLK North Levels					FLK Levels			FLK NN Levels			DK Levels		
/2 .1)	3 (12)	4 (13)	5 (14)	6 (15)	13 (16)	15 (17)	22 *Zinj* (18)	1 (19)	2 (20)	3 (21)	I-1 (22)	I-2 (23)	I-3 (24)
5.32	15.86	8.0	16.26	5.56	3.96	5.04	13.07	3.97	6.22	5.48	2.42	14.19	8.48
4.2	180.10	96.76	196.28	84.72	33.16	40.9	137.26	33.56	76.68	45.72	11.64	140.96	88.86
5.12	44.78	18.92	77.27	57.12	4.04	13.72	26.62	10.85	40.42	13.32	0	47.7	40.2
.27	.25	.20	.39	.67	.12	.34	.19	.32	.54	.29	0	.66	.55
7.68	380.64	192.0	390.24	133.44	95.04	120.96	313.68	95.28	149.28	131.52	58.08	340.56	203.52
.45	.53	.50	.50	.37	.65	.66	.56	.65	.49	.65	.80	.59	.56
3.49	200.54	95.24	193.96	48.72	61.88	80.06	176.72	61.72	72.6	85.8	46.44	199.66	114.66
1.17	54.11	32.72	51.08	7.94	19.45	16.94	25.58	4.05	23.66	29.04	0	11.92	11.3
'6	321	202	1039	85	42	113	2420	15	40	119		1454[c]	
3.14	11.36	12.1	12.06	15.24	8.37	8.12	13.63	8.45	12.33	8.34	4.81	9.93	10.48
0.86	12.64	11.9	11.94	8.76	15.63	15.8	10.38	15.55	11.67	15.66	19.19	14.07	13.52
3.71	3.41	4.09	3.14	1.43	4.84	3.36	1.96	1.02	3.8	5.3		.84	1.33
6.13	20.24	25.25	63.9	15.29	10.61	22.42	185.16	1.15	3.78	6.43	21.72		

ɔrmation from M. D. Leakey 1971: 256, Table 3, (indeterminate fragments).
tal for TK floors.
al for DK levels 1–3.

and complete bones and hence the proportions of complete bone articulator ends relative to those occurring as broken components within an assemblage. Both conditions would have the effect of displacing upward the distribution of cylinders per MNI relative to the number of articulator ends attached to complete bones.

These three comparisons (Figures 6.01–6.03) of the relationships between breakage indicators among the Olduvai assemblages with similar indicators among the control assemblages are quite revealing. First, the relationships in Figure 6.01 illustrate that some factors are conditioning the relationships between unidentified fragments and the number of missing articulator ends other than simple consumption of articulator ends, as was the case in the Alaskan materials. It is suggested that missing articulator ends in the Olduvai assemblages have either been transported away by predator–scavengers or introduced by the same from kill locations. All the Olduvai assemblages except FLK North Levels 1/2, FLK North Level 5, MNK Main, FC West, Floor and Tuff, and MNK H. 13 appear to be "selected" assemblages, in that parts have been either deleted from or selectively introduced to the location. Both BK II and the *Zinj* floor at FLK 22 stand out as extraordinary cases in that the production of unidentified fragments is very high relative to the number of articulator ends present; that is, abnormal destruction seems to have taken place at these sites.

In the analytical comparison illustrated in Figure 6.02, where the numbers of unidentified fragments is

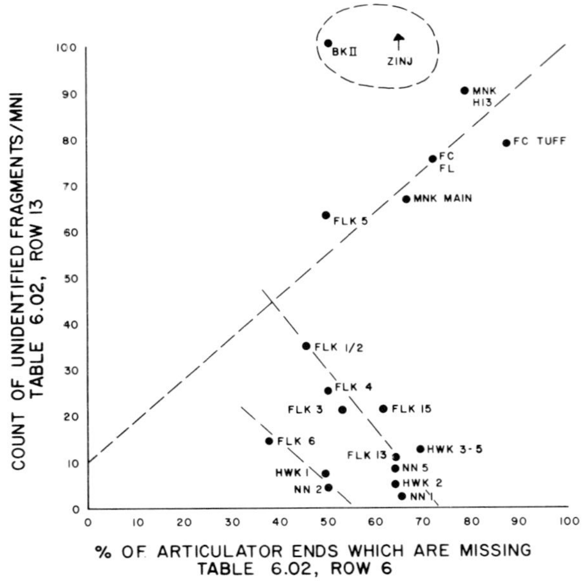

FIGURE 6.01. ***Relationship between the percentage of expected articulator ends which are missing and the count of unidentified fragments per MNI for sites in Olduvai Gorge.***

displayed against the number of limb-bone shafts, believed to be cylinders, a distribution somewhat different from that observed in the highly controlled wolf kill material is seen. From one perspective, there appears to be a set of Olduvai cases that actually follow the distribution seen in the wolf data; these include MNK H. 13, FC West, Tuff, MNK Main, FLK North Levels 1–6, FLK Level 13, FLK NN Level 1, and HWK East Levels 3–5. A subset of sites consisting of HWK East Levels 1 and 2, FLK NN Levels 2 and 3, and FLK Level 13 have relatively high frequencies of cylinders with corresponding low frequencies of unidentified splinters. This could occur if one was observing an animal-transported and gnawed assemblage minus a defecation area. Also standing out are the *Zinj* floor at FLK 22, BK II and FC West, Floor, where there are very high frequencies of unidentified fragments as well as relatively high frequencies of cylinders. This could occur if there was high breakage of nonlimb bone parts such as skulls, pelvic parts, and vertebrae, or if there was "hammer" breakage of some bones as well as some activity of gnawing animals as might occur in assemblages with low integrity. In any event, there does appear to be some internal differentiation among the Olduvai assemblages not seen among the control wolf kill assemblages, *suggesting more complicated formation processes.*

The final analytical display of the Olduvai materials (Figure 6.03) supports the foregoing conclusion but adds more information. First, there appear to be *several parallel* distributions that betray a similar relationship between variables but differing relative strengths for their operation in different situations.

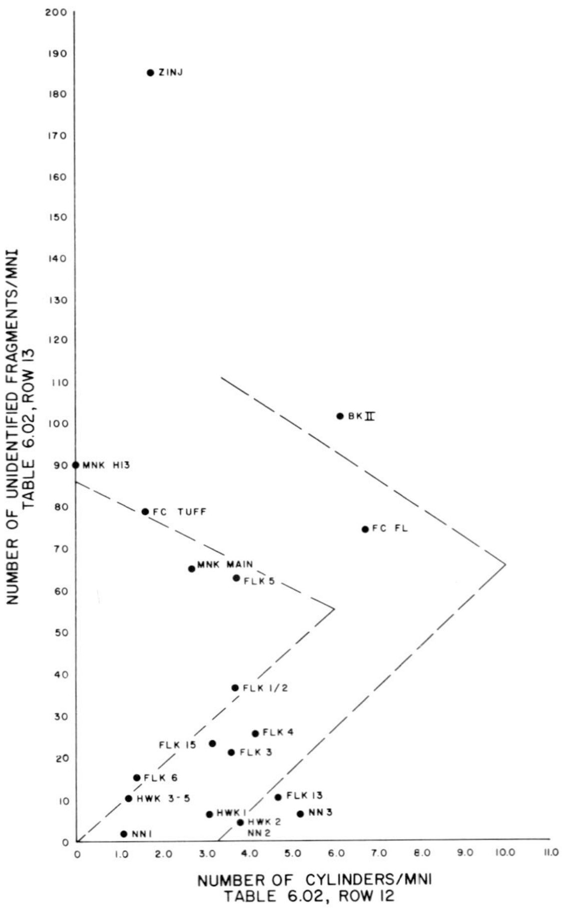

FIGURE 6.02. ***Relationship between the count of unidentified fragments per MNI and the number of cylinders per MNI for sites in Olduvai Gorge.***

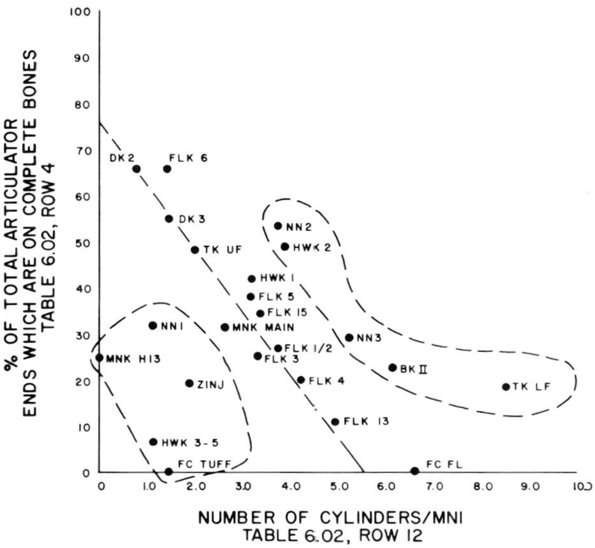

FIGURE 6.03. ***Relationship between the percentage of all articulator ends represented by complete bones and the number of cylinders per MNI for sites in Olduvai Gorge.***

Nevertheless, each is a negative relationship demonstrating that destruction of bones is related to cylinder production and, as noted earlier, this is only true for assemblages broken and gnawed by animals. The actual factors conditioning the relative "power" of this relationship as indicated by the parallel distributions are a matter for speculation at this point.

The *diagnostic* tests for bone-breakage patterning all support the view that the pattern of breakage (given that limb-bone shafts in Leakey's tables can be equated with cylinders) *is consistent with the patterning produced by gnawing animals*. Nevertheless, all tests show that the actual formation processes are more complicated than those observed among the wolf kills, with the clear implication that other conditions were affecting the relationships investigated. What these conditions might be was not clear from the tests. It was suggested that in the case of the comparison shown in Figure 6.01 transport of anatomical parts was modifying the distribution. In the case of the comparison shown in Figure 6.02, inordinate breakage of bones was indicated for some sites and high frequencies of cylinders for others, but no clear-cut "causes" were suggested. In the final comparison, animal destruction was clearly indicated but the levels of destruction were different among several subsets of the Olduvai sample. Nevertheless, the majority of the Olduvai sites fell on the same linear and negative distribution as did the wolf kill materials. These tests support the argument that the breakage patterning is consistent with that produced by gnawing animals. In the case of comparison 1 (Figure 6.01), the deviation away from the control materials is itself consistent with animal behavior, namely, transporting anatomical parts. Only in the case of comparisons 2 and 3 are factors indicated that are not recognizable at this point.

Evaluating the Degree of Destruction Suffered by the Olduvai Assemblages

In the previous comparisons, I was concerned with the degree to which patterns of bone destruction and breakage were indicative of the actions of nonhominid predator–scavengers versus tool-using hominids as the agents responsible for the assemblages. In this section I am concerned with the relative levels of destruction different assemblages have suffered. In the previous section it was found that comparisons between the proximal and distal ends of the humerus and the tibia permitted a fairly reliable estimate of assemblages that had suffered destruction. Figure 6.04 illustrates the relationships between proximal and distal ends of the humerus, and Figure 6.05 illustrates the frequency relationship between the proximal and distal tibia. It seems quite clear that most of the assemblages have suffered destruction. However, a few (DKI-1, FC West, Tuff, MNK H. 13, HWK East Level 2) seem to have suffered little if any destruction. FLK North Level 6 and *Zinj* appear to have suffered some but not really high levels of destruction as is characteristic of most assemblages. In the tibia comparisons, sites TK L.F. FLK NN Level 2, FLK North Level 6, DKI-3, and FLK Level 13 exhibit a similar profile of destruction, which seems much less than that of other assemblages. It is interesting that FLK NN Level 2 and

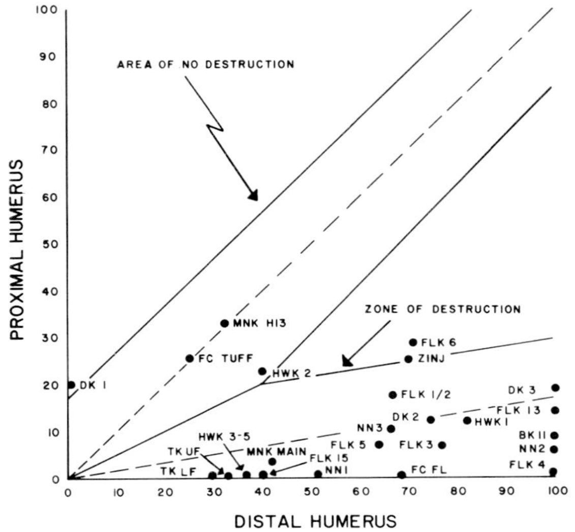

FIGURE 6.04. ***Relationship between the frequencies of proximal humerus and distal humerus for sites in Olduvai Gorge.***

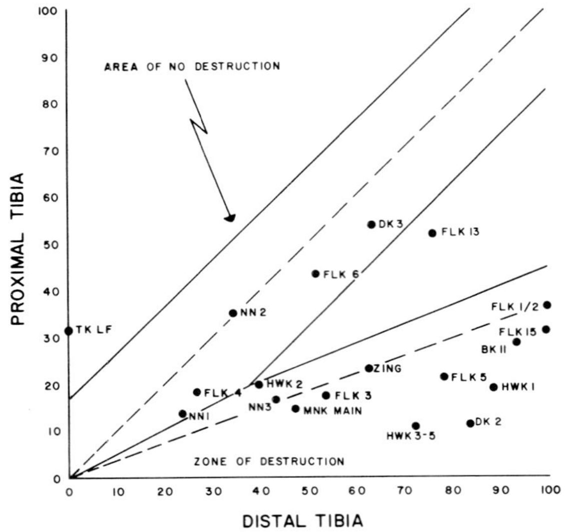

FIGURE 6.05. ***Relationship between the frequencies of proximal tibia and distal tibia for sites in Olduvai Gorge.***

FLK North Level 6 were noted earlier as sites where little evidence of hominids was present, and NN Level 3 was described as a site where undisturbed carcasses are seemingly well documented. Lower frequencies of distal tibiae in such situations could result from transport away of lower leg parts when the tibia is broken through the shaft. The estimates of destruction will help guide the need for correcting or "reconstructing" assemblages using survival percentages as a way of obtaining an impression of what was present prior to destruction.

Assemblage Composition—Olduvai Gorge

In treating the information from the important sites of Olduvai, several characteristics of these data must be cited as extremely limiting. For instance, frequency data on anatomical parts are given in Table 8 of Leakey (1971:276), but there is no information provided as to what was in fact counted. We do not know to what observations the tabulation refers. Is the information referable to the frequency of bone fragments assignable to the tabular classes, or is the frequency information a strict tabulation of the elements listed? For example, when the proximal humerus is listed, do the frequencies refer to the numbers of proximal humeri, or does it simply mean the numbers of fragments of proximal humeri? If the former, some analysis relative to the differential survival, consumption, and transport of anatomical parts is possible. If the latter, the situation is horribly confused, since breakage patterns condition the frequencies and no such direct analysis can be conducted. It is my guess that in fact the situation is even more confused, that for a category such as complete tibiae the number of complete tibiae was actually reported, but for a category like proximal femur the fragments of proximal femora were counted. There are other serious problems with the faunal information reported from Olduvai. There is one category that lists "isolated teeth," and two others, mandible and maxilla, which most certainly include teeth. Since I am all but certain that the latter two categories tabulate fragments of the anatomically designated parts, there is no way of knowing how many teeth remain seated in those elements. Since it seems likely that fragments are tabulated and the latter frequencies are conditioned by breakage as well as by the actual number of anatomical parts present, there is no way of directly converting the information given into meaningful statements about the frequencies of head parts present in the assemblages. That is, neither teeth, mandibles, nor maxillae may be meaningfully estimated from the data given.

Despite these truly impressive problems, I have transformed the data from Table 8 in Leakey (1971:276) into MNI estimates. That is, I have converted the percentage data back into counts and then divided the count by the number of parts for each category present in the anatomy of an ungulate.[4] Several steps were needed in many cases in order to accomplish this conversion. For instance, in obtaining a count for the proximal humerus I first had to multiply the total for the site by the percentage for proximal humerus; then I had to do the same thing for complete humerus and add the two results in order to obtain a total count for the proximal ends present. More difficult conversions were needed for teeth, where I have used estimators to approximate the number of teeth included in mandibular and maxillary fragments. To obtain the estimators, I consulted the complete inventories of anatomical parts that I made for the fauna from the Mousterian site of Combe Grenal in France. In these inventories each mandibular and maxillary fragment was listed together with an inventory of the teeth still remaining in the fragment. I found that for the site of Combe Grenal as a whole, including parts from reindeer (*Rangifer tarandus*) and red deer (*Cervus elaphas*), there was an average of 2.14 teeth remaining in each mandibular fragment and 2.50 teeth within maxillary sockets. Thus, each value of the mandible and maxilla was multiplied by 2.14 and 2.5 respectively to obtain an estimate of the number of teeth represented by such fragments. This total was then added to the total for isolated teeth to obtain an estimate for the total number of teeth present. The latter value was divided by 34, the total number of teeth characteristic of most ungulates, to obtain some estimate for the numbers of heads represented in the assemblage.

[4]For a discussion of MNI calculations, see L. R. Binford (1978b:69–72).

Table 6.03 summarizes all the information deemed usable for monitoring assemblages composition among the Olduvai sites, as expressed in MNI estimates. Each site has two columns. The first presents the MNI estimate for each anatomical part, and the second presents the percentage ratio (that is, the MNI for the part divided by the MNI for the most frequent part). Such values are arrayed between 0 and 100 but are not "percentages" in the sense that the array is constrained. That is, no degree of freedom is lost by expressing the relative frequencies in ratio form, but the problem of differential sample size is elminated since all cases are standardized as a ratio scale distributed between 0 and 100.

For comparative purposes, the mean values for large and small animals remaining at kills and introduced to dens (Table 5.08) was converted to the same format as the Olduvai data and multiplied by the survival percentages given in Table 5.04. This permits the direct comparison between populations of fauna modified by animals and those remaining in the deposits at Olduvai Gorge. (See Table 6.04.)

It was decided that, rather than proceed through a case-by-case comparison, some grouping strategy might be both timesaving and informative. For this reason, a factor analysis in both the "R" and "Q" modes was performed on the information presented in Table 6.03. To eliminate the problem of known curvilinear relationships, the data as presented in Table 6.03 were transformed to log values for each entry. In addition, all values less than 1 were raised to a value of 1 (this includes zeros) so that no negative log values appeared in the matrix. In the case of both "R" and "Q" mode analysis, squared multiple correlation coefficients were placed in the diagonal cells rather than leaving them unaltered. A standard eigenvalue cutoff value of 1.0 was used. A varimax rotation was performed and a Biomed program (BMD P4M, GREGION—195K) was used. Table 6.05 presents the factor loadings for the "R" mode of analysis for the four factors extracted.

A knowledge of the relationships between anatomical parts and their relative survival potential (see L. R. Binford and J. B. Bertram 1977) leads me to interpret the first factor as a *survivorship factor*; that is, the parts listed (scapula, proximal radio-cubitus, promimal metatarsal, proximal tibia, distal femur, distal tibia, proximal metacarpal, vertebrae, and ribs)

TABLE 6.03
Olduvai Gorge Data

	BK II		***TK*** Upper Floor		***TK*** Lower Floor		***FC West*** Tuff		***FC West*** Floor	
Anatomical part	MNI (1)	% (2)	MNI (3)	% (4)	MNI (5)	% (6)	MNI (7)	% (8)	MNI (9)	% (10
Teeth	15.69	100.0	2.12	.71	1.52	100.0	1.91	100.0	.74	100.
Pelvis	13.13	83.0	2.97	1.0	1.53	100.0	1.0	52.0	.5	68
Scapula	13.79	88.0	1.0	.34	1.0	65.0			.5	68.
Ribs	6.1	39.6	.61	.21	.15	10.0			.12	16
Vertebrae	3.59	23.0	.04	.1	.19	12.0	.04	2.0	.08	11.
Proximal humerus	1.3	8.0					.5	26.0		
Distal humerus	15.76	100.0	1.0	.34	.48	31.0	.5	26.0	.5	68.
Proximal radio-cubitus	8.54	54.0			.48	31.0			.5	68
Distal radio-cubitus	4.6	29.0	.5	.17	.48	31.0	.5	26.0		
Proximal metacarpal	10.51	67.0	1.5	.51	1.0	65.0				
Distal metacarpal	6.57	42.0	1.0	.34			1.0	57.0		
Proximal femur	4.6	29.0								
Distal femur	4.6	29.0	1.5	.51						
Proximal tibia	4.6	29.0			.48	31.0				
Distal tibia	14.45	92.0							.5	68
Proximal metatarsal	12.48	79.0	.5	.17	1.0	6.5			.5	68
Distal metatarsal	7.22	46.0	.5	.17	1.0	6.5	.5	26.0	.5	68

	FLK Levels 6		FLK Levels 13		FLK Levels 15		FLK Levels Zinj.	
Anatomical part	MNI (29)	% (30)	MNI (31)	% (32)	MNI (33)	% (34)	MNI (35)	% (3
Teeth	5.37	97.0	1.7	43.0	1.36	27.0	13.07	100
Pelvis	3.97	71.0	1.51	38.0	.44	9.0	11.26	86
Scapula	3.44	62.0	2.52	64.0	2.48	49.0	10.74	82
Ribs	3.38	61.0	.19	5.0	.54	11.0	8.97	69
Vertebrae	3.42	62.0	.66	17.0	.23	5.0	3.42	26
Proximal humerus	1.59	29.0	.51	13.02	0	0	3.58	27
Distal humerus	3.97	71.0	3.96	100.0	1.97	39.0	9.21	70
Proximal radio-cubitus	3.44	62.0	2.52	64.0	1.97	39.0	9.21	70
Distal radio-cubitus	3.71	67.0	0	0	1.02	20.0	3.07	23
Proximal metacarpal	4.5	81.0	2.52	64.0	3.94	78.0	5.63	43
Distal metacarpal	5.03	90.0	0	0	0	0	3.58	27
Proximal femur	1.59	29.0	0	0	0	0	3.07	23
Distal femur	3.18	57.0	1.01	26.0	0.51	10.0	4.09	31
Proximal tibia	2.38	43.0	2.02	51.0	1.54	31.0	3.07	23
Distal tibia	2.91	52.0	3.03	77.0	5.04	100.0	8.19	63
Proximal metatarsal	4.50	81.0	1.01	26.0	4.46	88.0	8.19	63
Distal metatarsal	5.56	100.0	0	0	0	0	7.68	59

ABLE 6.03—*Continued*

MNK				HWK East Levels						FLK North Levels							
Main		H.13		3–5		2		1		½		3		4		5	
MNI (11)	% (12)	MNI (13)	% (14)	MNI (15)	% (16)	MNI (17)	% (18)	MNI (19)	% (20)	MNI (21)	% (22)	MNI (23)	% (24)	MNI (25)	% (26)	MNI (27)	% (28)
.06	100.0	.92	61.0	3.14	78.0	6.45	100.0	3.71	47.0	13.93	39.0	8.05	38.0	6.78	85.0	10.78	66.0
.12	62.0	.49	33.0	.89	22.0	3.9	60.0	5.44	69.0	34.31	97.0	21.0	100.0	6.55	82.0	12.77	79.0
.68	82.0			4.02	100.0	1.3	20.0	2.44	31.0	27.25	77.0	6.07	29.0	4.36	55.0	9.87	61.0
.54	12.0	.23	15.0	.15	4.0	1.12	17.0	1.27	16.0	7.84	22.0	2.44	12.0	.34	4.0	1.38	8.0
.13	16.0	.12	8.0	.73	18.0	1.88	29.0	1.5	19.0	6.05	17.0	2.85	14.0	2.23	28.0	1.92	12.0
.43	3.0	.49	33.0	0	0	1.4	22.0	.94	12.0	6.06	17.0	1.4	7.0	0	0	1.16	7.0
.55	42.0	.49	33.0	1.45	36.0	2.6	40.0	6.38	81.0	23.21	66.0	15.86	76.0	8.0	100.0	10.45	64.0
.55	42.0	.49	33.0	1.45	36.0	2.08	32.0	5.70	72.0	22.2	63.0	7.93	38.0	4.73	59.0	8.13	50.0
.85	29.0			1.45	36.0	1.82	28.0	5.70	72.0	14.13	40.0	3.73	18.0	3.27	41.0	4.07	25.0
.83	52.0	.49	33.0	1.9	47.0	5.2	81.0	4.88	67.0	22.2	63.0	12.13	58.0	5.09	64.0	13.35	82.0
.13	39.0	1.5	100.0	1.9	47.0	3.64	56.0	3.94	50.0	19.17	54.0	8.87	42.0	6.55	82.0	7.55	46.0
.56	20.0			.89	22.0	.52	8.0	1.50	19.0	9.08	26.0	4.2	20.0	2.91	36.0	4.65	29.0
.14	16.0			.45	11.0	.52	8.0	.94	12.0	13.12	37.0	5.13	24.0	3.64	46.0	6.39	39.0
.14	16.0			.45	11.0	1.3	20.0	1.50	19.0	12.11	34.0	3.73	18.0	1.46	18.0	3.49	21.0
.98	46.0			2.90	72.0	2.6	40.0	6.94	88.0	35.32	100.0	11.2	53.0	2.18	27.0	12.77	79.0
.97	69.0	.49	33.0	1.45	36.0	2.6	40.0	7.88	100.0	34.31	97.0	8.87	42.0	7.64	96.0	16.26	100.0
.99	23.0			.89	22.0	3.12	48.0	2.07	26.0	21.19	60.0	7.0	33.0	2.91	36.0	9.87	61.0

FLK NN Levels						DK Levels					
1		2		3		I-1		I-2		I-3	
NI 7)	% (38)	MNI (39)	% (40)	MNI (41)	% (42)	MNI (43)	% (44)	MNI (45)	% (46)	MNI (47)	% (48)
91	73.0	3.64	59.0	4.37	80.0	2.42	100.0	14.19	100.0	7.77	92.0
1	63.0	2.41	39.0	3.33	61.0	.95	39.0	9.78	69.0	4.4	52.0
46	37.0	4.81	77.0	5.48	100.0	2.00	83.0	9.76	69.0	4.08	48.0
16	4.0	2.81	45.0	3.08	56.0	.81	33.0	2.79	20.0	2.27	27.0
38	22.0	.60	10.0	3.31	60.0	.73	30.0	2.63	19.0	1.93	23.0
	0	.40	6.0	.48	9.0	.48	20.0	1.63	11.0	1.57	19.0
93	51.0	6.22	100.0	3.57	65.0	0	0	10.84	76.0	8.48	100.0
46	37.0	5.82	94.0	5.00	91.0	.48	20.0	8.68	61.0	2.5	29.0
	0	2.41	39.0	1.43	26.0	.48	20.0	5.42	38.0	2.5	29.0
43	61.0	6.22	100.0	3.81	70.0	.48	20.0	8.13	57.0	4.09	48.0
97	24.0	5.42	87.0	1.43	26.0	0	0	7.05	50.0	2.52	26.0
46	37.0	0	0	1.43	26.0	.48	20.0	2.17	15.0	3.46	41.0
97	24.0	0.40	6.0	1.43	26.0	.95	39.0	3.79	27.0	5.34	62.0
9	12.0	2.21	36.0	0.95	17.0	0	0	1.63	11.0	4.4	52.0
97	24.0	2.21	36.0	2.38	43.0	0	0	11.92	84.0	5.34	62.0
97	100.0	5.22	84.0	0.95	17.0	1.52	63.0	5.42	38.0	5.65	67.0
0	76.0	1.81	29.0	0	0	0.95	39.0	3.8	27.0	4.71	56.0

TABLE 6.04
Behavioral Models for Olduvai Assemblages[a]

	Ravaged								Scavenged			
	Large prey, kill model		Large prey, den model		Small prey, kill model		Small prey, den model		Large prey, kill model		Transport model	
Anatomical part	MNI (1)	Col. 1 68.2 (2)	MNI (3)	Col. 3 39.7 (4)	MNI (5)	Col. 5 77.9 (6)	MNI (7)	Col. 7 70.3 (8)	MNI (9)	Col. 9 72.2 (10)	MNI (11)	Col. 11 89.0 (12)
Teeth	64.5	95.0	39.7	100.0	61.0	78.0	70.3	100.0	72.2	100.0	89.0	100.0
Pelvis	68.2	100.0	20.2	51.0	77.9	100.0	35.2	50.0	52.6	73.0	46.3	52.0
Scapula	27.9	41.0	12.7	32.0	45.1	58.0	21.3	30.0	38.3	53.0	30.8	35.0
Ribs	9.1	13.0	2.0	5.0	5.8	7.0	7.6	11.0	23.5	33.0	12.7	14.0
Vertebrae	23.4	34.0	5.0	13.0	14.5	19.0	8.9	13.0	41.0	57.0	24.4	27.0
Proximal humerus	6.2	9.0	5.0	13.0	3.2	4.0	3.9	6.0	33.1	46.0	11.9	13.0
Distal humerus	24.2	36.0	35.4	89.0	38.3	49.0	23.8	34.0	38.4	53.0	41.3	46.0
Proximal radio-cubitus	14.9	22.0	27.3	69.0	28.8	37.0	18.1	26.0	39.5	55.0	34.5	39.0
Distal radio-cubitus	12.9	19.0	24.5	62.0	14.6	19.0	14.2	20.0	27.4	38.0	41.5	47.0
Proximal metacarpal	10.7	16.0	30.5	77.0	5.5	7.0	17.1	24.0	35.2	49.0	33.4	38.0
Distal metacarpal	8.7	13.0	32.3	81.0	10.9	14.0	23.9	34.0	29.5	41.0	32.2	36.0
Proximal femur	11.6	17.0	5.6	14.0	19.3	25.0	6.9	10.0	.6	1.0	.7	1.0
Distal femur	11.0	16.0	13.6	34.0	10.3	13.0	15.9	23.0	0	0	0	0
Proximal tibia	9.5	14.0	8.2	21.0	14.9	19.0	14.8	21.0	35.9	50.0	18.3	21.0
Distal tibia	31.8	47.0	35.3	89.0	32.6	42.0	19.7	28.0	41.3	57.0	50.6	57.0
Proximal metatarsal	15.6	23.0	20.3	51.0	15.9	20.0	28.7	41.0	42.1	58.0	35.2	40.0
Distal metatarsal	12.5	18.0	22.5	57.0	11.3	15.0	23.8	34.0	33.8	47.0	33.8	38.0
Derivation												
Composition	$\bar{X}$ of large prey, kills, × SP		$\bar{X}$ of large prey, dens, × SP		$\bar{X}$ of small prey, kills, × SP		$\bar{X}$ of small prey, dens, × SP		$\bar{X}$ of large prey, kills × IGUI		IGUI (caribou) × SP	
Reference	Table 5.08, col. 2, × Table 5.04, col. 7		Table 5.08, col. 4, × Table 5.04, col. 7		Table 5.08, col. 6, × Table 5.04, col. 4		Table 5.08, col. 8, × Table 5.04 col. 4		Table 5.08, col. 2, × Table 5.08, col. 17		Table 5.08, col. 17, × Table 5.04, col. 7	

[a]SP = survival percentage; IGUI = inverse general utility index.

are all parts that are likely to vary considerably as a function of the level of destruction experienced by an assemblage and, in addition, are parts that roughly scale from high to low in the range of bone densities characteristic of young and old animals. Factor 2 appears to be those parts that most likely betray residual remains at an animal site, whereas factor 3 groups parts most likely to co-occur by virtue of having been transported away from animal kills—in short, parts diagnostic of transported assemblages. Factor 4 appears to be those elements that covary in the presence of the others listed that most consistently monitor large body size or old animals. This factor is interpreted as an age indicator for the prey animals represented in an assemblage.

A further warrant for the foregoing interpretation is provided by the factor scores as distributed among the cases that provided the data matrix. Table 6.06 summarizes the factor scores for the factor solution summarized in Table 6.05.

Of particular importance is that factor scores for factor 1 interpreted as a survivorship factor exhibit

TABLE 6.05
Sorted Rotated Factor Loadings (Pattern) for "R" Mode Analysis of Olduvai Sites

Anatomical part	Number	Factors 1	2	3	4
Scapula	23	.877	.351	0	0
Proximal radio-cubitus	28	.803	.396	0	.303
Proximal metatarsal	36	.746	.414	0	0
Proximal tibia	34	.729	0	0	.370
Distal femur	33	.718	0	.585	0
Distal tibia	35	.699	0	0	.364
Proximal metacarpal	30	.694	0	.264	.250
Vertebrae	25	.644	.527	.317	0
Ribs	24	.501	.421	0	0
Teeth	21	0	.837	0	0
Pelvis	22	.256	.808	.298	0
Distal metacarpal	31	0	0	.722	.436
Proximal femur	32	.478	0	.678	0
Distal radio-cubitus	29	.299	0	.518	0
Distal humerus	27	.414	.290	0	.848
Proximal humerus	26	0	.250	.443	0
Distal metapodial	37	0	.499	.327	0

no very consistent pattern with respect to the four known cases of destroyed assemblages included within the matrix, the den and kill models for large and small animals (rows 25–28, Table 6.05). It will be recalled that all four cases were modified by survival percentages prior to being used in the analysis. On the other hand, factor 2, designated a diagnostic for kill sites, exhibits some of the highest factor loadings on the two kill site cases and, equally interesting, shows low or negative values on the den assemblages. Likewise, factor 3, interpreted as diagnostic of transported assemblages, exhibits consistently high scores for the two control den assemblages and a negative value on the small kill assemblages. This nice fit is clouded by an ambiguous value on the large body-size kill. For the body size factor (4) the only positive score is on the large den sample, which is certainly the most distinctive body-size assemblage.

I am confident that the "R" mode analysis has recognized meaningful patterning and that the interpretation of the underlying structure is very close to being accurate. It now is of some interest to invert the data matrix and perform a "Q" mode analysis for purposes of clustering the cases and thereby permitting the recognition of assemblage types for the Olduvai sites. Table 6 07 summarizes the factor loadings that resulted from the "Q" mode analysis.

As can be seen, a five-factor solution was obtained with the majority of cases loading on the first factor. Not only do large numbers of Olduvai cases group on factor 1, but so do all the model data cases, the large and small body-size examples of both animal kills and dens. What the latter have in common is that they have been equally transformed by the survival percentages as a means of estimating what they would look like had they been ravaged by destructive agents. By extension, and by virtue of the "R" mode results factor 1 is interpreted as the survivorship cluster. However, a little more information is carried in that the model cases for dens (transported assemblages) exhibit consistently higher loadings than do the control cases for kills. This most certainly means that the Olduvai cases clustered with high loadings on factor 1 are transported; that is, they are most analogous to the animal den assemblages of the control cases. These are FLK North Levels, 3, 4, and 5, HWK East Levels 3–5, MNK Main, FLK North and ½, and BK II.

Factor 2 is the most interesting in that it exhibits moderate loadings on the den assemblage models and zero values on the kill models, suggesting that these are transported assemblages but differ from those of factor 1 in that they are not heavily modified by destructive agents. Grouped on this factor are HWK East Level 2, FLK NN Level 2, MNK H. 13, FLK North Level 6, and FC West, Tuff (compare this to undestroyed cases indicated in Figures 6.04 and 6.05).

Factor 3 exhibits no meaningful loadings on the control cases of large and small animal dens and kills. Assemblages from FLK Levels 13 and 15 and FLK NN Level 3 are most discretely loaded on this factor. No obvious meaning can be given to this grouping at this time; however, its potential significance is interesting since it obviously is telling us about something *not anticipated by the animal behavior models.*

Factor 4 exhibits moderate loadings on the models of animal kills with the highest loading on the large kill data, which are clearly most distinctive.

TABLE 6.06
Factor Scores for "R" Mode Analysis of Olduvai Sites [a]

	Number	Chi-square values			Factors			
		(df = 17)	(df = 4)	(df = 13)	1	2	3	4
HWK 3–5	1	1.068	.369	1.283	.921	−.422	−.726	−.201
HWK 2	2	.903	.321	1.082	−.668	.544	.712	−.512
HWK 1	3	.744	.237	9.901	.551	.037	−.379	.836
FLK ½	4	.650	.292	.761	1.035	.028	.244	−.161
FLK 3	5	.699	.490	.764	−.533	−.166	1.238	.462
FLK 4	6	1.146	.412	1.371	.530	−.099	.847	.782
FLK 5	7	.376	.119	.455	.576	−.004	.351	.035
FLK 6	8	.763	.524	.836	.737	.030	1.185	.212
FLK 13	9	1.428	1.489	1.410	1.421	−.865	−1.949	.606
FLK 15	10	1.310	.806	1.465	.412	−1.435	−1.023	.297
Zinj	11	.442	.136	9.536	.500	.315	.325	.283
NN 1	12	1.011	.044	1.308	.022	−.003	.377	−.232
NN 2	13	1.146	.242	1.424	.633	−.114	−.400	.778
NN 3	14	1.098	.124	1.398	.502	.039	.393	.266
DK I-1	15	1.452	3.975	.676	.217	1.117	−.136	−4.092
DK I-2	16	.422	.162	.502	.035	.100	.591	.540
DK I-3	17	.967	.611	1.076	.839	−.564	.409	1.175
BK II	18	.312	.219	.341	.356	.237	.601	.567
TK UF	19	1.471	5.377	.268	−1.691	−3.859	−.757	−1.495
TK LF	20	1.460	1.501	1.447	−.208	1.077	−2.116	.330
FC TUFF	21	1.402	3.456	.771	−3.580	.354	.765	.942
FC FL	22	1.452	2.617	1.094	−.416	1.356	−2.801	.446
MNK MAIN	23	.393	.222	.446	.319	.538	.196	−.740
MNK H13	24	1.414	1.273	1.458	−1.808	.533	−.657	.962
LG KILL MODEL	25	.638	.707	.617	−.613	.873	.889	−1.206
LG DEN MODEL	26	1.079	.339	1.307	.290	−.453	.463	.952
SM KILL MODEL	27	.982	.218	1.218	.170	.587	−.036	−.753
SM DEN MODEL	28	.770	.718	.786	−.548	.220	1.397	−1.082

[a] Estimated factor scores and Mahalanobis distances (chi squares) from each case to the centroid of all cases for original data (df = 17), factor scores (df = 4), and their difference (df = 13). Each chi-square value has been divided by its degree of freedom.

Olduvai sites grouped on this factor are FLK 22, or the *Zinj* floor, FC West, Floor, and DKI-1.

Factor 5, like factor 4, exhibits moderate loadings on the model kill assemblages, differing only in the slightly higher values for the small animal contributions. Only FLK NN Level 3 is loaded on this factor.

As a source of still further information, the factor scores for the above "Q" mode solution are summarized in Table 6.08.

Inspection of the factor scores supports the interpretations offered thus far. The suggestion that factor 3 represented patterning not modeled by the animal control data is further elucidated by recognizing that the scapula, ribs, proximal radio-cubitus, proximal metacarpal, and the bones of the tibia through the proximal metatarsal are contributing most to the definition of the factor. These are all parts that might be expected to ride if relatively large sections of leg were being transported or deleted from an assemblage. The scapula is one of the first bones to be disarticulated in a normal sequence and is one that is normally consumed relatively early in an animal consumption sequence.

The factor analysis has grouped both variables

and cases in ways that are consistent with the controls that were built into the analysis, that is, the model assemblages for prey of both large and small size remaining as residual parts at kills and transported by animals to dens. In the "Q" mode results, factor 1 is a grouping of assemblages that have suffered considerable attrition at the hands of bone-destructive agents, with a bias in this group in favor of assemblages that have been transported, or den forms of assemblage. Elements diagnostic of these assemblages are the extremities of the rear leg in particular. Factor 2 designates transported assemblages that have not suffered much attrition and are front-leg biased. Factor 3 is a grouping of assemblages in terms of some characteristic not modeled in the animal behavior cases. Factors 4 and 5 are kill or residual faunal assemblages that have not suffered heavy attrition. Factor 4 is biased toward lower rear leg parts whereas factor 5 is characterized by more upper front leg parts.

The factor solutions are robust, and the groupings of sites are, in my opinion, quite secure in terms of diagnostic criteria. This permits me to analyze several cases, each taken as representative of a class of assemblage isolated by the factor analysis. Analyzing assemblages in terms of compositional charac-

TABLE 6.07
Sorted Rotated Factor Loadings (Pattern) for "Q" Mode Analysis of Olduvai Sites

		Factors				
	Number	1	2	3	4	5
FLK 5	38	.950	0	0	0	0
FLK 4	37	.891	0	0	0	0
HWK 3–5	32	.863	0	0	0	0
FLK 3	36	.856	0	0	0	0
MNK MAIN	54	.852	0	.274	0	0
FLK ½	35	.810	0	.252	0	0
LG DEN MODEL	57	.806	.414	0	0	0
BK II	49	.803	0	.370	.306	0
SM DEN-MODEL	59	.799	.389	0	.304	0
DK I-2	47	.731	.423	.262	0	.261
NN 1	43	.726	0	0	.329	0
SM KILL MODEL	58	.701	0	0	.389	.469
DK I-3	48	.697	0	0	.284	0
HWK 1	34	.675	.484	.271	0	0
LG KILL MODEL	56	.573	0	0	.557	.517
HWK 2	33	.401	.809	0	0	0
NN 2	44	.360	.755	.428	0	0
MNK H13	55	0	.657	0	0	0
FLK 6	39	.529	.614	0	0	0
FLK 15	41	.369	0	.862	0	0
FLK 13	40	0	0	.683	0	0
NN 3	45	0	0	.643	0	.590
FC TUFF	52	0	.546	−627	0	.339
FLK 22	42	.394	.365	.325	.706	0
FC/FL	53	.418	0	0	.676	0
DK I-1	46	0	0	0	.499	0
TK UF	50	0	0	0	0	0
TK LF	51	.284	.389	.390	.366	0
	VP	10.508	3.764	3.076	2.417	1.279

TABLE 6.08
Factor Scores for "Q" Mode Analysis of Olduvai Sites

		Chi-square values			Factors				
	Number	(df = 16)	(df = 5)	(df = 11)	1	2	3	4	5
Teeth	1	.941	1.106	.866	.798	.927	−.313	1.466	1.206
Pelvis	2	.941	1.008	.911	.540	1.013	−.384	1.172	1.404
Scapula	3	.941	.781	1.014	.714	−1.138	.935	.871	−.060
Ribs	4	.941	1.553	.663	−2.033	.306	1.072	1.068	−.542
Vertebrae	5	.941	.594	.099	−1.063	−.103	−.392	.408	1.127
Proximal humerus	6	.941	1.360	.751	−2.353	.186	−.869	−.378	.098
Distal humerus	7	.941	.235	1.262	.753	.339	.360	−.064	.547
Proximal radio-cubitus	8	.941	.355	1.208	.185	.289	1.215	−.052	.081
Distal radio-cubitus	9	.941	.736	1.035	−.129	.480	−.461	−1.448	.774
Proximal metacarpal	10	.941	1.213	.818	.207	1.070	1.282	−1.308	−.644
Distal metacarpal	11	.941	1.244	.804	.593	.919	−1.478	−1.467	.052
Proximal femur	12	.941	1.594	.645	.036	−2.361	−.755	−.457	.529
Distal femur	13	.941	.490	1.146	.059	−1.378	−.170	−.314	−.468
Proximal tibia	14	.941	.339	1.215	−.461	−.791	.486	−.333	−.693
Distal tibia	15	.941	.435	1.171	−.854	−.200	.964	−.487	.197
Proximal metatarsal	16	.941	.780	1.015	.833	.109	.458	.367	−1.562
Distal metatarsal	17	.941	2.177	.379	.467	.333	−1.949	.955	−2.048

teristics as clues to both the identity of the agents and the behavior responsible for the assemblages is our next task.

FACTOR 1 ASSEMBLAGES

As has been said, these seem to have suffered considerable attrition and exhibit their greatest similarities to the den models, that is, *animal-transported assemblages*. Since the Olduvai assemblages load on the factor somewhat higher than do the den data, I immediately suspected that these were faunal scatters and were analogous to the assemblage from Alaska collected at Chandler rather than true animal dens or lairs as known from Swartklip. Inspection of Table 6.03 for the dominant characteristics of assemblages from FLK North Levels 4 and 5, the two most "typical" for the factor, shows that they are assemblages heavily dominated by lower limb parts. This fact further supports the impression that these are faunal scatters, or background distributions of anatomical parts removed from kills by animals to avoid competitive confrontations and to increase the comfort of eating. Experience with the Alaskan material showed that these separated parts were those most heavily gnawed and destroyed by animals. (See the description of wolf gnawing behavior at the time of the "second feeding" in Chapter 5.) Such heavy destruction is clearly indicated for the assemblages grouped in factor 1.

As discussed earlier, an assemblage transported by animals is most likely to be approximated by the inverse general utility index, and if subject to destruction, the latter must be modified by the survival percentages. Figures 6.06 and 6.07 illustrate the relationship between FLK North Level 5 and FLK North Levels 1/2 and the model described (values are summarized in Table 6.04) as modified by the survival percentages of caribou (medium-large sized animal).

Several things are of interest in these distributions. First, both are very similar in that the bulk of the assemblage is related to the model in a steeply linear fashion. Second, in both cases the femur and metapodials appear overrepresented, whereas teeth, representing head parts, are underrepresented. This type of situation, where a transported assemblage is biased in terms of lower extremities, was seen previously in the Chandler assemblage from Alaska.

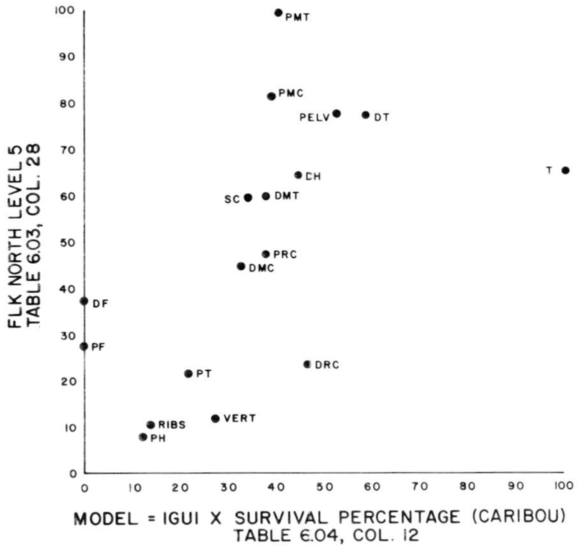

FIGURE 6.06. ***Relationship between the frequencies of anatomical parts from FLK N, Level 5, and modeled values.***

Figure 6.08 illustrates the relationship between FLK North Level 5 and the sample from Chandler in Alaska.

A strong correlation is indicated, which is positive and moderately curvilinear with only two parts falling off the curve as underrepresented in the FLK North Level 5 assemblage—the distal metatarsal and distal metacarpal. This is almost certainly related to breakage, as shown by the fact that the third highest entry for "metapodials, indeterminate" listed in the Olduvai tables (M. D. Leakey 1971:276) is for the sample from FLK 5. I interpret this to mean that broken parts of metapodials that could not be accurately assigned the front versus back legs are tabulated in the "indeterminate" category. If correct, it is not surprising to find these parts underrepresented in the "determinate" classes. I am quite confident, by virtue of facts of assemblage composition breakage and characteristic attrition, that the factor 1 faunal assemblages as grouped in Table 6.06 are *background scatters* or palimpsests of faunal elements dispersed largely by predator–scavengers. There is nothing present in these assemblages as known that suggests in any way distinctive activity by hominids.

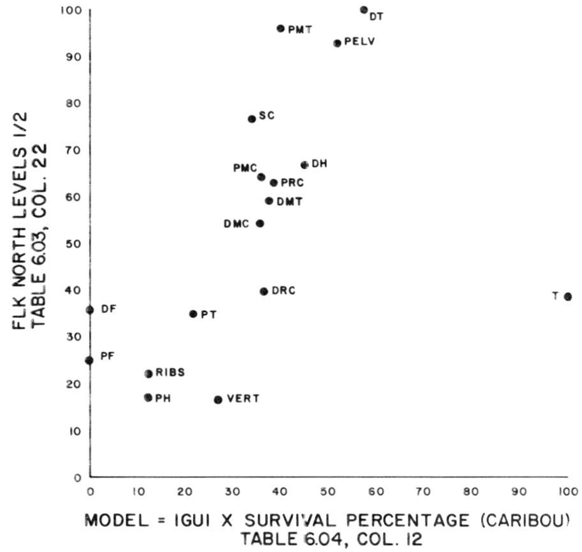

FIGURE 6.07. ***Relationship between frequencies of anatomical parts from FLK N, Levels 1 and 2, and modeled values.***

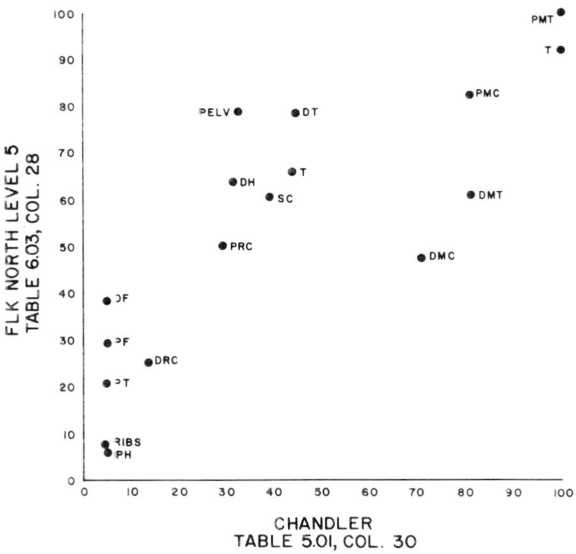

FIGURE 6.08. ***Relationship between frequencies of anatomical parts from FLK N, Level 5, and Alaskan assemblage from Chandler.***

It seems most likely that the bulk of the faunal material is behaviorally unrelated to the associated tools in these deposits.

FACTOR 2 ASSEMBLAGES

This is a most provocative grouping of assemblages. It indicates a strong similarity between both levels where relatively complete anatomical articulations were noted (FLK North Level 6 and HWK East Level 2) as well as *the only reported assemblage not associated with tools* (FLK NN Level 2). Two very small assemblages that seem to represent different contexts are also grouped on this factor: MNK H.13 and FC West, Tuff. I mention that they seem out of place in that the assemblages with the highest loadings are generally described as relatively unbroken, well-preserved faunas, whereas the latter two are described as poorly preserved and associated with many very small fragments of bone, presumably the by-product of bone weathering.

The overall character of the factor as judged by the factor scores relative to the anatomical parts indicates that these assemblages have not suffered heavy attrition by bone-destructive agents. The reports of poor preservation and many small splinters seem inconsistent with this at first consideration. However, if the fauna is poorly *preserved* as a function of exposure and weathering, then destruction is a function not necessarily of the strength of the anatomical parts but instead of their position on the surface and relative speed of burial. If the latter were in fact the circumstance, then there would be no necessary preservation related pattern of variability directly correlated with relative bone densities, as is the case when there is an active destruction of parts by gnawing animals and the like. In addition to the preservation-versus-destruction possibility, the two anomalous assemblages are very small indeed and could well be strongly biased by virtue of their small counts.

Returning to the large assemblages, they are most certainly representative of a "transported" condition. This is well illustrated by the truly remarkable fit between the assemblage from HWK East Level 2 and the inverse general utility index for medium- to large-sized animals (caribou) illustrated in Figure 6.09. The relationship is positive and curvilinear

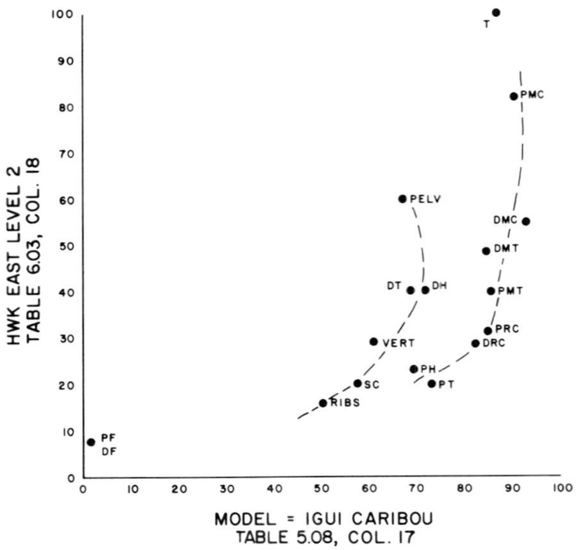

FIGURE 6.09. ***Relationship between frequencies of anatomical parts from HWK E, Level 2, and modeled values.***

such that parts of low to marginal utility are well represented, whereas parts of high to moderate utility are poorly represented.

This is precisely the model that anticipated the parts introduced to animal dens. It should also be emphasized that this model is very close to that which anticipates the parts abandoned by modern Eskimo hunters at kill sites after the "choice" parts have been removed to a base camp. In fact, in the HWK East Level 2 assemblage, the extremities of the lower legs are slightly underrepresented and, as I have indicated earlier, modern hunters often remove the lower leg parts in excess of their obvious utility as "riders" with more valuable upper leg parts. Even more provocative is the fact that these assemblages have not been ravaged. This is one of the marks of difference between an animal kill and a human kill: Animals consume at the kill and at least some destruction takes place there. The destruction may be minimal at the kill proper where mainly meat is consumed, whereas bones that are gnawed and subjected to heavy destruction are generally dragged away from the kill proper. Man using tools dismembers a kill and transports the most usable food away so that little actual destruction is effected on the parts remaining at a kill. Despite all these provocative observations that appear to warrant inferences of hominid-related activities, such inferences appear to be directly at odds with the excavators' observations:

1. Site FLK NN Level 2 yielded no stone tools or other evidence of hominid activity.
2. Site FLK North Level 6 is interpreted as a site "where an elephant was cut up by early man [Leakey 1971:64]."
3. The deposits at HWK East Level 2 were described as yielding tools "scattered at random" through 6½ feet of deposit (M. D. Leakey 1971:93). Large coprolites of hyena and lion-sized animals were regularly present. In the lower part of the deposit was a minor concentration of fauna largely representing the remains of an elephant together with articulated parts of a *Deinotherium*. In the upper part of the deposit was another concentration of related parts of a rhinoceros.

The faunal remains analyzed included the large mammal parts as well as the smaller bovids and equids that occurred in the deposits as background faunal remains. What is almost certainly represented here is the death site of several large mammals together with background scatters of other (mainly bovid) remains. The accumulation is a true palimpsest, differing from those of factor 1 in that the large-animal remains added to the scatter biased the total in the direction of "moderate" value anatomical elements (inspect M. D. Leakey 1971: Figures 32 and 47 and this bias becomes clear). In addition, the assemblages are not heavily ravaged, which is what most distinctively separates them from those of factor 1. Burial was relatively rapid and initial destruction at the kill–death site was minimal (as was the case in Hill's kill site materials from modern Africa).

Which of the many different events, agents, and episodes "recorded" in these geological palimpsests did man participate in? I think it is interesting that in all three distributions illustrated by Mary Leakey (1971: Figures 32 and 47) the stone tools are not concentrated or spatially clustered with the bones of the large animals that have been interpreted as the targets of the hominids' actions at the sites. In fact, the tools are clustered with the scatter of largely bovid remains in all cases. In short, the stone tools are parts of the same background palimpsest distribution (relative to other large-mammal death locations) within which the bovid remains occur. Both appear to be unrelated to the death assemblages of the large animals. It is provocative that the stone tool distribution at the now famous "hippo" site at Koobi Fora also exhibits a dispersion pattern such that the tools are distributed independently of the bones within the scatter recorded. Stated another way, where the tools are found there are few bones and where the bones are found in high density there are few, if any, tools (see Isaac 1978). It is tempting to see the two situations as similar—a death location of a large mammal superimposed on a vague scatter of both bones and tools in a situation of accretionary geological deposition. The result is a deposit of low integrity and low resolution, which provides no warranted reason for inferring behavioral connections between spatially associated things.

FACTOR 1 KILLS

It is my opinion that several sites loaded on factor 1 are grouped because they share a pattern derived from bone destruction. This is illustrated by the fact that the models for both kills and dens are loaded on factor 1. Four Olduvai assemblages, DKI-2, DK Level 3, FLK NN Level 1, and HWK East Level 1, are arranged with the kill site models on factor 1 and are therefore believed to be more like those models. Figure 6.10 illustrates the relationship between the ravaged model (Table 6.04, column 6) from small-animal kills and the part frequencies from the site of DKI-2. It is clear the fit is impressive, being a positive convex curvilinear relationship between the data and the model with only the pelvis being underrepresented in any anomalous manner.

FLK NN Level 1 is an important assemblage since it was recovered from one of the living floors (M. D. Leakey 1971:258). Yet only a very small tool assemblage was recovered (five tools and two flakes) and the living floor interpretation seems somewhat equivocal judging from Leakey's discussion (pp. 43–44). DKI-2 is clearly a palimpsest of moderate depth exhibiting no obvious internal structure (p. 25).

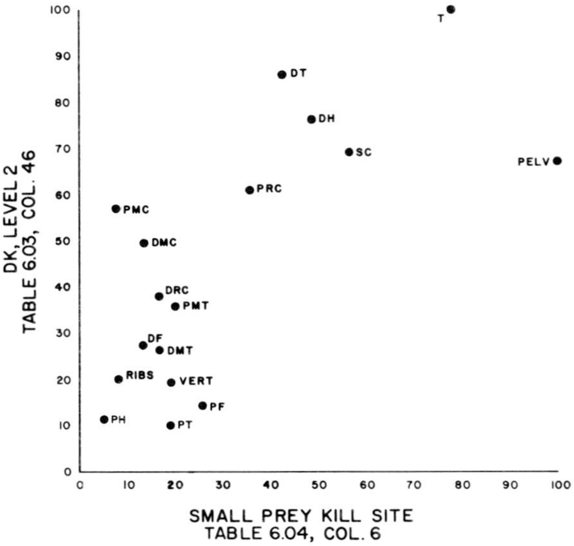

FIGURE 6.10. ***Relationship between frequencies of anatomical parts from DK, Level 2, and modeled values.***

FACTOR 3 ASSEMBLAGES

This Factor 3 groups two assemblages, FLK Levels 13 and 15, with a major loading also on FLK NN Level 3. This is an interesting factor, since there are no significant loadings from any of the model assemblages for animal kills or dens. This suggests that we are seeing some organizational facts that are outside the range of conditions currently documented for animal behavior. Inspections of Table 6.03 shows that these assemblages are not dominated either by lower legs, as is common with animal-transported assemblages, or by head and axial skeleton parts, as is common at animal kills. Instead, the distal humerus, distal tibia, and scapula appear as the most common elements on the three sites in order of their loadings. In addition to this difference, FLK Levels 13 and 15 exhibit an odd pattern of absence for distal metapodials. FLK NN Level 3 shares this property for the metatarsals. This is almost certainly a pattern of destruction not seen previously in either animal context or in modern human contexts. The analysis indicates that these assemblages are something special and share little with nonhominid animal assemblages. This view is supported by the excavators, who note that at FLK Level 15 "some of the mammalian limb bones show clear indications of battering and smashing [M. D. Leakey 1971:59]." *This is the only assemblage from Olduvai Gorge where such an observation was reported.* All of these observations, from both my analysis and the statements by the excavators, converge to suggest that we are seeing the faunal consequences of hominid behavior or at least some behavior not documented for other animals.

From my experience with the faunal materials of the Nunamiut Eskimo, I found that assemblages dominated by upper limb bone elements are generally second-order assemblages. That is, they represent a selection of parts from a population already strongly biased away from the proportions characteristic of a living animal. Such assemblages were found to be characteristic of human subsistence in hunting camps and stations where anatomical parts had been selected and transported from a kill and introduced to the camp, where a second selection was then made for use in the camp. There were many other contexts of Eskimo logistical–consumption activities where second-order assemblages were common. What might be some reasonable contexts where early hominid populations would behave in such a way as to produce a second-order assemblage? The answer is relatively clear: *If hominids scavenged usable parts from animal kills after the major predator–scavenger competition was over, the parts recovered would be a second-order assemblage.* In order to determine if the assemblages grouped by factor 3 could be anticipated by a "scavenger model," the mean kill assemblage for large animals as presented in Table 5.08, column 2, was multiplied by the IGUI (inverse general utility index) and standardized as presented in Table 6.04, column 10. Figure 6.11 illustrates the relationship between this "scavenger" model and the frequencies of parts recovered at site FLK Level 15.

I consider this distribution extremely *diagnostic* in that there are clearly two distributions, one low and linear and one high and convex curvilinear. The low and linear distribution is positive and groups teeth (which monitor the head), the pelvis, and the vertebrae. *These are the axial skeletal elements, which most commonly remain articulated at animal kills.* In the high and curvilinear distribution are all parts of the appendicular skeleton plus ribs, with the only meat-yielding part of any significance being the

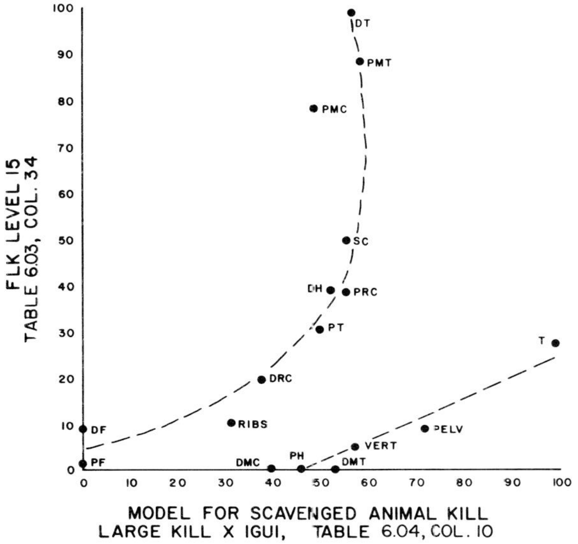

FIGURE 6.11. ***Relationship between frequencies of anatomical parts from FLK, Level 15, and modeled values.***

scapula. This is an extremely provocative pattern. It shows that the articulated axial parts typically remaining at animal kills are underrepresented in a correlated and systematic fashion, whereas the appendicular skeleton elements, most likely to be disarticulated and scattered about an animal kill, are overrepresented in a correlated and systematic fashion. I think it is highly unlikely that such a fit to a data-based model with such a diagnostic pattern could occur if the contextual identification was totally spurious. *I am quite confident that factor 3 represents anatomical parts selected from an already "consumed" and scavenged animal kill* as evidenced by the underrepresentation of the axial skeleton. This type of systematic secondary selection is not thus far documented for nonhominid predator–scavengers.

If early hominids were scavenging predator kills and natural death sites in a manner yielding the foregoing pattern, two conditions must have obtained:

1. The scavenging took place after most other predator–scavengers had already abandoned the site. This is shown by the model, which anticipates the data, as well as the apparent contrast between the biased removal of already disarticulated parts versus parts likely to have remained articulated at a kill.

2. The scavenging was the simple picking up of disarticulated parts and their transport to another location for consumption. No dismemberment by hominids seems indicated.

The second point seems warranted by the contrast between the axial and appendicular skeleton and the continuity between anatomical elements, with the possible exception of the distal metapodials whose absence appears anomalous. Put another way, the scavenging seems to have been of portable parts present at an animal kill rather than after the butchering of body segments to facilitate transport as in a "human" model. I will return to a discussion of some of the implications of these findings after I have completed discussing the Olduvai assemblages.

FACTOR 4 ASSEMBLAGES

Is a grouping of sites that share moderate loadings on both the large and small kills of prey with the large animal data most similar to the cases. It is likely that this is also a grouping of assemblages that are largely unravaged except as indicated by companion loadings on factor 1. The sites grouped by factor 4 are the *Zinj* floor (FLK 22), the floor at site FC, and level DKI-1. This is a very impressive grouping, placing in juxtaposition sites from the lowest part of Bed I to the middle part of Bed II. Not only is this the case, but the most important *Zinj* floor is included. The latter captured the imagination of both professionals and laymen almost from the first days of its discovery. This was the first demonstrated association between a hominid, tools, and fauna. This was the sought-after situation that, it was hoped, would provide the needed clues to the character of man's subsistence behavior and perhaps even something about other behaviors. Not only was the association present, but the material was from what appeared to be a single land surface. Everything appeared appropriate to interpret this as a living floor; there was even some speculation about the internal layout of the site and where the brush "houses" might have been placed (see M. D. Leakey 1971:260). In our earlier review of the observations by excavators, it was noted that there was at least a "lake margin" background of naturally occurring fauna, reptiles, and birds, in addition to concrete evidence for the action of relatively large predators, plus the tools evidencing the presence of hominid activities. We have a minimum of two contexts—generalized by-products of a lake margin habitat and specific episodes of action by both nonhominid and hominid agents. The number of events and the redundancy of the episodic contributions (the resolution of the assemblage) were not clear. That numbers of different situational contexts are represented on the *Zinj* floor is supported by the analysis summarized in Table 6.06 where it is to be noted that moderate factor loadings for the *Zinj* floor occur on factors 1, 2, and 3, in addition to a dominant loading on factor 4. This is the most heterogeneous or situationally mixed site, as judged from the factor analysis, within the studied set from Olduvai Gorge. Some subset of the assemblage is ravaged in a pattern suggestive of nonhominid consumption: A scavenged natural death of animals appears represented, as does some transported items scavenged, perhaps by hominids, from animal kills somewhere else. Most characteristic of the site seems to be an association of teeth, pelvis, ribs, distal metatarsal, and scapula (see Table 6.07), which, in their relative numerical frequencies, approximate an assemblage identified as animal kills of medium to large animals. The fact that the presence of the distal metatarsal is partially diagnostic of this factor when its absence is distinctive of factor 3 assemblages is interesting and perhaps of behavioral significance. If my interpretation of the analysis is correct, a considerable number of independent episodes of hominid activity are represented on the *Zinj* floor and related to an equally complicated history of nonhominid events, such as natural deaths and actions by nonhominid predators. The analysis clearly indicates that resolution is poor. Nevertheless, some observations (the loading of factor 3; the interesting positive loading for factor 4 on the distal metatarsal, a part that was anomalously negative on factor 3) are strongly suggestive that some faunal facts from the *Zinj* floor (not all of them, by any means) are potentially informative regarding early hominid behavior.

The kill site character of factor 4 raises the question of whether we are seeing the residuals at an animal kill or parts transported to a base camp for "sharing consumers" by human–hominid "hunter-scavengers." It will be recalled that what animals tend to leave at kill sites is generally what modern human hunters tend to transport to base camps. What happened on the *Zinj* floor?

Using the contrasts previously listed (see Tables 6.03 and 6.07), the *Zinj* floor is characterized by (*a*) high frequencies of heads, (*b*) more pelves than ribs, (*c*) low frequencies of the proximal humerus and the distal radio-cubitus, and (*d*) low frequencies of the femur relative to the tibia. These are without exception characteristics of *animal kills* rather than transported human assemblages. It is hard to imagine hominids dismembering a kill by eating through and therefore consuming the proximal humerus and the distal radio-cubitus. Similarly, it is difficult to picture the abandonment of the femur in favor of the tibia if hominids were transporting game to base camps. These facts are more impressive if it is recalled that heads are common but femora are rare, with the same being true for the pelvis–rib contrast, properties more characteristic of kills (see Figure 5.13). I think it can be said unequivocally that what is represented are animal kills that man may be partially exploiting after the other predator–scavengers have abandoned the location. Prior utilization is indicated by the general residual pattern, particularly by low frequencies of lower front legs relative to rear legs. It will be recalled that the front leg is the first to become dismembered and therefore is subject to early scattering by predator–scavengers. The frequencies at the *Zinj* site (see M. D. Leakey 1971: Table 53, column 36) show that such scattering had most likely taken place.

Factor 4 is recognized as a kill assemblage of anatomical parts with all the characteristics of carcasses either killed and consumed in addition to having been scavenged by nonhominid predator–scavengers. There is also a provocative linkage between factor 4 and factor 2 in the mirror-image patterning with respect to the distal metatarsal. This is taken as suggestive of hominid behavior and is therefore viewed as evidence for some utilization of the abandoned kills by hominids on the site.

FACTOR 5 ASSEMBLAGES

This is a factor defined by teeth, pelvis, and vertebrae with minor contributions from parts of the front leg. It is also assimilated to the kill site models (see Table 6.06) and appears to be a kill site variant where scapula and ribs as well as parts of the rear leg are less common relative to other parts of the axial skeleton as well as the upper front leg. Only one site, FLK NN Level 3 (also loaded heavily on factor 3) is designated as diagnostic. Inspection of the factor scores for factors 3 and 5 (Table 6.08) shows that they appear to be inverse images of one another. Figure 6.12 illustrates the relationships between the scores for factors 3 and 5. There appears to be an uncorrelated scatter made up of the ribs, scapula, proximal and distal ends of the humerus, proximal metacarpal, proximal radio-cubitus, as well as the distal tibia plus the proximal metatarsal, proximal tibia, and the distal femur. In addition, a positive relationship is indicated for the distal metatarsal and metacarpals,

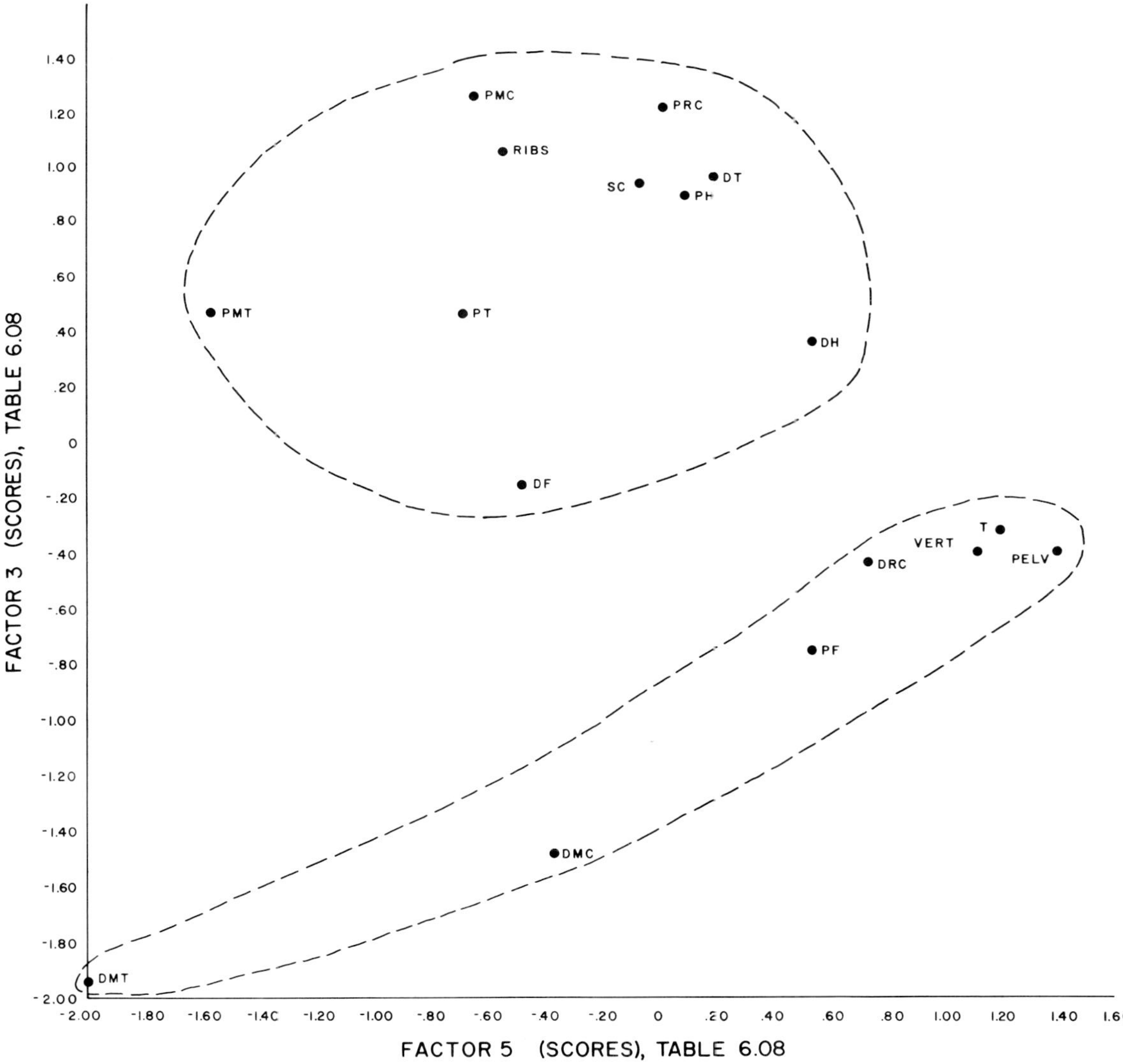

FIGURE 6.12. ***Relationship between factor scores on anatomical parts for factors 3 and 5.***

proximal femur, distal radio-cubitus, vertebrae, teeth, and pelvis. The latter is exactly the grouping within the low and linear relationship indicated between the data from FLK Level 15 (site diagnostic of factor 3 and our model for scavenged animal kills—see Figure 6.11). This similarity clinches the understanding of factor 5 as a remnant kill assemblage from which elements definitive of factor 3 were scavenged, presumably by hominids. The reason the two factors do not appear as the two ends of a single bipolar factor is most likely the destruction of parts by the hominids, particularly parts at the opposite ends of the long bones previously partially destroyed by animals, that is, the harder ends. The latter behavior would render a lack of correlation between hominid "ravaged" assemblages and the survival percentages. Thus they are not assimilated to factor 1 and after secondary destruction by hominids the source and the derivative are no longer complementary but appear as independent "factors."

I am most confident of the interpretations of cases grouped by analysis. In addition to the implications for the recognition of patterning that can be referred to behavior not characteristic of nonhuman predator–scavengers, the interaction between the patterning renders the contextual interpretation of behavior rather secure. For instance, the compound loading of factor 5 and factor 3 for the assemblage from FLK Level 15 answers the nagging question raised by the factor 3 interpretation: Did the hominid scavengers transport body parts back to a *base camp* for consumption? The co-occurrence of factor loadings for both 3 and 5 on the most diagnostic assemblages, FLK Level 15, illustrates nicely that the scavengers who "processed" the parts indicated by factor 3 did so at the same place where they obtained the parts (indicated by high loadings for factor 5), namely adjacent to the kill–death site abandoned by nonhominid predator–scavengers.

Summary

Criteria for diagnosis were employed using patterns of breakage and levels of destruction. Several characteristics of the Olduvai assemblages were recognized. Many of the assemblages were missing parts, not primarily because of destruction of parts in situ but rather because of parts having been introduced or deleted from assemblages through transport. A patterned arrangement of assemblages, from those relatively complete to those suffering attrition, was recognized as nearly identical to a distribution noted for wolf kill assemblages. That is, patterns of survivorship and destruction of long-bone elements were essentially identical to those of the control material, suggesting that the agent of destruction when active was an animal destroying bones through gnawing rather than a man breaking bones open for marrow.

As a clue to the relative levels of destruction the assemblages had suffered, tests were made (Figures 6.04 and 6.05) and it was recognized that assemblages DKI-1, FC West, Tuff, MNK H. 13, HWK East Level 2, FLK North Level 6 and the *Zinj* floor were not heavily destroyed in the manner characteristic of animal-related patterns of destruction, that is, destruction of parts in relation to the hardness of the bone. Given the information, it became clear that models of transported assemblages and reconstructions of ravaged models of most of the assemblages would be needed if successful comparisons were to be made between models based on known animal and human behavior and the Olduvai cases.

It was decided to perform a multivariate analysis aimed at recognizing a grouping of cases (a "Q" mode analysis) so that modeling could be carried out for representative cases taken as examples of sets of assemblages rather than modeling each case. A factor analysis in both modes was performed and interesting patterning was recognized, resulting in the recognition of five groups of cases (factors in the "Q" mode). As an aid to the interpretation of the factor analysis, the model assemblages for the modern data on animal dens and kills for both large and small prey as modified by survival percentages were included as "cases" in the original data matrix. It should be clear that three different approaches to analysis have been carried out on the same data base. Confidence in the conclusions could be increased by congruence or agreement between methods, as well as being influenced by the relative "power" of each approach taken individually.

As regards the relevance of analytical approaches developed in this book, the data from Olduvai had not been studied in detail for cut marks; breakage patterns were not reported; and even relative survivorship of some elements was not unambiguously reported. The most complete data derived from a modified list of anatomical parts usable for interassemblage comparisons; hence, the technique of interassemblage comparison developed in Chapter 5 was considered the most "powerful" procedure feasible. Any support that might be forthcoming from breakage patterning and measures of destruction would only be supportive and not conclusive in any way, given the minimal levels of reporting for these important classes of facts at present.

Some supporting conclusions were recognized using different methods. For instance, using the technique of comparison between the proximal and distal ends of the humerus as a clue to relative levels of destruction, basically two classes of assemblage were recognized: destroyed and not destroyed. In the "Q" mode factor analysis only factor 2 clearly indicated little or not destruction, given that factor 1 grouped assemblages that had suffered heavy attrition. Factors 3, 4, and 5 were somewhat ambiguous in that they assimilated kill site models or appeared as independent groupings. Since the kill and den models were modified as ravaged assemblages, there could be some highly patterned and expected forms of animal-related destruction (deletion of parts relative to bone density) and it would not appear as an independent diagnostic in the factor analysis. Table 6.09 illustrates the relationship between destruction as isolated in the factor analysis and as indicated by the simple humerus test. It is clear that the two approaches to analysis support one another and, therefore, render the accurate identification of relatively destroyed versus nonravaged assemblages more likely. This type of diagnostic analysis renders the modeling of assemblages easier, since important properties can be assessed in ways other than by trial and error.

Without any doubt, the most powerful analysis was the modeling of assemblages judged representative of the five and a half sets of cases isolated as internally similar by the "Q" mode factor analysis. The den-associated assemblages grouped by factor 1 were found to be best approximated by the inverse general utility index developed from anatomical

TABLE 6.09
Comparison of Destruction Estimates Derived from Factor Analysis and the Humerus Test

	Humerus test				
	Destruction		No destruction		
Factor analysis	Inventory	Count	Inventory	Count	Total
Maximum destruction (factor 1)	FLK ½ FLK 3 FLK 4 FLK 5 NN 1 DK I-2 DK I-2 BK II TK UF TK LF MNK MAIN	11	None	0	11
Minimal destruction (factors 3, 4, 5)	FC FL NN 3 FLK 13 FLK 15	4	DK I-1 ZINJ	2	6
No destruction (factor 2)	NN 2	1	FC TUFF MNK H13 HWK 2 FLK 6	4	5
Total		16		6	22

studies of sheep and caribou (L. R. Binford 1978b). This model had to be further modified by the survival percentages for bones of moderate- to large-sized animals (L. R. Binford and J. B. Bertram 1977). This modification was found to accommodate the data from FLK North Level 5, as well as FLK North Level 1/2, fairly well except there was an underrepresentation of heads and some overrepresentation of the femur and metapodials. The latter conditions were observed earlier in a wolf-transported assemblage from Alaska (Chandler). Comparison with the Chandler assemblage demonstrated a very good fit indeed, strongly supporting the inference from the model that these were transported assemblages that had suffered heavy attrition from gnawing. This is a common situation where bones are dragged off from a kill under conditions of competition at the kills. The Olduvai factor 1 cases are faunal palimpsests created by animals dragging bones of low food utility away from kill locations. If man was involved, he was behaving indistinguishably from other animals and no extraordinary "human-like" behaviors are justifiably referred to these assemblages.

Modeling factor 2 cases was interesting since no destruction was indicated and they were assimilated to den models by the factor analysis. This is particularly interesting since the only case judged to be a kill site by the excavators is grouped here, In addition, HWK East Level 2, a mixed deposit that contained parts of large animals in essentially articulated arrangement (a characteristic commonly cited as indicative of a kill site), is also included on this factor. Still more interesting is the fact that FLK NN Level 2, the only faunal assemblage reported from Olduvai that was not associated with stone tools, appears also in this grouping. The association of stone tools with largely articulated skeletal parts, as at site FLK North Level 6, is generally cited as evidence supporting the view that early Pleistocene hominids were regularly hunting.

> There are a small number of Plio-Pleistocene sites where the partly broken-up carcass of a single large animal has been found with artifacts scattered around. These are indistinguishable from sites termed butchery sites by the archaeologists. . . . I am inclined to think that the hominids were hunters, although they may have pirated and scavenged other animals' kills [Isaac 1976:499].

I think it is of extreme importance that for the sites included in this study the analysis shows that the bones largely ignored by the investigators, bovid bones in the case of FLK North Level 6 as well as the others in the sample, are demonstrably background assemblages that assimilate to a den model differing from those of factor 1 only in the relative degrees of destruction. The other fact of extreme importance is that the stone tools are physically associated with the background fauna and essentially inversely correlated with frequencies of bones remaining from recognizable megafauna. This lack of physical association is a striking characteristic of the so-called hippo butchery site at Koobi Fora (see Isaac 1978a). This new look provides a very different picture, namely that the speed of desposition and hence the preservation of fairly intact skeletal segments is a characteristic of the geological formation that also accounts for the lack of destruction of the background fauna as well. What we are seeing is the same depositional context indicated for the sites grouped by factor 1; only the rates of deposition and hence burial are much greater, resulting in the preservation of large animals in at least partial articulation and therefore inhibiting destruction of background fauna by both animals and weathering. The stone tools in these deposits are, however, no less components of palimpsests and appear to have been deposited most commonly with the background fauna or at least independently of the larger, more romantic types of animal remains. *These are not hominid kills of large animals.*

Factor 2 grouped three very interesting assemblages, all small, which might well indicate considerable integrity and resolution. Models for the fauna illustrate nicely that these are second-order faunal assemblages. Of even more interest, they are shown to be biased second-order assemblages; namely, two distributions are indicated when the data are plotted against a second-order model drawn from anatomical studies. The assemblage is biased against parts of the axial skeleton, the very parts that are most likely to remain articulated and the center of attention for other predator–scavengers. What were selected producing this second-order assemblage were the parts most quickly disarticulated and most likely to have become scattered around the site of a kill by nonhominid predators. The regular bias against axial parts demonstrates that butchering was not going on prior to the selection of parts for removal, presumably by hominids. The fact that the second-order assemblage model is based on the inverse general utility index means that the *biased selection was in favor of parts of least food utility.* These would, of course, be the parts generally ignored by other predator–scavengers. *The most common bones in these assemblages are those that yield only bone marrow as edible material.*

The factor analysis isolated these three assemblages as structurally important and very similar in composition. The modeling illustrated them to have properties that have never been observed among nonhuman predator–scavenger assemblages, thus qualifying them as very likely derivatives of hominid behavior. The analysis, very clean and unambiguous (see Figure 6.11) illustrates a bias that is only consistent with the removal of essentially marrow-yielding bones from the already ravaged kills of other predator–scavengers.

These observations are consistent with the fact that the only assemblage reported from Olduvai Gorge that was described as showing clear evidence of "battering and smashing" appears in this grouping. There is still further evidence as to the nature of the behavior that occurred at the factor 3 sites. Table 6.10 summarizes the tool assemblages recovered from the three sites grouped by this factor.

What is very obvious is that the assemblages are dominated by percussion tools or hammers; cutting tools are all but absent. I think it can be said with some confidence that we are seeing in Table 6.10 the materials used as hammers for breaking marrow bones. The degree to which these are tools in the sense of designed and produced items is questionable. All the chipping and modified edges could be the accidental by-product of using a hammer to break something else (bones). It is interesting in this regard that the assemblages from FLK NN Level 1, as well as HWK East Level 1 sites, grouped with ravaged kills on factor 1, also yielded assemblages with little debitage, dominated by large items seemingly used as hammers. This suggests that the behavior of the hominids at locations of animal kills was essentially the same as seen here when some bones have been dragged off from the kill after it has been exploited by other animals.

TABLE 6.10
Inventories of Artifacts Recovered from Factor 3 Sites, Olduvai Gorge[a]

	FLK				
	NN 3	15	13	Total	Group total
	(1)	(2)	(3)	(4)	(5)
Percussion tools					
Choppers	2	1	2	5	26
Anvils	0	2	2	4	
Polyhedron	1	0	1	2	
Hammerstones	1	2	0	3	
Nodules and blocks	5	2	4	11	
Heavy-duty scrapers	1	0	0	1	
Light duty flakes	2	0	0	2	11
Debitage	7	1	1	9	

[a]Data from M. D. Leakey 1971: 44, 59.

Both factors 4 and 5 were assimilated to predator kill sites in the factor analysis. Both exhibited properties that permitted their identification as residual populations at a kill site rather than as parts transported to a base camp in more "human" fashion.

The important assemblages from the *Zinj* floor, FC West, Floor, and level 1, at DKI were grouped on factor 4. As far as *Zinj* and FC West, Floor, are concerned, the factor analysis showed them to be moderatively distributed on several factors, strongly indicative of a palimpsest or a behaviorally complex assemblage. Observations by the excavators point clearly to evidence of many different agents being involved at the location and therefore tip the scales in favor of a complicated palimpsest with little integrity and resolution. There is clear evidence of hominid involvment in exploiting faunal elements presumably present as a consequence of the behavior of other agents, big cats and hyenas being most likely.

It seems to me that one major conclusion is justified from the foregoing analysis: The large, highly publicized sites as currently analyzed carry little specific information about hominid behavior. It might be possible, using detailed studies of breakage morphology and association, to isolate those clusters of material referable to hominid behavior on a floor such as *Zinj*; however, given its demonstrably low integrity and resolution, arguments about base camps, hominid hunting, sharing of food, and so forth are certainly premature and most likely wildly inaccurate. The only clear picture obtained is that of a hominid scavenging the kills and death sites of other predator–scavengers for abandoned anatomical parts of low food utility, primarily for purposes of extracting bone marrow. Some removal of marrow bones from kills is indicated, but there is no evidence of "carrying food home." Transport of the scavenged parts away from the kill site to more protected locations in a manner identical to that of all other scavengers is all that one need imagine to account for the unambiguous facts preserved in Olduvai.

The analysis of these assemblages through the applications of multivariate techniques in the context of a method of residues set of tactics illustrates perhaps more than anything else the potential of the approach advocated. We can now restudy these assemblages with some hope of recognizing patterning in breakage or surficial modification that most likely derives from hominid behavior uncomplicated by the actions of other agents. We have some chance here of recognizing patterning that might well contrast with breakage patterns produced by nonhominid agents, thereby providing the beginning clues in the important task of developing a "signature" for early hominid bone breakage. We cannot be certain, since our methods for isolating these assemblages are not totally unambiguous, but the study of these isolated assemblages is a better research bet than projecting Nunamiut bone-breaking behavior into the past as representative of "human" bone breakage, or of inferring "humanness" from assemblages for which there is no justification for attributing them to a single hominid agent. The potential feedback characteristic of applying a methodology and then investigating the archaeological record in new ways is well represented here.

Most of the patterning isolated by my tactics as well as my arguments as to its implications for the past have already been summarized. I must stress that this is not to be thought of as a conclusion but as the first step in the analytical treatment of remains from the dawn of our evolutionary beginnings. We must begin the important process of *interacting* with the materials remaining to us from the past. This interaction should be ongoing, so that as new information becomes available from actualistic studies that can be reasonably projected into the past, new pattern-recognition studies can be conducted. Intellectual anchors used to tease out patterning from the overall aggregate of facts remaining to us can be anticipated as more middle-range work is completed. The use of "knowns" to permit the isolation of residual patterning can then be studied uncomplicated by facts that reasonably appear to refer to actions or events of no direct relevance to hominids. Information gleaned from such controlled pattern-recognition studies may be then further studied both actualistically if possible and certainly distributionally and correlationally among the assemblages available from the past. If actualistic studies are possible, new methods may be developed; if only archaeological studies are possible, new patterning in need of methodological investigation will be uncovered.

This study does not even begin to answer all the important questions regarding the characteristic behaviors of our early hominid ancestors. It simply represents the beginnings of productive research into these questions. It should clear the air of many of the conventions that have misguided our views of the past for some time, and refocus our attention on the important issue of carving knowledge out of ignorance.

With regard to "cleaning the air" of conventions, I think a number of things can be stated with some certainty. *The famous Olduvai sites are not living floors.* They vary in the degree to which they represent stabilized land surfaces. The more stabilized the ancient land surfaces, the greater the likelihood that the integrity and resolution of materials occurring on those land surfaces are very low. In general this fact has not been appreciated.

The interpretation of deposits as living floors permitted the treatment of the contents of those deposits as if one were dealing with a monument; it was assumed that the contents of the provenience unit were there "by the hand of man." Placing the problem of identification at the level of the provenience unit ensures that the contents of "identified units" carry implications for hominid behavior.

I refer to this as a *configurational inference.* It is an inference made about a unit, a configuration of phenomena, believed to represent a derivative of past organized dynamics. It is an inference about a unit within which associations are assumed to be meaningful in an a priori sense regarding what one wants to know about the past. This may be both a logical and a tactical error. It is a logical error in that if we wish to learn when and under what conditions man began to live in base camps, assuming such camps as a methodological device ensures that we are involved in a hopeless tautology. Also, the criteria for recognition were not developed for the unit being identified, instead they are developed for one of the items within the unit—stone tools. The identification of the unit then becomes an inference from association of this identifiable class of things with a certain geological context (e.g., a paleosol). Tool use on a stabilized land surface does not ensure that man lived in base camps or even carried out localized activities on living floors!

It is on this point of identifying observational units that I am in tactical disagreement with Michael Schiffer (1976), who adovcates a strategy of "provenience identification." As pointed out, this can be very powerful if well informed by middle-range research, but the tactics Schiffer illustrates are ones of concept formation with little attention to the development of operational definitions for the concepts. For instance, are all unbroken pots indicative of "de facto garbage," and all deposits containing broken and typologically heterogeneous materials "secondary refuse"? Regardless of these problems, Schiffer's approach is directly linked with excavation in volumetric units rather than with removal of the matrix and the plotting of the provenience of things within the matrix. Excavation by units of matrix volume assumes that things associated in a given volume of similar matrix were deposited under similar "formation" conditions. This is establishing units within which associations are considered meaningful by stipulation.

In my opinion, where we do not have well-established methods for inference backed by substantial middle-range research, we must proceed with a pattern-recognition strategy, rather than a stipulative strategy. In terms of excavation procedure, a pattern-recognition strategy seeks to record properties of both matrix and inclusions particularistically; that is, in terms of point provenience so that later study can be conducted to determine if there is any spatial patterning between things and their matrices. Quite literally, we are documenting the archaeological record so as to permit the recognition of structural relationships between the matrix and its inclusions. Stipulative strategies assume the existence of patterning and also assume that the observer is skillful enough to isolate it unambiguously through the use of good field judgments.

This is a traditional approach where our understanding of the past hinges on the competence of the field excavator and his judgments rather than on methods derived from research directed toward understanding the archaeological record. The latter research would permit us to evaluate the utility of differing analytical and observational tactics quite independently of the subjective skills of the excavator.

The traditional stipulative strategy is the old approach of monument archaeology, where the matrix itself was considered to exist by the hand of man; therefore, its contents were referable to the same formation conditions as the matrix. Isolation of different formation episodes then becomes an art manifest in the excavator's ability to recognize the historically meaningful "provenience units" considered to be directly referable to past formation episodes. This is a classic paradigmatic exercise in which a series of post hoc arguments as to the meaning of matrix variability encountered during excavation are adopted as conventions for the recognition of provenience types. The meaning given to such units represents configurational inference, and conditions the meaning that may be reasonably assigned to the contents of such units (see Schiffer's [1976:133–139] discussion of proveniences and analytical units, particularly treating the problem of "floor" definition, [pp. 136–137]).

It was this very problem that led me to suggest that, at least during the early years of scientific development, methodological growth should be rooted in actualistic research where the bear and its footprint can be unambiguously related to one another. If we can develop robust methods for giving meaning to archaeological observations, then the use of such "secure" inferences as control for further pattern-recognition studies may be a powerful means of expanding our knowledge of the archaeological record. Until we have such robust methods however, using provenience identification tactics largely represents judgments and opinions as it is practiced. Our procedures for giving meaning to the archaeological record should be based on more reliable and less subjective tactics.

There is another tactic, which we might call "control once removed," which derives from tactics of provenience identification. This is the tactic that, it was argued, led to a misleading set of "methods" used in some Americanist research—that is, seeking a methodology through pattern-recognition studies conducted on materials about which an inferred meaning has already been assigned. The work of Frison at Glenrock (1970) was a good example. The configurational inference was made, namely that the deposit was the consequence of human behavior. Then a pattern-recognition study was conducted of bone modifications on inclusions and a very distinctive pattern was described. This pattern was then accepted as diagnostic of human behavior and as such was employed as a method for recognizing the hand of man in other assemblages. We have seen that the configurational inference was almost certainly wrong and the patterning recognized was almost certainly referable to the action of scavenging animals. *We must not fall into the trap of adopting a post hoc argument as a methodological convention and using it to ascribe meaning directly to the archaeological record.* This has been a major weakness of traditional archaeology, in that it recognized no explicit role for methodological research. Methods were simply conventions that were gradually adopted by virtue of having been used by "prominent" men.

The use of configurational inferences has not only misled the study of inclusions within identified proveniences, but has also had adverse effects on comparative studies making use of "assemblages" or the contents of "identified" proveniences.[5] For instance, if patterning was recognized among such "identified" units, then post hoc arguments as to the meaning of such patterning would be directly conditioned by the implications of the original "identification." For example, if I identify the assemblages from Olduvai as living floor units, then any patterning that might be recognized among such units would be necessarily modeled in terms of factors conditioning differences among living places. This is a fine example of the point made in Chapter 2, namely that we can never reason in a valid manner from premises to a conclusion that would contradict the premises with which we start. If we start with the premise that the condition of formation was a group of hominids living at a particular place, then we can never reason from that premise to a conclusion that would contradict the premise; our model must be about factors that differentially condition the behavior of hominids living at different places. This is exactly what was done by Speth and Davis (1976). They assumed that the composition of the Olduvai faunal assemblages was the consequence of hominid behavior. Given this assumption, they compared the assemblages summarized into relative frequencies of Carnivora, tortoises, and bovids to seasonally summarized ethnographic data from the !Kung and G/wi Bushman. They were able to show that in the Bushman assemblages, tortoises tended to occur in rainy season assemblages, whereas hunting of bovids with few tortoises present was a pattern characteristic of the African dry season. A very similar dichotomy was noted in the faunal assemblages from Olduvai Gorge. There were assemblages with substantial numbers of tortoises and there were others dominated by bovids. The latter type of assemblage was most common, leading the authors to the conclusion that most of the sites in Olduvai Gorge were dry season "occupations"!

> In conclusion, the archaeological record from Olduvai Gorge probably reflects only part of the total settlement system of these early hominids. . . . During the rainy season the Olduvai hominids either moved to nearby areas that have not been sampled, or more probably, as part of their seasonal round they abandoned the area of the Gorge entirely [Speth and Davis 1976:445].

Let us look at this situation slightly differently. *Are all assemblages of predominantly bovid remains*

[5]This problem was recognized by Schiffer (1976), but he viewed the solution as resting with his "insight" that man cleaned up his sites and did not always leave everything where he had used it. Schiffer proposed that we give up the "false principles" of the New Archaeology, which allegedly expected everything in the archaeological record to be distributed and associated as if the living system had been miraculously stopped in mid "process" by some catastrophic event (see Schiffer 1976:12–13). This is of course nonsense.

the consequences of hominid hunters? The answer is absolutely not. Bone assemblages from Africa in almost any context are predominantly bovid remains relative to tortoises and carnivores. Hill's (1976) fauna was predominantly bovid remains. Behrensmeyer's fauna was predominantly bovid remains, as was Klein's fauna at Swartklip, and so on. Thus there is no necessary relationship between bovid remains and dry season hunting by hominids. It remains for us to ask if there is any necessary relationship between tortoise remains and rainy season hominid hunting, although we know less about the context of occurrence of tortoise remains. One of the few observations on tortoises is by Richard Klein (1978a) with regard to the fauna from "Cutting 10" at Elandsfontein (Hopefield):

> Fossil tortoise remains are extremely abundant at Elandsfontein, and their frequency appears to vary independently of the frequencies of mammal bones and artifacts, suggesting that, by and large, the tortoise remains are present for different reasons. I think they probably represent mainly natural mortality [Klein 1978a:77–78].

In this judgment I would concur with Klein. For instance, examination of the map of finds supplied by Mary Leakey (1971: Figure 20) for FLK NN Level 3 reveals the tortoise remains to be distributed in essentially nine separate localizations, strongly suggesting the death sites of different tortoises. There is no demonstrable association between tortoise localizations and tools, or other bones for that matter. This "floor" also has a localization of bovid remains in the northern projection of the excavations, consisting mainly of ribs, mandibular and maxillary parts, as well as some vertebrae. This seems indicative of the remains of a kill or natural death site of a bovid. In both the tortoise and bovid remains the distribution has the character of a plain set of differing episodes with the natural deaths of several tortoises being situationally unrelated to the other faunal remains or to the few associated "tools." This is the pattern indicated by Klein for Elandsfontein. Unfortunately the other site in Olduvai from which large numbers of tortoise remains are reported, MNK skull site of H. 13 (see M. D. Leakey 1971:257), was not mapped so as to document the location of the tortoise remains (see M. D. Leakey 1971: Figure 58). Nevertheless, bovids and tortoises are part of the natural world of African fauna. Bovids can be expected to be common in almost any type of faunal assemblage. The presence of tortoises together with bovids does not necessarily imply hominid behavior. Such an argument from ethnographic analogy to the Olduvai assemblages only makes sense if you make the prior assumption that the Olduvai assemblages were hominid produced. The fact that there is such a good fit between the Olduvai data and the Bushman data in no way proves that the causes were the same. This is like the situation with any other post hoc model—if the fit had *not* been good, then the analogy would not have been "impressive" or appeared plausible. The fit must be good for the argument from analogy to appear plausible, but plausibility does not prove causation. This example perhaps demonstrates the awesome "power" of a paradigm. If we accept the initial identification of a provenience unit, then there is only a limited and logically constrained set of implications to the "facts" summarized from the provenience unit. The character of the implications are established not by the observations, but by the assumptive identification of the unit from which the facts are summarized from observations. This technique of using paradigmatic tactics to identify a unit in the archaeological record as to the conditions of its formation is what I previously referred to as the strategy of configurational inference. If one is accurate in the use of this strategy, it can be very powerful in that recognition of the behavioral conditions under which a deposit is formed permits the contents of that deposit to be considered in an integrated manner. Each component of the deposit can be evaluated as to its functional role in the processes that conditioned the formation of the deposit. In the aforementioned case the bones became food "facts," the variability in species became "facts" of seasonal variability in diet, and so on. In a very real sense the context of relevance was established for the material contents of the deposits, hence the implicative contexts were established and therefore the character of the inference that could be drawn were basically established. In order to do this correctly we must have comprehensive knowledge of the various contexts of formation for both archaeological and geo-paleontological deposits as well as the combinations of such processes. Although this may be a goal of middle-range research, it is certainly not a capability that we currently possess. Attempting to use the tactic of configurational inference at this stage of our ignorance regarding depositional or formation processes (see Schiffer [1971] for the initial usage of this term) seems to me to be self-defeating and certain to contribute to the generation of more myths.

Given our current state of knowledge regarding the archaeological record, our excavation procedures must be consistent with analytical approaches and therefore must consist of point provenience strategies until we know something of the range of structural patterning in the archaeological record that may in fact carry "configurational" meanings. Similarly, we must use analytical tactics in middle-range research and the evaluation of deposits in attempting to gain an understanding of patterning in different constituents of the deposits considered quite independently of the associations that might exist within the deposits. This approach recognizes that man may frequently contribute to deposits, but not necessarily create them.

Postscript 1

Between the time of my writing and the time of my review of the page proofs (February 6, 1981), two important articles have appeared (Bunn, *et al.* 1980, and Lewin 1981) which bear directly on the arguments in this chapter. In addition I have had privileged access to two additional articles which are currently in press and will most likely be published around the time of the appearance of this book. These are "Archaeological Evidence for Meat Eating by Plio-Pleistocene Hominids from Koobi Fora, Kenya and Olduvai Gorge, Tanzania" by Henry T. Bunn and "Cutmarks Made by Stone Tools on Bones from Olduvai Gorge, Tanzania" by R. Potts and P. Shipman, both submitted to *Nature* for publication.

It should be pointed out that this is an extraordinary level of publication and activity illustrating nicely how exciting the field of early hominid archaeology is at the present time.

These publications represent the first materials made available on the important subject of surficial marking on bones from the important sites at Olduvai Gorge and at Koobi Fora. Actualistic studies were conducted by both H. Bunn and Potts and Shipman to provide recognition criteria useful in distinguishing cutmarks made with stone tools from tooth scoring and other forms of modification. Both researchers report that the unambiguous recognition of stone tool cutmarks is possible as is the recognition of tooth scoring produced by gnawing animals.

Potts and Shipman studied 12 levels from Olduvai Gorge, while Henry Bunn studied nine assemblages from the Koobi Fora area and four from Olduvai Gorge with a particular emphasis on FXJj 50 at Koobi Fora and FLK 22 Zinj at Olduvai Gorge.

Bunn examined bones both for surficial marks but also for the traditionally controversial properties of bone breakage.

> At FLK Zinj . . . over 400 bones show damage *attributable to chewing by medium sized carnivores,* and this damage potentially obscures signs of prior hammerstone breakage by hominids. This sequence of events is suggested because over 500 bone flakes, some of which probably resulted from breakage by hominids have been identified in the assemblage [Bunn, in press, emphasis mine].

This seems to be a feeble attempt to salvage or save the conventional interpretation that all the faunal remains were deposited at the Zinj location by the hominids. The ad-hoc auxiliary hypothesis (see Popper 1965:80–84) that carnivore activity must be subsequent to hominid activity is certainly holding onto the conventional view of this as a living floor produced by hominids by a very thin thread indeed. One wonders why carnivores should consistently chew bones previously broken for marrow by hominids? I would imagine that the chewing by the carnivores would most likely be related to their attempts to obtain food!

Particularly exciting is the illustration of what I consider to be unambiguous dismembering marks on the medial face of a distal humerus from Koobi Fora site FXJj 50 dated to approximately 1.5 million years ago. Bunn's illustrations (Bunn, *et. al.*, 1980, Figure 4b; Bunn, in press, Figure 2) should be compared to Figures 4.15 and 4.30 in this book. The form of the marks and their placement are in my opinion unquestionably referrable to hominids using tools to dismember the joint in question. Also quite interesting is the observation that most of the marks interpreted as inflicted by tools *are not dismemberment marks.* They are seemingly mostly transverse marks on the diaphysis of shafts of limb bones. It is hard to imagine these as deriving from dismemberment. It is much more likely that they arose during the breaking of already dismembered bones for marrow. When lower limb bones or bones remaining from meat consumption by carnivores are exploited for marrow, one of the first problems is the partially dried and very tough adhering tissue. In addition, the periostieum is very tough and increases in strength with the size of the animal. Even on fresh bones cleaned of skin and tendon, the periostieum can be difficult to pull apart after bones are broken. Much more commonly this is cut and scraped off, at least along the intended impact surface before attempting to break the bone. This "cleaning" of the bone results in incisions if lower limbs are skinned and may also result in scrape marks across the surface of periostieum removal. It is quite provocative that, of the 23 marks interpreted by Potts and Shipman as tool inflicted, only 4 are on articular ends *and 17 are on long bone shaft fragments.* In addition, marks were identified on a scapula blade and on a rib fragment. This bias in favor of marked diaphysis indicates to me that most use of tools was in the task of marrow recovery and only a very minimal use for dismemberment. Some few marks may be the result of scavenging remnant morsels of meat from bones remaining on predator kills and sites of natural deaths. The biased distribution of tool-inflicted marks was summarized at a recent national meeting as follows:

> "They've also found extensive cutmarks on fossil horse limbs from Olduvai at a point where virtually no meat would be present, only tendons and skin," he added. "Either these creatures weren't smart enough to know where the meat was, which is unlikely, or they were interested in something other than meat. One can only speculate what they were after" [Lewin 1981:373, quoting Isaac].

The arguments presented in this chapter anticipate just such a distribution. This is most encouraging.

It is equally gratifying to hear from the researchers currently ' closest ' to the early hominid materials the following:

> [B]oth hominids and other animals modified the bones in these assemblages. Therefore we cannot attribute the observed patterns of faunal and skeletal representation solely to hominid activity [Potts & Shipman, 1981, p. 5].
>
> It seems clear that the formation of these early archaeological sites involved a complex interplay of several key factors . . . the excavated materials cannot therefore be viewed as simple, complete records of hominid activities at the site [Bunn, 1981, p. 6].

Postscript 2

In November, 1980, I had the opportunity to see some of the faunal remains from the important Russian site on the island of Jersey, La Cotte de Saint-Brelade, excavated over an extended period of time by Charles McBurney (see McBurney & Callow, 1971). On the occasion of my visit to the archaeology laboratories at Cambridge University, Mrs. McBurney, Katharine Scott, and Peter Callow all shared with me a very exciting morning looking at the massive bones of young mammoth and rhinoceros recovered from the site. I must admit I had gone to the laboratory prepared to see the telltale evidence of predator–scavengers. I was *both surprised and absolutely fascinated when I observed no evidence of nonhominid predator–scavengers.* While at the same time I was able to point to very convincing evidence of both dismemberment of the megafauna (a cut in the same place, as was illustrated by Bunn, *et. al.*, 1980, Figure 4b, and in this volume, Figures 4.15 and 4.30, was observed on the distal humerus of a young mammoth as well as other dismemberment marks) and filleting of meat. (There were long-filleting marks down the length of the massive rhinoceros scapula.) These filleting marks were aligned much as those illustrated in Figure 4.06 of this volume. Judging from the disposition of the bones on the site, it would appear that after filleting the bones were stacked in an out-of-the-way place (see Scott 1980) near the location where processing of the meat had occurred as a kind of insurance "cache." While these observations are of interest, the major point of mentioning these materials is because of the nature of the cut marks themselves. My impression was of very wide marks, almost as if made by a chisel-edged tool, quite unlike the small "V" shaped "slicing marks" described by Potts and Shipman (in press) as characteristic of cut marks produced by stone tools. The marks on the Cotte de Saint-Brelade fauna should be studied in some detail since it is my impression that they would not "fit into" the diagnostic scheme proposed thus far for recognizing the marks of stone tools, yet, judging from their placement, they were almost certainly produced by tool-using hominids.

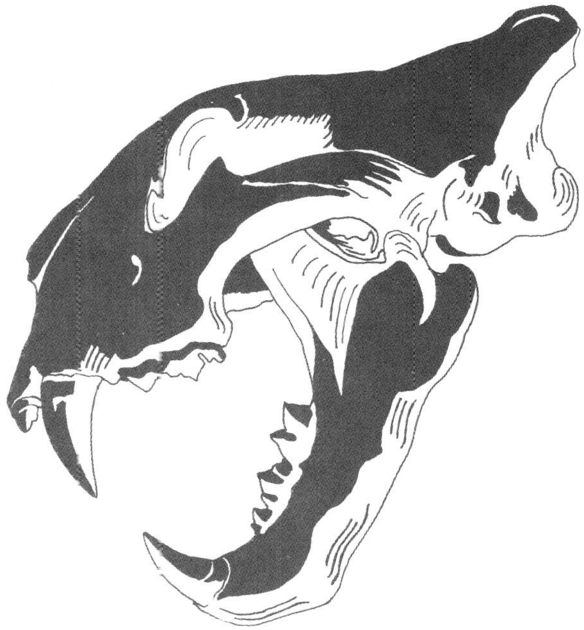

Chapter 7
General conclusions

This book has been about methodology. It has been primarily concerned with documenting empirical conditions that are thought to be redundant in their patterning and general in their relevance. The underlying message goes beyond a strict concern for methodology, however: *A science that lacks robust methodology cannot operate as a science.* In the absence of reliable methods it cannot evaluate the ideas that are set forth about the subject matter of the field, and that is, of course, its function.

I have touched upon the ideas of the Abbé Breuil regarding the historical significance of his observations on the fauna of Choukoutien. Similarly, I have treated the ideas of Raymond Dart regarding the significance for the past of his observations on the fauna from Makapansgat. In like fashion, I have discussed ideas set forth by George Frison as to the significance (in terms of past behavior) of certain patterned observations he made on fauna from the Glenrock Buffalo Jump and later extended to other data sets. Finally, I have attempted a fairly systematic application of both knowledge and methods developed here to the important ideas set forth by Glynn Isaac and heralded by Richard Leakey as the last word in the interpretation of our early hominid ancestors' behavior. In these and other examples where I have been critical of the *meanings* that archaeologists have given to observations made on the archaeological record, I have been able to do so because of the development of both a body of knowledge and an understanding of process that, when consistently utilized, permitted me to (*a*) see patterning others had not seen or (*b*) understand the significance of patterning as referable to processes not imagined or considered relevant by those whose ideas I criticized.

I have suggested before that meanings come from us and, as I see it, there are only two basic ways for this to happen: (*a*) by convention and (*b*) by appeal to a highly confirmed body of knowledge about the nature of the world and how it works. The fine line between the two is sometimes hard to recognize since the warranting arguments offered for the adoption of a *convention* are frequently phrased *as if* the justification was based on a theoretically supported body of understanding. The development of robust methodology has been advocated as the need for *middle-range theory* (L. R. Binford 1977a). Middle-range theory provides us with the tools for giving meaning to observations of interest. For instance, when an astronomer exposes a special film through a telescope aimed at a distant star and obtains a spectral array of the different colors, the patterning and relative dominance of the spectrum by different colors representing different wavelengths permit the astronomer to estimate the rate at which a star or other body in the galaxy or beyond is moving away from us, something of its distance away, and so on. How can we give such meaning to these observations of color and arrangement? We can do so only because we understand the processes as summarized by theories about the conditions and interactions between masses and energy sources that *cause* spectral change. *Insofar as our understanding of these processes is accurate and complete, we are able to give meaning accurately and unambiguously to the spectral observations we make.* This is an example of the use of middle-range theory. It is not middle range because it is second rate, particularistic, or unimportant, but because the propositions that make up the facilitating theory are independent of the argument under investigation. Thus, middle-range theory plays no role in the *explanations* offered for the variability in the subject of interest. That is, the causes of celestial movement and dynamics are not to be explained by a knowledge of the relationships between light and motion. The latter are only symptomatic of the dynamics of the universe and the phenomena we use for diagnosing its conditions.

The reader may well ask why I have not offered an example from the science of archaeology, for after all I have previously urged the development of explanatory theories (Binford 1968b), which of course argues for there being a necessary connection between specified causes and effects. Insofar as we can independently monitor the causes and effects, we can evaluate the claims of "necessity" quite directly. This situation would of course also allow us to evaluate the accuracy of the uniformitarian assumption directly. Although this strategy is feasible it presupposes the prior existence of instruments for measuring or monitoring both the alleged causes and the effects. That is, it presupposes an operational set of instruments for measurement specific to the properties cited as both causes and effects by the theory. In short, this is a strategy of a *mature* science and archaeology has not yet reached such levels of maturity in most important areas. As an example of this strategy the development of C-14 analysis is noteworthy. There was a robust body of theory regarding the relationships between living organisms, atmospheric C-14 and rates of degeneration of the radioactive isotope. There are clear implications for measuring time elapsed since an organism was alive. Given such a theory and a means of measuring the amount of the radioactive isotope of carbon remaining in a derivative of a living organism, one could estimate the time elapsed since the death of the organism. One could infer this elapsed time by measuring the radioisotope quantitatively. The bristlecone pine (see Renfrew 1973) provided a means of directly evaluating one of the uniformitarian assumptions necessary to the original development of the technique and it was found to be inadequate, but at the same time it provided the basis for estimating the past variability in atmospheric C-14, rendering the uniformitarian assumption regarding that particular constant unnecessary. Most archaeologists use C-14 as a means of "dating," rarely viewing it as an inferential means of estimating elapsed time. In this case the product of theoretically based scientific research has moved into a paradigmatic function for the archaeologist. Given that the research was sound and the theory robust, a major advance in method was achieved for those interested in understanding the past.

Of importance in this example is the fact that the theory that serves to warrant the inference of elapsed time is intellectually independent of arguments that may be made about the age of some remains, the duration of some historical episode, and so on. This is a crucial characteristic of middle-range theories; they must be intellectually independent of the arguments about the phenomena they are used to enlighten. In addition to intellectual independence from theories of historical process, middle-range research must be strenuously tested, since inaccuracies can lead to spurious problem recognition and inaccurate description of the past.

Ad hoc middle-range theory is never acceptable and rarely ever correct. For instance, the arguments generated by Frison as to the actions responsible for patterning in the faunal remains observed at Plains bison sites were invented to accommodate a particular set of observations. The behavioral reconstructions offered by Isaac for the facts from the early hominid sites in Africa were also invented to accommodate a particular set of observations. All such post hoc accommodations remain particular, and if extended, as in the case of interpretations of faunal materials found at Lubbock Lake (Johnson 1977, 1978), become conventions untested but accepted as true because of their plausibility and sometimes simply because of respect for the person proposing them. Obviously they will be plausible if they are in fact invented to account for given facts. If done by a person whose judgment we respect, as in the case of Frison, then they become integrated into the methodology of the field and become commonly acknowledged conventions for giving meaning to observations. In the case of Isaac's arguments, for example, if we accept his convention that an association between stone tools and heterogeneous faunal remains means a base camp (see Isaac 1975a:17) and a base camp means sharing, then all archaeological sites (locations where there is an association of tools and heterogeneous fauna) will confirm Isaac's belief that sharing was the critical factor that made us men! This type of tautology results from the adoption of a conventional methodology rather than one that demands a well-established understanding of the processes that bring into being the observations, which are in turn assigned meanings.

A major source of conventions has been the post hoc arguments of prominent researchers. Under the canons of traditional archaeology, reconstructions of the past were considered to be the consequences of intelligence and competence on the part of archaeologist. The past as known through archaeological research was then the past as imagined by a series of men considered to be both competent and intellectually qualified to interpret their findings. Since familiarity with the data was considered to ensure good interpretation, rarely were the data of a prominent person reanalyzed or otherwise argued from by other workers. After all, who was more familiar with the facts? This resulted in the growth over time of a body of understanding that can only be described as an accumulation of "myths by prominent men" regarding the past. Men were evaluated, not ideas. Perhaps the most striking example of this situation is the commonly accepted understanding of the site of Choukoutien. Most any textbook tells us that the men of Choukoutien cooked meat over hearths, predominantly ate venison, but also ate elephants, rhinoceroses, beavers, bison, and wild pigs, and on occasion one another.

Evidence for the use of fire was summarized by Teilhard de Chardin in Black (1933:57) as so extensive as to need no further comment! In this judgment I must agree but the evidence hardly justifies the picture one gets from textbooks, of early man seated around his hearth roasting meat and carrying on a fireside chat.

> The thickest layer of ash, in the upper-middle part of the cave deposit, *is up to six meters deep.* Stone tools and fossilized small vertebrates—rats and bats—were numerous in this layer, *sometimes indeed forming their own layers.* The ash deposit here is not in piles, but spread out in even layers, apparently the result of water movement. In the lower-middle part of the cave deposit, the ash layer is thicker near the south wall. *At its maximum, it is four meter deep.* It was around the fringe of this ash layer that most of the human fossils and stone tools were unearthed [Lan-Po 1975:34; emphasis added].

That man was a cannibal as evidenced by patterns of bone destruction at Choukoutien is a standard part of the myth of the past that is passed on to each student generation:

> All the human skulls from Choukoutien have been carefully opened through the base, presumably to extract the brain. It seems likely that the Choukoutien people were

> cannibals and perhaps headhunters. We tend to think of cannibalism as a bestial and inhuman practice, but in fact nothing better demonstrates the humanity of the Choukoutien people. Among living peoples, cannibalism is never a matter of nutrition, no animal, human or nonhuman, eats its dead for food. Rather, it is a solemn ritual act, sometimes to express family piety toward the deceased or magically to impart the deceased's spirit and qualities to the living. We may be confident that the atmosphere of Choukoutien during the cannibal meal was closer to mass than to McDonald's [Jolly and Plog 1976:169].

> One other aspect of Choukoutien deserving comment is the evidence of cannibalism. Not a single skull from this site had an attached face, most of the cranial remains were mixed up with ashes, animal bones, and other refuse, and there is evidence that the base of each skull had been pried open to get at the brain [Wenke 1980:153].

The cannibalism interpretation was offered despite the fact that at the time the following facts were known:

> On top of the reddish mud deposit is a layer of coarse sand, evidence of a big flood in the area. When the flood receded, cave-dwelling Chinese hyenas, an extinct species, were the first visitors. This sandy layer has yielded a large number of hyena skeletons and a layer of coprolites, or fossilized dung. These hyenas had the distinctive habit of excreting at a fixed spot . . . their powerful jaws and teeth could crack and chew even the hardest animal bones . . . Peking man took over the cave not long after these hyenas. His bones and those of other animals, as well as his tools of bone and stone and traces of his use of fire *are to be found in this layer* [Lan-Po 1975:9; emphasis added].

I have already indicated how gnawing animals typically chew out the face of prey early in the consumption sequence of the skull. In fact, in cervids the remains of a cranial section analogous to the skulls of Peking man is considered *diagnostic* of nonhominid animal skull destruction (see Figures 3.24 and 3.26). For a particularly graphic illustration of what gnawing animals can do to a human face, see Beard (1979:277), for the picture of a human killed and partially consumed by animals.

In light of the arguments presented here the interpretations of Choukoutien should stand out as some of the more imaginative myths created about early man. That the investigators were most likely wrong and developed post hoc arguments from very little information and a great deal of romantic imagination is not necessarily to be regretted or even criticized. What *is* regrettable, and open to strong criticism, is that 45 years have passed and research aimed at evaluating their ideas has not been initiated. Instead, the arguments cited as warranting their interpretations have sometimes served as "methods" for others seeking to draw inferences about the past. Dart (1957) cited Breuil's work on alleged bone tools as justification for his interpretations. Blanc (1950) cited Weidenreich's work at Choukoutien as support for his interpretation of cannibalism in European sites. The myths go on.

One of the great virtues of science is that it is considered to be self-corrective:

> It appeals to no special revelation or authority. . . . It claims no infallibility, but relies upon the methods of developing and testing hypotheses for assured conclusions. The canons of inquiry are themselves discovered in the process of reflection, and many themselves become modified in the course of study. The method makes possible the noting and correction of errors by continued application of itself [Cohen and Nagel 1934:395–396].

It should be generally clear that no such self-corrective characteristics have been evident in the study of early man. Advances have generally been made through discoveries of new materials and not by virtue of the successive growth of knowledge that generally accompanies the use of scientific methods. We desperately need to abandon the technique of evaluating men and adopt the strategy of evaluating ideas.

I have attempted to describe in relevant detail the consequences of animal behavior for the distribution and association of anatomical parts, breakage, and surviving morphology of bones remaining as by-products of predator–scavenger behavior. I have contrasted, when possible, this behavior with that of humans in accomplishing similar tasks. By this approach I have sought to recognize *diagnostic* characteristics that can be used to identify faunal assemblages that derive from nonhominid behavior and distinguish them from assemblages with properties referable to the behavior of man and by inference hominids. During the course of these investigations, I have attempted to point to inaccurate, or at best ambiguous, interpretations of faunal facts offered by other researchers. This has not been done out of churlishness but as an object lesson to archaeologists. It is so very easy to misread the facts of the archaeological record and thereby distort our ideas of man and the past. Such distortions have been common in the past and are continuously being made by archaeologists operating in the absence of a justifiable methodology for giving meaning to archaeological observations.

In demanding a processual understanding of how the record of the past was constituted, I have emphasized the fact that deposits yielding traces of human–hominid behavior can be expected to be heterogeneous in the identity of the agents producing the traces preserved, as well as in the events or episodes recorded in the deposit. It was suggested that archaeological deposits are relatively rare and are commonly a derivative of recent complex cultural systems. For most of man's history as a hunter–gatherer, we can expect archaeological remains to occur as inclusions within geological deposits. Under such conditions it is totally unwarranted to assume that what is associated with undisputed archaeological remains is also a derivative of human–hominid behavior and/or derives from the same events or episodes that resulted in the deposition of the archaeological remains. Establishing and warranting that associations signify integrated behavior is one of the more difficult tasks facing the researcher of early man. I have briefly touched upon many ideas and arguments concerning the character of life in the ancient past that are almost certainly false by virtue of the failure of the persons offering the interpretation to treat seriously the problem that correlation (association) does not necessarily imply causation (behavioral integration in this case).

It is relatively easy to demonstrate that the role of the nonhominid predator–scavengers as contributors to deposits in which occur traces of hominid behavior has been generally overlooked and in many cases unrealistically denied. The cases of bone tools from Choukoutien, Makapansgat, and many German caves and rockshelters are good examples. The fanciful interpretations of deposits such as Torralba, Cueva Morin, and Pin Hole Cave in Great Britain should stand as reminders that in the absence of processual understanding accommodative fantasy has been passed off as knowledge. All the arguments for cannibalism among early hominids are almost certainly based on associations and morphological modifications of bones made primarily by nonhominid predator–scavengers. This is almost certain for the case of Choukoutien (see Lan-Po 1975:9). When one reflects on statements such as the following, one can only wish that the investigators had been more concerned with details of context and association and less with imagined romance:

> At La Caune de l'Arago, the anterior part of a human skull lay upside down on a very rich prehistoric living site, skull-cap down, jaw bone in the air. It has been left by man in the midst of a heap of rhinoceros, horse, auroch, deer and ibex bones. The flesh had been stripped away before it was left there and the back part of the skull had been removed. A flake was found in the skull cavity. It is not impossible that the skull may have been opened in order to extract the brain perhaps during ritual proceedings [DeLumley 1975: p. 799].

Almost certainly most of the arguments about the ritual significance of cave bears in the life of Neanderthal man are equally fanciful. The "evidence" has been considered by recent authorities (see Kurten 1976; Koby 1953) with the conslusion that natural conditions accounted for the modifications on bear bones that had been cited as evidence for much undisciplined speculation leading to the modern myths of bear cults and Middle Paleolithic religiosity.

We have a great deal to learn, and from a methodological perspective, the biggest lesson is to give up our anthropocentrism. We must critically seek understanding as to the role of nonhominid agents and conditions as contributors to the geological palimpsests within which the earliest evidence of hominid behavior occurs.

As a trial demonstration of both the application of my methods and the need for such systematic applications, I have analyzed some of the important data reported from Olduvai Gorge. I have demonstrated what will most certainly become a point of controversy—a new and perhaps "unflattering" view regarding the nature of the tool use among the hominids of the Plio-Pleistocene boundary. This view was made possible by the application of a *methodology* to a body of observations. There was no simple accommodation of ideas to the observations. A good methodology must be uncompromising. Only then is the quality of the results referable to the quality of the methodology. When "just-so stories" and intellectual sleight of hand are the basis for making statements as to what the past was like, one can only attack the logic and the persons making such arguments. When a methodology is the basis for statements about the past, it is possible to research the methodology itself and to uncover its weaknesses and perfect its strengths. Progress is made under such conditions, whereas in the absence of methodology only persons' reputations are reshuffled.

The methodology outlined has led to certain conclusions about the character of the past, importantly about the behavior of our early hominid ancestors. The picture one gains from the analysis of the Olduvai materials is a far cry from many of the romantic pictures that have been advanced (see, for instance, Ardrey 1976; R. Leakey and R. Lewin 1977; 1978). This analysis seems to justify several statements about our early hominid ancestors:

1. They were scavenging the consumed kills and scavenging death sites of animals after most of the other predator–scavengers had abandoned the carcass and scattered some of its parts.
2. The parts scavenged were primarily leg bones that appear to have already had the meat removed, or they were lower leg bones that had little meat present to begin with.
3. The major, or in many cases the only, usable or edible parts consisted of bone marrow.
4. Hominids were using hammer tools to break open the leg bones and thereby expose the usable marrow.

All of the foregoing statements are essentially post hoc accommodative arguments offered to account for patterning isolated in this study. As such they are not endpoints of research but should be points of beginning. These suggestions regarding the character of the past must be evaluated and the evaluation must rest with the perfection of methods for unambiguously recognizing the products of hominids as distinct from other agents. It must also rest with the development of means for evaluating alternative arguments that could be offered to account for the same facts. For instance, some of the patterning isolated in my analysis, though clearly outside the range of the currently documented animal-produced assemblages, could be the consequences of water transport (see Behrensmeyer 1975a; Voorhies 1969) biasing animal-generated bone populations, or perhaps it could be the result of other processes not yet considered.

Although my suggestions as to the significance of patterning recognized in this study may be viewed as starting points for further research, there have been some arguments presented in evaluation of inferences previously made that appear to be rather conclusive.

There is no evidence supporting the idea that the hominids were removing food from the location of procurement to a base camp for consumption. In fact, the covariant patterning among anatomical parts shows that the parts selected by hominids for use were taken from already consumed and abandoned carcasses: the coincidence of both the residual components of such an animal kill and the modified elements used by the hominids at the same site demonstrates nicely that consumption was at the place of procurement. *No evidence for base camps exists.* Similarly, *the argument that food was shared is totally unsupported.* It is hard to imagine the sharing of tiny morsels of bone marrow. For sharing to become a regular behavior, the food package must at least exceed in size the amount a single individual might consume at a single sitting. I am firmly convinced that sharing was important in human evolution and I am equally convinced that it was a consequence of shifts to the hunting of animal foods, particularly animals of moderate to large size. *There is no evidence supporting the argument that the hominids at Olduvai Gorge were hunting.* The most commonly cited evidence, such as the elephant "kill" at FLK North Level 6, the *Deinotherium* "kill" at FLK, and perhaps the large semiarticulated animals at HWK East Level 2, as well as the "hippo" location at Koobi Fora, are all most likely natural death locations. As pointed out, the tools at the "kills" in Olduvai and Koobi Fora are localized largely independently of the bones from animals said to have been killed. In fact, the tools are most commonly associated with the bones of bovids, and it is very hard to argue that these bones were in the site by virtue of the same events that resulted in the presence of the large animal carcasses.

For many years the problem of what the selective context for the initial use of tools was like and in turn what the earliest tools were used for has received some speculative attention. Arguments recently advanced regarding early tool use have been accommodated to the fanciful word pictures that Ardrey, Isaac, and Leakey and Lewin have painted of our early ancestors.

Robert Ardrey, the advocate of the hunting hypothesis, states,

> so every selective pressure favored those of us who most reliably could stand erect and move on two feet, just as it favored those newly freed hands that could most ably wield a weapon. Perhaps as our capacities increased we used our weapons not entirely to defend ourselves, but on occasion to knock down a small, slow animal and so supplement our diet of fruit with odd bits of meat [Ardrey 1976:43].

Glynn Isaac, the advocate of the sharing hypothesis, pictures the context of tool use as follows:

> What about stone tools? Our ancestors, like ourselves, could probably break up the body of a small animal, as chimpanzees do, with nothing but their hands and teeth. It is hard to visualize them or us, however, eating the meat of an elephant, a hippopotamus or some other large animal without the aid of a cutting implement. As the archaeological evidence demonstrates abundantly, the protohumans of East Africa not only knew how to produce such stone flakes by percussion, but also found them so useful that they carried the raw materials needed to make the implements with them from place to place [Isaac 1978:103].

Richard Leakey and Roger Lewin, popularizers of other people's ideas, re-create the context of early stone tool use as follows:

> Because different sticks are more effective when they are relatively sharp, the early hominids probably soon learned to whittle branches to a point, using stone flakes they happened to come across. . . . From there it is but a short intellectual leap to manufacture a cutting edge by breaking one stone in half by striking it against another or, more crudely, by smashing it to the ground. The point here though is that the first stone-tool technology . . . was very probably needed to fashion other tools of wood [R. Leakey and R. Lewin 1978:142].

We are told that the first tools were weapons for defense and killing prey, knives for cutting up large animals, or tools used for shaping more important tools of wood. All these arguments were invented to fit the author's beliefs. Ardrey believes in the "killer-apes"; Isaac in a kind of middle-class genteel protohuman who shared his food, took care of his family, and was on his way to being emotionally and intellectually "human"; for Leakey and Lewin, hominid life was an impoverished projection into the past of Richard Lee's ideas about the !Kung Bushmen.

What behavioral context for the earliest use of tools is indicated by the faunal analysis performed here?

One of the interesting generalizations that biologists often make about the ecological relationships among species living in the same habitat is that no two species occupy identical niches. Another commonly noted phenomenon is that species occupying nearly identical habitats are related to one another in a sometimes intriguingly complex "food web." In most situations where we have multiple species occupying very similar habitats, we find that they eat complementary components of basically the same food sources. For instance, some grazers feed on the seed heads, others feed on the large leaves, and still others crop the young growth of the same grass species. When we view such situations from an ecological perspective, most successful adaptive diversifications among species are accomplished by one species living on the entropy from another species' adaptation. The long and complicated succession from top predator to the lowly Diptera larva is essentially a story of who eats what another missed, ignored, or abandoned. Almost all earlier discussions of early hominid adaptations have essentially seen a shift to either scavenging or hunting as a direct competitive confrontation between hominids and other predator–scavengers. Rather than eating what they were not, hominids were viewed as "taking their food away" from others. For instance, Robert Ardrey, in what he later refers to as the "death of the scavenger hypothesis " put things this way:

> If we were incapable of killing large prey animals such as wildebeest and waterbuck, then how were we capable of stealing their remains from their rightful and more dangerous killers? If we had been concerned with only a few stray bones, then luck could account for it. But the

> impressive accumulations at early hominid living sites must indicate either that we had been even more adept thieves than we are today, or that the great carnivores in those times were unaccountably lazy at guarding their kills [Ardrey 1976:17].

The analysis of the materials remaining from our ancestral adaptations suggests no such confrontations or need for postulating "lazy" carnivores. Our ancestors were simply taking advantage of matter abandoned and/or ignored by other animals. No confrontations need have occurred, for our ancestors were simply moving in on "waste" and "entropy" from the predator–scavenger adaptations already in operation. The small morsels of such waste would have brought our ancestors into situations of face-to-face competition with nothing more fierce than perhaps a Diptera larva! The opposable thumb and the configuration of the hand made possible the use of a stone as a hammer to gain access to abandoned tiny morsels of food. In fact, access to bone marrow by hammer-wielding hominids was probably more direct and effective than any other, save perhaps the legendary jaws of the hyena. In competition for bone marrow, early hominids had an advantage. Ecological success is bestowed upon those with advantages.

The setting one can imagine for early tool use is just that *tool use* in which available stones were used to break bones. The percussion, however, also broke the stones, particularly when used in conjunction with an anvil. Over the millenia our hominid ancestors learned something of percussion techniques and eventually used this knowledge in tool manufacture and for the production of tools other than hammers. I strongly doubt that tool manufacture as opposed to patterned breakage was a characteristic of hominid modification of lithics until we begin to see design characteristics in lithic items in Lower Bed II at Olduvai Gorge. I suspect that increased scavenging of meat from carcasses is a behavioral trend and may well relate to increases in the "flake tool" component of the early industries, as well as increases in the frequency of larger animals seen through the Olduvai sequence. When one employs a regular set of scavenging strategies aimed at recovering meat, it is quite likely that there would be a numerical shift toward larger animals, in that the larger the body size the greater the likelihood of meat remaining available to a larger number of scavengers, including hominids.

I am convinced that hunting as an important contribution to a human adaptation is a part of our history that must be understood in terms of the radiation of "men" out of Africa. I can see no selective context for hunting to have arisen as an organized component of a hominid adaptation within the African "homeland." In fact, I would be very surprised to find that hunting played an important role in most African adaptations until after the appearance of the bow and arrow. Although these "hunches" are interesting, this is not the place for a return to sheer speculation.

The questions I have raised regarding the currently accepted ideas about early man are perhaps sufficent to justify the need for the development of robust middle-range theory and to follow up the research leads clarified by the applications of a *methodology*. The imaginative researcher should appreciate that many of the arguments presented here that stem directly from my attempt to apply a nascent methodology demand further types of research, which have not thus far been conducted. For instance, no systematic analyses on the internal spatial structure of the distributions carefully plotted by Mary Leakey have been carried out. It is true that Milla Ohel (1977) has made some provocative observations isolating some locations of probable hominid activity, but the details of the faunal elements represented in such localities are not available. Similarly, the comparative study of associations between debitage clusters on the floor and scatters of larger tools might well yield interesting facts, yet these studies have not been made. Even the minor attempts that have been made proceeded from the assumption that internal differentiations noted within the distributions referred to internal differentiations and specialized space within hominid sites (see Ohel 1977). It appears much more likely that most such differential distributions and the apparent structure on these "floors" derive from the compounding of many different episodes and agents, each contributing slightly different materials having slightly different spatial properties. For instance, the concentration of broken bone on the *Zinj* floor (see M. D. Leakey 1971: Figure 24) is most likely referable to hominids. *Zinj* himself is more likely the scavenged body of a hominid deposited independently of the marrow-cracking episode indicated by the bone and lithic debris concentration. Similarly, the scattered bones frequently referred to as peripheral to the site are probably a normal paleontological background of altered bones from natural death sites and kills by predators. We need to pay much more attention to the actual structure of spatial patterning among elements that are fairly reliably referred to hominid behavior. It is in such patterning that diagnostic evidence for important behaviors such as sharing and the use of home bases will be developed. Methodological development is needed in the domain of site structure to aid both in the separation of the different episodes and agents that may contribute to a deposit and in the recognition of evolutionarily important behaviors.

In addition to development of new domains of research such as site structure, there are other needs. Almost all the control populations, and control studies (such as those on bone density) remain in the "encouraging" stage. That is, enough work has been done for us to be encouraged that we are on the track of valuable and useful knowledge, but not enough has been done to render our current knowledge very secure. Much more basic research is needed to perfect methods for knowing the past. We have had far too much of what I tend to think of as the *National Geographic* approach to research. That is the view that progress is made through discoveries (preferably photogenic ones) that are treated as self-evident in their meaning. In fact just the reverse is the case. Basic research makes possible the reliable assignment of meaning to observations. Without such research, discoveries simply serve as the stimulus for modern myth making. We have had quite enough of that. It is time we got down to the difficult and perhaps not so photogenic task of carrying out the basic research that will make possible the move from the age of myths to the era of understanding.

References

Abel, O.

1926 How Neanderthal man hunted cave bears in the Dragon's Cave near Mixnitz, Austria. *Natural History 26*:252–256.

Agenbroad, L. D.

1978 *The Hudson–Meng site: An Alberta bison kill in the Nebraska High Plains.* Washington, D.C.:University Press of America.

Altuna, J.

1971 Los mamiferos del yacimiento prehistorico de Morin. In *Cueva Morin Excavaciones 1966–1968,* by G. Echegaray *et al.* Santander: Patronato de las Cuevas Prehistóricas. Pp. 371–397.

Amsterdamski, S.

1975 Between experience and metaphysics. *Boston Studies in the Philosophy of Science 35.* Boston:Reidel.

Andree, J.

1939 *Der eiszeitliche Mensch in Deutschland und seine kulturen.* Stuttgart:Enke.

Archer, M., M. Crawford, and D. Merrilees

1980 Incisions, breakages, and charring, some probably man-made, in fossil bones from Mammoth Cave, Western Australia. *Alcheringa 4* (1–2):115–131.

Ardrey, R.

1976 *The hunting hypothesis: A personal conclusion concerning the evolutionary nature of man.* New York:Atheneum.

Bächler, E.

1921 *Das Drachenloch ob Vättis im Tamintale.* 2445 M.U.M. und seine Bedeutung als paläontologische funstätte und prähistorische Niederlassumg aus der Altsteinzeit (Pälaolithikum) in Schweizerlande. Switzerland:St. Gall. Pp. 1–144.

1940 *Das Alpine Pälaolithikum der Schweiz.* Monographien zur Ur-und Frühgeschichte der Schweiz. Bd. II.

Battaglia, R.

1953 Le ossa lavorate della caverna Pocala Nella Venezia Guiulia e il problema del "Mosteriano Alpino." *Bulletino di Palentologia Italiana 8*(6):5–31.

Beard, P. H.

1979 *The end of the game.* Garden City, N.Y.:Doubleday.

Behrensmeyer, A. K.

1975a Taphonomy and paleoecology in the hominid fossil record. In *Yearbook of Physical Anthropology 19,* edited by J. Buettner-Janusch. Pp. 36–50.

1975b The taphonomy and paleoecology of Plio-Pleistocene vertebrate assemblages east of Lake Rudolf, Kenya. *Bulletin of the Museum of Comparative Zoology 146*(10):473–578.

1978 Taphonomic and ecologic information from bone weathering. *Paleobiology 4*(2):150–162.

Behrensmeyer, A. K., and D. E. Dechant-Boaz

1980 The recent bones of Amboseli Park, Kenya, in relation to East African paleoecology. In *Fossils in the making: Vertebrate taphonomy and paleoecology,* edited by A. K. Behrensmeyer and A. P. Hill. Chicago:University of Chicago Press. Pp. 72–92.

Behrensmeyer, A. K., D. Western, and D. E. Dechant-Boaz
1979 New perspectives in vertebrate paleoecology from a recent bone assemblage. *Paleobiology* 5(1):12–21.

Berckhemer, F.
1934 Der Steinheimer Urmensch und die Tierwelt seines Lebensgebietes. *Naturwiss Monatschrift der Deutschen Naturkundever 47*(4). Stuttgart.

Berounioux, F. M.
1958 "Spiritualité" de l'homme de Néandertal. In *Hundert Jahre Neanderthaler 1856–1956*, edited by G. H. R. von Koenigswald. Utrecht, The Netherlands:Wenner Gren Foundation for Anthropological Research. Pp. 151–165.

Biberson, P., and E. Aguirre
1965 Experiences de taille d'outils préhistoriques dan des os d'elephant. *Quaternaria 7*:165–183.

Binford, L. R.
1967 Smudge pits and hide smoking: The use of analogy in archaeological reasoning. *American Antiquity 32*(1):1–12.
1968a Archaeological perspectives. In *New prespectives in archaeology*, edited by S. R. Binford and L. R. Binford. Chicago:Aldine. Pp. 5–32.
1968b Archaeological theory and method. In *New perspectives in Archaeology*, edited by S. R. Binford and L. R. Binford. Chicago:Aldine. Pp. 1–3.
1968c Post-Pleistocene adaptations. In *New perspectives in archaeology*, edited by S. R. Binford and L. R. Binford. Chicago:Aldine. Pp. 313–341.
1968d Some comments on historical versus processual archaeology. *Southwestern Journal of Anthropology 24*:267–275.
1972a Archaeological reasoning and smudge pits—revisited. In *An archaeological perspective*, by L. R. Binford. New York:Seminar Press. Pp. 52–58.
1972b Contemporary model building: Paradigms and the current state of Paleolithic research. In *Models in archaeology*, edited by D. L. Clarke. London:Methuen. Pp. 19–166.
1973 Interassemblage variability—the Mousterian and the "functional argument." In *The explanation of culture change*, edited by C. Renfrew. London:Duckworth. Pp. 227–254.
1976 Forty-seven trips: A case study in the character of some formation processes of the archaeological record. In *Contributions to anthropology: The interior peoples of northern Alaska*, edited by E. S. Hall, Jr. *National Museum of Man, Mercury Series* (Paper No. 49). Ottawa:National Museum of Canada. Pp. 299–351.
1977a General introduction. In *For theory building in archaeology*, edited by L. R. Binford. New York:Academic Press. Pp. 1–10.
1977b Historical archaeology: Is it historical or archaeological? In *Historical archaeology and the importance of material things*, edited by L. Ferguson. Society for Historical Archaeology, Special Publication No. 2. Pp. 13–32.
1977c Olorgesailie deserves more than the usual book review. *Journal of Anthropological Research 33*(4):493–502.
1978a Dimensional analysis of behavior and site structure: Learning from an Eskimo hunting stand. *American Antiquity 43*(3):330–361.
1978b *Nunamuit ethnoarchaeology*. New York : Academic Press.
1979 Organization and formation processes, looking at curated technologies. *Journal of Anthropological Research* 35(3):255–273.
1980 Willow smoke and dog's tails: Hunter-gatherer settlement systems and archaeological site formation. *American Antiquity 45*(1):4–20.

Binford, L. R., and J. B. Bertram
1977 Bone frequencies—and attritional processes. In *For theory building in archaeology*, edited by L. R. Binford. New York:Academic Press. Pp. 77–153.

Binford, L. R., and S. R. Binford
1966 A preliminary analysis of functional variability in the Mousterian of Levallois facies. *American Anthropologist 68*(2):238–295.

Binford, L. R., and W. J. Chasko
1976 Nunamiut demographic history: A provocative case. In *Demographic anthropology*, edited by E. B. W. Zubrow. Albuquerque:University of New Mexico Press. Pp. 63–143.

Binford, S. R.
1968a Ethnographic data and understanding the Pleistocene. In *Man the hunter*, edited by R. B. Lee and I. DeVore. Chicago:Aldine. Pp. 274–275.
1968b A structural comparison of disposal of the dead in the Mousterian and Upper Paleolithic. *Southwestern Journal of Anthropology 24*(2):139–154.

Binford, S. R., and L. R. Binford (Editors)
1968 *New perspectives in archaeology*. Chicago:Aldine.

Black, D. (Editor)
1933 Fossil man in China: The Choukoutien cave deposits with a synopsis of our present knowledge of the Late Cenozoic in China. *Memoir of the Geological Survey of China Series A, 11*.

Blanc, A.
1939 L'Homme fossile du Monte Circeo. *L'Anthropologie 44*:253–264.
1950 Was Neanderthal man a ritual cannibal? Evidence from the Cave of Circeo which points to a grim practice common to the head-hunters of today

and the cave-men of prehistory. In *Illustrated London News 216*(5784):292.

1958 Torre in Pietra, Saccopastore, Monte Circeo: On the position of the Mousterian in the Pleistocene sequence of the Rome area. In *Hundert Jahre Neanderthaler 1856–1956*, edited by G. H. R. von Koenigswald. Utrecht, The Netherlands:Wenner Gren Foundation for Anthropological Research. Pp. 167–174.

1961 Some evidence for the ideologies of early man. In *Social life of early man*, edited by S. L. Washburn. Viking Fund Publications in Anthropology 31. Pp. 119–136.

Boaz, N. T., and A. K. Behrensmeyer

1976 Hominid taphonomy: Transport of human skeletal parts in an artificial fluviatile environment. *American Journal of Physical Anthropology 45*(1):53–60.

Bond, G.

1969 The geology of the Kalambo Falls prehistoric site. In *Kalambo Falls Prehistoric Site*, Vol. 1, by J. D. Clark. Cambridge:Cambridge University Press. Pp. 197–213.

Bonnichsen, R.

1973 Some operational aspects of human and animal bone alteration. In *Mammalian osteo-archaeology: North America*, edited by M. Gilbert. Columbia, Miss.:Missouri Archaeological Society. Pp. 9–24.

1975 On faunal analysis and the australopithecines. *Current Anthropology 16*(4):635–636.

1978 Critical arguments for Pleistocene artifacts from the Old Crow Basin, Yukon: A preliminary statement. In *Early man in America, from a circum-Pacific perspective*, edited by A. L. Bryan. Edmonton:Archaeological Researchers International. Pp. 102–118.

1979 Pleistocene bone technology in the Beringian Refugium. *Archaeological Survey of Canada Paper, Mercury Series* (Paper No. 89). Ottawa:National Museum of Canada.

Bordes, F.

1961 *Typologie du Paléolithique ancien et moyen.* Publications de l'Institut de Préhistoire de l'Université de Bordeaux, Memoir No. 1. Bordeaux:Imprimeries Delmas.

1968 *The Old Stone Age.* New York:World University Library, McGraw-Hill.

1969 Os percé Mousterien et os gravé Acheuléen du Pech de l'Azé II. *Quaternaria 11*:1–6.

1972 *A tale of two caves.* New York:Harper & Row.

Boucher de Perthes, M.

1849 *Antiquites celtiques and Antediluviennes*, Vol. 1. Paris.

Bourliere, F.

1963 Specific feeding habits of African carnivores. *African Wild Life 17*(1):21–27.

1966 Observations on the ecology of some large African mammals. In *African ecology and human evolution*, edited by F. C. Howell and F. Bourliere. Chicago:Aldine. Pp. 43–54.

Brain, C. K.

1967a Bone weathering and the problem of bone pseudo-tools. *South African Journal of Science 63*:97–99.

1967b Hottentot food remains and their meaning in the interpretation of fossil bone assemblages. *Scientific Papers of the Namib Desert Research Station* No. 32:1–11.

1968 Who killed the Swartkrans ape-man? *South African Museum Bulletin* No. 9:127–139.

1969 The contribution of Namib Desert Hottentots to an understanding of australopithecine bone accumulations. *Scientific Papers of the Namib Desert Research Station* No. 39:13–22.

1970 New finds at the Swartkrans australopithecine site. *Nature 225*:112–1119.

1976a A re-interpretation of the Swartkrans site and its remains. *South African Journal of Science 72*:141–146.

1976b Some principles in the interpretation of bone accumulations associated with man. In *Human origins: Louis Leakey and the East African evidence*, edited by G. Ll. Isaac and B. R. McCown. Menlo Park, Calif.:Benjamin. Pp. 97–116.

1980 Some criteria for the recognition of bone-collecting agencies in African caves. In *Fossils in the making: Vertebrate taphonomy and paleoecology*, edited by A. K. Behrensmeyer and A. P. Hill. Chicago:University of Chicago Press. Pp. 107–130.

Breuil, H.

1932 Le feu et l'industrie de pierre et d'os dans le gisement du "Sinanthropus" à Chou Kou Tien. *L'Anthropologie 42*:1–17.

1938 The use of bone implements in the Old Paleolithic period. *Antiquity 12*:56–67.

1939 Bone and antler industry of the Choukoutien *Sinanthropus* site. *Palaeontologia Sinica*, n. s. D., No. 6, Peking.

Brooks, R. H.

1967 A comparative analysis of bone from locality 2 (C1-245), Tule Springs, Nevada. *Nevada State Museum Anthropological Papers* No. 13:402–411.

Brumley, J. H.

1973 Quantitative methods in the analysis of butchered

faunal remains: A suggested approach. *Archaeology in Montana* 14(1):1–40.

Bryan, A. L.
1978 An overview of Paleo-American prehistory from a circum-Pacific perspective. In *Early man in America, from a circum-Pacific perspective,* edited by A. L. Bryan. Edmonton:Archaeological Researchers International. Pp. 306–327.

Buckland, W.
1823 *Reliquiae Diluvianae; or, observations on the organic remains contained in caves, fissures, and diluvial gravel, and on other geological phenomena attesting the actions of a universal deluge.* London: John Murray. New York:Arno Press. (Reprinted 1978).

Bunge, M.
1979 *Causality and modern science* (3rd rev. ed.). New York:Dover.

Bunn, H. T.
In press Archaeological evidence for meat-eating by Plio-Pleistocene hominids from Koobi Fora, Kenya and Olduvai Gorge, Tanzania. Submitted to *Nature*

Bunn, H. T., J. Harris, G. L. Isaac, Z. Kaufulu, E. Kroll, K. Schick, N. Toth & A. K. Behrensmeter
1980 FxJj 50: An Early Pleistocene Site in Northern Kenya. *World Archaeology* 12(2):109–136.

Campbell, J. B., and C. Sampson
1971 A new analysis of Kent's Cavern, Devonshire, England. *University of Oregon Anthropological Papers* No. 3.

Canby, T. Y.
1979 The search for the first Americans. *National Geographic* 156(3):330–363.

Chamberlin, T. C.
1965 The method of multiple working hypotheses. *Science* 148 (3671):754–759.

Chaplin, R. E.
1969 The use of non-morphological criteria in the study of animal domestication from bones found in archaeological sites. In *The domestication and exploitation of plants and animals,* edited by P. J. Ucko and G. W. Dimbleby. Chicago:Aldine. Pp. 231–245.
1971 *The study of animal bones from archaeological sites.* New York:Seminar Press.

Childe, V. G.
1956 *Piecing together the past.* New York:Praeger.

Clark, G.
1977 *World prehistory in new perspective* (3rd ed). Cambridge:Cambridge University Press.

Clark, J. D.
1969 *Kalambo Falls prehistoric site,* Vol. I. Cambridge:Cambridge University Press.

Clark, J. D., and C. V. Haynes
1970 An elephant butchery site at Mwanganda's village, Karonga, Malawi, and its relevance for Palaeolithic archaeology. *World Archaeology* 1:390–411.

Clarke, D. L.
1972 Review of explanation in archaeology: An explicitly scientific approach. *Antiquity* 46:237–239.
1973 Archaeology: The loss of innocence. *Antiquity* 47:6–18.

Cleveland, G. C.
1976 Experimental replication of butchered Artiodactyla bone with special reference to archaeological features at 45 FR 5. In *Preliminary archaeological investigations at the Miller site, Strawberry Island, 1976,* by G. C. Cleveland *et al.* Project Report No. 46. Pullman, Washington:Washington Archaeological Research Center.

Coe, M.
1978 The decomposition of elephant carcasses in the Tsavo (East) National Park, Kenya. *Journal of Arid Environments* 1(1):71–86.
1980 The role of modern ecological studies in the reconstruction of paleoenvironments in sub-Saharan Africa. In *Fossils in the making: Vertebrate taphonomy and paleoecology,* edited by A. K. Behrensmeyer and A. P. Hill. Chicago:University of Chicago Press. Pp. 55–67.

Cohen, M. R., and E. Nagel
1934 *An introduction to logic and scientific method.* New York:Harcourt, Brace.

Coles, J. M., and E. S. Higgs
1969 *The archaeology of early man.* New York:Praeger.

Coon, C.
1962 *The origin of races.* New York:Knopf.

Crader, D. C.
1974 The effects of scavengers on the bone material from a large mammal: An experiment conducted among the Bisa of the Luangwa Valley, Zambia. In *Ethnoarchaeology,* edited by C. B. Connan and C. W. Clewlow, Monograph 4. Los Angeles:Institute of Archaeology, University of California. Pp. 161–173.

Daly, P.
1969 Approaches to faunal analysis. *American Antiquity* 34(2):146–153.

Daniel, G.
1968 One hundren years of Old World prehistory. In *One hundred years of anthropology,* edited by J. O. Brew. Cambridge, Mass.:Harvard University Press, Pp. 57–93.

Dart, R. A.
1949 The predatory implemental technique of *Aus-*

tralopithecus. American Journal of Physical Anthropology 7(1):1–39.
1956 The myth of the bone-accumulating hyena. *American Anthropologist 58*(1):40–62.
1957 The osteodontokeratic culture of *Australopithecus prometheus. Memoir of the Transvaal Museum* No. 10.
1958 The minimal bone-breccia content of Makapansgat and the australopithecine predatory habit. *American Anthropologist 60*(5):923–931.
1959a An "australopithecine" scoop from Herefordshire. *Nature 183*:844.
1959b Further light on australopithecine humeral and femoral weapons. *American Journal of Physical Anthropology 17*(1):87–94.
1960 The bone tool-manufacturing ability of *Australopithecus prometheus. American Anthropologist 62*(1):134–143.
1961 An australopithecine scoop made from a right *Australopithecus* upper arm bone. *Nature 191*:372–373.
1962 From cannon-bone scoops to skull bowls at Makapansgat. *American Journal of Physical Anthropology 20*(3):287–295.
1967 Mousterian osteodontokeratic objects from Geula Cave (Haifa, Israel). *Quaternaria 9*:105–140.

David, F.
1972 Annexe III, Témoins Osseux. In *Fouilles de Pincevent,* essai d'analyse ethnographique d' un habitat Magdalénien (La section 36), by A. Leroi-Gourhan et M. Brézillon. Paris:Editions du Centre National de la Recherche Scientifique. Pp. 295–320.

Davis, L. B., and M. Wilson (Editors)
1978 Bison procurement and utilization: A symposium. *Plains Anthropologist Memoir* No. 14.

Dawkins, W. B.
1874 *Cave hunting, researches on the evidence of caves respecting the early inhabitants of Europe.* London:Macmillan.

DeLumley, H.
1969 *Une cabane Acheuléenne dans la Grotte du Lazaret.* Memoirs de la Société Préhistorique Française, Tome 7. Paris:Société Préhistorique Française.
1975 Cultural evolution in France in its paleoecological setting during the Middle Pleistocene. In *After the australopithecines,* edited by K. W. Butzer and G. Ll. Isaac. The Hague:Mouton. Pp. 745–808.

De Mortillet, G. Et. A.
1900 Le Préhistorique, Origine et Antiquité de L'Homme (3rd ed.). Paris:Schleicher Freres.

Dibble, D. S., and D. Lorrain
1968 Bonfire Shelter: A stratified bison kill site, Val Verde County, Texas. *Texas Memorial Museum, Miscellaneous Papers* No. 1.

Douglas-Hamilton, I., and O. Douglas-Hamilton
1975 *Among the elephants.* New York:Viking.

Ducos, P.
1969 Methodology and results of the study of the earliest domesticated animals in the Near East (Palestine). In *The domestication and exploitation of plants and animals,* edited by P. J. Ucko and G. W. Dimbleby. Chicago:Aldine. Pp. 265–275.

Ehrenberg, K.
1959 Die urzeitlichen Fundstellen und Funde in der Salzofenhöhle, Steiermark. *Archaeologia Austriaca 25*:8–24.
1962 Über weitere urzeitliche Fundstellen und Funde in der Salzofenhöhle, Steiermark. *Archaeologia Austriaca 32*:1–23.

Einstein, A.
1947 The problem of space, ether, and the field of physics. In *Man and the universe: The philosophers of science,* edited by S. Commins and R. N. Linscott. New York:Random House. Pp. 471–482.

Elton, G. R.
1969 *The practice of history.* London:Collins.

Evans, J. C.
1978 *An introduction to environmental archaeology.* Ithaca, N.Y.:Cornell University Press.

Fagan, B. M.
1967 *Iron Age cultures in Zambia (Kolomo and Kangela),* Vol. 1. London:Chatto and Windus.

Fagan, B. M., D. W. Phillipson, and S. G. H. Daniels
1969 *Iron Age cultures in Zambia (Dambwa, Ongombe, Ilede, and Tonga),* Vol. 2. London:Chatto and Windus.

Feustel, R.
1969 Zur problematik der "Protolithischen Knochencultur" und der "Osteodontokeratic Culture." *Alt-Thuringen 10*:1–7.
1970 Comment on article by Donald L. Wolberg. *Current Anthropology 11*(1):32.

Freeman, L. G.
1968 A theoretical framework for interpreting archaeological materials. In *Man the hunter,* edited by R. B. Lee and I. DeVore. Chicago:Aldine. Pp. 262–267.
1971 El Hueso Trabajado Musteriense de Cueva Morin. In *Cueva Morin Excavaciones 1966–1968,* by G. Echegaray *et al.* Santander:Patronato de las Cuevas Prehistóricas. Pp. 135–161.
1973 El Musteriense. In *Cueva Morin Excavaciones*

1969, by G. Echegaray *et al.* Santander:Patronato de las Cuevas Prehistóricas. Pp. 15–140.

1975 Acheulian sites and stratigraphy in Iberia and the Mahgreb. In *After the australopithecines*, edited by K. W. Butzer and G. Ll. Isaac. The Hague:Mouton. Pp. 661–743.

1978 Mousterian worked bone from Cueva Morin (Santander, Spain): A preliminary description. In *Views of the past*, edited by L. G. Freeman. The Hague:Mouton. Pp. 29–51.

Freeman, L. G., and K. W. Butzer

1966 The Acheulean station at Torralba (Spain): A progress report. *Quaternaria* 8:9–21.

Friedman, K.

1980 Stone or bone? In *Odyssey*, edited by M. Ambrosino. Boston:Public Broadcasting Associates. P. 7.

Frison, G.

1970 The Glenrock Buffalo Jump, 48CO304. *Plains Anthropologist Memoir* No. 7.

1971 Shoshonean antelope procurement in the Upper Green River Basin, Wyoming. *Plains Anthropologist* 16(54):258–284.

1978 *Prehistoric hunters of the High Plains*. New York:Academic Press.

Frison, G. (Editor)

1974 *The Casper site: A Hell Gap bison kill on the High Plains*. New York:Academic Press.

Frison, G., M. Wilson, and D. J. Wilson

1976 Fossil bison and artifacts from an early Altithermal period arroyo trap in Wyoming. *American Antiquity* 41(1):28–57.

Gamble, C.

1978 Resource exploitation and the spatial patterning of hunter–gatherers: A case study. In *Social organization and settlement: Contributions from anthropology, archaeology and geography*, edited by D. Green, C. Haselgrove, and M. Spriggs. Oxford:British Archaeological Reports. Pp. 153–185.

1979 Hunting strategies in the Central European Paleolithic. *Proceedings of the Prehistoric Society* 45:35–52.

Gifford, D. P.

1977 Observations of modern human settlements as an aid to archaeological interpretation. Ph.D. dissertation, University of California, Berkeley.

1978 Ethnoarchaeological observations of natural processes affecting cultural materials. In *Explorations in ethnoarchaeology*, edited by Richard A. Gould. Albuquerque:University of New Mexico Press. Pp. 77–101.

1980 Ethnoarchaeological contributions to the taphonomy of human sites. In *Fossils in the making: Vertebrate taphonomy and paleoecology*, edited by A. K. Behrensmeyer and A. P. Hill. Chicago:University of Chicago Press. Pp. 93–106.

1981 Taphonomy and paleoecology: A critical review of archaeology's sister disciplines. In *Advances in archaeological method and theory*, Vol. 4, edited by M. B. Schiffer. New York:Academic Press.

Gilbert, B. M.

1969 Some aspects of diet and butchering techniques among prehistoric Indians in South Dakota. *Plains Anthropologist* 14(46):277–294.

Gorjanovic-Kramberger, K.

1906 Der diluviale Mensch von Krapina in Kroatien. *Ein Beitrag zur Palaonthropologie*. Studien uber die entwicklungsmechanik des primatenskelettes mit besond. Berucks Anthropologie un Descendenzlehre. Lief. I–XI. Wiesbaden:Herausgeg von O Walkhoff. Pp. 59–277.

Gould, R. A.

1978 Beyond analogy in ethnoarchaeology. In *Explorations in ethnoarchaeology*, edited by R. A. Gould. Albuquerque:University of New Mexico Press.

1980 *Living archaeology*. Cambridge:Cambridge University Press.

Gould, S. J.

1977 *Ever since Darwin: Reflections in natural history*. New York:Norton.

Guilday, J.

n.d. *The Buffalo site (46PU31), Putnam County, West Virginia*. Pittsburgh, Pa.:Carnegie Museum (on file).

Guilday, J. E., P. Parmalee, and D. Tanner

1962 Aboriginal butchering techniques at the Eschelman site (36LA12), Lancaster County, Pennsylvania. *Pennsylvania Archaeologist* 32:59–83.

Guilday, J. E., and D. P. Tanner

1962 Animal remains from the Quaker State Rockshelter (36VE27) Venango County, Pennsylvania. *Pennsylvania Archaeologist* 32(3–4), Whole No.

Hammond, N. (Editor)

1979 *Analytical Archaeologist:* Collected papers of David L. Clarke. London and New York:Academic Press.

Harington, C. R., R. Bonnichsen, and R. E. Morlan

1975 Bones say man lived in Yukon 27,000 years ago. *Canadian Geographical Journal* 91:42–48.

Haynes, G.

1978a Excavation of a small Virginia cave. *Bulletin of the Archaeological Society of Virginia*. 32(3): 37–41.

1978b Morphological damage and alteration to bone: Laboratory experiments, field studies, and zoo studies. *AMQUA: Abstracts of the Fifth Biennial Meeting*. Edmonton. P. 210.
1980 Evidence of carnivore gnawing on Pleistocene and Recent mammalian bones. *Paleobiology* 6(3):341–351.

Hempel, C. G.
1965 *Aspects of scientific explanation*. New York:Free Press.
1977 Formulation and formalization of scientific theories. In *The Structure of Scientific Theories* (2nd Ed.), edited by F. Suppe. Urbana:University of Illinois Press. Pp. 244–265.

Hendey, Q. B., and R. Singer
1965 The faunal assemblages from the Gamtoos valley shelters. *South African Archaeological Bulletin 20* (80):206–213.

Henschel, J. R., R. Tison, and F. von Blottnitz
1979 Implications of a spotted hyaena bone assemblage in the Namib Desert. *South African Archaeological Bulletin 34*:127–131.

Higgs, E. S., and M. R. Jarman
1972 The origins of animal and plant husbandry. In *Papers in Economic Prehistory*, edited by E. S. Higgs. Cambridge:Cambridge University Press. Pp. 3–13.

Hill, A. P.
1975 Taphonomy of contemporary and Late Cenozoic East African vertebrates. Ph.D. dissertation, University of London, London, England.
1976 On carnivore and weathering damage to bone. *Current Anthropology 17*(3):335–336.
1979 Butchery and Natural disarticulation: an investigatory Technique. *American Antiquity 44*(4): 739–744.
1980 Early postmortem damage to the remains of some contemporary East African mammals. In *Fossils in the making: Vertebrate taphonomy and paleoecology*, edited by A. K. Behrensmeyer and A. P. Hill. Chicago:University of Chicago Press. Pp. 131–152.

Hole, F., K. V. Flannery, and J. A. Neely
1969 Prehistory and human ecology of the Deh Luran Plain, an early village sequence from Khuzistan, Iran. *Memoirs of the Museum of Anthropology, University of Michigan* No. 1.

Hooijer, D. A.
1951 On the supposed evidence of early man in the Middle Pleistocene of southwest China. *Southwestern Journal of Anthropology 7*(1):77–81.

Hörmann, K.
1923 Die Petershöhle bei Veldan in Mittelfranken. *Naturhistorische Gesellschaft zu Nuernberg; Abhandlungen, Grabungsberichte der Anthropologischen Sektion 21*:125–153.

Howell, F. C.
1965 *Early man*. New York:Time-Life Books.
1966 Observations on the earlier phases of the European Lower Paleolithic. *American Anthropologist 68*(2):88–201.

Huckell, B. B.
1979 Of chipped stone tools, elephants, and the Clovis hunters. *Plains Anthropologist 24*(85):177–189.

Hughes, A. R.
1954 Hyenas versus australopithecines as agents of bone accumulation. *American Journal of Physical Anthropology 12*:467–486.

Humphrey, R. L.
1979 Forward. In *Pre-Llano cultures of the Americas: Paradoxes and possibilities*, edited by R. L. Humphrey and D. Stanford. Washington, D.C.:Anthropological Society of Washington. Pp. v–ix.

Irving, W. N.
1978 Pleistocene archaeology in eastern Beringia. In *Early man in America, from a circum-Pacific perspective*, edited by A. L. Bryan. Edmonton:Archaeological Researchers International. Pp. 96–101.

Irving, W. N., and C. R. Harington
1973 Upper Pleistocene radio-carbon dated artifacts from the northern Yukon: Man in Beringia 27,000 years ago. *Science 179*:335–340.

Isaac, G. Ll.
1971 The diet of early man: Aspects of archaeological evidence from Lower and Middle Pleistocene sites in Africa. *World Archaeology 2*:278–298.
1975a Early hominids in action: A commentary on the contribution of archaeology to understanding the fossil record in East Africa. *Yearbook of Physical Anthropology 19*, edited by J. Buettner-Janusch.
1975b Early stone tools—an adaptive threshold? In *Problems in economic and social archaeology*, edited by G. de G. Sieveking, I. H. Longworth, and K. E. Wilson. London:Duckworth. Pp. 39–47.
1975c Stratigraphy and cultural patterns in East Africa during the middle ranges of Pleistocene time. In *After the australopithecines*, edited by K. W. Butzer and G. Ll. Isaac. The Hague:Mouton. Pp. 469–542.
1976 Stages of cultural elaboration in the Pleistocene: Possible archaeological indications of the development of language capabilities. In *Origins and evolution of language and speech*, edited by S. R. Harnad, H. D. Steklis, and J. Lancaster. Annals of the New York Academy of Sciences, Vol. 280. Pp. 275–288.

1977 *Olorgesailie: Archaeological studies of a Middle Pleistocene lake basin in Kenya.* Chicago and London:University of Chicago Press.

1978a The food-sharing behavior of protohuman hominids. *Scientific American 238*(4):90–106.

1978b Food sharing and human evolution: Archaeological evidence from the Plio-Pleistocene of East Africa. *Journal of Anthropological Research 34*(3):311–325.

Isaac, G. Ll., and B. Isaac

1975 Africa. In *Varieties of culture in the Old World,* edited by Robert Stigler. New York:St. Martin's. Pp. 8–48.

Johnson, E.

1977 Animal food resources of PaleoIndians. *The Museum Journal, Vol. XVII (PaleoIndian Lifeways).* Lubbock, Texas:West Texas Museum Association. Pp. 65–77.

1978 PaleoIndian bison procurement and butchering patterns on the Llano Estacado. *Plains Anthropologist Memoir* No. 14:98–105.

Jolly, C. J., and F. Plog

1976 *Physical anthropology and archaeologv.* New York:Knopf.

Jones, P. R.

1980 Experimental butchery with modern stone tools and its relevance for Palaeolithic archaeology. *World Archaeology 12*(2):153–165.

Jones, R. (Editor)

1875 *Reliquiae Aquitanicae: by E. Lartet and H. Christy. Being contributions to the archaeology and paleontology of Perigord and the adjoining provinces of Southern France.* Paris:R. Jones.

Kehoe, T. F.

1967 The Boarding School Bison Drive site. *Plains Anthropologist Memoir* No. 4.

Keith, Sir A.

1931 *The antiquity of man,* Vol. 2. Philadelphia:Lippincott.

Kitching, J. W.

1963 *Bone, tooth, and horn tools of Palaeolithic man: An account of the osteodontokeratic discoveries in Pin Hole Cave, Derbyshire.* Manchester England:Manchester University Press.

Kitts, D. B.

1977 *The structure of geology.* Dallas:Southern Methodist University Press.

Klein, R. G.

1973 *Ice-Age hunters of the Ukraine.* Chicago:University of Chicago Press.

1975 Paleoanthropological implications of the nonarchaeological bone assemblage from Swartklip I, southwestern Cape Province, South Africa. *Quaternary Research 5*:275–288.

1976a The mammalian fauna of the Klasies river mouth sites, southern Cape Province, South Africa. *South African Archaeological Bulletin 31*:75–98.

1976b A preliminary report on the "Middle Stone Age" open-air site of Duinefontein 2 (Melkbosstrand, southwestern Cape Province, South Africa). *South African Archaeological Bulletin 31*:12–20.

1977 The mammalian fauna from the Middle and Later Stone Age (Later Pleistocene) levels of Border Cave, Natal Province, South Africa. *South African Archaeological Bulletin 32*:14–27.

1978a The fauna and overall interpretation of the "Cutting 10" Acheulean site at Elandsfontein (Hopefield), southwestern Cape Province, South Africa. *Quaternary Research 10*:69–83.

1978b Preliminary analysis of the mammalian fauna from the Redcliff Stone Age cave site, Rhodesia. *Occasional Papers of the National Museums and Monuments of Rhodesia Series A. Human Sciences 4*(2):74–80.

1978c A preliminary report on the larger mammals from the Boomplass Stone Age cave site, Cango Valley, Oudtshoorn District, South Africa. *South African Archaeological Bulletin 33*:66–75.

1978d The vertebrate fauna from the Buffelskloof rock shelter. *South African Archaeological Bulletin 33*:35–38.

1980 The interpretation of mammalian faunas from Stone Age archaeological sites, with special reference to sites in the southern Cape Province, South Africa. In *Fossils in the making,* edited by A. K. Behrensmeyer and A. Hill. Chicago:University of Chicago Press. Pp. 223–246.

Klima, B., R. Musil, and J. Jelinek

1962 Der erforschung der Höhle Švédův stůl, 1953–1955. *Anthropos 13.* Brno: Krajske.

Koby, F. E.

1941 Le "charriage a sec" des ossements dans les cavernes. *Ecologie und Geology Helvet 34.*

1953 Les soi-disant instruments osseux de paleolithique alpin et la charriage a sec des os de l'ours des cavernes. *Verhandlg. d. Natsurforsch.* Gesellschift in Basel, Vol. 54.

Krause, W.

1937 Die eiszeitlichen Knochenfunde von Meiendorf. In *Das altsteinzeitliche Rentierjägerlager Meiendorf,* by A. L. Rust. Neumünster:Karl Wachholtz. Pp. 48–61.

Krause, W., and W. Kollau

1943 Die steinzeitlichen Wirbeltierfaunen von Stellmoor in Holstein. In *Die alt-und mittelstein-*

zeitlichen Funde von Stellmoor, by A. L. Rust. Neumünster:Karl Wachholtz. Pp. 49–59.

Kruuk, H.
1970 Interactions between populations of spotted hyenas (*Crocuta crocuta* erxben) and their prey species. In *Animal populations in relation to their food resources,* edited by A. Watson. Oxford:Blackwell. Pp. 359–374.
1972 *The spotted hyena: A study of predation and social behavior.* Chicago:University of Chicago Press.

Kuhn, T. S.
1962 The historical structure of scientific discovery. *Science 136*:760–764.
1977 Second thoughts on paradigms. In *The structure of scientific theories* (2nd ed.), edited by F. Suppe. Urbana:University of Illinois Press. Pp. 459–482.

Kurten, B.
1968 *Pleistocene mammals of Europe.* Chicago:Aldine.
1976 *The cave bear story: Life and death of a vanished animal.* New York:Columbia University Press.

Lan-Po, C.
1975 *The cave home of Peking man.* Peking:Foreign Languages Press.

Leakey, L. S. B.
1946 A prehistorian's paradise in Africa: Early Stone Age sites at Olorgesailie. *Illustrated London News,* October 5.
1968 Bone smashing by Late Miocene Hominidae. *Nature 218*:528–530.

Leakey, M. D.
1971 *Olduvai Gorge,* Vol. 3: *Excavations in Beds I and II 1960–1963.* Cambridge:Cambridge University Press.
1979 Footprints in the ashes of time. *National Geographic 155* (4):446–457.

Leakey, R., and R. Lewin
1977 *Origins.* New York:Dutton.
1978 *People of the lake: Mankind and its beginning.* Garden City, N.Y.:Anchor Press/Doubleday.

Lee, R., and I. DeVore (Editors)
1968 *Man the hunter.* Chicago:Aldine.

Leechman, D.
1951 Bone grease. *American Antiquity 16*(4):355–356.

Leonardi, F.
1958 Temoiganges de l'homme de Neandertal dans l'Italie de Nord. In *Hundert Jahre Neanderthaler 1856–1956,* edited by G. H. R. von Koenigswald. Utrecht, the Netherlands:Wenner Gren Foundation for Anthropological Research. Pp. 231–252.

Leroi-Gourhan, A., and M. Brézillon
1972 *Fouilles de Pincevent,* essai d'analyse ethnographique d'un habitat Magdalénien (La section 36). Paris:Editions du Centre National de la Recherche Scientifique.

Lewin, R.
1981 AAAS briefings: Ancient cut marks reveal work of prehuman hands. *Science 211*:372–373.

Lopez, B. H.
1978 *Of wolves and men.* New York:Scribner's.

Luttschwager, J., and A. C. von Bemmel
1962 Uber von Hominiden bearbeitete knochenfunde aus dem oberen Villafranchien in den Niederlanden. *Beaufortia 9*:139–144.

Lyman, R. L.
1978 Prehistoric butchering techniques in the lower Granite Reservoir, southeastern Washington. *Tebiwa 13*:1–25.

Lyon, P. J.
1970 Differential bone destruction: An ethnographic example. *American Antiquity 35*(2):213–215.

Magoun, A. J.
1979 Summer scavenging activity in northeastern Alaska. In *Proceedings of the First Conference on Scientific Research in the National Parks.* (Volume 1), edited by Robert M. Linn. Washington: National Park Service, Department of Interior. Pp. 335–340.

Margalef, R
1968 *Perspective in ecological theory.* Chicago:University of Chicago Press.

Marshack, A.
1972 *The roots of civilization.* New York:McGraw-Hill.
1975 *Exploring the mind of Ice Age man.* National Geographic *147*(1):65–89.
1976 Some implications of the Paleolithic symbolic evidence for the origin of language. In *Origins and evolution of language and speech,* edited by S. R. Harnad, H. D. Steklis, and J. Lancaster. Annals of the New York Academy of Sciences, Vol. 280. Pp. 213–239.

Martin, H.
1906 Présentation d'ossement de renne pertant des lésions d'origine humaine et animale. *Bulletin Société Préhistorique de France 3*:385–397.
1907–1910 *Recherches sur l'evolution du Mousterien dans le gisement de la Quina (Charente).* Volume—Industrie Osseuse. Paris:Schleicher Freres.
1910 La percussion osseuse et les esquilles qui en dérivent, expérimentation. *Bulletin Société Préhistorique de France 7*:299–304.

Mason, O. T.
1893 The birth of invention. In *Annual Report, Smith-*

sonian Institution for 1892. Washington, D.C.:Smithsonian Institution.

Mason, R. J., R. A. Dart, and J. W. Kitching
1958 Bone tools at the Kalkbank Middle Stone Age site and the Makapansgat australopithecine locality, central Transvaal. *Archaeological Bulletin* *8*(51):85–116.

Masterman, M.
1970 The nature of a paradigm. In *Criticism and the growth of knowledge*, edited by I. Lakatos and A. Musgrave. Cambridge:Cambridge University Press. Pp. 59–89.

McBurney, C. B. M. and P. Callow
1971 The Cambridge excavations at La Cotte de Saint-Brelade, Jersey—A preliminary report. *Proceedings of the Prehistoric Society 37*(2):167–207.

McCullough, D. R.
1970 Secondary productions of birds and mammals. In *Analysis of temperate forest ecosystems*, edited by D. E. Reichle. Ecological Studies, Vol. 1. Heidelberg. Pp. 107–130.

Mech, L. D.
1966 *The wolves of Isle Royale.* Fauna of the National Parks of the United States, Fauna Series 7. Washington, D.C.:U.S. Government Printing Office.

Menghin, O.
1931 *Weltgeschichte der Steinzeit.* Vienna:Anton Schroll.

Merriam, J. C.
1906 Recent cave exploration in California. *American Anthropologist* n.s. *8*(2):221–228.

Mill, J. S.
1850 *A system of logic.* New York:Harper.

Miller, G. J.
1969 A study of cuts, grooves, and other marks on recent and fossil bone, 1: Animal tooth marks. *Tebiwa 12*(1):20–26.
1975 A study of cuts, grooves and other marks on recent and fossil bone, 2: Weathering cracks, fractures, splinters and other similar natural phenomena. In *Lithic technology: Making and using stone tools*, edited by E. H. Swanson. Chicago:Mouton. Pp. 211–228.

Mills, M. G. L. and Mills, M. E. J.
1979 An analysis of bones collected at hyaena breeding dens in the Gemsbok National Park. In *Annals of the Transvaal Museum, 30*(14):145–153.

Morlan, R. E.
1978 Early man in northern Yukon Territory: Perspectives as of 1977. In *Early man in America, from a circum-Pacific perspective*, edited by A. L. Bryan. Edmonton:Archaeological Researchers International. Pp. 78–95.
1979 A stratigraphic framework for Pleistocene artifacts from Old Crow River, northern Yukon Territory. In *Pre-Llano cultures of the Americas: Paradoxes and possibilities*, edited by R. L. Humphrey and D. Stanford. Washington, D.C.:Anthropological Society of Washington. Pp. 125–145.
1980a Evidence for stone artifacts in the Early Wisconsin of northern Yukon Territory. Paper delivered at the 13th annual meeting of the Canadian Archaeological Association, Saskatoon.
1980b Taphonomy as an aid to archaeological reconnaissance and interpretation: An example from northern Yukon Territory. *AMQUA: Abstracts of the Sixth Biennial Meeting.* Orono.
1981 Taphonomy and archaeology in the Upper Pleistocene of northern Yukon Territory: A glimpse of the peopling of the New World. *Archaeological Survey of Canada, Mercury Series* (Paper No. 94). Ottawa:National Museum of Canada.

Mottl, G. von M.
1951 Die repolust-hohle be Puggau (Steiermark) und ihre eiszeitlichen Bewohner. *Archaeologia Austriaca 8*:1–81.

Movius, H. M., Jr.
1953 The Mousterian cave of Teshik-Tash, southeastern Uzbekistan, central Asia. *American School of Prehistoric Research Bulletin 17*:11–71.

Müller, A. H.
1963 *Lehrbuch der Paläozoologie.* Band I, *Allegemeine Grundlagen.* Jena:Gustav Vischer Verlag.

Myers, T. P., M. R. Voorhies, and R. G. Corner
1980 Spiral fractures and bone pseudotools at paleontological sites. *American Antiquity 45*(3):483–490.

Nagel, E.
1961 *The structure of science.* New York:Harcourt, Brace and World.

Nelson, N.C.
1928 Pseudo-artifacts from the Pliocene of Nebraska. *Science 67*:316–317.

Nicolaescu-Plopsor, C. S., and D. Nicoloaescu-Plopsor
1963 The possible existence of the proto-hominids in Rumania's Villafranchian. *Dacia 7*:9–25.

Noe-Nygaard, N.
1977 Butchering and marrow fracturing as a taphonomic factor in archaeological deposits. *Paleobiology 3*:218–237.

Ohel, M. Y.
1977 Patterned concentrations on living floors at Old-

uvai, Beds I and II: An experimental study. *Journal of Field Archaeology* 4:423–433.

Owens, D. D., and M. J. Owens
1979 Communal denning and clan associations in brown hyenas (*Hyena brunnea, Thunberg*) of the central Kalahari Desert. *African Journal of Ecology* 17:35–44.

Ozegovic, F.
1958 Die bedeutung der Enteckumg des Diluvialen Menchen von Krapina in Kroatien. In *Hundert Jahre Neanderthaler 1856–1956,* edited by G. H. R. von Koenigswald. Utrecht, The Netherlands:Wenner-Gren Foundation for Anthropological Research. Pp. 27–32.

Park, E.
1978 The Ginsberg caper: Hacking it as in the Stone Age. *Smithsonian* 9(4):85–96.

Parmalee, P. W.
1965 The food economy of Archaic and Woodland peoples at the Tick Creek Cave site, Missouri. *Missouri Archaeologist 27* (1):1–34.

Pei, W. C.
1938 Le Role des animaux et des causes naturelles dans le cassure des os. *Palaeontologica Sinica,* n.s. D., No. 7. Peking.

Perkins, D., and P. Daly
1968 A hunters' village in Neolithic Turkey. *Scientific American* 219(5):97–106.

Pienaar, U. de V.
1969 Predator–prey relationships amongst the larger mammals of the Kruger National Park. *Koedoe* 12:108–176.

Pillaert, E. E.
1969 Faunal remains from the Millville site (47GT53), Grant County, Wisconsin. *Wisconsin Archaeologist* 50(2):93–108.

Platt, J. R.
1964 Strong inference. *Science* 146 (3642):347–353.

Popper, K. R.
1965 *The logic of scientific discovery.* New York: Harper and Row.

Posnansky, M.
1959 A Hope Fountain site at Olorgesailie, Kenya Colony. *South African Archaeological Bulletin* 14(55):83–90.

Potts, R. and P. Shipman
In press Cutmarks made by stone tools on bones from Olduvai Gorge, Tanzania. Submitted to *Nature.*

Putnam, F. W.
1906 Evidence of the work of man on objects from Quaternary caves in California. *American Anthropologist,* n.s. 8(2):229–235.

Quine, W. V., and J. S. Ullian
1978 *The web of belief* (2nd ed.). New York:Random House.

Rathje, W. L.
1979 Modern material culture studies. In *Advances in archaeological method and theory,* Vol. 2, edited by M. B. Schiffer. New York:Academic Press. Pp. 1–37.

Read, C. E.
1971 Animal bones and human behavior: Approaches to faunal analysis in archaeology. Ph.D. dissertation, University of California, Los Angeles.

Read-Martin, C. E., and D. W. Read
1975 Australopithecine scavenging and human evolution: An approach from faunal analysis. *Current Anthropology* 16(3):359–368.

Renfrew, C.
1973 *Before civilization: The radiocarbon revolution and prehistoric Europe.* New York:Knopf.

Riek, G.
1934 *Die Eiszeitjaegerstation am Vogelherd im Lonetal, Vol. I: Die Kulturen.* Leipzig:Curt Kabitzsch.

Runes, D. D. (Editor)
1962 *Dictionary of philosophy.* Totowa, N.J.:Littlefield, Adams.

Rust, A. L.
1937 Das altsteinzeitliche Rentierjägerlager Meiendorf. Neumünster:Karl Wachholtz.
1943 *Die atl-und mittelsteinzeitlichen Funde von Stellmoor.* Neumünster:Karl Wachholtz.
1958 *Die jungpaläolithischen Zeltanlagen von Ahrensburg.* Neumünster:Karl Wochholtz.

Rutot, A.
1907 Le cannibalisme a l'époque des cavernes en Belgique. *Bulletin Société Préhistorique de France* 4:138–327.

Sackett, J. R.
1968 Method and theory of Upper Paleolithic archaeology in southwestern France. In *New perspectives in archaeology,* edited by S. R. Binford and L. R. Binford. Chicago:Aldine. Pp. 61–83.

Sadek-Kocros, H.
1972 Primitive bone fracturing: A method of research. *American Antiquity* 37(3):369–382.
1975 Intentional fracturing of bone: Description of criteria. In *Archaeozoological Studies,* edited by A. T. Clason. Amsterdam:North Holland Publishing. Pp. 139–150.

Schaller, G. B.
1972 *The Serengeti lion: A study of predator–prey relations.* Chicago:University of Chicago Press.

Schaller, G. B., and G. Lowther
1969 The relevance of carnivore behavior to the study of early hominids. *Southwestern Journal of Anthropology 25*:307–341.

Schiffer, M. B.
1972 Archaeological context and systemic context. *American Antiquity 37*:156–165.
1976 *Behavioral archaeology*. New York:Academic Press.
1978 Methodological issues in ethnoarchaeology. In *Explorations in ethnoarchaeology*, edited by R. A. Gould. Albuquerque:University of New Mexico Press. Pp. 229–247.

Schmerling, D.
1833 *Recherches sur les ossements fossiles decouverts dans les cavernes de la province de Liege*. Liege:privately published.

Schmidt, R. R.
1908 Die neuen palaeolithischen Kulturstaetten der Schwaebischen Alb. *Archiv für Anthropologie, Völkerforschung und kolonialen Kulturwandel 235*:Pp. 62–71.
1912 *Die diluviale Vorzeit Deutschlands*. Stuttgart:E. Schweizerbartsche Verlags Naegele & Dr. Sproesser.

Schmidt, R. R., and P. Wernert
1910 Die archaeologischen Einschluesse der Loessstation Achenheim, 1: Elsass und die palaeolithischen kulturen des Rheintalloesses. *Prehistorische Zeitschrift 1*(3/4):339–346. Suedende-Berlin:Verlag der Praehistorischen Zeitschrift.

Scott, K.
1980 Two hunting episodes of Middle Paleolithic age at La Cotte de Saint-Brelade, Jersey (Channel Islands). *World Archaeology 12*(2):137–152.

Semenov, S. A.
1964 *Prehistoric technology* (translated by M. W. Thompson). Bath, England:Adams and Dart.

Shipman, P.
1975 Implications of drought for vertebrate fossil assemblages. *Nature 257*:667–668.

Shipman, P., and J. E. Phillips
1976 On scavenging by hominids and other carnivores. *Current Anthropology 17*(1):170–172.
1977 Hominid tool-making versus carnivores' scavenging. *American Journal of Physical Anthropology 46*(1):77–86.

Singer, R.
1956 The "bone tools" from Hopefield. *American Anthropologist 58*:1127–1134.

Sollas, W. J.
1924 *Ancient hunters and their modern representatives* (3rd ed.). London:Macmillan.

Speth, J. D., and D. D. Davis
1976 Seasonal variability in early hominid predation. *Science 192*:441–445.

Spiess, A. E.
1979 *Reindeer and caribou hunters: An archaeological study*. New York and London:Academic Press.

Stanford, D.
1979a Bison kill by Ice Age hunters. *National Geographic 155*(1):114–119.
1979b The Selby and Dutton sites: Evidence for a possible pre-Clovis occupation of the High Plains. In *Pre-Llano cultures of the Americas: Paradoxes and possibilities*, edited by R. L. Humphrey and D. Stanford. Washington, D.C.:Anthropological Society of Washington. Pp. 101–123.

Stephenson, R. O., and L. Johnson
1972 *Wolf report Volume X*. Project Progress Report, Federal Aid in Wildlife Restoration, Project W-17-3. Juneau, Alaska:Alaska Department of Fish and Game.

Stiles, D.
1977 Ethnoarchaeology: A discussion of methods and applications. *Man 12*:87–103.

Sullivan, A. P.
1978 Inference and evidence in archaeology: A discussion of the conceptual problems. In *Advances in archaeological method and theory*, Vol. 1, edited by M. B. Schiffer. New York:Academic Press, Pp. 183–222.

Suppe, F.
1977 Afterword—1977. In *The structure of scientific theories* (2nd ed.), edited by F. Suppe. Urbana:University of Illinois Press. Pp. 617–730.

Sutcliffe, A. J.
1970 Spotted hyena: Crusher, gnawer, digester and collector of bones. *Nature 227*:1110–1113.
1973 Similarity of bones and antlers gnawed by deer to human artifacts. *Nature 246*:428–430.

Taylor, W. W.
1948 A study of archaeology. *American Anthropologist Memoir* No. 69.

Thomas, D. H.
1971 On distinguishing natural from cultural bone in archaeological sites. *American Antiquity 36*(3):366–371.

Thompson, R. H.
1956 The subjective element in archaeological inference. *Southwestern Journal of Anthropology 12*(3):327–332.

Toots, H.
1965 Sequence of disarticulation in mammalian skeletons. *University of Wyoming Contributions in Geology 4*(1):37–39.

Trigger, B. G.
1978 *Time and traditions.* New York:Columbia University Press.

Turnbull, P. F., and C. A. Reed
1974 The fauna from the Terminal Pleistocene of Palegawra Cave, a Zarzian occupation site in northeastern Iraq. *Fieldiana Anthropology* 63(3):81–146.

Uerpmann, H. P.
1973 Animal bone bonds and economic archaeology: A critical study of "osteoarchaeological" method. *World Archaeology* 4(3):307–332.

Veyrier, M., and J. Combier
1952 L'industrie osseuse Mousterienne de la Grotte Neron a Soyons (Ardeche). *L'Anthropologie* 56:383–385.

Vincent, A. S.
1978 East meets West: An analysis of the development of early man studies in East Africa 1919–1978. Honours Thesis—Bachelor of Arts, Australian National University, Canberra. ACT.

Von Daniken, E.
1969 *Chariots of the gods?* New York:Bantam.

Von Den Driesch, A. and J. Boessneck
1975 Schnittspuren en neolithischen Tierknochen. *Germania* 53:1–23.

Voorhies, M. R.
1969 Taphonomy and population dynamics of an Early Pliocene vertebrate fauna, Knox County, Nebraska. *University of Wyoming Contributions to Geology,* Special Paper No. 1. Pp. 1–69.

Vrba, E. S.
1975 Some evidence of chronology and paleoecology at Sterkfontein, Swartkraans, and Kromdraai from the fossil Bovidae. *Nature* 254:301–304.
1980 The significance of bovid remains as indicators of environment and predation patterns. In *Fossils in the making: Vertebrate taphonomy and paleoecology,* edited by A. K. Behrensmeyer and A. P. Hill. Chicago:University of Chicago Press. Pp. 247–271.

Walker, P. L., and J. C. Long
1977 An experimental study of the morphological characteristics of tool marks. *American Antiquity* 42(4):605–616.

Washburn, S. L.
1957 Australopithecines: The hunters or the hunted? *American Anthropologist* 59:612–614.

Watson, P. J., S. A. LeBlanc, and C. L. Redman
1971 *Explanation in archaeology: An explicitly scientific approach.* New York and London:Columbia University Press.

Webb, W. S., and D. L. DeJarnette
1942 An archaeological survey of Pickwick Basin in the adjacent portions of the states of Alabama, Mississippi, and Tennessee. *Bureau of American Ethnology Bulletin* No. 129. Washington, D.C.:Smithsonian Institution.

Weidenreich, F.
1928 *Der Schädelfund von Weimer-Ehringsdorf.* Jena.
1941 The extremity bones of *Sinanthropus pekinensis. Palaeontologia Sinica,* n.s. D., No. 5. Peking.

Welbourne, R. G.
1975 Tautswe Iron Age site: Its yield of bones. *Botswana Notes and Records* 7:1–16.

Wenke, R. J.
1980 *Patterns in prehistory.* New York and Oxford:Oxford University Press.

Wernert, P.
1948 Le culte des Crânes à l'Époque Paléolithique. In *Histoire Générale des Religions,* Vol. 1. sous le direction de M. Gorce et R. Mortier. Paris:A. Quillet.

Wetzel, R.
1969 Kulturknochen und Knochenkultur. In *Die Bocksteinchmiede im Loheta,* by R. Wetzel and G. Bosinski. Veroffentichungen des stattichen Amtes fur Denkmalpflege, Reihe A. Heft 15, Teil I. Stuttgart:Verlag Muller and Graff. Pp. 75–132.

Wheat, J. B.
1967 A Paleo-Indian bison kill. *Scientific American* 216(1):44–52.
1972 The Olsen–Chubbuck Site: A Paleo-Indian bison kill. *Society for American Archaeology, Memoirs* No. 26.
1979 The Jurgens site. *Plains Anthropologist Memoir* No. 15.

White, T. E.
1952 Observations on the butchering techniques of some aboriginal peoples, 1. *American Antiquity* 17(4):337–338.
1953a A method of calculating the dietary percentage of various food animals utilized by aboriginal peoples *American Antiquity* 18 (4):396–398.
1953b Observations on the butchering technique of some aboriginal peoples, No. 2. *American Antiquity* 19(2):160–164.
1954 Observations on the butchering technique of some aboriginal peoples, Nos. 3, 4, 5, and 6. *American Antiquity* 19(3):254–264.
1955 Observations on the butchering techniques of some aboriginal peoples, Nos. 7, 8, and 9. *American Antiquity* 21(2):170–178.

Whitehead, A. N.

1957 *The concept of nature.* Ann Arbor:University of Michigan Press.

1967 *Science and the modern world: Lowell Lectures, 1925.* New York:Free Press.

Willey, G. R., and J. A. Sabloff

1980 *A history of American archaeology* (2nd ed.). San Francisco:Freeman.

Wolberg, D. L.

1970 The hypothesized osteodontokeratic culture of the Australopithecinae: A look at the evidence and the opinions. *Current Anthropology* *11*(1):23–37.

Wood, W. R.

1962 Notes on the bison bone from the Paul Brave, Huff, and Demery sites (Oahe reservoir). *Plains Anthropologist* *7*(17):201–204.

1968 Mississippian hunting and butchering patterns: Bone from the Vista Shelter, 23SR20, Missouri. *American Antiquity* *33*(2):170–179.

Woodburn, J. C.

1968 Comment. In *Man the hunter,* edited by R. B. Lee and I. DeVore. Chicago:Aldine. P. 342.

Yellen, J. E.

1977a *Archaeological approaches to the present: Models for reconstructing the past.* New York:Academic Press.

1977b Cultural patterning in faunal remains: Evidence from the !Kung Bushmen. In *Experimental archaeology,* edited by D. Ingersoll, J. E. Yellen, and W. Macdonald. New York:Columbia University Press. Pp. 271–331.

Zapfe, M.

1939 Lebensspuren der eiszeitlichen Höhlenhyäne: Die urgenschictliche Bedeutung der Lebensspuren knochenfressender Raubtiere. *Paleobiologica* *7*:111–146.

Zierhut, N. W.

1967 Bone breaking activities of the Calling Lake Cree. *Alberta Anthropologist* *1*:33–36.

Zotz, L. F.

1951 *Altsteinzeitkunde Mitteleuropas.* Stuttgart:Enke.

Index

STUDIES IN ARCHAEOLOGY

Consulting Editor: Stuart Struever

Department of Anthropology
Northwestern University
Evanston, Illinois

Charles R. McGimsey III. **Public Archeology**

Lewis R. Binford. **An Archaeological Perspective**

Muriel Porter Weaver. **The Aztecs, Maya, and Their Predecessors: Archaeology of Mesoamerica**

Joseph W. Michels. **Dating Methods in Archaeology**

C. Garth Sampson. **The Stone Age Archaeology of Southern Africa**

Fred T. Plog. **The Study of Prehistoric Change**

Patty Jo Watson (Ed.). **Archeology of the Mammoth Cave Area**

George C. Frison (Ed.). **The Casper Site: A Hell Gap Bison Kill on the High Plains**

W. Raymond Wood and R. Bruce McMillan (Eds.). **Prehistoric Man and His Environments: A Case Study in the Ozark Highland**

Kent V. Flannery (Ed.). **The Early Mesoamerican Village**

Charles E. Cleland (Ed.). **Cultural Change and Continuity: Essays in Honor of James Bennett Griffin**

Michael B. Schiffer. **Behavioral Archeology**

Fred Wendorf and Romuald Schild. **Prehistory of the Nile Valley**

Michael A. Jochim. **Hunter-Gatherer Subsistence and Settlement: A Predictive Model**

Stanley South. **Method and Theory in Historical Archeology**

Timothy K. Earle and Jonathon E. Ericson (Eds.). **Exchange Systems in Prehistory**

Stanley South (Ed.). **Research Strategies in Historical Archeology**

John E. Yellen. **Archaeological Approaches to the Present: Models for Reconstructing the Past**

Lewis R. Binford (Ed.). **For Theory Building in Archaeology: Essays on Faunal Remains, Aquatic Resources, Spatial Analysis, and Systemic Modeling**

James N. Hill and Joel Gunn (Eds.). **The Individual in Prehistory: Studies of Variability in Style in Prehistoric Technologies**

Michael B. Schiffer and George J. Gumerman (Eds.). **Conservation Archaeology: A Guide for Cultural Resource Management Studies**

Thomas F. King, Patricia Parker Hickman, and Gary Berg. **Anthropology in Historic Preservation: Caring for Culture's Clutter**

Richard E. Blanton. **Monte Albán: Settlement Patterns at the Ancient Zapotec Capital**

R. E. Taylor and Clement W. Meighan. **Chronologies in New World Archaeology**

Bruce D. Smith. **Prehistoric Patterns of Human Behavior: A Case Study in the Mississippi Valley**

Barbara L. Stark and Barbara Voorhies (Eds.). **Prehistoric Coastal Adaptations: The Economy and Ecology of Maritime Middle America**

Charles L. Redman, Mary Jane Berman, Edward V. Curtin, William T. Langhorne, Nina M. Versaggi, and Jeffery C. Wanser (Eds.). **Social Archeology: Beyond Subsistence and Dating**

Bruce D. Smith (Ed.). **Mississippian Settlement Patterns**

Lewis R. Binford. **Nunamiut Ethnoarchaeology**

J. Barto Arnold III and Robert Weddle. **The Nautical Archeology of Padre Island: The Spanish Shipwrecks of 1554**

Sarunas Milisauskas. **European Prehistory**

Brian Hayden (Ed.). **Lithic Use-Wear Analysis**

William T. Sanders, Jeffrey R. Parsons, and Robert S. Santley. **The Basin of Mexico: Ecological Processes in the Evolution of a Civilization**

David L. Clarke. **Analytical Archaeologist: Collected Papers of David L. Clarke. Edited and Introduced by His Colleagues**

Arthur E. Spiess. **Reindeer and Caribou Hunters: An Archaeological Study**

Elizabeth S. Wing and Antoinette B. Brown. **Paleonutrition: Method and Theory in Prehistoric Foodways.**

John W. Rick. **Prehistoric Hunters of the High Andes**

Timothy K. Earle and Andrew L. Christenson (Eds.). **Modeling Change in Prehistoric Economics**

Thomas F. Lynch (Ed.). **Guitarrero Cave: Early Man in the Andes**

Fred Wendorf and Romuald Schild. **Prehistory of the Eastern Sahara**

Henri Laville, Jean-Philippe Rigaud, and James Sackett. **Rock Shelters of the Perigord: Stratigraphy and Archaeological Succession**

Duane C. Anderson and Holmes A. Semken, Jr. (Eds.). **The Cherokee Excavations: Holocene Ecology and Human Adaptations in Northwestern Iowa**

Anna Curtenius Roosevelt. **Parmana: Prehistoric Maize and Manioc Subsistence along the Amazon and Orinoco**

Fekri A. Hassan. **Demographic Archaeology**

G. Barker. **Landscape and Society: Prehistoric Central Italy**

Lewis R. Binford. **Bones: Ancient Men and Modern Myths**

Richard A. Gould and Michael B. Schiffer (Eds.). **Modern Material Culture: The Archaeology of Us**

in preparation

Muriel Porter Weaver. **The Aztecs, Maya, and Their Predecessors: Archaeology of Mesoamerica, 2nd edition**

Ross H. Cordy. **A Study of Prehistoric Social Change: The Development of Complex Societies in the Hawaiian Islands**